# Concepts and Applications of Organic Chemistry

# Concepts and Applications of Organic Chemistry

Editor: Saul Rowen

NY RESEARCH
P R E S S

New York

Published by NY Research Press
118-35 Queens Blvd., Suite 400,
Forest Hills, NY 11375, USA
www.nyresearchpress.com

Concepts and Applications of Organic Chemistry
Edited by Saul Rowen

International Standard Book Number: 978-1-63238-551-2 (Hardback)

The publisher's policy is to use permanent paper from mills that operate a sustainable forestry policy. Furthermore, the publisher ensures that the text paper and cover boards used have met acceptable environmental accreditation standards.

**Trademark Notice:** Registered trademark of products or corporate names are used only for explanation and identification without intent to infringe.

**Cataloging-in-Publication Data**

Concepts and applications of organic chemistry / edited by Saul Rowen.
   p. cm.
Includes bibliographical references and index.
ISBN 978-1-63238-551-2
1. Chemistry, Organic. 2. Chemistry. I. Rowen, Saul.
QD251.3 .C64 2017
547--dc23

Printed in the United States of America.

# Contents

Preface...................................................................................................................................IX

Chapter 1    **Self-catalysed aerobic oxidization of organic linkerin porous crystal for on-demand regulation of sorption behaviours**.........................................1
Pei-Qin Liao, Ai-Xin Zhu, Wei-Xiong Zhang, Jie-Peng Zhang, Xiao-Ming Chen

Chapter 2    **Nitrogen-embedded buckybowl and its assembly with $C_{60}$**.........................9
Hiroki Yokoi, Yuya Hiraoka, Satoru Hiroto, Daisuke Sakamaki, Shu Seki, Hiroshi Shinokubo

Chapter 3    **Synthesis of tetrasubstituted 1-silyloxy-3- aminobutadienes and chemistry beyond Diels–Alder reactions**.............................................................................18
Xijian Li, Siyu Peng, Li Li, Yong Huang

Chapter 4    **Total synthesis of (+)-gelsemine via an organocatalytic Diels–Alder approach**.........27
Xiaoming Chen, Shengguo Duan, Cheng Tao, Hongbin Zhai, Fayang G. Qiu

Chapter 5    **The key role of the scaffold on the efficiency of dendrimer nanodrugs**.........33
Anne-Marie Caminade, Séverine Fruchon, Cédric-Olivier Turrin, Mary Poupot, Armelle Ouali, Alexandrine Maraval, Matteo Garzoni, Marek Maly, Victor Furer, Valeri Kovalenko, Jean-Pierre Majoral, Giovanni M. Pavan, Rémy Poupot

Chapter 6    **Targeting bacteria via iminoboronate chemistry of amine-presenting lipids**.........44
Anupam Bandyopadhyay, Kelly A. McCarthy, Michael A. Kelly, Jianmin Gao

Chapter 7    **Macroscopic ordering of helical pores for arraying guest molecules noncentrosymmetrically**.....................................................................................53
Chunji Li, Joonil Cho, Kuniyo Yamada, Daisuke Hashizume, Fumito Araoka, Hideo Takezoe, Takuzo Aida, Yasuhiro Ishida

Chapter 8    **Chemical reaction mechanisms in solution from brute force computational Arrhenius plots**.......................................................................................62
Masoud Kazemi, Johan Åqvist

Chapter 9    **Iron-catalysed cross-coupling of organolithium compounds with organic halides**.........69
Zhenhua Jia, Qiang Liu, Xiao-Shui Peng, Henry N.C. Wong

Chapter 10    **Induction and control of supramolecular chirality by light in self-assembled helical nanostructures**.................................................................................76
Jisung Kim, Jinhee Lee, Woo Young Kim, Hyungjun Kim, Sanghwa Lee, Hee Chul Lee, Yoon Sup Lee, Myungeun Seo, Sang Youl Kim

Chapter 11    *De novo* branching cascades for structural and functional diversity in small
molecules.................................................................................................................84
Miguel Garcia-Castro, Lea Kremer, Christopher D. Reinkemeier,
Christian Unkelbach, Carsten Strohmann, Slava Ziegler, Claude Ostermann,
Kamal Kumar

Chapter 12    Cellular delivery and photochemical release of a caged inositol-pyrophosphate
induces PH-domain translocation *in cellulo*..............................................................97
Igor Pavlovic, Divyeshsinh T. Thakor, Jessica R. Vargas, Colin J. McKinlay,
Sebastian Hauke, Philipp Anstaett, Rafael C. Camuña, Laurent Bigler, Gilles Gasser,
Carsten Schultz, Paul A. Wender, Henning J. Jessen

Chapter 13    Self-assembly of dynamic orthoester cryptates............................................................105
René-Chris Brachvogel, Frank Hampel, Max von Delius

Chapter 14    Unique distal size selectivity with a digold catalyst during alkyne homocoupling.............112
Antonio Leyva-Pérez, Antonio Doménech-Carbó, Avelino Corma

Chapter 15    Transient signal generation in a self-assembled nanosystem fueled by ATP.......................120
Cristian Pezzato, Leonard J. Prins

Chapter 16    Tunable solid-state fluorescent materials for supramolecular encryption...........................128
Xisen Hou, Chenfeng Ke, Carson J. Bruns, Paul R. McGonigal, Roger B. Pettman,
J. Fraser Stoddart

Chapter 17    Discovery and enantiocontrol of axially chiral urazoles via organocatalytic
tyrosine click reaction......................................................................................137
Ji-Wei Zhang, Jin-Hui Xu, Dao-Juan Cheng, Chuan Shi, Xin-Yuan Liu, Bin Tan

Chapter 18    Fluorescence microscopy as an alternative to electron microscopy for microscale
dispersion evaluation of organic–inorganic composites....................................................147
Weijiang Guan, Si Wang, Chao Lu, Ben Zhong Tang

Chapter 19    Putting pressure on aromaticity along with *in situ* experimental electron density
of a molecular crystal......................................................................................154
Nicola Casati, Annette Kleppe, Andrew P. Jephcoat, Piero Macchi

Chapter 20    Practical carbon–carbon bond formation from olefins through nickel-catalyzed
reductive olefin hydrocarbonation..........................................................................162
Xi Lu, Bin Xiao, Zhenqi Zhang, Tianjun Gong, Wei Su, Jun Yi, Yao Fu, Lei Liu

Chapter 21    Atom-economic catalytic amide synthesis from amines and carboxylic acids
activated *in situ* with acetylenes.........................................................................170
Thilo Krause, Sabrina Baader, Benjamin Erb1 & Lukas J. Gooßen

Chapter 22    Organic–inorganic supramolecular solid catalyst boosts organic reactions in water...........177
Pilar García-García, José María Moreno, Urbano Díaz, Marta Bruix, Avelino Corma

Chapter 23    Bio-based polycarbonate as synthetic toolbox................................................................186
O. Hauenstein, S. Agarwal, A. Greiner

Chapter 24   **Merging rhodium-catalysed C–H activation and hydroamination in a highly selective [4+2] imine/alkyne annulation** ........................................................................ 193
Rajith S. Manan, Pinjing Zhao

Chapter 25   **Highly regio- and enantioselective multiple oxy- and amino-functionalizations of alkenes by modular cascade biocatalysis** ........................................................... 204
Shuke Wu, Yi Zhou, Tianwen Wang, Heng-Phon Too, Daniel I.C Wang, Zhi Li

Chapter 26   **Metal-free intermolecular formal cycloadditions enable an orthogonal access to nitrogen heterocycles** ................................................................................................. 217
Lan-Gui Xie, Supaporn Niyomchon, Antonio J. Mota, Leticia González, Nuno Maulide

**Permissions**

**List of Contributors**

**Index**

# Preface

This book has been an outcome of determined endeavour from a group of educationists in the field. The primary objective was to involve a broad spectrum of professionals from diverse cultural background involved in the field for developing new researches. The book not only targets students but also scholars pursuing higher research for further enhancement of the theoretical and practical applications of the subject.

Since twentieth century, organic chemistry emerged and has been rapidly expanding throughout without deceleration in the progress of this field. Foundations of medicine and biotechnology are built on organic compounds. It is contributing to the research in medicinal chemistry, polymer chemistry, as well as several aspects of materials science. This book discusses the fundamentals and concentrates on the applications of organic chemistry. It presents the concepts and applications in a concise manner for a better understanding of the subject. Students, researchers, experts and all associated with organic chemistry will benefit alike from this book.

It was an honour to edit such a profound book and also a challenging task to compile and examine all the relevant data for accuracy and originality. I wish to acknowledge the efforts of the contributors for submitting such brilliant and diverse chapters in the field and for endlessly working for the completion of the book. Last, but not the least; I thank my family for being a constant source of support in all my research endeavours.

**Editor**

# Self-catalysed aerobic oxidization of organic linker in porous crystal for on-demand regulation of sorption behaviours

Pei-Qin Liao[1], Ai-Xin Zhu[1,2], Wei-Xiong Zhang[1], Jie-Peng Zhang[1] & Xiao-Ming Chen[1]

Control over the structure and property of synthetic materials is crucial for practical applications. Here we report a facile, green and controllable solid–gas reaction strategy for on-demand modification of porous coordination polymer. Copper(I) and a methylene-bridged bis-triazolate ligand are combined to construct a porous crystal consisting of both enzyme-like $O_2$-activation site and oxidizable organic substrate. Thermogravimetry, single-crystal X-ray diffraction, electron paramagnetic resonance and infrared spectroscopy showed that the methylene groups can be oxidized by $O_2$/air even at room temperature via formation of the highly active $Cu(II)\text{-}O_2^{\cdot-}$ intermediate, to form carbonyl groups with enhance rigidity and polarity, without destroying the copper(I) triazolate framework. Since the oxidation degree or reaction progress can be easily monitored by the change of sample weight, gas sorption property of the crystal can be continuously and drastically (up to 4 orders of magnitude) tuned to give very high and even invertible selectivity for $CO_2$, $CH_4$ and $C_2H_6$.

[1] MOE Key Laboratory of Bioinorganic and Synthetic Chemistry, School of Chemistry and Chemical Engineering, Sun Yat-Sen University, Guangzhou 510275, P. R. China. [2] Faculty of Chemistry and Chemical Engineering, Yunnan Normal University, Kunming 650092, P. R. China. Correspondence and requests for materials should be addressed to J.-P.Z. (email: zhangjp7@mail.sysu.edu.cn).

Compared with conventional adsorbents, porous coordination polymers (PCPs) are unique for their diversified and tailorable coordination frameworks[1-11]. While tailoring structure/property generally refers to design/synthesis of new ligands and frameworks, some PCP prototypes can adopt several different metal ions and/or ligands, providing a rational strategy for adjusting the structure/property, albeit only in a limited degree[5,8,12-14]. A few of these prototypes can even form mixed-component (solid-solution) crystals with variable concentration/ratio of functional building blocks, which in principle allows the structure/property to be adjusted more continuously and precisely[15,16]. However, so far there is no rational strategy to directly monitor/control the composition of solid-solution frameworks during the synthesis process because complicated reaction environments involving solvents and/or liquid reactants are generally required for known direct-synthesis and post-synthetic modification (PSM)[3,7,17-21] methods. It should be noted that the reaction time/feeding ratio can be hardly used as a parameter for precise control over the reaction progress/framework composition, since the two variables have complicated relationships. Single-crystal X-ray diffraction is straightforward for direct visualization of the physical adsorption and chemical reaction events in PCPs[22], but retaining the sample single crystallinity after physical/chemical changes is always a great challenge[22-26], and crystallography is not a technique for quantitative analysis. If a PCP crystal could react with a gas ($O_2$ is a good yet difficult candidate) in the absence of assistant solvent/liquid, the sample weight or gas pressure can serve as an easily measurable parameter directly and linearly associated with its framework composition.

The selective and 'green' oxidation of organic molecules by air or dioxygen ($O_2$) is fundamental in the biosystem and has long been pursued in industry and chemical sciences[27-29]. Because the energy barrier for electron transfer from singlet organic substrates to the triplet $O_2$ is generally very high[30], harsh reaction conditions and/or efficient catalysts are usually necessary to activate the $O_2$ molecule for aerobic oxidization, so that the earth atmosphere can maintain a high $O_2$ concentration. Natural enzymes are well known as highly selective and efficient catalysts under mild conditions. For instance, the highly efficient $O_2$ activation centres in various copper proteins, such as galactose oxidase, tyrosinase and dopamine $\beta$-monooxygenase, have been extensively studied or used as structural models for synthetic catalysts.

Considering that PCPs generally lack sufficient robustness, reactivity and/or $O_2$-activation ability to allow aerobic oxidation of themselves, we designed and synthesized a porous metal azolate framework (MAF)[4] on the basis of Cu(I) and a methylene-bridged bis-triazolate ligand (Fig. 1). Although Cu(I)-based PCPs are very scarce because Cu(I) can be easily oxidized as Cu(II) by air to destroy the original metal–ligand connectivity[31], azolate derivatives have been demonstrated as suitable ligands for highly stable PCPs even with Cu(I) (ref. 4), and the coordination between Cu(I) and triazolate can be expected to give low-coordinated metal centres similar to those in the $O_2$-activating copper proteins. Furthermore, the methylene bridge in a diarylmethane-type ligand is obviously flexible and oxidizable (activated by two aromatic rings)[32], which could be oxidized in suitable catalytic conditions to form a more rigid and polar ketone group.

## Results

### Preparation and characterization of the porous crystal.
Colourless crystals of the titled compound, MAF-42, were synthesized in its large-pore (*lp*) form as $[Cu_4(btm)_2] \cdot C_6H_6$

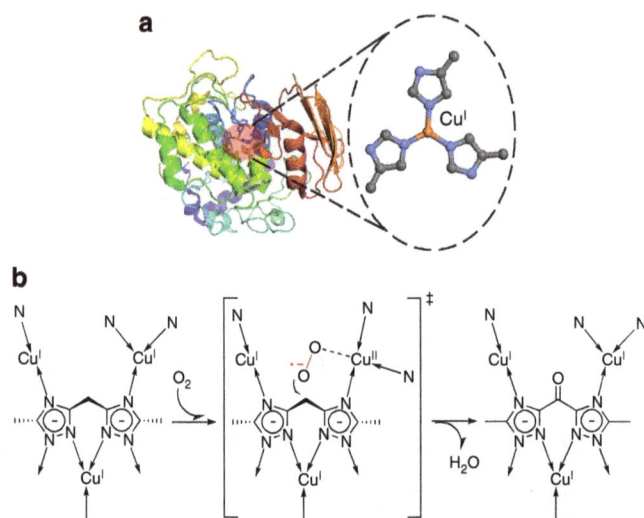

**Figure 1 | Bioinspired aerobic oxidation strategy for multifunctional porous frameworks.** (**a**) The $O_2$-activating site in a typical copper protein. (**b**) The possible coordination structure and reactivity of an extended network solid (porous crystal) consisting of Cu(I) and a methylene-bridged bistriazolate ligand.

(denoted as $C_6H_6$@MAF-42-*lp*, $H_2$btm = bis(5-methyl-1,2,4-triazolate-3-yl)methane) by solvothermal reaction of $H_2$btm and $[Cu(NH_3)_2]OH$ in a mixed aqueous ammonia/methanol/benzene solvent. Single-crystal X-ray diffraction analysis (Supplementary Table 1) showed that MAF-42-*lp* is a three-dimensional (3D) porous coordination framework composed of five independent (two of them locate at twofold axes with occupancies of 1/2) Cu(I) ions and two independent, fully deprotonated btm$^{2-}$ ligands in a 2:1 molar ratio. As expected, all btm$^{2-}$ ligands are six-coordinated and the average coordination number of Cu(I) ions is three[4]. Nevertheless, each Cu(I) ion either adopts the linear, distorted T-shaped or tetrahedral coordination geometry (Supplementary Fig. 1). The two independent ligands have different surrounding environments. The methylene group (C4) of one ligand is adjacent to a two- and two three-coordinated Cu(I) ions, while that (C11) of another ligand is only adjacent to two three- and a four-coordinated Cu(I) ions (Supplementary Table 2). The two triazolate rings of btm$^{2-}$ are not coplanar because they are linked by a $sp^3$ hybridized methylene C atom. The 3D coordination framework can be regarded as a cross-packing structure of ribbon-like fragments (Supplementary Fig. 2), which retains large 1D rhombic channels (void 37.8%, cross-section size $4.8 \times 7.1$–$6.0 \times 10.8$ Å$^2$) with disordered benzene molecules filled inside. It should be noted that the Cu(I) ions and methylene groups are fully exposed on the pore surface (Fig. 2a).

Thermogravimetry (TG) and powder X-ray diffraction (PXRD) measurements of $C_6H_6$@MAF-42-*lp* showed that the benzene molecules can be completely removed below 260 °C to form the small-pore (*sp*) phase $[Cu_4(btm)_2]$ (denoted as MAF-42-*sp* or simplified as MAF-42), which can be stable up to 410 °C (Supplementary Figs 3 and 4). Single-crystal structure of MAF-42-*sp* (Supplementary Table 1) showed significantly contracted unit cell (–22%) and channel size (void 14.1%, cross-section size $2.4 \times 2.6$–$3.5 \times 3.7$ Å$^2$) because the cross angle of packed ribbons reduced from 72° to 53° (Fig. 2b)[5,8]. In MAF-42-*sp*, the Cu(I) ions and methylene groups are less exposed on the pore surface; however, their separations are similar to those in MAF-42-*lp* (Supplementary Table 2). While the coordination bond lengths were changed very little ($\Delta_{max} = 0.022$ Å), the structural variation mainly occurred on the ligand conformations and coordination

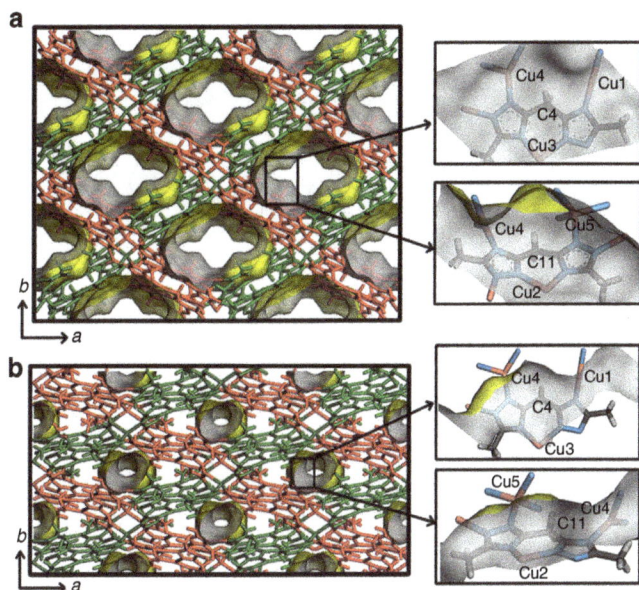

**Figure 2 | Guest-induced framework breathing.** (**a**) $C_6H_6$@MAF-42-*lp* (**b**) MAF-42-*sp*. The ribbon-like fragments of the coordination framework running along two directions are highlighted in green and red, respectively, and the pore surfaces of 1D channels are highlighted as yellow/grey curved planes. Desolvation leads to -28%, +10% and -22% changes in *a*, *b* and *V*, respectively.

geometries of the Cu(I) ions serving as the joints between the ribbon-like fragments ($\Delta_{max} = 7.3°$; Supplementary Tables 2 and 3). Actually, the ligand-bending directions are reversed, that is, the average torsion angle between two triazolate rings of $btm^{2-}$ changes from 167.6° in MAF-42-*lp* to − 165.1° in MAF-42-*sp* (Supplementary Fig. 5). The structure transformation between $C_6H_6$@MAF-42-*lp* and MAF-42-*sp* can be reversibly triggered by adsorption/desorption of benzene vapour (Supplementary Fig. 4).

**Self-catalysed aerobic oxidation.** In air, colourless $C_6H_6$@MAF-42-*lp* turns brown and then black quickly (Fig. 3a), which should not be originated from Cu(I) to Cu(II) oxidation because the Cu(II) complex is usually blue or green. Electrospray ionization mass spectrometry of the demetalated samples showed that fresh $C_6H_6$@MAF-42-*lp* has only a signal at $m/z = 179$ ($H_3btm^+$), while a new peak at $m/z = 193$ corresponding to the expected oxidation product bis(5-methyl-1,2,4-triazol-3-yl)methanone ($H_3btk^+$) appeared after the sample was exposed in air (Supplementary Fig. 6). The infrared spectrum of the oxidized sample exhibits a strong band at $1641 cm^{-1}$ in the characteristic region of $1,750 \pm 150 cm^{-1}$ for carbonyl groups (Supplementary Fig. 7). The single-crystal structure of a black crystal obtained by prolonged exposure (1 month) of $C_6H_6$@MAF-42-*lp* in air was measured (Supplementary Table 1). A residual electron peak appeared near one of the two crystallographically independent methylene groups that adjacent to the two-coordinated Cu(I) ion, which can be refined as an oxygen atom without any restriction, giving an occupancy of 0.28(5) and a C=O bond length of 1.20(5) Å (Supplementary Fig. 8)[33]. These observations demonstrated that the $btm^{2-}$ ligands in $C_6H_6$@MAF-42-*lp* can be readily oxidized by air at room temperature, although the reaction rate is quite slow. The oxidation of MAF-42-*sp* (almost no colour change after exposed in air at room temperature for several days) is much slower, which may be ascribed to the even smaller pore size and less exposed active sites.

The TG analysis showed that heating microcrystalline MAF-42-*sp* in an $O_2$ flow at 418 K could be an optimized reaction condition, in which complete oxidation can be achieved after ~ 8,000 min (Fig. 3b). Meanwhile, mass spectrometry analysis of the effluent showed that water is produced as a byproduct during the oxidation process (Supplementary Fig. 9). Therefore, on the basis of the simple reaction equation (Fig. 1b), the oxidation degree of the crystals can be simply calculated and controlled by the heating time or by monitoring the sample weight. The PXRD pattern of the fully oxidized sample is similar to that of MAF-42-*sp* (Supplementary Fig. 10), indicating the retention of the original framework connectivity. X-ray photoelectron spectroscopy showed that, even on the particle surface of the oxidized sample, only very small amounts of Cu(I) ions have been oxidized to Cu(II) (Supplementary Fig. 11). While some oxidation reactions have been used to modify the organic linkers in PCP crystals, strong and environment-unfriendly oxidants (such as nitric acid and dimethyldioxirane), as well as organic solvents are necessary[17,18,34–36]. Compared with the conventional oxidation methods, the solvent-free aerobic oxidation is obviously much greener yet more difficult. At the same conditions, the free ligand $H_2btm$ cannot be oxidized (Supplementary Fig. 12), highlighting the difficulty of covalent redox PSM and crucial role of enzyme-like Cu(I) centres. Compared with known direct-synthesis and PSM methods, the aerobic oxidation is also noteworthy for its controllable, solvent-free and solid–gas reaction mechanism, which can be conveniently used to tailor the adsorbent properties. Interestingly, the oxidized samples can be completely reduced back to $C_6H_6$@MAF-42-*lp* by hydrazine (Supplementary Fig. 13).

The structure of a highly oxidized single crystal $[Cu_4(btm)_{0.7}(btk)_{1.3}]$ (denoted as **O65** to reflect the oxidation degree, 65%) has been successfully measured (Supplementary Table 1). There are residual electron peaks near the two independent methylene groups, which were refined as oxygen atoms to give occupancies of 0.73(2) and 0.57(2) and C=O bond lengths of 1.236(10) and 1.234(11) Å, respectively (Fig. 3c,d). Again, the oxidation degree of methylene near the two-coordinated Cu(I) is higher. In **O65**, weak Cu(I)-carbonyl interactions can be observed (Supplementary Table 2), which are consistent with the relatively low wave number of carbonyl absorption in the infrared spectrum. Such unambiguous crystallography evidence is noteworthy because the harsh reaction condition for covalent PSM generally degrades crystallinity so that the modification degree is not high enough to be determined by crystal-structure analysis[23,37,38]. **O65** has a slightly smaller and distorted unit cell compared with MAF-42-*sp* ($\Delta V/V_1 = − 0.6\%$, $\Delta\beta = − 1.0°$), which can be ascribed to the generation of polar carbonyl groups on the pore surface and dipole–dipole attractive interactions within the host framework of **O65**. While the unit-cell volume follows $C_6H_6$@MAF-42-*lp* $\gg$ MAF-42-*sp* > **O65**, their organic ligands deviate from the planar conformation by 12.4, − 14.9 and − 12.9°, respectively (Supplementary Fig. 5). Obviously, oxidation enhances the conjugation degree and planarity of ligands, which are disadvantageous for framework contraction. To further shrink from MAF-42-*sp* to **O65**, the coordination bonding lengths ($\Delta_{max} = 0.044$ Å) and angles ($\Delta_{max} = 9.6°$) are forced to change more (Supplementary Tables 2 and 3). On the other hand, framework contraction also reduces the planarity of ligands in **O65**.

To gain more insight into the aerobic oxidation process, we tried to capture some reaction intermediates. In the single-crystal structure of MAF-42 loaded with $O_2$ (denoted as $O_2$@MAF-42) at low temperature of − 140 °C (Supplementary Table 1), although $O_2$ molecules are highly disordered, it can be seen that the primary adsorption sites are close to the two- and three-

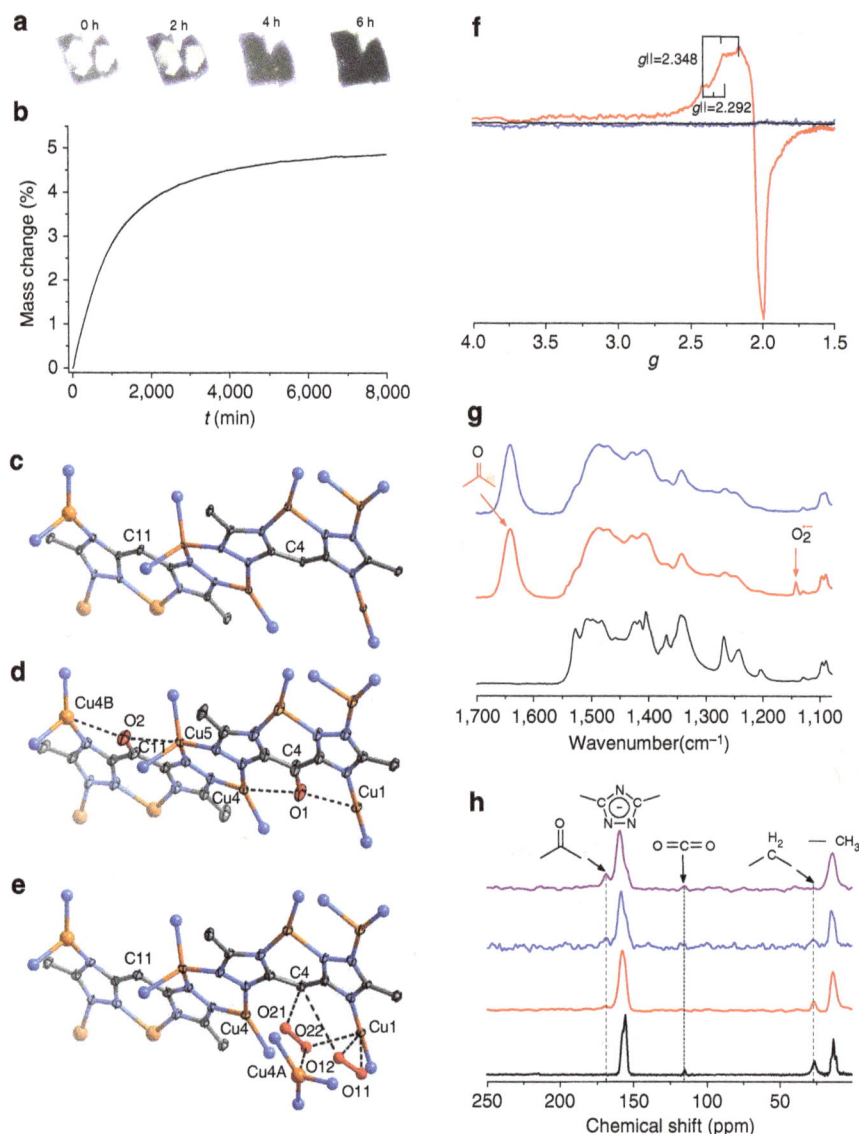

**Figure 3 | Controlling the reaction degree and revealing the catalytic mechanism of aerobic oxidation. (a)** Photographs of $C_6H_6$@MAF-42-*lp* exposed in air at room temperature for different time. **(b)** TG of MAF-42 in an $O_2$ flow at 418 K. **(c-e)** X-ray single-crystal structures of MAF-42, **O65** and $O_2$@MAF-42 (probability drawn at 30% for independent atoms; symmetric codes: $A = 1-x, -y, 1-z$; $B = x, 1-y, -0.5+z$). The black dashed lines are drawn to highlight relatively short nonbonding interactions (Cu1···O11 3.04(4) Å, Cu1···O12 3.28(3) Å, Cu4···O22 3.73(6) Å), C4···O12 4.47(6) Å, C4···O21 3.46(6) Å, Cu1···O1 2.61(1) Å, Cu4···O2 2.79(1) Å). **(f)** EPR spectra of MAF-42 in $O_2$ measured at 298 K (black) and then at 473 K (red) and then back to 298 K (blue). **(g)** DR-FTIR spectra of MAF-42 measured at 473 K under a variable atmosphere from helium (black) to oxygen (red) and then back to helium (blue). **(h)** Solid-state $^{13}$C NMR spectra of MAF-42 (black), **O53** (red), **O74** (blue) and **O100** (purple).

coordinated Cu(I) ions and the methylene group having a preference for oxidation (Fig. 3e and Supplementary Table 2). The presence of Cu(II) species during the aerobic oxidation reaction was confirmed by the axially symmetric signal with $g_{\parallel} = 2.292 \sim 2.348$ and the vertical signal with $g_{\perp} = 2.089$ in the *in situ* electron paramagnetic resonance (EPR) spectroscopy measured at 200 °C (Fig. 3f)[39]. Because the peak intensity at $g \sim 2.0$–$2.3$ originated from the Cu(II) ion is vastly greater than that of the radical $O_2^{\cdot-}$ at $g \sim 2$, the latter is difficult to assign. On the other hand, *in situ* diffuse reflectance-Fourier transform infrared spectroscopy showed absorption bands at $1,140\,cm^{-1}$ characteristic for the $O_2^{\cdot-}$ species (Fig. 3g) and $458\,cm^{-1}$ characteristic for the Cu–O coordination bond (Supplementary Fig. 14)[40]. The observation of physical adsorption at low temperature and chemical adsorption at high temperature is similar with some other

heterogeneous reactions involving gas reactants and solid catalysts[41]. Therefore, some key stages of the aerobic oxidation mechanism have been observed, in which $O_2$ molecules first attack the low-coordinated Cu(I) ions, and then form the highly active intermediate Cu(II)-$O_2^{\cdot-}$ reactive enough to break the hydrocarbon C–H bond, and finally returned to Cu(I) and produced the carbonyl product and water. The reaction can be unambiguously assigned to the Cu(II)/Cu(I) redox couple and metal-centred four-electron oxidation mechanism, which are commonly expected for natural enzymes and small molecular complexes but have been scarcely confirmed[18,19]. The catalytic activity of the low-coordinated Cu(I) sites in MAF-42 for aerobic oxidation of guest reactants was further confirmed by its effectiveness and low activation energy for the reaction of $CO + O_2 = CO_2$ (Supplementary Methods, Supplementary Figs 4 and 15).

## Regulation of gas adsorption properties.

To reveal the structure–property relationship, MAF-42 was oxidized to different degrees by the above established aerobic oxidation method. By virtue of the solid–gas reaction mechanism, the products were obtained in quantitative yields and were used without further workup procedure. Solid-state $^{13}C$ nuclear magnetic resonance (NMR) spectra showed that the methylene peak at 27 p.p.m. and carbonyl peak at 169 p.p.m. gradually reduces and increases, respectively, as the oxidation degree increases (Fig. 3h). On the basis of the peak areas of the methylene (2.8 p.p.m.) and methyl (0.8 p.p.m.) groups observed in the solution $^1H$ NMR spectra of the DCl-digested samples (Supplementary Fig. 16), the oxidation degrees of the oxidized samples (hereafter denoted as O53, O74 and O100, respectively) were calculated as 53%, 74% and 100%. PXRD and TG showed that oxidization enhances the hydrophilicity, decreases the thermal stability and reduces the unit-cell volumes of the samples (Supplementary Figs 17 and 18), being consistent with the generation of more polar carbonyl groups[4].

Single-component $CO_2$ isotherms were measured for MAF-42, O53, O74 and O100 at 195 K to characterize their porosity and framework flexibility/rigidity, which all exhibited hysteresis (Fig. 4a and Supplementary Fig. 19a). The isotherm of MAF-42 showed a direct transition from the nonporous (np) phase ($P/P_0 < 0.25$) to the lp phase ($P/P_0 > 0.40$). The absence of the sp phase in the isotherm of MAF-42 can be explained by its extremely small aperture size ($2.2 \times 2.6 \text{Å}^2$) and inert pore surface. Differently, O53 and O74 exhibited the expected sp-to-lp transition, as their first-step saturation uptakes are coincident with the theoretical values calculated from the crystal structures of MAF-42 and O65 (Supplementary Table 4). The increased $CO_2$-binding ability of the oxidized samples is consistent with their enhanced pore surface polarity. The second-step saturation uptake and the corresponding pore volume follow O74 (128 and $0.23 \text{cm}^3 \text{g}^{-1}$) > O53 (116 and $0.21 \text{cm}^3 \text{g}^{-1}$) > MAF-42 (105 and $0.19 \text{cm}^3 \text{g}^{-1}$), indicating that oxidization can expand the lp phase of the host frameworks, which was further quantified by the unit-cell volumes of guest-saturated samples (Supplementary Fig. 20). Framework expansion of the lp phase could be ascribed to the enhanced rigidity of the oxidized ligand, which usually increases the bridging length. In the lp phase, the dipole–dipole

attraction effect (causing the shrinkage of the sp phase[42]) can be eliminated by insertion of guest in the channels[5,8].

More interestingly, when the oxidation degree increases, the gate-opening/closing pressure (defined as the intermediate pressure between two isotherm steps) increases during adsorption ($P/P_0 = 0.32$, 0.35 and 0.74 for MAF-42, O53 and O74, respectively) but decreases during desorption ($P/P_0 = 0.25$, 0.21 and 0.14 for MAF-42, O53 and O74, respectively) to widen the hysteresis loop, which demonstrates that oxidization increases the ligand rigidity and difficulty of phase transition and the oxidized samples are solid-solution frameworks rather than mechanical mixtures[15,43]. Further, O100 showed one-step sorption isotherm with a small saturation uptake of ca $30 \text{cm}^3 \text{g}^{-1}$ (Fig. 4a and Supplementary Fig. 19a), confirming that it's sp phase has the most shrunk structure compared with the counterparts with lower oxidation ratios, and is too rigid to undergo the sp-to-lp transition. Therefore, the oxidized framework exhibited larger breathing amplitude, energy change and energy barrier during sp-to-lp transition. Only a couple of examples have demonstrated the control of framework flexibility by PSM of PCPs[42,44], in which different functional building blocks were added into the coordination framework, which always reduced the pore volume and the regulation is uncontrollable and discontinuous. In our case, the aerobic oxidation has negligible size effect (similar for methylene and carbonyl) but directly increases the ligand/framework rigidity and the pore volume of the lp phase, so that the oxidized framework exhibits enhanced adsorption capacity. More importantly, the solid–gas reaction mechanism allows modification ratio and gas sorption property to be continuously and conveniently monitored/controlled.

High-pressure single-component $CO_2$, $C_2H_6$ and $CH_4$ sorption isotherms were measured at 298 K (Fig. 4b–d and Supplementary Fig. 19b–d). The room-temperature $CO_2$ sorption isotherms are similar to those measured at 195 K, except that MAF-42 shows the sp-to-lp transition at room temperature. The abnormal temperature-dependent sorption behaviour of the sp phase of MAF-42 can be ascribed to thermal expansion (Supplementary Fig. 4) and/or decreased diffusion barrier[45]. The gradually increased gate-opening pressure (16, 18, 28 and >40 bar for

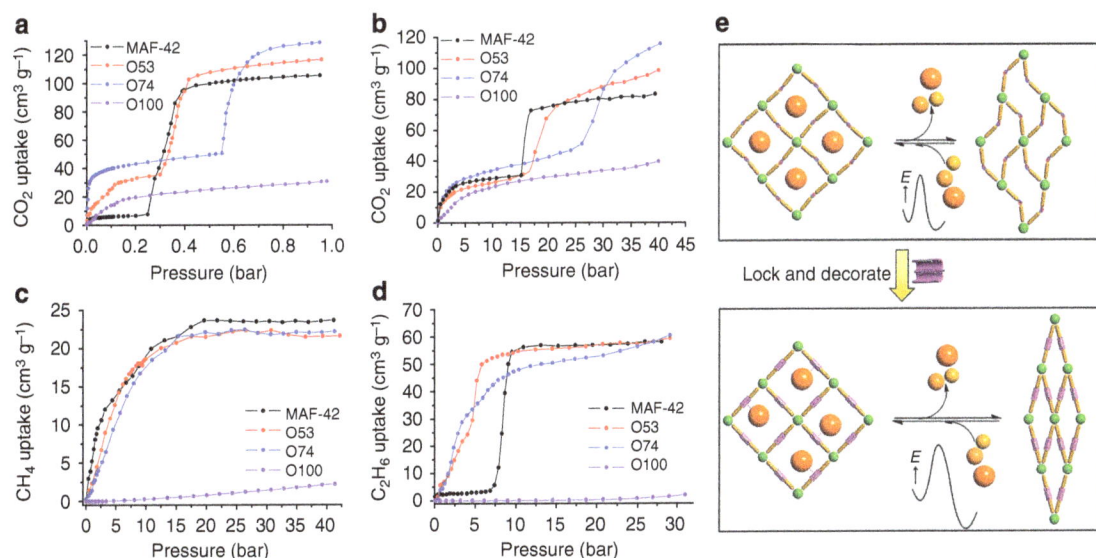

**Figure 4 | Modulation of single-component gas adsorption properties.** (a) $CO_2$ at 195 K. (b) $CO_2$ at 298 K. (c) $CH_4$ at 298 K. (d) $C_2H_6$ at 298 K. Lines are drawn to guide eyes. (e) Elucidation of the gas adsorption mechanism. The framework flexibility and pore surface polarity of a porous crystal are modulated by chemical modification of the organic linker. Specifically, for MAF-42, the hydrophobic and flexible methylene bridges are transformed to hydrophilic and rigid carbonyl ones via aerobic oxidation, and the transformation ratio is controlled by the oxidation degree.

MAF-42, **O53**, **O74** and **O100**, respectively) and decreased isotherm slope of the *sp*-to-*lp* transition are consistent with the increased framework rigidity of the oxidized samples (Fig. 4e).

For $CH_4$, MAF-42, **O53** and **O74** show type-I adsorption isotherms corresponding to the *sp* phase (Supplementary Table 4). Compared with $CO_2$, $CH_4$ has very low boiling point and polarity (Supplementary Table 5), meaning that it interacts with the host framework very weakly and can hardly open the narrow channel or induce the *sp*-to-*lp* transition. The variation of $C_2H_6$ sorption isotherms of MAF-42, **O53** and **O74** resembles that of $CO_2$ at 195 K. The gate-opening pressure is 8 bar for the *np*-to-*lp* transition of MAF-42, which is higher than those of 5 and 6 bar for the *sp*-to-*np* transitions of **O53** and **O74**, respectively. The nonporous *sp* phase of MAF-42 for $C_2H_6$ can be explained by the very large molecular size of the guest (Supplementary Table 5). Because of the over-shrinkage of the host framework, **O100** is virtually nonporous for the low-polarity gases $CH_4$ and $C_2H_6$.

The finely/drastically tunable gas sorption properties suggest their usefulness for on-demand gas separation. Mixed gas adsorption isotherms were measured to reveal the real gas adsorption selectivities of selected samples with promising isotherms. The $CO_2/CH_4$ selectivities of MAF-42 and **O100** at 1–10 bar (measured by mixed $CO_2/CH_4$ with 40:60 molar ratio) were observed as 28–14 and 700–600, respectively (Supplementary Methods, Fig. 5a–c), meaning that the performance of the crystal can be significantly improved by the aerobic oxidation treatment. The extremely high $CO_2/CH_4$ selectivity of **O100** should be suitable for purifying $CH_4$ from biogases ($CH_4$ 45~65%, $CO_2$ 30~50%)[46] and landfill gases ($CH_4$ 35~55%, $CO_2$ 40~45%)[47] by selective adsorption of $CO_2$.

More interestingly, the real $CH_4/C_2H_6$ selectivities (measured by mixed $CH_4/C_2H_6$ with 20:80 molar ratio) were observed to be 500–200 and 1/24–1/17 for MAF-42 and **O74**, respectively, indicating the catalytic oxidation can drastically invert the $CH_4/C_2H_6$ selectivity up to 4 orders of magnitude, which has not been realized by other adsorbents or methods (Supplementary Methods, Fig. 5d–f). The selective adsorption of $CH_4$ by MAF-42 at low pressure is due to its nonporous nature for $C_2H_6$

according to the molecular sieving effect. On the other hand, **O74** prefers adsorption of $C_2H_6$ because it is porous to both gases, and $C_2H_6$ has the larger molecular weight and quadrupole moment (Supplementary Table 5). This property may be useful for on-demand purification of different mixture gases (for example, natural gases usually contain $CH_4$ 69~96% and $C_2H_6$ 1~14%)[2,48].

Conventional adsorbents generally exhibit Langmuir-type isotherm and monotonically decreased (versus pressure) adsorption selectivity, which is not suitable for some gas separation applications operating at high pressures[49]. Some flexible PCPs showing gate-opening phenomena and stepped isotherms may be used to solve this problem; however, the real selectivities are usually much lower than those predicted by the single-component isotherms because the non-preferred guest can also enter the opened channel above the gate-opening pressure[50]. We observed that the real $C_2H_6/CH_4$ selectivity (measured by mixed $C_2H_6/CH_4$ with 75:25 molar ratio) of **O53** significantly increases from 92 to 140 at $C_2H_6$ partial pressure from 4.7 to 5.5 bar, respectively, and remains 113 at $C_2H_6$ partial pressure of 8.1 bar (Supplementary Methods and Supplementary Fig. 21). The abrupt increase in real adsorption selectivity occurs around the gate-opening pressure of single-component $C_2H_6$ isotherm at 4.7–5.2 bar. This observation indicates that the gate-opening is forced by filling the $C_2H_6$ molecules in the newly generated space. In other words, there is no space left for $CH_4$, preventing the commonly observed co-adsorption effect.

## Discussion

In summary, we demonstrated that Cu(I) ions with enzyme-like $O_2$ activation ability can cooperate with a flexible and oxidizable bis-triazolate ligand to fabricate a porous crystal reactive towards molecular oxygen even at room temperature. The methylene bridge of the bis-triazolate ligand provides the key reactivity to molecular oxygen, as well as modifiable framework flexibility and pore surface polarity. Although the free ligand is inert to oxygen, the flexible methylene bridge near the low-coordinated Cu(I) centres in the coordination framework is reactive enough to be

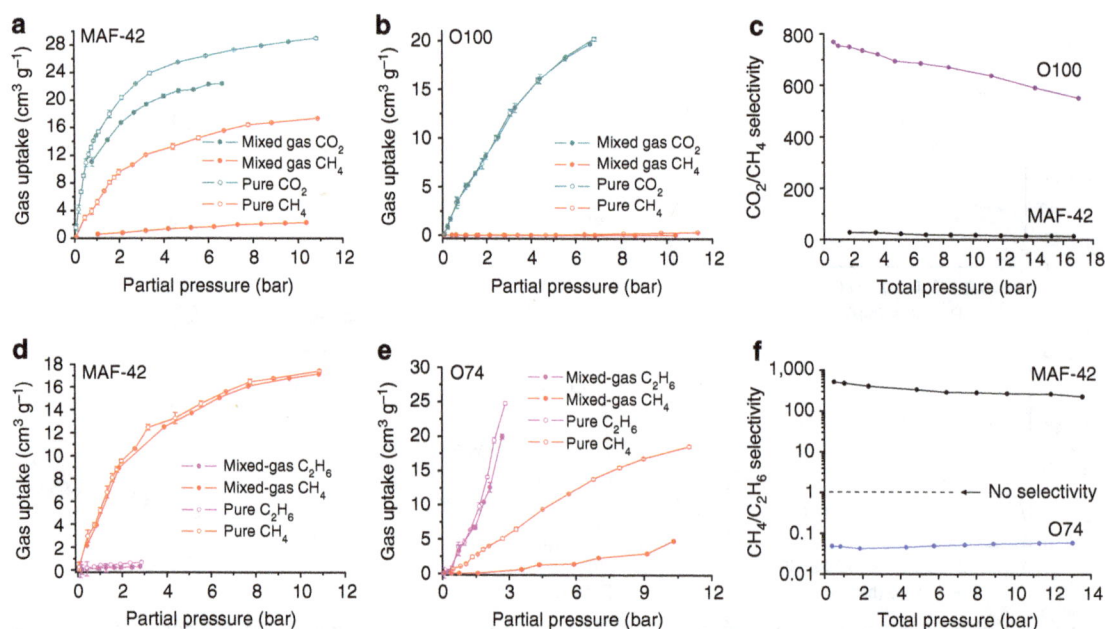

**Figure 5 | Modulation of mixed-gas adsorption selectivity.** (**a–c**) A ratio of 40:60 $CO_2/CH_4$ for MAF-42 and **O100**, as well as the corresponding $CO_2/CH_4$ selectivity. (**d–f**) A ratio of 80:20 $CH_4/C_2H_6$ for MAF-42 and **O74**, as well as the corresponding $CH_4/C_2H_6$ selectivity. All gas adsorption isotherms were measured at 298 K. Lines are drawn to guide eyes.

oxidized as a more rigid and polar carbonyl group. *In situ* structural and spectroscopic analyses confirmed that the low-coordinated Cu(I) centres behave like those in copperproteins during the aerobic oxidation. Compared with known material synthesis and processing methods, the solvent-free aerobic oxidation reaction has a series of advantages, such as green, quantitative yield, easy and precise monitor/control of modification degree, work-up procedure free, and so on. While conventional adsorbents separate adsorbates with fixed selectivities by either sorption affinity difference or molecular sieving effect, the porous crystal consisting of a reactive organic linker with modifiable flexibility and polarity offers a possibility to utilize both mechanisms, as demonstrated by the very high, drastically tunable and even switchable gas sorption selectivities. These results may enlighten future design and construction of multifunctional and controllable porous materials.

## Methods

**Materials.** Commercially available reagents and solvents were used without further purification. The ligand H$_2$btm was synthesized according to the literature method[51].

**Measurements.** Infrared spectra were obtained from KBr pellets on a Bruker TENSOR 27 FT IR spectrometer in the 400- to 4,000-cm$^{-1}$ region. Diffused reflection Fourier transform infrared spectroscopy (DR-FTIR) was performed on a Bruker VERTEX 70 spectrometer in the 400- to 4,000-cm$^{-1}$ region. TG-mass spectra were performed on a hyphenated apparatus of NETZSCH STA 449 F3 Jupiter and NETZSCH QMS 403C Aedo. Elemental analyses (C, H and N) were performed with a Vario EL elemental analyser. TG analyses were performed by using a TA Q50 system. Solid-state $^{13}$C NMR measurements were carried out at ambient temperature on a Bruker AVANCE 400 spectrometer. Solution $^1$H NMR measurements were carried out at ambient temperature on a VARIAN Mercury-Plus 300 spectrometer, for which ~20 mg of solid samples were digested with sonication in 550 μl of DCl (20 wt% in D$_2$O). PXRD patterns were collected (0.02° per step, 0.06 s per step except for otherwise stated) on a Bruker D8 Advance diffractometer (Cu Kα) at room temperature. Mass spectra were measured by a SHIMADZU LCMS-2010A equipment using an electrospray ionization source with MeOH as the mobile phase. Electron paramagnetic resonance measurements were performed at 9.7 GHz (X-band) using a Bruker BioSpin A300 spectrometer. The spin concentrations in the samples were determined from the second integral of the spectra using CuSO$_4$·5H$_2$O as a standard. The as-synthesized sample (weight of ~100–200 mg) was placed in the self-made quartz tube and dried for 8 h at 260 °C under high vacuum to remove the remnant solvent molecules, sealed with back-filled O$_2$ before measurements.

**Synthesis of materials.** A mixture of aqueous ammonia (25%, 4 ml) solution of [Cu(NH$_3$)$_2$]OH (0.025 mol l$^{-1}$), H$_2$btm (0.009 g, 0.05 mmol), methanol (3.0 ml) and benzene (1.0 ml) was sealed in a 15-ml Teflon-lined stainless reactor, which was heated in an oven at 160 °C for 72 h. The oven was cooled to room temperature at a rate of 5 °C h$^{-1}$. The resulting colourless block crystals were filtered, washed by ethanol and dried in air to give C$_6$H$_6$@MAF-42-*lp* (yield *ca* 83%). Guest-free MAF-42 was obtained by heating C$_6$H$_6$@MAF-42-*lp* at 260 °C under high vacuum for 24 h. The oxidized samples were obtained by heating MAF-42 at 418 K in the O$_2$ flow for different times.

**X-ray crystallography.** Single-crystal diffraction intensities were collected on a Bruker Apex CCD diffractometer with graphite-monochromated Mo Kα radiation or a Oxford Gemini S Ultra CCD diffractometer using mirror-monochromated Cu Kα radiation. Absorption corrections were applied by using the multiscan programme SADABS. The structures were solved by the direct method and refined with a full-matrix least-squares technique with the SHELXTL 6.10 programme package. The occupancies of carbonyl oxygen atoms and guest molecules were obtained by free refinement. Anisotropic thermal parameters were applied to all non-hydrogen atoms of host frameworks except for the oxygen atom in structure obtained by prolong exposure of C$_6$H$_6$@MAF-42-*lp* in air. Hydrogen atoms were generated geometrically. Crystal data for the complexes were summarized in Supplementary Table 1.

PXRD data for Pawley refinement were collected in 0.02° per step and 3 s per step. Indexing and Pawley refinement of the PXRD patterns was carried out by using the Reflex module of Material Studio 5.0. The patterns were indexed by the TREOR90 method with the aid of unit-cell parameters from single-crystal data. Pawley refinements were carried out with the cell parameters obtained from indexing in space group C2/c. Peak profiles, zero-shifts and unit-cell parameters were refined simultaneously. The peak profiles were refined by the Pseudo-Voigt function with Berar–Baldinozzi asymmetry correction parameters.

**Gas sorption measurement and calculation.** The single-component gas sorption isotherms were measured with automatic volumetric adsorption apparatuses (BEL-SORP-max and BELSORP-HP). The mixed gas adsorption isotherms were measured by volumetric adsorption – gas chromatograph instrument BELSORP-VC with a maximum measurement pressure of 15 bar. The as-synthesized sample (weight of ~500–800 mg) was placed in the quartz tube and dried for 8 h at 260 °C to remove the remnant solvent molecules before measurements. CO$_2$ (99.999%), CH$_4$ (99.999%) and C$_2$H$_6$ (99.99%) were used for all measurements. The temperatures were controlled by an acetone – dry ice bath (195 K) or a water bath (298 K).

## References

1. Herm, Z. R. *et al.* Separation of hexane isomers in a metal-organic framework with triangular channels. *Science* **340**, 960–964 (2013).
2. Li, J.-R., Kuppler, R. J. & Zhou, H.-C. Selective gas adsorption and separation in metal-organic frameworks. *Chem. Soc. Rev.* **38**, 1477–1504 (2009).
3. Cohen, S. M. Postsynthetic methods for the functionalization of metal–organic frameworks. *Chem. Rev.* **112**, 970–1000 (2012).
4. Zhang, J.-P., Zhang, Y.-B., Lin, J.-B. & Chen, X.-M. Metal azolate frameworks: from crystal engineering to functional materials. *Chem. Rev.* **112**, 1001–1033 (2012).
5. Férey, G. & Serre, C. Large breathing effects in three-dimensional porous hybrid matter: facts, analyses, rules and consequences. *Chem. Soc. Rev.* **38**, 1380–1399 (2009).
6. Bunck, D. N. & Dichtel, W. R. Mixed linker strategies for organic framework functionalization. *Chem. Eur. J.* **19**, 818–827 (2013).
7. Burrows, A. D. Mixed-component metal-organic frameworks (MC-MOFs): enhancing functionality through solid solution formation and surface modifications. *CrystEngComm* **13**, 3623–3642 (2011).
8. Horike, S., Shimomura, S. & Kitagawa, S. Soft porous crystals. *Nat. Chem.* **1**, 695–704 (2009).
9. Nugent, P. *et al.* Porous materials with optimal adsorption thermodynamics and kinetics for CO$_2$ separation. *Nature* **495**, 80–84 (2013).
10. Rabone, J. *et al.* An adaptable peptide-based porous material. *Science* **329**, 1053–1057 (2010).
11. Liao, P.-Q. *et al.* Strong and dynamic CO$_2$ sorption in a flexible porous framework possessing guest chelating claws. *J. Am. Chem. Soc.* **134**, 17380–17383 (2012).
12. Henke, S., Schneemann, A., Wütscher, A. & Fischer, R. A. Directing the breathing behavior of pillared-layered metal–organic frameworks via a systematic library of functionalized linkers bearing flexible substituents. *J. Am. Chem. Soc.* **134**, 9464–9474 (2012).
13. Schneemann, A. *et al.* Flexible metal-organic frameworks. *Chem. Soc. Rev.* **43**, 6062–6096 (2014).
14. Shekhah, O. *et al.* Made-to-order metal-organic frameworks for trace carbon dioxide removal and air capture. *Nat. Commun.* **5**, 4228 (2014).
15. Fukushima, T. *et al.* Solid solutions of soft porous coordination polymers: fine-tuning of gas adsorption properties. *Angew. Chem. Int. Ed.* **49**, 4820–4824 (2010).
16. Deng, H. *et al.* Multiple functional groups of varying ratios in metal-organic frameworks. *Science* **327**, 846–850 (2010).
17. Bernt, S., Guillerm, V., Serre, C. & Stock, N. Direct covalent post-synthetic chemical modification of Cr-MIL-101 using nitrating acid. *Chem. Commun.* **47**, 2838–2840 (2011).
18. Burrows, A. D., Frost, C. G., Mahon, M. F. & Richardson, C. Sulfur-tagged metal-organic frameworks and their post-synthetic oxidation. *Chem. Commun.* **0**, 4218–4220 (2009).
19. Jiang, H.-L., Feng, D., Liu, T.-F., Li, J.-R. & Zhou, H.-C. Pore surface engineering with controlled loadings of functional groups via click chemistry in highly stable metal–organic frameworks. *J. Am. Chem. Soc.* **134**, 14690–14693 (2012).
20. Karagiaridi, O. *et al.* Opening ZIF-8: a catalytically active zeolitic imidazolate framework of sodalite topology with unsubstituted linkers. *J. Am. Chem. Soc.* **134**, 18790–18796 (2012).
21. Li, T., Kozlowski, M. T., Doud, E. A., Blakely, M. N. & Rosi, N. L. Stepwise ligand exchange for the preparation of a family of mesoporous MOFs. *J. Am. Chem. Soc.* **135**, 11688–11691 (2013).
22. Zhang, J.-P., Liao, P.-Q., Zhou, H.-L., Lin, R.-B. & Chen, X.-M. Single-crystal X-ray diffraction studies on structural transformations of porous coordination polymers. *Chem. Soc. Rev.* **43**, 5789–5814 (2014).
23. Sato, H., Matsuda, R., Sugimoto, K., Takata, M. & Kitagawa, S. Photoactivation of a nanoporous crystal for on-demand guest trapping and conversion. *Nat. Mater.* **9**, 661–666 (2010).
24. Kawamichi, T., Kodama, T., Kawano, M. & Fujita, M. Single-crystalline molecular flasks: chemical transformation with bulky reagents in the pores of porous coordination networks. *Angew. Chem. Int. Ed.* **47**, 8030–8032 (2008).
25. Yang, S. *et al.* Selectivity and direct visualization of carbon dioxide and sulfur dioxide in a decorated porous host. *Nat. Chem.* **4**, 887–894 (2012).

26. Vaidhyanathan, R. *et al.* Direct observation and quantification of CO$_2$ binding within an amine-functionalized nanoporous solid. *Science* **330**, 650–653 (2010).

27. Punniyamurthy, T., Velusamy, S. & Iqbal, J. Recent advances in transition metal catalyzed oxidation of organic substrates with molecular oxygen. *Chem. Rev.* **105**, 2329–2364 (2005).

28. Que, L. & Tolman, W. B. Biologically inspired oxidation catalysis. *Nature* **455**, 333–340 (2008).

29. Li, L. *et al.* A crystalline porous coordination polymer decorated with nitroxyl radicals catalyzes aerobic oxidation of alcohols. *J. Am. Chem. Soc.* **136**, 7543–7546 (2014).

30. Piera, J. & Bäckvall, J.-E. Catalytic oxidation of organic substrates by molecular oxygen and hydrogen peroxide by multistep electron transfer—a biomimetic approach. *Angew. Chem. Int. Ed.* **47**, 3506–3523 (2008).

31. Grzywa, M. *et al.* CFA-2 and CFA-3 (Coordination Framework Augsburg University-2 and -3); novel MOFs assembled from trinuclear Cu(I)/Ag(I) secondary building units and 3,3',5,5'-tetraphenyl-bipyrazolate ligands. *Dalton Trans.* **42**, 6909–6921 (2013).

32. Barbara, E. *Ullmann's Encyclopedia of Industrial Chemistry* (Wiley-VCH Verlag GmbH & Co. KGaA, Weinheim, 2000).

33. Orpen, A. G. *et al.* Tables of bond lengths determined by X-ray and neutron diffraction. Part 2. Organometallic compounds and co-ordination complexes of the d- and f-block metals. *J. Chem. Soc. Dalton Trans.* S1–83 (1989).

34. Sippel, P. *et al.* Dielectric relaxation processes, electronic structure, and band gap engineering of MFU-4-type metal-organic frameworks: towards a rational design of semiconducting microporous materials. *Adv. Funct. Mater.* **24**, 3885–3896 (2014).

35. Yoon, J. W. *et al.* Controlled reducibility of a metal–organic framework with coordinatively unsaturated sites for preferential gas sorption. *Angew. Chem. Int. Ed.* **49**, 5949–5952 (2010).

36. Liu, T.-F. *et al.* Stepwise synthesis of robust metal–organic frameworks via postsynthetic metathesis and oxidation of metal nodes in a single-crystal to single-crystal transformation. *J. Am. Chem. Soc.* **136**, 7813–7816 (2014).

37. Deshpande, R. K., Minnaar, J. L. & Telfer, S. G. Thermolabile groups in metal–organic frameworks: suppression of network interpenetration, post-synthetic cavity expansion, and protection of reactive functional groups. *Angew. Chem. Int. Ed.* **49**, 4598–4602 (2010).

38. Wang, Z. & Cohen, S. M. Postsynthetic covalent modification of a neutral metal – organic framework. *J. Am. Chem. Soc.* **129**, 12368–12369 (2007).

39. Smeets, P. J., Woertink, J. S., Sels, B. F., Solomon, E. I. & Schoonheydt, R. A. Transition-metal ions in zeolites: coordination and activation of oxygen. *Inorg. Chem.* **49**, 3573–3583 (2010).

40. Vaska, L. Dioxygen-metal complexes: toward a unified view. *Acc. Chem. Res.* **9**, 175–183 (1976).

41. Otsuka, K. & Wang, Y. Direct conversion of methane into oxygenates. *Appl. Catal. A* **222**, 145–161 (2001).

42. Wang, Z. & Cohen, S. M. Modulating metal – organic frameworks to breathe: a postsynthetic covalent modification approach. *J. Am. Chem. Soc.* **131**, 16675–16677 (2009).

43. Coudert, F. X., Jeffroy, M., Fuchs, A. H., Boutin, A. & Mellot-Draznieks, C. Thermodynamics of guest-induced structural transitions in hybrid organic – inorganic frameworks. *J. Am. Chem. Soc.* **130**, 14294–14302 (2008).

44. Volkringer, C. & Cohen, S. M. Generating reactive MILs: isocyanate- and isothiocyanate-bearing mils through postsynthetic modification. *Angew. Chem. Int. Ed.* **49**, 4644–4648 (2010).

45. Wei, Y.-S. *et al.* Turning on the flexibility of isoreticular porous coordination frameworks for drastically tunable framework breathing and thermal expansion. *Chem. Sci.* **4**, 1539–1546 (2013).

46. Grande, C. A. & Rodrigues, A. E. Biogas to fuel by vacuum pressure swing adsorption i. behavior of equilibrium and kinetic-based adsorbents. *Ind. Eng. Chem. Res.* **46**, 4595–4605 (2007).

47. Pires, J., Bestilleiro, M., Pinto, M. & Gil, A. Selective adsorption of carbon dioxide, methane and ethane by porous clays heterostructures. *Sep. Purif. Technol.* **61**, 161–167 (2008).

48. He, Y., Zhou, W., Krishna, R. & Chen, B. Microporous metal-organic frameworks for storage and separation of small hydrocarbons. *Chem. Commun.* **48**, 11813–11831 (2012).

49. Horike, S., Inubushi, Y., Hori, T., Fukushima, T. & Kitagawa, S. A solid solution approach to 2D coordination polymers for CH$_4$/CO$_2$ and CH$_4$/C$_2$H$_6$ gas separation: equilibrium and kinetic studies. *Chem. Sci.* **3**, 116–120 (2012).

50. Hamon, L. *et al.* Co-adsorption and separation of CO$_2$ − CH$_4$ mixtures in the highly flexible MIL-53(Cr) MOF. *J. Am. Chem. Soc.* **131**, 17490–17499 (2009).

51. Bahceci, S., Yuksek, H. & Serdar, M. Reactions of amidines with some carboxylic acid hydrazides. *Indian J. Chem. B* **44**, 568–572 (2005).

## Acknowledgements

This work was supported by the '973 Project' (2012CB821706 and 2014CB845602) and NSFC (21225105, 21290173 and 21473260). We thank Professor Hongbing Ji, Dr Zebao Rui and Mr Huayao Chen for the DR-FTIR measurement and Mr Guping Hu for the EPR measurement.

## Author contributions

J.-P.Z. designed the research. P.-Q.L. and A.-X.Z. performed syntheses and measurements. P.-Q.L. and W.-X.Z. performed PXRD analyses. P.-Q.L., J.-P.Z. and X.-M.C. wrote the manuscript.

# 2

# Nitrogen-embedded buckybowl and its assembly with C$_{60}$

Hiroki Yokoi[1], Yuya Hiraoka[1], Satoru Hiroto[1], Daisuke Sakamaki[2], Shu Seki[2] & Hiroshi Shinokubo[1]

Curved $\pi$-conjugated molecules have attracted considerable interest because of the unique properties originating from their curved $\pi$ surface. However, the synthesis of such distorted molecules requires harsh conditions, which hamper easy access to heteroatom-containing curved $\pi$ systems. Here we report the synthesis of a $\pi$-extended azacorannulene with nitrogen in its centre. The oxidation of 9-aminophenanthrene provides tetrabenzocarbazole, which is converted to the azabuckybowl through palladium-catalysed intramolecular coupling. The electron-donating nature and curved $\pi$ surface of the azabuckybowl enable its tight association with C$_{60}$ in solution and solid states. High charge mobility is observed for the azabuckybowl/C$_{60}$ assembly. This compound may be of interest in the fields of curved $\pi$ systems as fullerene hosts, anisotropic $\pi$ donors and precursors to nitrogen-containing nanocarbon materials.

[1] Department of Applied Chemistry, Graduate School of Engineering, Nagoya University, Nagoya 464-8603, Japan. [2] Department of Molecular Engineering, Graduate School of Engineering, Kyoto University, Kyoto 615-8510, Japan. Correspondence and requests for materials should be addressed to S.H. (email: hiroto@apchem.nagoya-u.ac.jp) or to H.S. (email: hshino@apchem.nagoya-u.ac.jp).

Curved π-conjugated molecules have captivated numerous scientists[1-4]. Curving a π system induces a large displacement from a plane to construct three-dimensional structures[5-10]. Besides their figurative beauty, the curved π surface generates unique functions such as chiroptical properties, anisotropic electron transitions and dynamic motion in solution and solid states[11-13]. To enhance these characteristics, the introduction of heteroatoms is an effective strategy. However, the synthesis of heteroatom-containing curved π systems remains a challenge[14,15]. The preparation of distorted π systems requires harsh reaction conditions that do not tolerate heterocyclic molecules.

Buckybowls, that is, bowl-shaped molecules such as corannulenes[16] and sumanenes[17] represent important curved π-conjugated molecules, which can be precursors for the bottom-up synthesis of fullerenes and nanotubes. In 1966, Barth and Lawton[16] reported the first chemical synthesis of corannulene. Since Scott and Siegel's groups[18-21] developed their straightforward synthesis of corannulene, numerous bowl-shaped hydrocarbons have been synthesized. On the other hand, the nitrogen-embedded bowl-shaped molecules have been sought as model compounds for azafullerenes and nitrogen-doped carbon nanotubes[22-24]. Furthermore, dramatic changes in electronic structures of buckybowls by nitrogen are expected. However, the synthesis of buckybowls with internal nitrogen atoms has been still challenging[25]. Oxidative fusion approaches are not compatible with electron-rich nitrogen because of its less tolerant nature for oxidation.

Recently, our group reported the oxidative dimerization of aminoarenes to distorted π-conjugated molecules in a one-step operation[26,27]. We here disclose that the oxidation of 9-aminophenanthrenes affords tetrabenzocarbazoles in good yields. Furthermore, consecutive fusion reactions of tetrabenzocarbazole 2 through palladium-catalysed C–H/C–Cl and C–H/C–Br coupling achieve the synthesis of nitrogen-embedded buckybowl 5, that is, 'azabuckybowl'[28] under mild conditions. Owing to the electron-donating nature of the nitrogen atom, azabuckybowl 5 strongly interacts with $C_{60}$ to furnish an inclusion complex, which exhibits a substantially high charge-carrier mobility in the solid state.

## Results

### Synthesis of nitrogen-embedded buckybowl. 
The synthesis of nitrogen-embedded buckybowl 5 started with the oxidative dimerization of 1 (Fig. 1)[26]. 9-Aminophenanthrene 1 was oxidized to tetrabenzocarbazole 2 in 94% yield. Reaction of 2 with Pd(OAc)$_2$/tricyclohexylphosphine provided singly fused product 3 in 63% yield[29]. The twisted conformation of 3 was unambiguously elucidated by X-ray diffraction analysis (Supplementary Fig. 18). The bromination of 3 with bromine afforded tribrominated product 4 in 56% yield. Finally, the palladium-catalysed double C–H/C–Br coupling furnished nitrogen-embedded buckybowl 5 in 46% yield. The proton nuclear magnetic resonance ($^1$H NMR) spectrum of 5 exhibited six proton signals in the aromatic region, indicating the formation of a fused and symmetrical molecule.

### Structural elucidation and characteristics of azabuckybowl. 
The bowl-shaped structure of 5 was unambiguously elucidated by X-ray diffraction (Fig. 2). In the crystal, one asymmetric unit contained two independent molecules of 5. The bowl depth, which is defined as the distance between the mean plane that consisted of five carbons at the edge and the centroid of the pyrrole ring, was 1.65 and 1.70 Å. The bowl depth of the central azacorannulene core was 0.90 and 0.92 Å, which is slightly greater

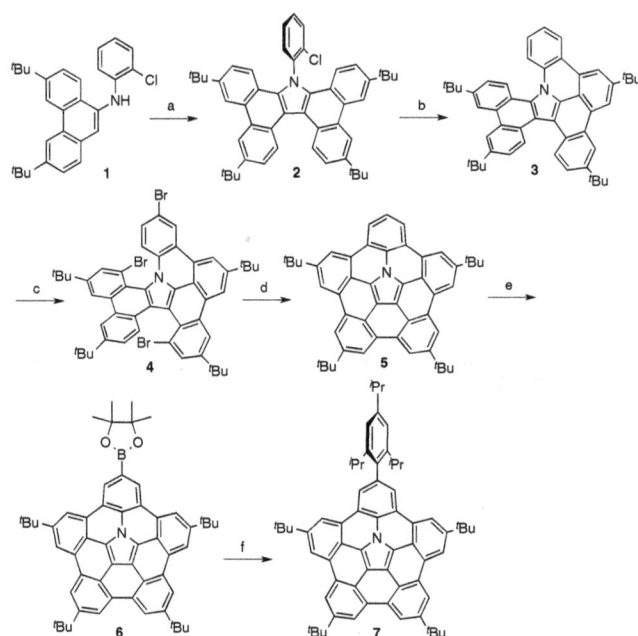

**Figure 1 | Synthesis of nitrogen-embedded buckybowl 5 from phenanthrene 1.** Conditions: (a) 2,3-dichloro-5,6-dicyanobenzoquinone (DDQ), TFA, toluene, room temperature, 1 h, 94% yield. (b) Pd(OAc)$_2$, PCy$_3$·HBF$_4$, K$_2$CO$_3$, DMA, 130 °C, 43 h, 63% yield. (c) Br$_2$, CCl$_4$, 70 °C, 12.5 h, 56% yield. (d) Pd(OAc)$_2$, PCy$_3$·HBF$_4$, K$_2$CO$_3$, DMA, 130 °C, 16 h, 46% yield. (e) bis(pinacolato)diboron, [Ir(OMe)cod]$_2$, 4,4'-di-*tert*-butyl-2,2'-bipyridyl, octane, 10.5 h, 110 °C, 80% yield. (f) 2-bromo-1,3,5-triisopropylbenzene, PdCl$_2$(dppf)•CH$_2$Cl$_2$, Cs$_2$CO$_3$, 1,4-dioxane, 13 h, 100 °C, 55% yield.

than that of corannulene (0.86 Å). The curvature of 5 was further evaluated by Haddon's π-orbital axis vector (POAV) angles[30]. As shown in Fig. 2a, the POAV angles around the central pyrrole ring are in the range of 7.2°–9.3°. These values are comparable with that of corannulene (9.1°). It is noteworthy that the molecules constructed a one-dimensional chain stacking structure in the crystal (Fig. 2c). Distances between the two closest molecules were 3.25 and 3.41 Å, indicating the existence of a π–π interaction.

Bowl-to-bowl inversion of the azabuckybowl was investigated. Azabuckybowl 5 was further functionalized by iridium-catalysed C–H borylation to provide 6 in 80% yield[31,32]. The Suzuki–Miyaura cross-coupling reaction of 6 with 2-bromo-1,3,5-triisopropylbenzene furnished the corresponding coupling product 7 in 55% yield. The $^1$H NMR spectrum of 7 in 1,2-dichlorobenzene-$d_4$ at room temperature exhibited three doublet peaks for methyl protons of isopropyl groups at 1.43, 1.39 and 1.13 p.p.m. This non-symmetric feature indicates that 7 shows no bowl-to-bowl inversion at room temperature. As the temperature was raised, two proton signals at 1.43 and 1.13 p.p.m. were gradually broadened (Fig. 2d). Even at 170 °C, these signals were not coalesced. Accordingly, the bowl-to-bowl inversion energy ($\Delta G^{\ddagger}$) was measured by two-dimensional exchange spectroscopy (2D EXSY) experiments. At 393 K, $\Delta G^{\ddagger}$ was determined to be 23.3 kcal mol$^{-1}$ in 1,2-dichlorobenzene-$d_4$ (Supplementary Figs 32 and 33). This value is higher than that of the parent sumanene ($\Delta G^{\ddagger} = 19.7$–20.4 kcal mol$^{-1}$)[33]. The high bowl-inversion energy of 7 was also supported by theoretical calculations. The inversion barrier of 7 was calculated to be 19.9 kcal mol$^{-1}$ by density functional theory (DFT) calculations at the B3LYP/cc-pVDZ level, which is higher than those of sumanene (18.2 kcal mol$^{-1}$) and corannulene (9.1 kcal mol$^{-1}$) calculated at the same level of theory[34].

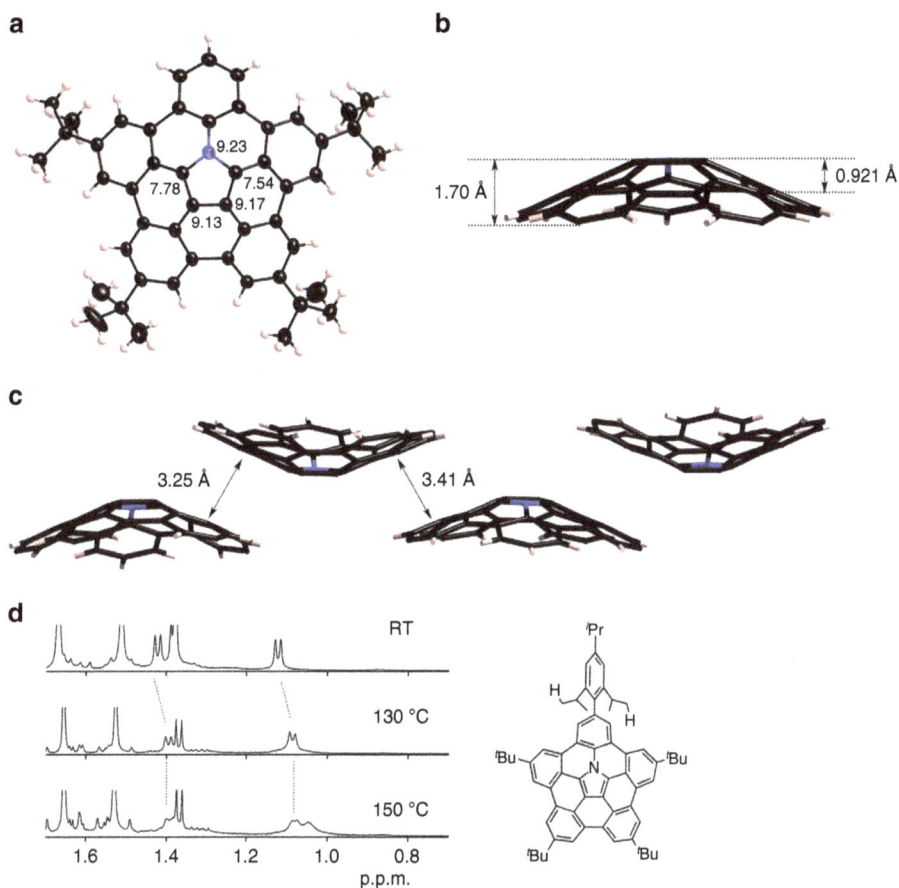

**Figure 2 | Structural features of azabuckybowls.** (**a**) Top view and POAV pyramidalization angles, (**b**) side view of **5** and (**c**) packing structure of **5** in the crystal. Thermal ellipsoids in **a** are scaled at 50% probability level and t-butyl groups are omitted for clarity in **b** and **c**. (**d**) Temperature-dependent NMR spectra of **7** in 1,2-dichlorobenzene-$d_4$. POAV angles and bowl depths in one of two molecules in the crystal are displayed in **a** and **b**. Solvent molecules (o-xylene) in the crystal structure of **5** were omitted for clarity. RT, room temperature.

**Optical and electrochemical properties of azabuckybowl.** Figure 3 shows ultraviolet–visible absorption and emission spectra of **3** and **5** in $CH_2Cl_2$. The lowest energy bands shifted to the low-energy region as the degree of fusion increased. All compounds exhibited fluorescence in the visible region (Fig. 3b). The emission quantum yield of **5** was 17%, which is the highest among buckybowls[35]. The Stokes shifts of **3** ($2,800 \, cm^{-1}$) and **5** ($1,500 \, cm^{-1}$) were relatively larger than that of a planar molecule. This reflects their excited state dynamics, owing to their distorted characteristics.

The electronic structures of **3** and **5** were further investigated by an electrochemical analysis (Supplementary Fig. 22 and Supplementary Table 2). Reversible oxidation waves were observed for all compounds. The first oxidation potentials were lowered in the order of **2** > **3** > **5**, indicating effective electron donation from the nitrogen atom to the entire π system.

We then examined the protonation behaviour of **5**, because **5** was expected to have higher basicity than planar amines. The addition of trifluoroacetic acid (TFA) to a dichloromethane solution of **5** induced a dramatic change in its absorption spectrum (Fig. 3c). Interestingly, the same change was observed on the addition of a one-electron oxidant, tris(4-bromophenyl) aminium hexachloroantimonate (BAHA) (Fig. 3d). We also monitored the electro-oxidative absorption spectrum of **5** in $CH_2Cl_2$, which exhibited essentially the same change (Fig. 3e). These facts strongly indicate that the addition of TFA resulted in the generation of the radical cation species rather than simple protonation. The formation of the radical cation was confirmed by electron spin resonance (ESR) measurements (Supplementary

Fig. 26). The solution of **5** in the presence of TFA exhibited a distinct signal at $g = 2.002$, as was the case of the oxidation of **5** with BAHA. The conversion of **5** to the radical cation was almost quantitative under air atmosphere but was substantially lower under argon (Supplementary Fig. 27). The radical cation generation is likely due to electron transfer between **5** and protonated **5** involving air oxidation[36]. The facile generation of the radical cation from **5** would allow the investigation of the effect of oxidative doping on solid-state properties. This phenomenon also implies that nitrogen-doped electron-rich nanocarbons may undergo a similar radical cation generation by protonation.

**Association behaviour of 5 with $C_{60}$.** The effect of nitrogen also appeared in the association behaviour of **5** with $C_{60}$. For hydrocarbon buckybowls, their association constants with $C_{60}$ were very low to be measured[37–40]. The incorporation of electron-rich nitrogen in buckybowls should enable tighter binding with electron-deficient fullerenes. The electrochemical analysis revealed its much lower oxidation potential (0.20 V) when compared with corannulene (1.57 V) (Fig. 3f)[41]. The addition of $C_{60}$ into an 1,2-dichlorobenzene solution of **5** induced a change in the ultraviolet–visible absorption and emission spectra (Fig. 4b,c). In particular, the appearance of broad absorption bands in the near-infrared region suggests intermolecular charge-transfer interactions between **5** and $C_{60}$. The association behaviour was also monitored by $^1H$ NMR analysis. On the addition of $C_{60}$ into a toluene-$d_8$ solution of **5**, all aromatic proton signals were upfield

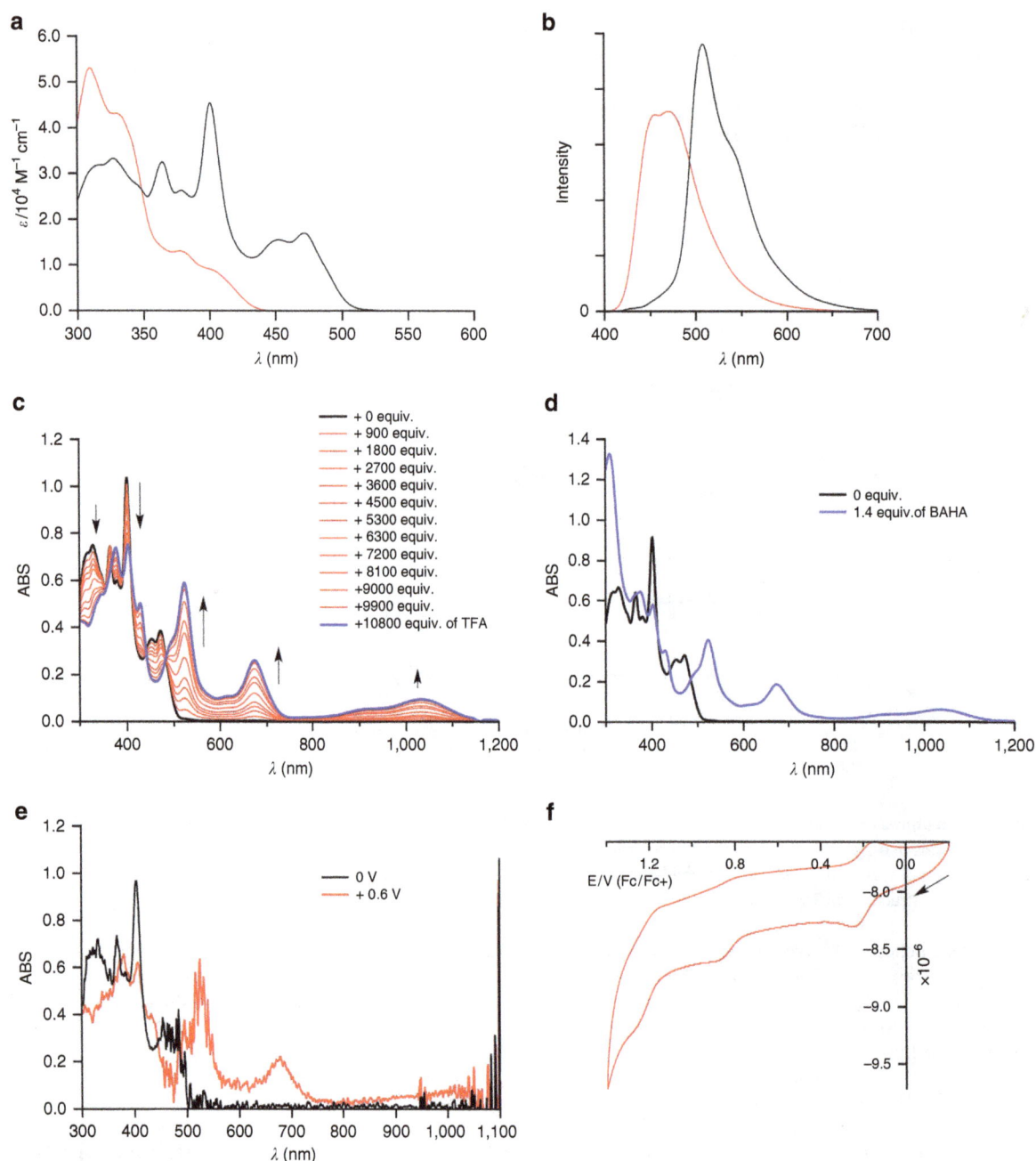

**Figure 3 | Physical properties of 5.** (**a**) Ultraviolet-visible absorption spectra in $CH_2Cl_2$ and (**b**) emission spectra of **3** (red) and **5** (black) in $CH_2Cl_2$ (concentration: $7.6 \times 10^{-7}$ M). (**c**) Spectral changes in absorption spectra of **5** on the addition of TFA into a dichloromethane solution of **5**. (**d**) Absorption spectra of **5** in $CH_2Cl_2$ before and after the addition of 1.4 equiv. of BAHA. (**e**) Spectroelectrochemical analysis of **5** in $CH_2Cl_2$. (**f**) Cyclic voltammogram of **5** measured in $CH_2Cl_2$ with tetra-n-butylammonium hexafluorophosphate as the electrolyte.

shifted (Fig. 4a). This indicates that **5** and $C_{60}$ interacted in a convex–concave manner. This was revealed by the X-ray crystallographic analysis and showed that $C_{60}$ was located above the centre of **5** (Fig. 4d,e). The penetration depth of $C_{60}$ into **5** measured from the centroid of the pyrrole ring to the centroid of $C_{60}$ is 6.82 Å, and that measured from the shortest distance from the concave surface of **5** to a $C_{60}$ surface is 3.29 Å, whereas the depths of $C_{60}$ into the corannulene/$C_{60}$ complex are 6.94 and 3.75 Å. The short distance between **5** and $C_{60}$ indicates the presence of attractive interactions between them. Judging from the relatively long distance ($>3.74$ Å), the CH–$\pi$ interaction between tert-butyl groups and $C_{60}$ was not essential. The binding constant was determined to be 3,800 M$^{-1}$ by titration with absorption and fluorescence spectra (Supplementary Figs 23–25).

This value is approximately three times larger than that of perthiolated corannulene. The existence of intermolecular charge-transfer interactions between **5** and $C_{60}$ was indicated by the quenching behaviour of the emission on the addition of $C_{60}$ to a 1,2-dichlorobenzene solution of **5** (Fig. 4c). The DFT optimization of $5 \supset C_{60}$ afforded nearly the same structure as the crystal structure. The highest occupied molecular orbital was spread over the entire surface of **5**, whereas the lowest unoccupied molecular orbital was delocalized on $C_{60}$ (Supplementary Fig. 29). In addition, oscillator strengths of the absorption of $5 \supset C_{60}$ were simulated by the time-dependent DFT method (Supplementary Fig. 30). The broad lowest-energy band in the near-infrared region was assigned as the highest occupied molecular orbital–lowest unoccupied molecular orbital transition. These results

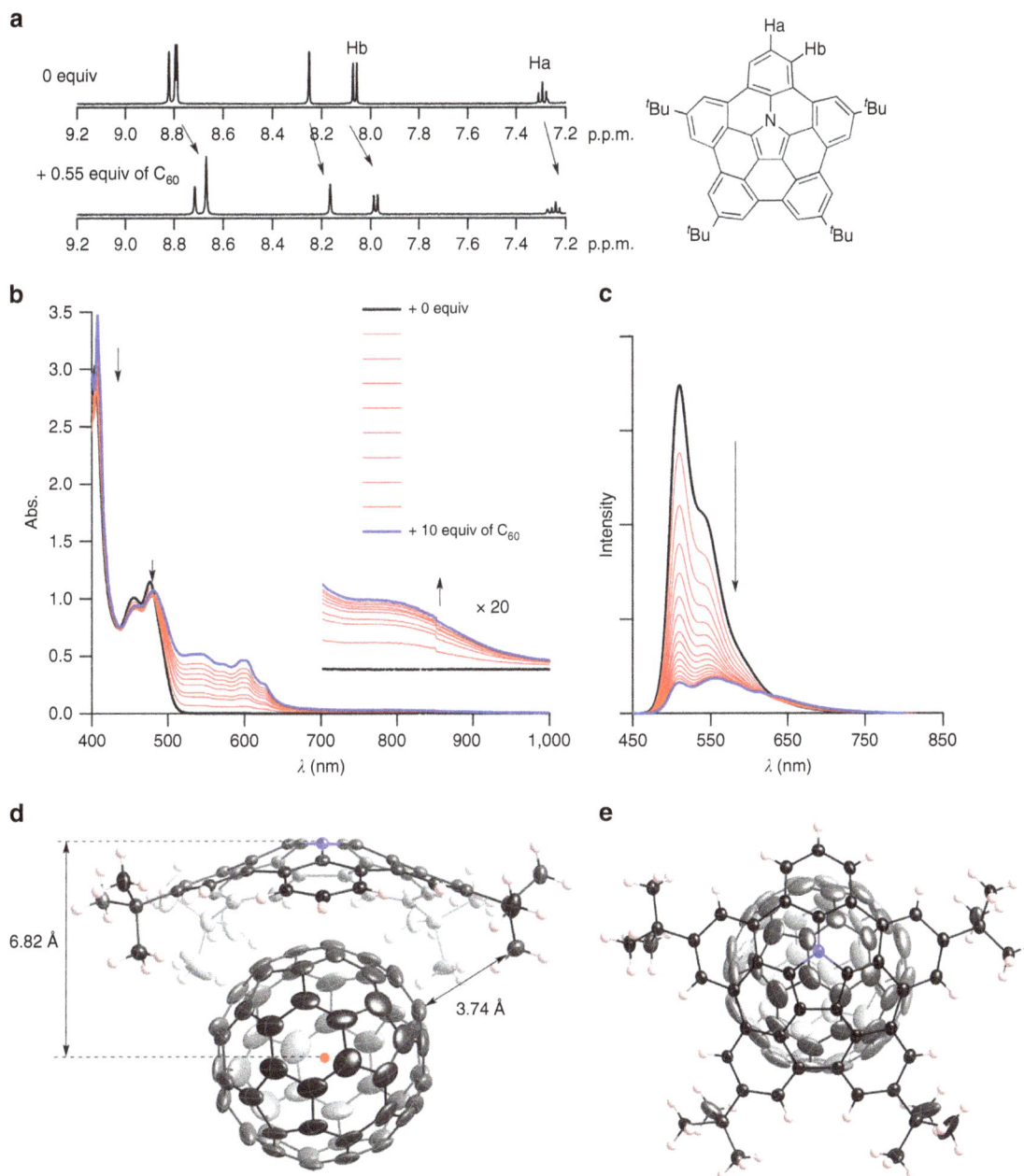

**Figure 4 | C$_{60}$ binding behaviour of 5 in solution and solid.** (**a**) $^1$H NMR spectra before (top) and after (bottom) addition of 0.55 equiv. of C$_{60}$ into a toluene-$d_8$ solution of **5**. (**b**) Ultraviolet–visible absorption spectra of addition of 0–10 equiv. of C$_{60}$ into a 1,2-dichlorobenzene solution of **5**. (**c**) Fluorescence spectra of addition from 0–150 equiv. of C$_{60}$ into a 1,2-dichlorobenzene solution of **5**. (**d**) Side view and (**e**) top view of X-ray crystal structure of **5 ⊃ C$_{60}$**. Thermal ellipsoids are scaled at 50% probability level. Solvent molecules (toluene) in the crystal structure of **5 ⊃ C$_{60}$** were omitted for clarity.

supported the conclusion that an intermolecular charge-transfer interaction exists between **5** and C$_{60}$.

Finally, we investigated the effect of the association with C$_{60}$ of **5** on the charge-carrier mobility of **5** by flash-photolysis time-resolved microwave conductivity (FP-TRMC) measurements[42]. The maximum transient conductivity ($\phi\Sigma\mu$) of **5** was measured to be $1.5 \times 10^{-5} \, \text{cm}^2 \, \text{V}^{-1} \, \text{s}^{-1}$ (Supplementary Fig. 28). For the co-crystal of **5 ⊃ C$_{60}$**, the mobility was enhanced to $2.4 \times 10^{-4} \, \text{cm}^2 \, \text{V}^{-1} \, \text{s}^{-1}$ (Fig. 5a,b). The charge-carrier generation efficiency ($\phi$) was determined to be $4.4 \times 10^{-3}$ by the transient absorption spectroscopy measurement. Accordingly, the local charge mobility of **5 ⊃ C$_{60}$** was $0.17 \, \text{cm}^2 \, \text{V}^{-1} \, \text{s}^{-1}$. Such a large mobility of **5 ⊃ C$_{60}$** should originate from effective charge separation caused by an electronic interaction between **5** and C$_{60}$. Furthermore, the alignment of **5**

and C$_{60}$ in the co-crystal may contribute to mobility enhancement. Both C$_{60}$ and **5** construct one-dimensional chain alignments in the co-crystal (Fig. 5c).

## Discussion

In summary, we have achieved the synthesis of a nitrogen-embedded buckybowl under mild conditions. The total yield of azabuckybowl **5** was 11% from 9-bromophenanthrene. We also found that the protonation of **5** resulted in the efficient generation of radical cation species. The nitrogen-embedded buckybowl was sufficiently electron-rich to assemble tightly with C$_{60}$ in solution and solid states. The molecular assembly of **5** with C$_{60}$ exhibited a significantly high charge mobility ($0.17 \, \text{cm}^2 \, \text{V}^{-1} \, \text{s}^{-1}$). The nitrogen-embedded buckybowl can be a novel molecular entity

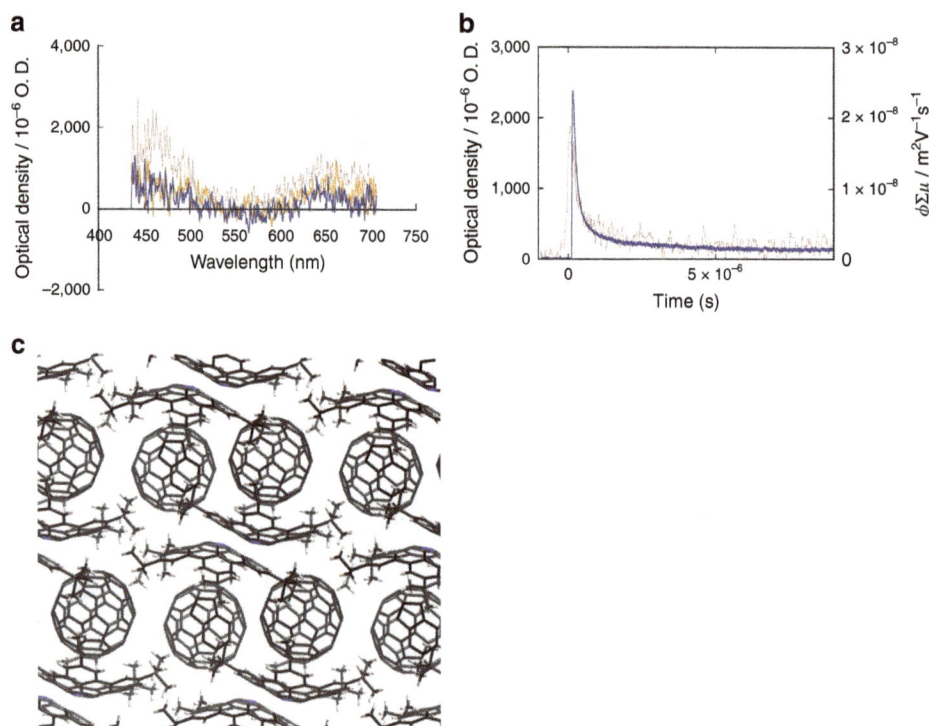

**Figure 5 | Charge-carrier mobility of 5 ⊃ C₆₀.** (**a**) Transient absorption spectra of **5 ⊃ C₆₀** on exposure to 355 nm laser pulses at $3.8 \times 10^{16}$ photons per cm$^2$. Spectra wire observed immediately after pulse exposure (red), 2 ms (orange) and 7 ms (blue) after pulse exposure. All spectra were recorded at room temperature under air-saturated atmosphere. (**b**) Kinetic traces of a photoconductivity transient (blue) recorded by FP-TRMC measurements and transient optical absorption at 650 nm for **5 ⊃ C₆₀** on exposure to 355 nm laser pulses at $9.1 \times 10^{15}$ photons per cm$^2$ (conductivity) and $3.8 \times 10^{16}$ photons per cm$^2$ (optical). (**c**) Side view of crystal packing of **5 ⊃ C₆₀**. Solvent molecules (toluene) were omitted for clarity.

in the field of curved π systems as fullerene hosts, anisotropic π donors and precursors to nitrogen-containing nanocarbon materials.

## Method

**Materials and characterization.** $^1$H NMR (500 MHz) and $^{13}$C NMR (126 MHz) spectra were recorded using a Bruker AVANCE III HD spectrometer. Chemical shifts were reported at the delta scale in p.p.m. relative to CHCl$_3$ ($\delta = 7.260$ p.p.m.), CH$_2$Cl$_2$ ($\delta = 5.320$ p.p.m.), toluene-$d_8$ ($\delta = 7.000$ p.p.m.), acetone-$d_6$ ($\delta = 2.05$ p.p.m.) and 1,2-dichlorobenzene-$d_4$ ($\delta = 6.930$ p.p.m.) for $^1$H NMR and CDCl$_3$ ($\delta = 77.0$ p.p.m.) for $^{13}$C NMR. $^1$H and $^{13}$C NMR spectra are provided for all compounds; see Supplementary Figs 1–17. Ultraviolet–visible–near infrared absorption spectra were recorded using a Shimadzu UV-2550 or JASCO V670 spectrometer. Emission spectra were recorded using a JASCO FP-6500 spectrometer and absolute fluorescence quantum yields were measured by the photon-counting method using an integration sphere. Mass spectra were recorded using a Bruker microTOF by electrospray ionization (ESI) methods. Unless otherwise noted, materials obtained from commercial suppliers were used without further purification.

**Synthesis of 3,6-Di-*tert*-butyl-9-bromophenanthrene.** 3,6-Di-*tert*-butylphenanthrene (0.540 g, 1.86 mmol) was dissolved in CCl$_4$ (11 ml) in a two-necked flask equipped with a dropping funnel. Br$_2$ (0.10 ml, 1.95 mmol) and CCl$_4$ (11 ml) were added into the dropping funnel. The solution was heated to 50 °C and then the bromine solution was added slowly over 1 h. After the addition was completed, the mixture was stirred for additional 30 min. The reaction mixture was cooled to room temperature and then quenched with aqueous Na$_2$S$_2$O$_3$. The resulting mixture was extracted with CH$_2$Cl$_2$ and the organic layer was washed with aqueous Na$_2$S$_2$O$_3$, dried over Na$_2$SO$_4$ and concentrated *in vacuo*. Purification by silica-gel column chromatography (cyclohexane as eluent) afforded the title compound (0.645 g, 1.75 mmol) in 94% yield as a white solid. $^1$H NMR (500 MHz) (CDCl$_3$): $\delta = 8.70$ (d, $J = 1.5$ Hz, 1H), 8.66 (s, 1H), 8.30 (d, $J = 8.5$ Hz, 1H), 8.02 (s,1H), 7.77 (dd, $J_1 = 8.5$ Hz, $J_2 = 1.5$ Hz, 1H), 7.74 (d, $J = 8.5$ Hz, 1H), 7.68 (dd, $J_1 = 8.5$ Hz, $J_2 = 1.5$ Hz, 1H), 1.53 (s, 9H), 1.52 (s, 9H) p.p.m.; $^{13}$C NMR (126 MHz) (CDCl$_3$): $\delta = 150.0$, 149.6, 131.1, 130.4, 129.5, 129.3, 128.5, 127.8, 127.5, 125.6, 125.4, 120.7, 118.2, 118.1, 35.25, 35.14, 31.42 p.p.m.; high-resolution atmospheric pressure chemical ionization–MS (APCI–MS): $m/z = 368.1124$, calcd for $(C_{22}H_{25}Br)^+ = 368.1134$ [$M^+$].

**Synthesis of compound 1.** A Schlenk tube containing 3,6-di-*tert*-butyl-9-bromophenanthrene (0.200 g, 0.542 mmol), Cs$_2$CO$_3$ (0.265 g, 0.812 mmol), Pd$_2$dba$_3$ · CHCl$_3$ (28.0 mg, 27.1 µmol) and Xantphos (31.3 mg, 54.0 µmol) was flushed with N$_2$ three times. To the tube, 2-chloroaniline (86 µl, 0.812 mmol) and dry 1,4-dioxane (2.0 ml) were added. The mixture was stirred for 46 h at 100 °C. The resulting mixture was cooled to room temperature, passed through a pad of Celite and concentrated *in vacuo*. Purification by silica-gel column chromatography (hexane/CH$_2$Cl$_2$) afforded 1 (0.158 g, 0.380 mmol) in 70% yield as a pale yellow solid. $^1$H NMR (500 MHz) (CDCl$_3$): $\delta = 8.76$ (d, $J = 2.0$ Hz, 1H), 8.68 (d, $J = 1.0$ Hz, 1H), 8.04 (d, $J = 8.5$ Hz, 1H), 7.74 (d, $J = 8.5$ Hz, 1H), 7.69 (dd, $J_1 = 8.5$ Hz, $J_2 = 2.0$ Hz, 1H), 7.66 (dd, $J_1 = 8.5$ Hz, $J_2 = 1.5$ Hz, 1H), 7.61 (s, 1H), 7.41 (dd, $J_1 = 8.0$ Hz, $J_2 = 1.5$ Hz, 1H), 7.04 (ddd, $J_1 = 8.5$ Hz, $J_2 = 7.0$ Hz, $J_3 = 1.5$ Hz, 1H), 6.88 (dd, $J_1 = 8.5$ Hz, $J_2 = 1.5$ Hz, 1H), 6.77 (ddd, $J_1 = 8.0$ Hz, $J_3 = 1.5$ Hz, 1H), 6.36 (s, 1H), 1.53 (s, 9H), 1.52 (s, 9H) p.p.m.; $^{13}$C NMR (126 MHz) (CDCl$_3$): $\delta = 149.6$, 148.6, 142.4, 134.6, 131.4, 130.2, 129.5, 128.5, 127.7, 127.6, 127.1, 125.1, 125.0, 122.8, 120.5, 119.4, 118.9, 118.6, 117.9, 115.6, 35.18, 31.54, 31.48 p.p.m.; high-resolution APCI–MS: $m/z = 416.2141$, calcd for $(C_{28}H_{31}ClN)^+ = 416.2140$ [$(M + H)^+$].

**Synthesis of compound 2.** A flask-containing compound 1 (0.100 g, 0.241 mmol) was flushed with N$_2$ three times. To the flask, a dry and degassed toluene/CF$_3$COOH (10 ml, 33 µl) solution was added. To the solution, a solution of DDQ (0.109 g, 0.482 mmol) in dry and degassed toluene/CF$_3$COOH (10 ml, 33 µl) was added and the mixture was stirred for 1 h at room temperature. The reaction mixture was quenched with aqueous NaHCO$_3$ and aqueous Na$_2$S$_2$O$_3$, and extracted with CH$_2$Cl$_2$. The organic layer was washed with water, dried over Na$_2$SO$_4$ and concentrated *in vacuo*. Purification by silica-gel column chromatography (hexane/CH$_2$Cl$_2$) afforded compound 2 (79.6 mg, 0.113 mmol) in 94% yield as a pale yellow solid. $^1$H NMR (500 MHz) (CDCl$_3$): $\delta = 9.00$ (d, $J = 8.5$ Hz, 2H), 8.80 (d, $J = 1.5$ Hz, 2H), 8.75 (d, $J = 1.5$ Hz, 2H), 7.85 (dd, $J_1 = 8.5$ Hz, $J_2 = 1.5$ Hz, 1H), 7.69–7.77 (m, 4H), 7.62 (ddd, $J_1 = J_2 = 7.5$ Hz, $J_3 = 1.5$ Hz, 1H), 7.32 (dd, $J_1 = 9.0$ Hz, $J_2 = 1.5$ Hz, 2H), 7.08 (dd, $J = 9.0$ Hz, 2H), 1.58 (s, 18H), 1.48 (s, 18H) p.p.m.; $^{13}$C NMR (126 MHz) (CDCl$_3$): $\delta = 147.3$, 146.8, 140.7, 135.7, 132.6, 132.2, 131.3, 131.1, 130.6, 128.9, 127.8, 126.6, 125.8, 124.4, 123.6, 121.7, 120.5, 119.7, 119.4, 116.5, 35.02, 34.91, 31.61, 31.40 p.p.m.; ultraviolet–visible (CH$_2$Cl$_2$): $\lambda_{max}$ ($\varepsilon [M^{-1} cm^{-1}]) = 342$ (22,000), 359 (23,000), 376 (21,000) nm; high-resolution APCI–MS: $m/z = 702.3872$, calcd for $(C_{50}H_{53}ClN)^+ = 702.3861$ [$(M + H)^+$].

**Synthesis of compound 3.** A Schlenk tube containing compound 2 (30.1 mg, 42.8 µmol), K$_2$CO$_3$ (35.5 mg, 0.257 mmol), Pd(OAc)$_2$ (9.56 mg, 42.6 µmol) and

$PCy_3 \cdot HBF_4$ (31.5 mg, 85.5 μmol) was flushed with $N_2$ three times. To the tube, dry and degassed DMA (1.5 ml) was added. The mixture was stirred for 43 h at 130 °C. The resulting mixture was cooled to room temperature, passed through a pad of Celite and concentrated *in vacuo*. Purification by silica-gel column chromatography (hexane/$CH_2Cl_2$) afforded compound **3** (17.8 mg, 26.8 μmol) in 63% yield as a yellow solid. $^1H$ NMR (500 MHz) ($CDCl_3$): $\delta = 9.15$ (d, $J = 9.0$ Hz, 1H), 9.13 (d, $J = 10$ Hz, 1H) 8.92 (d, $J = 1.0$ Hz, 1H), 8.89 (s, 1H), 8.86 (d, $J = 0.5$ Hz, 1H), 8.83 (d, $J = 0.5$ Hz, 1H), 8.61 (dd, $J_1 = 8.0$ Hz, $J_2 = 1.5$ Hz, 1H), 8.59 (s, 1H), 8.51 (d, $J = 8.5$ Hz, 1H), 8.40 (d, $J = 8.0$ Hz, 1H), 7.88 (d, $J = 8.5$ Hz, 1H), 7.84 (d, $J = 8.5$ Hz, 1H), 7.66 (dd, $J_1 = 8.5$ Hz, $J_2 = 1.5$ Hz, 1H), 7.50–7.56 (m, 2H), 1.71 (s, 9H), 1.64 (s, 9H), 1.63 (s, 9H), 1.60 (s, 9H) p.p.m.; $^{13}C$ NMR (126 MHz) ($CDCl_3$): $\delta = 148.9$, 147.7, 147.7, 146.5, 135.2, 132.2, 129.3, 129.1, 129.0, 128.9, 127.5, 127.3, 127.2, 126.9, 126.6, 126.1, 125.4, 124.7, 124.6, 124.2, 124.2, 124.0, 123.4, 123.1, 122.8, 121.9, 120.1, 119.7, 119.6, 119.5, 118.4, 118.1, 116.0, 110.8, 35.84, 35.12, 35.10, 35.01, 32.00, 31.65, 31.64, 31.56 p.p.m.; ultraviolet–visible ($CH_2Cl_2$): $\lambda_{max}$ ($\varepsilon$[$M^{-1}$ $cm^{-1}$]) = 310 (53,000), 330 (43,000), 378 (13,000) and 404 (8,700) nm; fluorescence ($CH_2Cl_2$, $\lambda_{ex} = 378$ nm): $\lambda_{em} = 456$ and 471 nm ($\Phi_f = 0.18$); high-resolution APCI–MS: $m/z = 666.4088$, calcd for $(C_{50}H_{52}N)^+ = 666.4094$ [$(M+H)^+$].

**Synthesis of compound 4.** Compound **3** (40.1 mg, 60.3 μmol) was dissolved in $CCl_4$ (6.0 ml) in a two-necked flask equipped with a dropping funnel. A solution of $Br_2$ (0.10 ml, 2.0 mmol) in $CCl_4$ (3.0 ml) was added to the dropping funnel. The mixture was heated to 70 °C and then the bromine solution was added slowly over 15 min. After the addition was complete, the mixture was stirred for an additional 12.5 h. The reaction mixture was cooled to room temperature and then quenched with aqueous $Na_2S_2O_3$. The resulting mixture was extracted with $CH_2Cl_2$ and the organic layer was washed with aqueous $Na_2S_2O_3$, dried over $Na_2SO_4$ and concentrated *in vacuo*. Purification by silica-gel column chromatography (hexane only) afforded compound **4** (30.3 mg, 33.6 μmol) in 56% yield as a yellow solid. $^1H$ NMR (500 MHz) ($CDCl_3$): $\delta = 8.83$ (s, 1H), 8.80 (s,1H), 8.75 (s, 1H), 8.59 (d, $J = 1.5$ Hz, 1H), 8.58 (s, 1H), 8.47 (s, 1H), 8.41 (d, $J = 8.5$ Hz, 1H), 8.05 (s, 1H), 7.96 (d, $J = 9.0$ Hz, 1H), 7.91 (s, 1H), 7.68 (dd, $J_1 = 7.8$ Hz, $J_2 = 1.5$ Hz, 1H), 7.49 (dd, $J_1 = 9.0$ Hz, $J_2 = 2.0$ Hz, 1H), 1.66 (s, 9H), 1.60 (s, 9H), 1.59 (s, 9H), 1.52 (s, 9H) p.p.m.; $^{13}C$ NMR (126 MHz) ($CDCl_3$): $\delta = 149.7$, 149.0, 147.9, 147.9, 135.7, 132.3, 132.1, 131.0, 130.6, 130.2, 129.9, 129.4, 129.4, 127.8, 127.6, 127.1, 126.8, 125.5, 124.9, 124.6, 124.3, 124.2, 123.4, 121.0, 120.3, 119.9, 119.7, 118.9, 118.8, 118.8, 118.1, 117.4, 117.1, 108.8, 35.91, 35.13, 35.05, 34.99, 31.93, 31.54, 31.51 p.p.m.; high-resolution APCI–MS: $m/z = 900.1403$, calcd for $(C_{50}H_{49}Br_3N)^+ = 900.1410$ [$(M+H)^+$].

**Synthesis of compound 5.** A Schlenk tube containing compound **4** (20.3 mg, 22.5 μmol), $K_2CO_3$ (24.8 mg, 0.180 mmol), $Pd(OAc)_2$ (10.7 mg, 47.7 μmol) and $PCy_3 \cdot HBF_4$ (33.2 mg, 90.0 μmol) was flushed with $N_2$ three times. To the tube, dry and degassed DMA (2.6 ml) was added. The mixture was stirred for 16 h at 130 °C. The resulting mixture was cooled to room temperature and extracted with ethyl acetate. The organic layer was washed with water, dried over $Na_2SO_4$ and concentrated *in vacuo*. Purification by silica-gel column chromatography (hexane only) afforded compound **5** (6.81 mg, 10.3 μmol) in 46% yield as a yellow solid. $^1H$ NMR (500 MHz) ($CDCl_3$): $\delta = 8.61$ (s, 2H), 8.54 (s, 2H), 8.52 (s, 2H), 8.23–8.24 (m, 4H), 7.50 (t, $J = 8.0$ Hz, 1H), 1.63 (s, 18H), 1.60 (s, 18H) p.p.m.; $^{13}C$ NMR (126 MHz) ($CDCl_3$): $\delta = 148.8$, 147.6, 140.1, 135.3, 132.5, 131.1, 130.0, 129.0, 128.6, 127.7, 126.1, 123.4, 123.0, 122.5, 120.5, 120.0, 119.4, 117.9, 35.92, 35.84, 32.30, 32.17 p.p.m.; ultraviolet–visible ($CH_2Cl_2$): $\lambda_{max}$ ($\varepsilon$[$M^{-1}cm^{-1}$]) = 400 (35,000), 453 (12,000), 472 (13,000) nm; fluorescence ($CH_2Cl_2$, $\lambda_{ex} = 400$ nm): $\lambda_{em} = 508$ and 542 nm ($\Phi_f = 0.17$); high-resolution APCI–MS: $m/z = 662.3748$, calcd for $(C_{50}H_{48}N)^+ = 662.3781$ [$(M+H)^+$].

**Synthesis of compound 6.** A Schlenk tube containing compound **5** (30.5 mg, 46.0 μmol), bis(pinacolato)diboron (117 mg, 0.461 mmol), $[Ir(OMe)(cod)]_2$ (30.5 mg, 46.0 μmol) and 4,4'-di-*tert*-butyl-2,2'-bipyridyl (25.1 mg, 93.4 μmol) was flushed with $N_2$ three times. To the tube, dry and degassed octane (1.5 ml) was added. The mixture was stirred for 10.5 h at 110 °C. The resulting mixture was cooled to room temperature and concentrated *in vacuo*. Purification by silica-gel column chromatography afforded compound **6** (29.0 mg, 36.8 μmol) in 80% yield as a yellow solid. $^1H$ NMR (500 MHz) ($CDCl_3$): $\delta = 8.68$ (s,2H), 8.62 (s, 2H), 8.55 (s, 2H), 8.52 (s, 2H), 8.34 (s, 2H), 1.64 (s, 18H), 1.64 (s, 18H), 1.49 (s, 12H) p.p.m.; $^{13}C$ NMR (126 MHz) ($CDCl_3$): $\delta = 148.9$, 147.7, 140.0, 137.1, 132.6, 130.9, 130.1, 128.9, 128.5, 127.7, 125.4, 123.0, 120.4, 120.0, 119.4, 118.4, 84.21, 36.01, 35.84, 32.29, 32.24, 24.99 p.p.m.; high-resolution ESI–MS: $m/z = 787.4571$, calcd for $(C_{56}H_{58}BNO_2)^+ = 787.4564$ [$(M)^+$].

**Synthesis of compound 7.** A Schlenk tube containing compound **6** (9.48 mg, 12.0 μmol), $PdCl_2dppf \cdot CH_2Cl_2$ (4.97 mg, 6.09 μmol) and $Cs_2CO_3$ (9.80 mg, 30.1 μmol) was flushed with $N_2$ three times. To the tube, 2,4,6-triisopropylbromobenzene (56.2 mg, 0.199 mmol) and dry and degassed 1,4-dioxane (1.0 ml) were added. The mixture was stirred for 13 h at 100 °C. The resulting mixture was cooled to room temperature and concentrated *in vacuo*. Purification by silica-gel column chromatography afforded compound **7** (5.75 mg, 6.66 μmol) in 55% yield as a yellow solid. $^1H$ NMR (500 MHz) (acetone-$d_6$): $\delta = 8.85$ (s, 2H), 8.79 (s, 2H), 8.79 (s, 2H), 8.65 (d, $J = 1.0$ Hz, 2H), 8.36 (s, 2H), 7.25 (d, $J = 1.5$ Hz, 1H), 7.14 (d, J = 1.5 Hz, 1H), 3.13 (sext, $J = 7.0$ Hz, 1H), 3.01 (sext, $J = 7.0$ Hz, 1H), 2.37 (sext, $J = 7.0$ Hz, 1H), 1.62 (s, 18H), 1.57 (s, 18H), 1.35 (d, $J = 7.0$ Hz, 6H), 1.31 (d, $J = 7.0$ Hz, 6H), 0.88 (d, $J = 7.0$ Hz, 6H) p.p.m.; $^1H$ NMR (500 MHz) (1,2-dichlorobenzene-$d_4$): $\delta = 8.75$ (s, 2H), 8.74 (s, 2H), 8.72 (s, 2H), 8.25 (s, 2H), 8.21 (s, 2H), 7.35 (s, 1H), 7.28 (s, 1H), 3.35 (sext, $J = 7.5$ Hz, 1H), 3.02 (sext, $J = 7.0$ Hz, 1H), 2.85 (sext, $J = 7.0$ Hz, 1H), 1.66 (s, 18H), 1.50 (s, 18H), 1.41 (d, $J = 7.0$ Hz, 6H), 1.37 (d, 6H), 1.11 (d, $J = 7.0$ Hz, 6H) p.p.m.; $^{13}C$ NMR (126 MHz) ($CDCl_3$): $\delta = 148.9$, 148.5, 147.6, 147.1, 147.0, 140.4, 137.3, 135.9, 134.3, 132.5, 131.2, 130.1, 129.3, 128.5, 127.9, 125.9, 123.9, 123.3, 121.1, 120.7, 120.5, 119.9, 119.3, 117.9, 35.89, 35.84, 34.38, 32.31, 32.10, 30.66, 30.10, 24.36, 24.14, 24.06 p.p.m.; high-resolution ESI–MS: $m/z = 863.5438$, calcd for $(C_{65}H_{69}N)^+ = 863.5425$ [$(M)^+$].

**X-ray diffraction analysis.** X-ray data were obtained using a Bruker D8 QUEST X-ray diffractometer with an IμS microfocus X-ray source and a large-area (10 cm × 10 cm) CMOS detector (Photon 100) for **3** and **4**, and using a Rigaku CCD diffractometer (Saturn 724 with MicroMax-007) with Varimax Mo optics using graphite monochromated Mo-Kα radiation ($\lambda = 0.71075$ Å) for **5** and **5** ⊃ $C_{60}$. For ORTEP structures of **3**, **4**, **5** and **5** ⊃ $C_{60}$, see Supplementary Figs 18–21. Crystallographic details are given in CIF files (Supplementary Data 1–4). A fine crystal of **5** for the X-ray diffraction analysis was obtained by the vapour diffusion of methanol into its *o*-xylene solution. For the X-ray crystal structure of **5** ⊃ $C_{60}$, a fine crystal for the X-ray diffraction analysis was obtained by the vapour diffusion of methanol into a toluene solution with a 1:1 mixture of **5** and $C_{60}$. The molecule $C_{60}$ was significantly disordered and refined as two disordered rigid bodies by restraining with DFIX, DANG, DELU and SIMU commands as generally used for the refinement of $C_{60}$. The toluene solvent molecules were assigned as two disordered units by using minus part number, because it was located at the special position (Supplementary Fig. 21). The resolution and data were sufficiently suitable to determine the binding manner in the crystal ($R_{int} = 0.0252$, 22,169 total reflections and 11,448 unique reflections were observed). The detailed crystallographic data for all compounds are listed in Supplementary Table 1.

**Electrochemical analysis.** The cyclic voltammogram and differential-pulse voltammogram of **5** were recorded using an ALS electrochemical analyser 612C. Measurements were performed in freshly distilled dichloromethane with tetrabutylammonium hexafluorophosphate as the electrolyte. A three-electrode system was used. The system consisted of a platinum working electrode, a platinum wire and $Ag/AgClO_4$ as the reference electrode. The scan rate was 100 $mVs^{-1}$. The measurement was performed under nitrogen atmosphere. All potentials are referenced to the potential of ferrocene/ferrocenium cation couple. The data are listed in Supplementary Table 2. The electro-oxidative absorption of **5** was recorded under argon atmosphere with a BAS SEC-F spectroelectrochemical flow cell kit equipped with a DH-2000-BAL as the ultraviolet–visible–near infrared light source and an HR4000CG-ultraviolet–near infrared spectrometer.

**Determination of binding constant.** The binding constant ($K_a$) of $C_{60}$ with compound **5** was determined by ultraviolet–visible absorption and emission spectral analysis on the titration of $C_{60}$ into the 1,2-dichlorobenzene solution of **5**. The fitting was performed with the correlation between the change of absorbance or fluorescence intensity ($\Delta X$) at 700 and 508 nm, and the initial concentration of the guest ($[G]_0$) using the equation as follows:

$$1/\Delta X = 1/(b\Delta\varepsilon[G]_0[H]_0K_a) + 1/(b\Delta\varepsilon[H]_0), \quad (1)$$

where $\Delta\varepsilon$ is the gap of molar coefficients between guest and complex, and $[H]_0$ is the initial concentration of the host (Supplementary Figs 23 and 24). The estimated $K_a$ values by ultraviolet–visible spectral analysis were $3.9 \times 10^3\ M^{-1}$ for the first attempt and $3.7 \times 10^3\ M^{-1}$ for the second attempt. The $K_a$ was also estimated by the emission spectral analysis to be $3.8 \times 10^3\ M^{-1}$. The average $K_a$ is $3.8 \times 10^3\ M^{-1}$.

**ESR measurement.** ESR spectra were recorded at room temperature using a Bruker E500 spectrometer with 2.6$\phi$ quartz sample tubes. A sample solution of **5** was prepared under air and the ESR tube was sealed. Other samples were prepared by the addition of the degassed solution of TFA and BAHA in $CH_2Cl_2$ to the solution of **5**.

**Time-resolved microwave conductivity measurement.** Transient photoconductivity was measured by FP-TRMC[43]. A resonant cavity was used to obtain a high degree of sensitivity in the conductivity measurement. The resonant frequency and microwave power were set at ~9.1 GHz and 3 mW, respectively, such that the electric field of the microwave was sufficiently small not to disturb the motion of charge carriers. The conductivity value is converted to the product of the quantum yield $\phi$ and the sum of charge-carrier mobilities $\Sigma\mu$ by $\phi\Sigma\mu = \Delta\sigma\ (eI_0F_{light})^{-1}$, where $e$, $I_0$, $F_{light}$ and $\Delta\sigma$ are the unit charge of a single electron, incident photon density of excitation laser (photons per $m^2$), a correction (or filling) factor ($m^{-1}$) and a transient photoconductivity, respectively. The sample was set at the highest electric field in a resonant cavity. FP-TRMC experiments were performed at room temperature. The measurements of **5** and **5** ⊃ **C₆₀** were performed for crystalline samples covered with a polyvinyl alcohol film on a quartz substrate.

**Theoretical calculations.** All calculations were performed using the Gaussian 09 programme[44]. The geometry of **5**$^{+\bullet}$, in which all *tert*-butyl groups were replaced with hydrogen, was optimized by the DFT method using the B3LYP[45,46] functional and the 6-31G(d) basis set. The geometry of **5** ⊃ **C₆₀** was optimized by Zhao's M06-2X functional[47] and the 6-31G(d) basis set. The oscillator strengths of **5**$^{+\bullet}$ and **5** ⊃ **C₆₀** were calculated by the time-dependent DFT method at the B3LYP/6-31G(d) level (Supplementary Figs 30 and 31). For calculations of the bowl-to-bowl inversion energy, the ground and transition state geometries of **7** were optimized at the B3LYP/cc-pVDZ level. Zero-point energy and thermal energy corrections were conducted for the optimized structures. The calculation results are summarized in Supplementary Tables 3–7.

**Determination of bowl-to-bowl inversion energy by 2D EXSY measurement.**
The bowl-to-bowl inversion barrier of **7** was measured by 2D EXSY using the signals for methine protons of isopropyl groups at approximately $\delta = 3.3$ and 2.8 p.p.m. (Supplementary Fig. 32)[48]. 2D EXSY measurements were performed in 1,2-dichlorobenzene-$d_4$ at 393 K with a phase-sensitive nuclear Overhauser effect spectroscopy pulse sequence. The mixing time was increased from 50 to 300 ms. The rate constant ($k$) was determined using equation as follows:

$$k = (1/\tau_m)\ln((r+1)/(r-1)), \qquad (2)$$

where $\tau_m$ is the mixing time and $r$ is defined by the equation as follows:

$$r = (I_{AA} + I_{BB})/(I_{AB} + I_{BA}), \qquad (3)$$

where $I_{AB}$ and $I_{BA}$ are the intensities of the cross-peaks between two exchangeable signals A and B, and $I_{AA}$ and $I_{BB}$ are the intensities of the diagonal signals (Supplementary Fig. 33). The free energy ($\Delta G^{\ddagger}$) of the bowl-to-bowl inversion was finally obtained using the Eyring equation.

# References

1. Haley, M. M. & Tykwinski, R. R. *Carbon-Rich Compounds: From Molecules to Materials* (Wiley-VCH, Weinheim, 2006).
2. Petrukhina, M. A. & Scott, L. T. *Fragments of Fullerenes and Carbon Nanotubes: Designed Synthesis, Unusual Reactions, and Coordination Chemistry* (Wiley, Hoboken, 2012).
3. Kroto, H. W., Heath, J. R., O'Brien, S. C., Curl, R. F. & Smalley, R. E. C₆₀: buckminsterfullerene. *Nature* **318**, 162–163 (1985).
4. Iijima, S. Helical microtubules of graphitic carbon. *Nature* **354**, 56–58 (1991).
5. Kawasumi, K., Zhang, Q., Segawa, Y., Scott, L. T. & Itami, K. A grossly warped nanographene and the consequences of multiple odd-membered-ring defects. *Nat. Chem.* **5**, 739–744 (2013).
6. Shen, Y. & Chen, C.-F. Helicenes: synthesis and applications. *Chem. Rev.* **112**, 1463–1535 (2012).
7. Golder, M. R. & Jasti, R. Syntheses of the smallest carbon nanohoops and the emergence of unique physical phenomena. *Acc. Chem. Res.* **48**, 557–566 (2015).
8. Amaya, T. & Hirao, T. Chemistry of sumanene. *Chem. Rec.* **15**, 310–321 (2015).
9. Scott, L. T. *et al.* Geodesic polyarenes with exposed concave surfaces. *Pure Appl. Chem.* **71**, 209–219 (1999).
10. Feng, C. -N., Kuo, M. -Y. & Wu, Y. -T. Synthesis, structural analysis, and properties of [8]circulenes. *Angew. Chem. Int. Ed.* **52**, 7791–7794 (2013).
11. Field, J. E., Muller, G., Riehl, J. P. & Venkataraman, D. Circularly polarized luminescence from bridged triarylamine helicenes. *J. Am. Chem. Soc.* **125**, 11808–11809 (2003).
12. Wakamiya, A. *et al.* On-top π-stacking of quasiplanar molecules in hole-transporting materials: Inducing anisotropic carrier mobility in amorphous films. *Angew. Chem. Int. Ed.* **53**, 5800–5804 (2014).
13. Miyajima, D. *et al.* Liquid crystalline corannulene responsive to electric field. *J. Am. Chem. Soc.* **131**, 44–45 (2009).
14. Imamura, K., Takimiya, K., Aso, Y. & Otsubo, T. Triphenyleno[1,12-*bcd*:4,5-*b'c'd*':8,9-*b"c"d*"]trithiophene: the first bowl-shaped heteroaromatic. *Chem. Commun.* 1859–1860 (1999).
15. Tan, Q., Higashibayashi, S., Karanjit, S. & Sakurai, H. Enantioselective synthesis of a chiral nitrogen-doped buckybowl. *Nat. Commun.* **3**, 891 (2012).
16. Barth, W. E. & Lawton, R. G. Dibenzo[*ghi,mno*]fluoranthene. *J. Am. Chem. Soc.* **88**, 380–381 (1966).
17. Sakurai, H., Daiko, T. & Hiraoka, T. A synthesis of sumanene, a fullerene fragment. *Science* **301**, 1878 (2003).
18. Scott, L. T., Hashemi, M. M., Meyer, D. T. & Warren, H. B. Corannulene. A convenient new synthesis. *J. Am. Chem. Soc.* **113**, 7082–7084 (1991).
19. Tsefrikas, V. M. & Scott, L. T. Geodesic polyarenes by flash vacuum pyrolysis. *Chem. Rev.* **106**, 4868–4884 (2006).
20. Butterfield, A. M., Gilomen, B. & Siegel, J. S. Kilogram-scale production of corannulene. *Org. Process. Res. Dev.* **16**, 664–676 (2012).
21. Wu, Y.-T. & Siegel, J. S. Aromatic molecular-bowl hydrocarbons: Synthetic derivatives, their structures, and physical properties. *Chem. Rev.* **106**, 4843–4867 (2006).
22. Vostrowsky, O. & Hirsch, A. Heterofullerenes. *Chem. Rev.* **106**, 5191–5207 (2006).
23. Jang, J. W., Lee, C. E., Lyu, S. C., Lee, T. J. & Lee, C. J. Structural study of nitrogen-doping effects in bamboo-shaped multiwalled carbon nanotubes. *Appl. Phys. Lett.* **84**, 2877–2879 (2004).
24. Gong, K., Du, F., Xia, Z., Durstock, M. & Dai, L. Nitrogen-doped carbon nanotube arrays with high electrocatalytic activity for oxygen reduction. *Science* **323**, 760–764 (2009).
25. Gao, X., Zhang, S. B., Zhao, Y. & Nagase, S. A nanoscale jigsaw-puzzle approach to large π-conjugated systems. *Angew. Chem. Int. Ed.* **49**, 6764–6767 (2010).
26. Goto, K. *et al.* Intermolecular oxidative annulation of 2-aminoanthracenes to diazaacenes and aza[7]helicenes. *Angew. Chem. Int. Ed.* **51**, 10333–10336 (2012).
27. Ito, S. *et al.* Synthesis of highly twisted and fully π-conjugated porphyrinic oligomers. *J. Am. Chem. Soc.* **137**, 142–145 (2015).
28. Ito, S., Tokimaru, Y. & Nozaki, K. Benzene-fused azacorannulene bearing an internal nitrogen atom. *Angew. Chem. Int. Ed.* **54**, 7256–7260 (2015).
29. Nishihara, Y., Suetsugu, M., Saito, D., Kinoshita, M. & Iwasaki, M. Synthesis of cyclic 1-alkenylboronates via Zr-mediated double functionalization of alkynylboronates and sequential Ru-catalyzed ring-closing olefin metathesis. *Org. Lett.* **15**, 2418–2421 (2013).
30. Haddon, R. C. Comment on the relationship of the pyramidalization angle at a conjugated carbon atom to the σ bond angles. *J. Phys. Chem. A* **105**, 4164–4165 (2001).
31. Mkhalid, I. A. I., Barnard, J. H., Marder, T. B., Murphy, J. M. & Hartwig, J. F. C–H activation for the construction of C–B bonds. *Chem. Rev.* **110**, 890–931 (2010).
32. Eliseeva, M. N. & Scott, L. T. Pushing the Ir-catalyzed C–H polyborylation of aromatic compounds to maximum capacity by exploiting reversibility. *J. Am. Chem. Soc.* **134**, 15169–15172 (2012).
33. Amaya, T., Sakane, H., Muneishi, T. & Hirao, T. Bowl-to-bowl inversion of sumanene derivatives. *Chem. Commun.* 765–767 (2008).
34. Wu, T. -C., Hsin, H. -J., Kuo, M. -Y., Li, C. -H. & Wu, Y. -T. Synthesis and structural analysis of a highly curved buckybowl containing corannulene and sumanene fragments. *J. Am. Chem. Soc.* **133**, 16319–16321 (2011).
35. Wu, Y. -T. *et al.* Multiethynyl corannulenes: synthesis, structure, and properties. *J. Am. Chem. Soc.* **130**, 10729–10739 (2008).
36. Rathore, R. & Kochi, J. K. Acid catalysis vs. electron-transfer catalysis via organic cations or cation-radicals as the reactive intermediate. Are these distinctive mechanisms? *Acta Chem. Scand.* **52**, 114–130 (1998).
37. Mizyed, S. *et al.* Embracing C₆₀ with multiarmed geodesic partners. *J. Am. Chem. Soc.* **123**, 12770–12774 (2001).
38. Georghiou, P. E., Tran, A. H., Mizyed, S., Bancu, M. & Scott, L. T. Concave polyarenes with sulfide-linked flaps and tentacles: new electron-rich hosts for fullerenes. *J. Org. Chem.* **70**, 6158–6163 (2005).
39. Dawe, L. N. *et al.* Corannulene and its penta-*tert*-butyl derivative co-crystallize 1: 1 with pristine C₆₀-fullerene. *Chem. Commun.* **48**, 5563–5565 (2012).
40. Filatov, A. S. *et al.* Bowl-shaped polyarenes as concave–convex shape complementary hosts for C₆₀- and C₇₀-fullerenes. *Crystal Growth Design* **14**, 756–762 (2014).
41. Seiders, T. J., Baldridge, K. K., Siegel, J. S. & Gleiter, R. Ionization of corannulene and 1,6-dimethylcorannulene: photoelectron spectra, electrochemistry, charge transfer bands and *ab* initio computations. *Tetrahedron Lett.* **41**, 4519–4522 (2000).
42. Seki, S., Saeki, A., Sakurai, T. & Sakamaki, D. Charge carrier mobility in organic molecular materials probed by electromagnetic waves. *Phys. Chem. Chem. Phys.* **16**, 11093–11113 (2014).
43. Saeki, A., Koizumi, Y., Aida, T. & Seki, S. Comprehensive approach to intrinsic charge carrier mobility in conjugated organic molecules, macromolecules, and supramolecular architectures. *Acc. Chem. Res.* **45**, 1193–1202 (2012).
44. Frisch, M. J. *et al. Gaussian 09, Revision D.01* (Gaussian, Inc., Wallingford, CT, 2013).
45. Becke, A. D. Density-functional exchange-energy approximation with correct asymptotic behavior. *Phys. Rev. A* **38**, 3098–3100 (1988).
46. Lee, C., Yang, W. & Parr, R. G. Development of the Colle-Salvetti correlation-energy formula into a functional of the electron density. *Phys. Rev. B* **37**, 785–789 (1988).
47. Zhao, Y. & Truhlar, D. G. The M06 suite of density functionals for main group thermochemistry, thermochemical kinetics, noncovalent interactions, excited states, and transition elements: two new functionals and systematic testing of four M06-class functionals and 12 other functionals. *Theor. Chem. Acc.* **120**, 215–241 (2008).
48. Miyawaki, A., Kuad, P., Takashima, Y., Yamaguchi, H. & Harada, A. Molecular puzzle ring: *pseudo*[1]rotaxane from a flexible cyclodextrin derivative. *J. Am. Chem. Soc.* **130**, 17062–17069 (2008).

## Acknowledgements

This work was supported by a Grant-in-Aid for Scientific Research on Innovative Areas 'pi-System Figuration' (26102003) and 'Science of Atomic Layers' (26107519), by the Program for Leading Graduate Schools 'Integrative Graduate Education and Research in Green Natural Sciences' and a Grant-in-Aid for Scientific Research(C) (25410039) from the Ministry of Education, Culture, Sports, Science and Technology, Japan. H.S. acknowledges the Asahi Glass Foundation for financial support. H.Y. appreciates the Japan Society for the Promotion of Science, Research Fellowship for Young Scientists.

## Author contributions

S.H. and H.S. designed and conducted the project. H.Y. performed the synthesis and characterization, and measured the optical and electrochemical properties. S.H. performed X-ray diffraction analysis and DFT calculations. Y.H. measured and analysed the ESR spectra. D.S. and S.S. measured and analysed the charge-carrier mobility by using the TRMC method. S.H. and H.S. prepared the manuscript.

# 3

# Synthesis of tetrasubstituted 1-silyloxy-3-aminobutadienes and chemistry beyond Diels–Alder reactions

Xijian Li[1], Siyu Peng[1], Li Li[1] & Yong Huang[1]

Electron-rich dienes have revolutionized the synthesis of complex compounds since the discovery of the legendary Diels–Alder cycloaddition reaction. This highly efficient bond-forming process has served as a fundamental strategy to assemble many structurally formidable molecules. Amino silyloxy butadienes are arguably the most reactive diene species that are isolable and bottleable. Since the pioneering discovery by Rawal, 1-amino-3-silyloxybutadienes have been found to undergo cycloaddition reactions with unparalleled mildness, leading to significant advances in both asymmetric catalysis and total synthesis of biologically active natural products. In sharp contrast, this class of highly electron-rich conjugated olefins has not been studied in non-cycloaddition reactions. Here we report a simple synthesis of tetrasubstituted 1-silyloxy-3-aminobutadienes, a complementarily substituted Rawal's diene. This family of molecules is found to undergo a series of intriguing chemical transformations orthogonal to cycloaddition reactions. Structurally diverse polysubstituted ring architectures are established in one step from these dienes.

[1] Key Laboratory of Chemical Genomics, Peking University, Shenzhen Graduate School, Shenzhen 518055, China. Correspondence and requests for materials should be addressed to Y.H. (email: huangyong@pkusz.edu.cn).

Amino silyloxy butadienes are a class of highly reactive compounds that have demonstrated tremendous utility in natural product synthesis[1–8] and asymmetric catalysis[9–13]. In 1997, Rawal and colleagues[14,15] reported the first general synthesis of 1-amino-3-silyloxybutadienes from vinylogous amides. This family of highly reactive species showed significantly enhanced reactivity compared with the other electron-rich dienes in a number of cycloaddition reactions[16,17]. The extraordinary reactivity of the Rawal's diene is a result of an exceptionally high and polarized electron density along the four carbon conjugation system. Besides the Rawal's diene, other amino silyloxydienes were also reported, most of which bear the 1-amino-3-silyloxy substitution pattern[18–20]. In comparison, the complementarily substituted 1-silyloxy-3-aminodienes, which have a similar electronic property as their 1-amino-3-silyloxy counterparts, have been much less studied[21,22]. Schlessinger and Tsuge reported that 1-silyloxy-3-aminobutadienes could be prepared by γ-silylation of vinylogous amides or carbamates, and they underwent facile Diels–Alder reactions. Among these

amino silyloxy butadienes, tetrasubstituted analogues are very rare. In addition, amino silyloxy butadienes have received very little attention outside the cycloaddition paradigm. Considering the versatile functionalities embedded in these highly electron-rich molecules, developing efficient synthesis of 1-silyloxy-3-aminodienes and exploring transformations beyond cycloadditions would lead to significantly broadened synthetic utility of these versatile intermediates.

Herein, we report a synthesis of tetrasubstituted 1-silyloxy-3-aminobutadienes from readily available allenyl aldehydes and secondary amines. The reaction occurs under mild conditions in the absence of any catalyst and is highly atom economical. These dienes are discovered to undergo a number of captivating transformations other than cycloadditions that lead to a wide range of heavily substituted ring structures.

## Results

**Serendipitous synthesis of 1-silyloxy-3-aminobutadiene.** We recently developed a one-step synthesis of tri- and

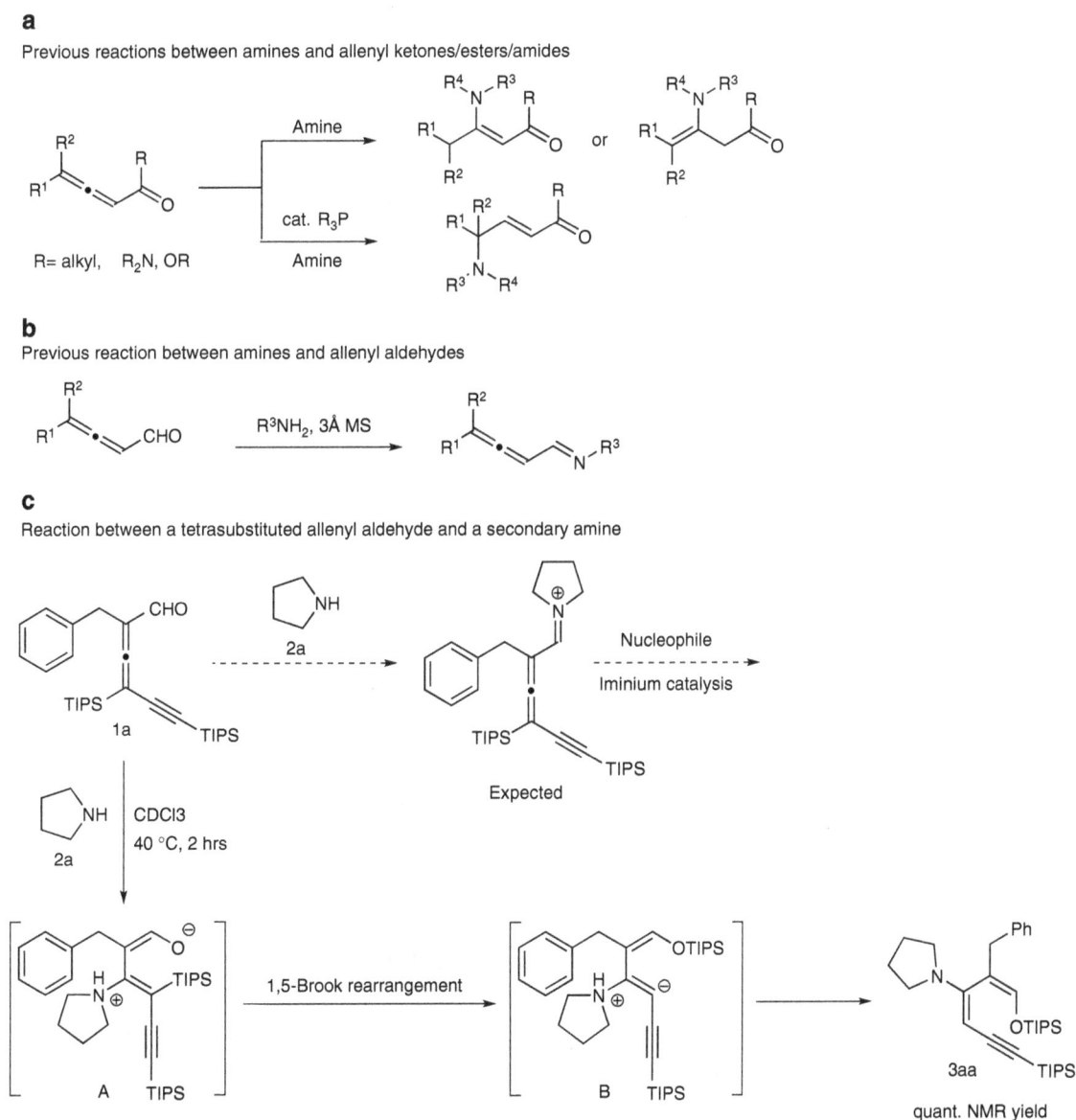

**a**
Previous reactions between amines and allenyl ketones/esters/amides

**b**
Previous reaction between amines and allenyl aldehydes

**c**
Reaction between a tetrasubstituted allenyl aldehyde and a secondary amine

**Figure 1 | Chemistry of allenyl carbonyls and amines. (a)** Conjugate addition of amines and phosphines to allenyl ketones, esters and amides are well established in the literature[25,26]. **(b)** Trisubstituted allenyl aldehydes react with primary amines to give the imine products. **(c)** Reaction between the tetrasubstituted allenyl aldehyde **1a** and pyrrolidine **2a** generated substituted 1-amino-3-silyloxybutadiene **3aa** in high yield via a conjugate addition, 1,5-Brook rearrangement and protonation sequence. TIPS, triisopropylsilyl.

**Table 1 | Substrate scope of allenyl aldehyde.**

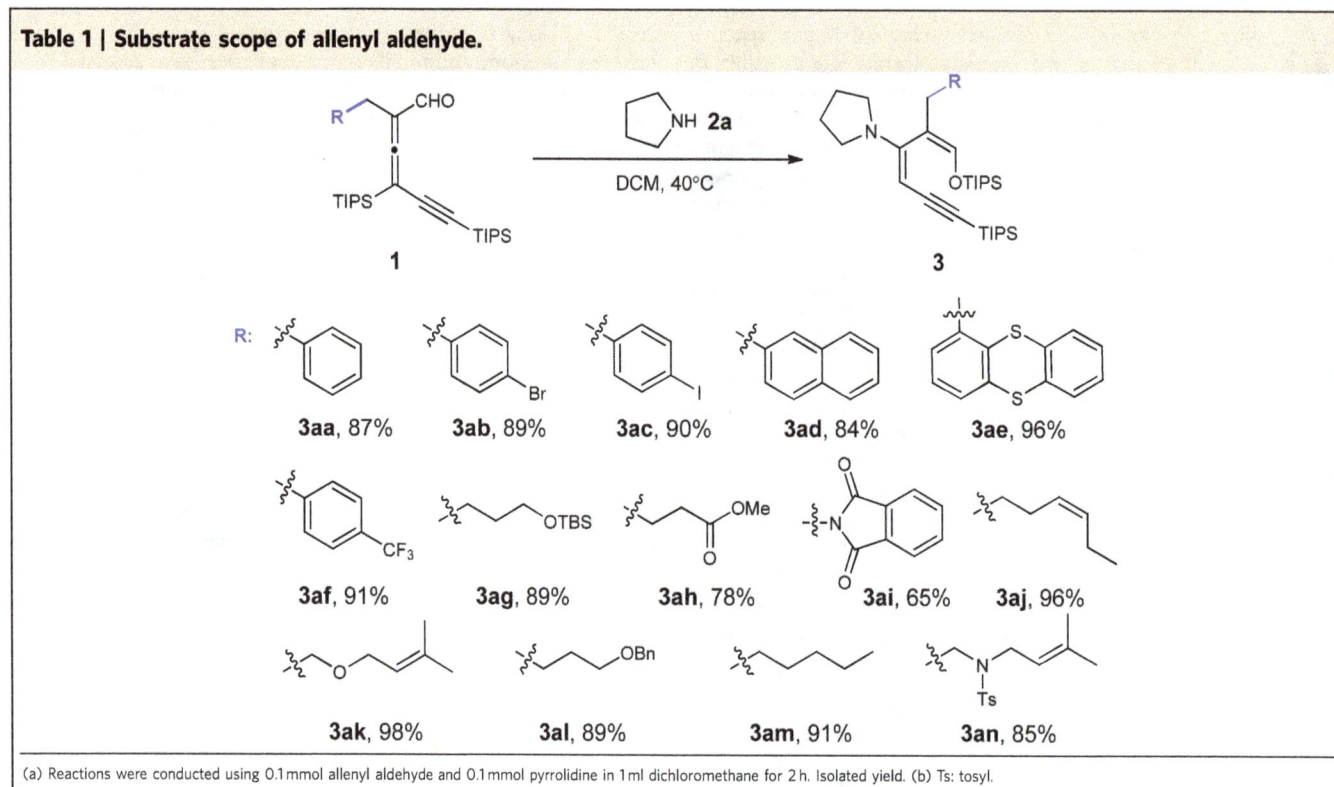

| | | | | |
|---|---|---|---|---|
| **3aa**, 87% | **3ab**, 89% | **3ac**, 90% | **3ad**, 84% | **3ae**, 96% |
| **3af**, 91% | **3ag**, 89% | **3ah**, 78% | **3ai**, 65% | **3aj**, 96% |
| **3ak**, 98% | **3al**, 89% | **3am**, 91% | **3an**, 85% | |

(a) Reactions were conducted using 0.1 mmol allenyl aldehyde and 0.1 mmol pyrrolidine in 1 ml dichloromethane for 2 h. Isolated yield. (b) Ts: tosyl.

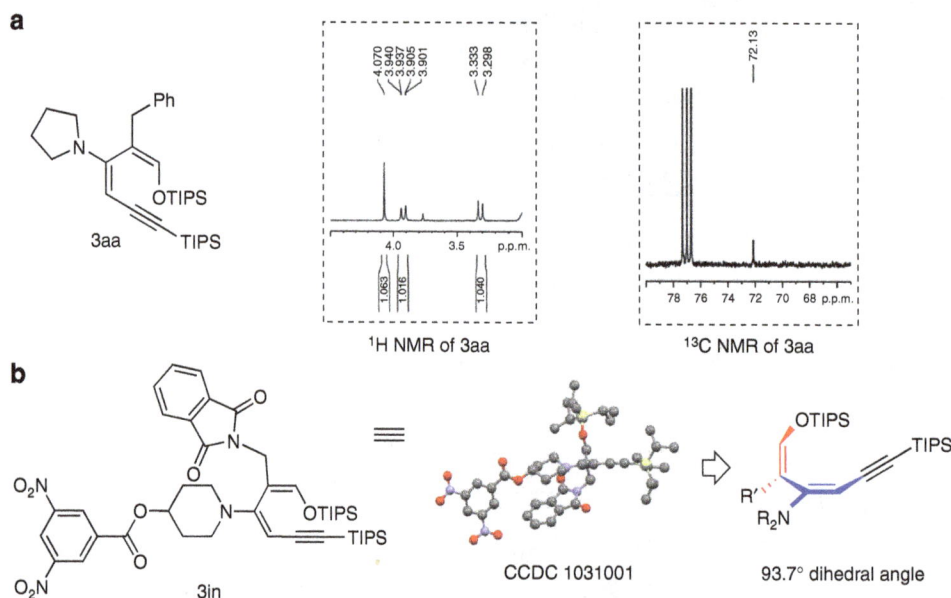

**Figure 2 | Structural information of the 1-silyloxy-3-aminobutadiene.** (a) NMR indicated highly eletron-rich nature of the enamine moiety. (b) X-ray structure revealed an unusual twisting of the diene olefinic carbons.

tetrasubstituted allenyl aldehydes from simple aldehydes and an electrophilic alkynylation reagent using gold/amine synergistic catalysis[23]. We found that a trisubstituted allenyl aldehyde reacted readily with an amine to form an ynenamine intermediate, which could be intercepted by various electrophiles including molecular oxygen[24]. Subsequently, we decided to explore a reaction between a tetrasubstituted allenyl aldehyde and a secondary amine. We were hoping that they would form a highly reactive allenyl iminium species that might promote novel chemical transformations with nucleophiles. Surprisingly, although the conjugate addition of amines to the

sp carbon of allenyl ketones, esters and amides were well known in the literature (Fig. 1a)[25,26], reactions involving the much more labile allenyl aldehydes were much less explored. There is only one report in which allenyl aldehydes readily formed the corresponding imines (1,2-addition) when treated with primary amines (Fig. 1b)[27]. No information was found for reactions involving allenyl aldehydes and secondary amines.

Surprisingly, the desired iminium intermediate was not observed when allenyl aldehyde **1a** was treated with 1 equiv. pyrrolidine **2a** under various conditions. Instead, a diene product **3aa** was formed in quantitative yield (Fig. 1c). Nuclear

**Table 2 | Substrate scope of amine.**

3ba, 85%    3ca, 89%    3da, 84%    3ea, 84%    3fa, 85%    3ga, 99%    3ha, 79%

3ia, 96%    3ja, 95%    3ka, 84%    3la, 90%    3ma, 91%    3na, 70%    3oa, <5%

(a) Reactions were conducted with 0.1 mmol allenyl aldehydes and 0.1 mmol pyrrolidine in 1 ml solvent for 2 h. Isolated yield. (b) For products **3fa**, **3ha** and **3na**, 2.0 equiv. amine was used and the reaction time was 20 h. (c) Boc: *t*-Butyloxycarbonyl.

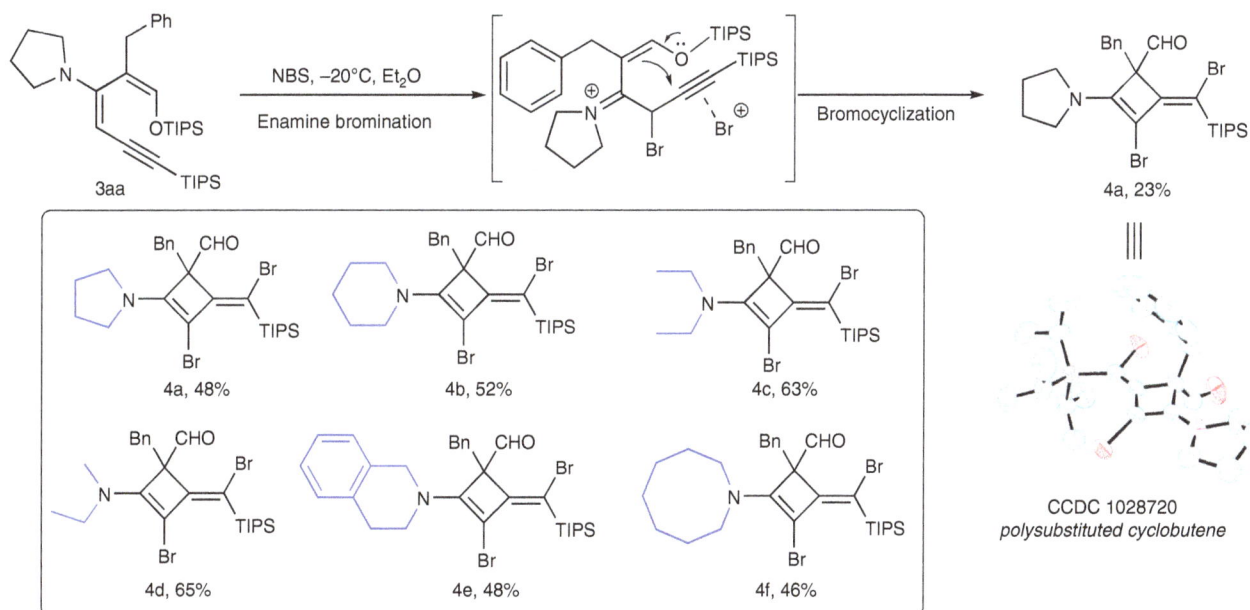

4a, 23%

4a, 48%    4b, 52%    4c, 63%

4d, 65%    4e, 48%    4f, 46%

CCDC 1028720
*polysubstituted cyclobutene*

**Figure 3 | Formation of highly substituted cyclobutenes.** The reaction is believed to proceed through an α-bromination/cyclization pathway. The structure of the product was assigned by X-ray. The products in the box were synthesized using 0.05 mmol dienes and 0.1 mmol N-bromosuccinimide (NBS) in 1 ml dichloromethane for 2 h; isolated yield.

Overhauser Effect studies showed that the silyl enol ether existed as the *Z*-isomer. On the basis of this information, we proposed that **3aa** was formed by the conjugate addition of pyrrolidine **2a** to allenyl aldehyde **1a**, followed by a facile 1,5-Brook rearrangement[28]. This reaction proceeded well in a number of aprotic solvents. Although prolonged exposure to silica gel led to decomposition, the zero-byproduct nature of this reaction made the work-up extremely easy. Quick filtration through a plug of neutral $Al_2O_3$ afforded the product in good purity. In some cases, simple removal of solvent gave essentially pure product.

**Reaction scope of allenyl aldehyde.** The scope of allenyl aldehyde was examined next. Various substrates were prepared directly from commercially available aldehydes and were subjected to pyrrolidine in dichloromethane (DCM) at 40 °C (Table 1). β-Aryl substituents were well tolerated. Good to excellent yields were obtained uniformly. Allenyl aldehydes from simple alkyl aldehydes were also explored. Alkyl groups bearing various functional groups (ether, ester, imide, olefin and sulfamide and so on) were excellent substrates. The reaction became slow when a sterically hindered R group was introduced. For example, only a trace amount of the corresponding 1-silyloxy-3-aminobutadiene was observed at elevated temperature after overnight when a secondary alkyl group is attached to the α-carbon of the allenyl aldehyde.

**Reaction scope of amine.** The scope of amine was also investigated (Table 2). A number of cyclic and acyclic secondary amines worked well for this reaction. Double bond, hydroxyl, carbamate and tertiary amine functionalities did not interfere with the diene formation. However, the reaction rate slowed down when hindered secondary amines were employed (Table 2, **3fa**, **3ha**, **3na**, **3oa**). Excess amine (2 equiv.) was used along with prolonged reaction time for those substrates. The reaction did not proceed

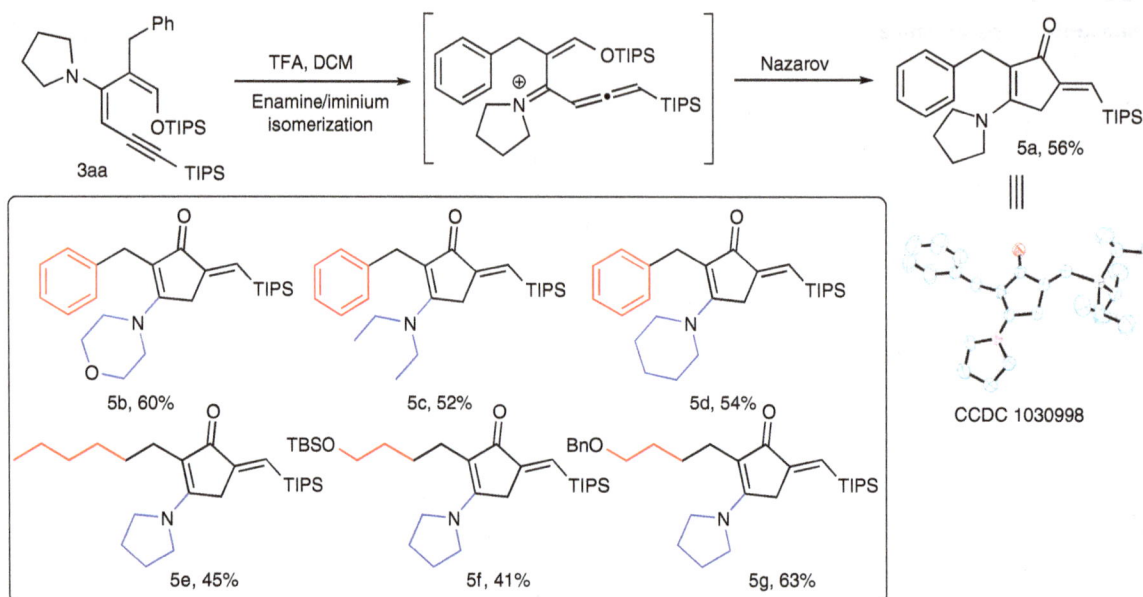

**Figure 4 | Formation of the exo-olefinic cyclopentenone product. (a)** Reactions were conducted with 0.05 mmol dienes in 1 ml solvent; isolated yield. **(b)** It is interesting that the closing ends of the iminium intermediates are both electron deficient. **(c)** Condition screening: MeSO$_3$H (1.1 equiv.), DCM, 30%; CF$_3$SO$_3$H (1.1 equiv.), DCM, 35%; trifluoroacetic acid (TFA, 1.1 equiv.), DCM, 56%; TFA (1.1 equiv.), Et$_2$O, 49%; TFA (1.1 equiv.), tetrahydrofuran (THF), 33%; TFA (0.5 equiv.), DCM, 30%; TFA (2.0 equiv.), DCM, 28%; TFA (5.0 equiv.), DCM, trace. **(d)** The products in the box were synthesized using 0.05 mmol dienes and 0.055 mmol TFA in 1 ml dichloromethane; isolated yield.

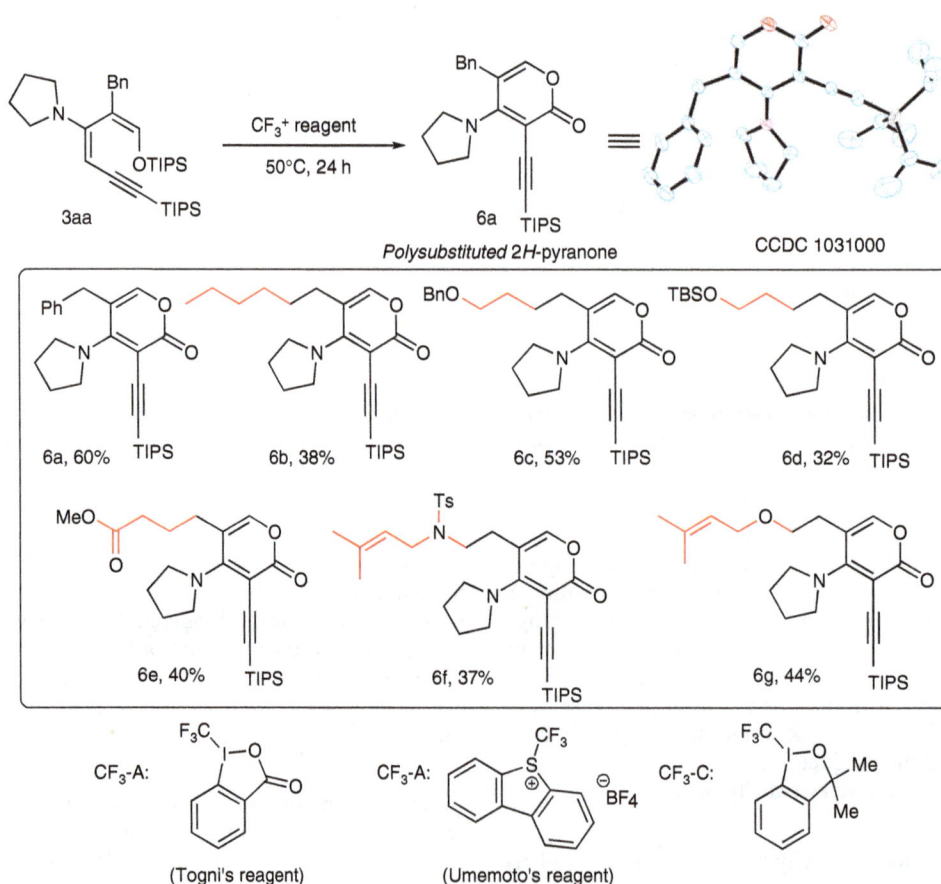

**Figure 5 | Formation of pyranone and condition screening. (a)** Condition screening: CF$_3$-A, Et$_2$O, no product; CF$_3$-A, tetrahydrofuran (THF), trace; CF$_3$-A, Toluene; trace; CF$_3$-A, DCE, trace; CF$_3$-A, CHCl$_3$, 22%; CF$_3$-A, CH$_2$Cl$_2$, 60%; CF$_3$-B, CH$_2$Cl$_2$, 17%; CF$_3$-C, CH$_2$Cl$_2$, trace. **(b)** The products in the box were synthesized using 0.05 mmol diene and 0.05 mmol CF$_3$-A in 1 ml CH$_2$Cl$_2$ at 50 °C for 24 h; isolated yield. **(c)** For product **6f**, the reaction temperature was 90 °C.

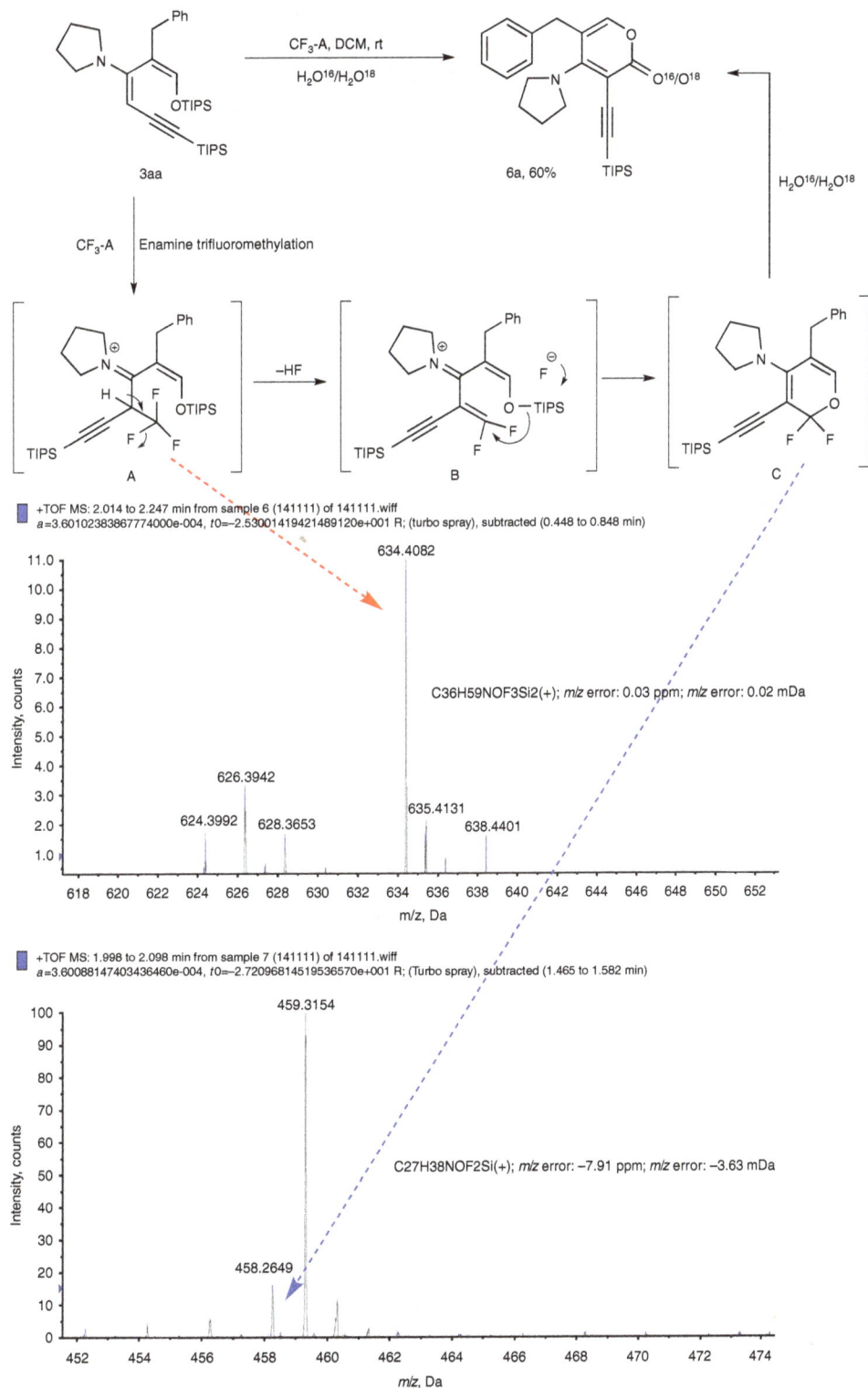

**Figure 6 | Proposed mechanism for the formation of six-membered ring. (a)** For the isotope experiment, dichloromethane was shaken rigorously with $H_2^{18}O$ and subsequently used for the reaction. **(b)** The enamine functionality is the primary reason for the facile conversion of the $CF_3$ group to the carbonyl, as it pushes an electron pair towards $CF_3$ and triggers the elimination of a fluoride anion. The alkyne moiety is also believed to be important for stabilizing the adjacent enamine.

for very bulky dialkyl amines such as diisopropylamine. Although no iminium formation was observed for secondary amines, the 1,2-addition was the major reaction pathway for primary amines.

**Structure of the 1-silyloxybutadiene and its implication.** 1-Silyloxy-3-aminobutadiene **3aa** showed an HNMR singlet

signal at 4.07 and a CNMR signal at 72.13, an indication of a highly electron-rich enamine olefinic carbon (Fig. 2a). The structure of the diene was subsequently determined by X-ray using a phthalyl-substituted analogue **3in**. Interestingly, the diene was severely distorted out of conjugation in this structure, with the two double bonds nearly perpendicular to each other

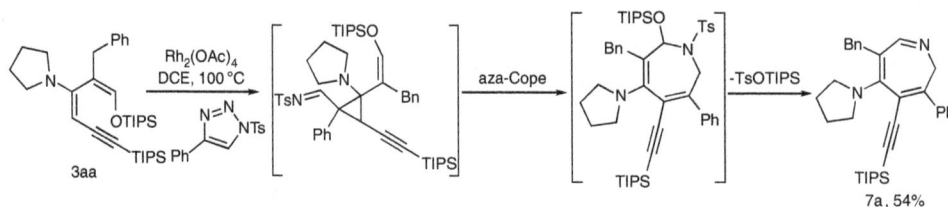

**Figure 7 | Formation of seven-membered ring.** Condition: **3aa** (0.2 mmol), 4-phenyl-1-tosyl-1H-1,2,3-triazole (0.1 mmol), Rh₂(OAc)₂ (0.01 mmol), DCE (2 ml), 100 °C, 48 h; isolated yield.

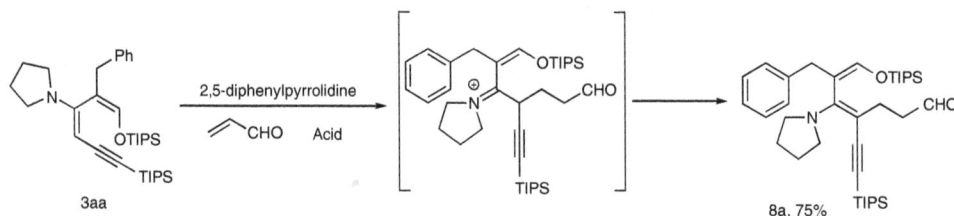

**Figure 8 | Differentiation of enamine and silyl enol ether.** Condition: **3aa** (0.1 mmol), acrolein (0.15 mmol), *rac-trans*-2,5-diphenylpyrrolidine (0.02 mmol), 4-Nitrobenzoic acid (0.02 mmol), dichloromethane (2 ml), room temperature, 2 h; isolated yield.

(Fig. 2b). This unexpected structure of the diene suggested that these dienes might not be suitable for the Diels–Alder reactions due to twisted conjugation. Various dienophiles were examined and no cycloadduct was formed under either thermal or Lewis acid catalysis conditions. In view of the electron-rich silyl enol ether and enamine present in these molecules, we decided to explore opportunities to affect novel transformations of **3aa** using electrophiles. Several intriguing transformations were discovered and a series of highly substituted small ring structures were generated.

**Cyclization to form cyclobutenes.** When **3aa** was treated with N-bromosuccinimide (NBS) at low temperature in ether, a dibromo cyclobutenyl aldehyde **4aa** was formed in 23% yield (Fig. 3). Cyclobutenes are quite unstable due to high ring strain and preparation of highly substituted cyclobutenes structures have been very challenging. Presumably, bromination of the enamine generated an α-bromo iminium intermediate that would rotate the π*-orbital of the alkyne to overlap with π-electron pair of the silyl enol ether to engage in a facile electrophilic ring closure. The exocyclic olefin geometry of the product was fully controlled as the corresponding E-isomer.

This reaction was optimized using different solvents. Very little product was formed in tetrahydrofuran, toluene and ethyl acetate. Decent conversion was obtained when the reaction was carried out at − 20 °C in dichloromethane. The desired product was isolated in 48% yield. This reaction was general for various amines and exo-olefinic cyclobutenes bearing different nitrogen substituents were prepared in moderate yields (Fig. 3). Bromo-cyclization reactions are well known for heteroatom nucleophiles, with five- and six-membered rings being the most common products[29–33]. Our reaction represents a rare example where a four-membered ring is formed using a carbon nucleophile. The intriguing structure of **4a** offers potential chemical viability to novel cyclobutane scaffolds.

**Imino-Nazarov cyclization to form cyclopentenone.** On treatment with an acid, the ynenamine moiety of diene **3aa** was converted to allenyl iminium by γ-protonation[34], which experienced a cationic 4-π electrocyclization to give **5aa** as a single isomer (Fig. 4). This type of imino-Nazarov cyclization

involving allene is very rare in the literature, and often occurs with poor selectivity and yield[35]. We subsequently found that the reaction was very sensitive to the acid used. Methanesulfonic acid and trifluoromethanesulfonic acid afforded poor yields. TFA was later found to be the best proton source. Solvent and acid stoichiometry were examined next. A catalytic amount of TFA led to an incomplete conversion and excess acid led to diene decomposition. Eventually, 1.1 equiv. TFA afforded the exo-olefinic cyclopentenone product in 56% isolated yield. The key exo-olefinic cyclopentenone moiety in product **5** is often found in methylenomycin and prostaglandin families of natural products and drugs[36]. The substrate scope was investigated in a broad manner (Fig. 4). Various 2-alkyl and 3-amino groups were well tolerated. The reaction condition was mild enough to retain acid labile functionalities, such as TBS ether. The TIPS group could be removed readily using TBAF. The existence of many orthogonal functional groups in these products offers opportunities for versatile structural manipulation.

**Carbonylation to form pyranone.** When diene **3aa** was treated with electrophilic trifluoromethylating reagent, we were very surprised to find that a six-membered pyranone product **6a** was formed (Fig. 5). This reaction was optimized immediately. We found that neutral hypervalent trifluoromethyl iodide (Togni's reagent) performed better than CF₃ salts, such as the Umemoto reagent. The reaction exhibited a strong solvent effect. No or trace product was observed in ether, tetrahydrofuran, toluene and dichloroethane. Interestingly, there is a large discrepancy between chloroform and dichloromethane. The conversion was much higher in dichloromethane, with 60% isolated yield. With the optimized reaction condition in hand, we examined the generality of the pyranone formation using 1-silyloxy-3-aminobutadienes bearing various substituents. It was found that alkyl group at the second position of the dienes were particularly well tolerated. Common functionality groups, such as ether, ester, olefin, silicon and sulfonamide, were successfully incorporated in the pyranone product (Fig. 5). Further investigation of this transformation using simple substrates is currently underway to better understand the structural requirement for such a delicate transformation.

The formation of pyranone **6** is very interesting and highly unexpected. Careful control experiments showed that the reaction was very sensitive to the amount of water in the system. In the absence of water, very little product was observed. With excess water, the starting material decomposed quickly. HRMS showed that when the solvent was saturated with $H_2^{18}O$, the product was formed with significant heavy isotope enrichment (Fig. 6). In addition, subjecting the reaction mixture to HRMS revealed the existence of a trifluoromethylated intermediate **A** and a difluoro cyclic ether species **C** (Fig. 6). On the basis of these data, we propose the following reaction mechanism: first, trifluoromethyl-ation of the enamine generates α-$CF_3$ iminium intermediate **A**. This species has a highly acidic carbon with three electron-withdrawing groups attached—iminium, alkyne and $CF_3$, which likely results a rapid elimination of hydrogen fluoride to give difluoro olefin **B**. The release of HF would deprotect the TIPS group, followed by a concurrent cyclization of the enolate to the electron-deficient difluoroalkene to give cyclic difluoroether **C**. Although hydrolysis of $CF_2$ to a carbonyl group requires harsh conditions[37], we believe that the adjacent enamine greatly accelerates this conversion by promoting rapid elimination of fluorides. Notably, this represents the first example that the Togni's reagent is used as a carbonyl precursor.

### Formal [4 + 3] cycloaddition to form azepine.

Diene **3aa** was also tested for formal [4 + 3] cycloaddition reactions. Under rhodium(II) iminocarbene condition, tetrasubstituted azepine **7a**, was obtained in 54% yield (Fig. 7). The structure of this product differed from a recent report using simple dienes (Tang and colleagues[38]) by containing one extra double bond in the already strained seven-member ring as a result of *in situ* elimination of TsOTIPS. No dihydropyrrole product was found in our reaction, which was formed on prolonged heating in Tang's report. The reaction was sensitive to the catalyst and reaction temperature. When changed to Rh(II) complexes other than $Rh_2(OAc)_4$, much lower yields were observed. Although both enamine and silyl enol ether are electron rich, the first cyclopropanation occurred exclusively at the enamine double bond, leading to a single isomer. The subsequent aza-Cope rearrangement was companied by simultaneous loss of TsOTIPS. The versatile functionalities contained in this tetrasubstituted azepine product offered great opportunities for structural diversification.

### Selective functionalization of enamine.

Finally, the nucleophilic reactivity of the silyl enol ether and the enamine could be differentiated using amine catalysed conjugate addition reaction. When diene **3aa** was treated with acrolein in the presence of a hindered secondary amine, the Michael addition product **8aa** was obtained in 75% yield (Fig. 8)[39]. Although 4-nitrobenzoic acid was used, both the enamine and the silyl enol ether moieties were not hydrolysed, suggesting the intrinsic stability of this pentasubstituted diene.

In summary, we developed a general synthesis of highly substituted 1-silyloxy-3-aminobutadienes using a very mild condition from readily available allenyl aldehydes. This unusual diene formation proceeded through an unprecedented conjugate addition of a secondary amine to an allenyl aldehyde, followed by a 1,5-Brook rearrangement. In sharp contrast to other electron-rich dienes, these 1-silyloxy-3-aminobutadienes exhibit extraordinary synthetic versatility for non-cycloaddition reactions. Several unexpected transformations were discovered and studied in detail. Structurally sophisticated four-, five-, six- and seven-membered ring systems bearing multiple functionalities were assembled in one step from these dienes. We expect that this chemistry will stimulate the design of novel synthetic strategies towards complex structures using amino silyloxy butadienes.

## Methods

**General methods and materials.** Solvents for reactions were distilled according to general practice before use. All reagents were purchased and used without further purification unless specified otherwise. Allenyl aldehyde substrates were prepared according to the literature-reported procedure[23]. Solvents for chromatography were technical grade and distilled before use. Flash chromatography was performed using 200–300 mesh silica gel with the indicated solvent system according to standard techniques. Analytical thin-layer chromatography was performed using Huanghai silica gel plates with HSGF 254. Qingdao Haiyang Chemical HG/T2354-92 silica gel was used for silica gel flash chromatography. Visualization of the developed chromatogram was performed by ultraviolet absorbance (254 nm) or appropriate stains. $^1$HNMR data were recorded on Bruker nuclear resonance spectrometers (300, 400 or 500 MHz) unless specified otherwise. Chemical shifts ($\delta$) are reported as quoted relative to the residual signals of chloroform ($^1$H 7.26 p.p.m. or $^{13}$C 77.16 p.p.m.). Multiplicities are described as: s (singlet), bs (broad singlet), d (doublet), t (triplet), q (quartet), m (multiplet); and coupling constants ($J$) are reported in Hertz (Hz). $^{13}$C NMR spectra were recorded on Bruker spectrometers (75, 101 or 126 MHz) with total proton decoupling. HRMS (electrospray ionization) analysis was performed by The Analytical Instrumentation Center at Peking University, Shenzhen Graduate School and (HRMS) data were reported with ion mass/charge ($m/z$) ratios as values in atomic mass units. $^1$H NMR, $^{13}$C NMR and HRMS are provided for all compounds; see Supplementary Figs 1–102. For ORTEP structures of **3in**, **4a**, **5a**, **6a**, see Supplementary Figs 103–106. See Supplementary Methods for the characterization data for all compounds. See Supplementary Data sets 1–4 for X-ray CIF files of compounds **3in**, **4a**, **5a**, **6a** (CCDC 1028720, 1030998, 1031000, 1031001). See Supplementary Data set 5 for a list of structurally novel compounds.

*General procedure for the synthesis of 3.* Aldehyde **1** (0.1 mmol, 1.0 equiv.) was dissolved in DCM (1.0 ml, 0.1 M) in an oven-dried 8-ml vial equipped with a magnetic stir bar and a rubber septum. Secondary amine **2** (0.1 mmol, 1.0 equiv.) was added and the reaction mixture was stirred at 40 °C for 2 h. Solvent was removed under *vacuo* and the residue was purified by flash neutral $Al_2O_3$ column chromatography (eluent: petroleum ether/ethyl ether = 100: 1) to afford the desired 1-silyloxy-3-aminobutadiene **3**.

*General procedure for the synthesis of 4.* Diene **3** (0.05 mmol, 1.0 equiv.) was dissolved in DCM (1 ml). The solution was stirred at − 20 °C for 5 min. N-bromosuccinimide (0.2 mmol, 2.0 equiv.) was added and the reaction mixture was stirred at − 20 °C for 2 h. Solvent was evaporated and the residue was purified by silica gel column chromatography ($Et_2O$: petroleum ether = 1: 10) to afford the desired product **4**.

*General procedure for the synthesis of 5.* Diene **3** (0.05 mmol, 1.0 equiv.) was dissolved in DCM (1 ml). The solution was stirred at − 20 °C for 5 min. A solution of trifluoroacetic acid (0.11 mmol, 1.1 equiv.) in DCM (0.2 ml) was added dropwise and the reaction mixture was stirred at − 20 °C overnight. The reaction was then warmed to room temperature and stirred for another 1 h. Solvent was evaporated and the residue was purified by silica gel column chromatography (ethyl acetate: petroleum ether = 1: 4) to afford the desired product **5**.

*General procedure for the synthesis of 6.* To a solution of diene **3** (0.05 mmol, 1.0 equiv.) in DCM (1 ml) was added the Togni's reagent (0.05 mmol, 1.0 equiv.). The reaction mixture was stirred at 50 °C for 24 h. Solvent was evaporated and the residue was purified by silica gel column chromatography (ethyl acetate: petroleum ether = 1: 4) to afford the desired product **6**.

*Procedure for the synthesis of 7a.* To a solution of **3aa** (113 mg, 0.2 mmol) in DCE (2 ml) was added 4-phenyl-1-tosyl-1$H$-1,2,3-triazole (30 mg, 0.1 mmol) and $Rh_2(OAc)_4$ (4.4 mg, 0.01 mmol). The reaction mixture was stirred at 100 °C for 48 h. Solvent was evaporated and the residue was purified by silica gel column chromatography (ethyl acetate: petroleum ether = 1: 8) to afford the desired product **7a** as a yellow oil.

*Procedure for the synthesis of 8a.* To the solution of diene **3aa** (56 mg, 0.1 mmol), *rac-trans*-2,5-diphenylpyrrolidine (4.5 mg, 0.02 mmol) and 4-nitrobenzoic acid (3.3 mg, 0.02 mmol) in DCM (2 ml) was added acrolein (10 μl, 0.15 mmol). The reaction mixture was stirred at room temperature for 2 h. Solvent was evaporated and the residue was purified by silica gel column chromatography (ethyl acetate: petroleum ether = 1: 25) to afford the desired product **8a** as a light yellow oil.

## References

1. Kozmin, S. A. & Rawal, V. H. A general strategy to aspidosperma alkaloids: efficient, stereocontrolled synthesis of tabersonine. *J. Am. Chem. Soc.* **120**, 13523–13524 (1998).
2. Kozmin, S. A., Iwama, T., Huang, Y. & Rawal, V. H. An efficient approach to aspidosperma alkaloids via [4 + 2] cycloadditions of aminosiloxydienes: Stereocontrolled total synthesis of ( ± )-Tabersonine. Gram-scale catalytic asymmetric syntheses of ( + )-Tabersonine and ( + )-16-Methoxytabersonine. Asymmetric syntheses of ( + )-Aspidospermidine and ( − )-Quebrachamine. *J. Am. Chem. Soc.* **124**, 4628–4641 (2002).
3. Smith, III A. B., Basu, K. & Bosanac, T. Total synthesis of ( − )-Okilactomycin. *J. Am. Chem. Soc.* **129**, 14872–14874 (2007).
4. Hayashida, J. & Rawal, V. H. Total synthesis of ( ± )-Platencin. *Angew. Chem. Int. Ed.* **47**, 4373–4376 (2008).

5.  You, L.-F. *et al.* An enantioselective synthesis of the ABD tricycle for ( − )-Phomactin A featuring Rawal's asymmetric Diels–Alder cycloaddition. *Adv. Synth. Catal.* **350**, 2885–2891 (2008).

6.  Nicolaou, K. C., Scott Tria, G., Edmonds, D. J. & Kar, M. Total syntheses of ( ± )-Platencin and ( − )-Platencin. *J. Am. Chem. Soc.* **131**, 15909–15917 (2009).

7.  Petronijevic, F. R. & Wipf, P. Total synthesis of ( ± )-Cycloclavine and ( ± )-5-*epi*-Cycloclavine. *J. Am. Chem. Soc.* **133**, 7704–7707 (2011).

8.  Mukai, K., Urabe, D., Kasuya, S., Aoki, N. & Inoue, M. A convergent total synthesis of 19-Hydroxysarmentogenin. *Angew. Chem. Int. Ed.* **52**, 5300–5304 (2013).

9.  Huang, Y., Iwama, T. & Rawal, V. H. Highly enantioselective Diels − Alder reactions of 1-amino-3-siloxy-dienes catalyzed by Cr(III)-Salen complexes. *J. Am. Chem. Soc.* **122**, 7843–7844 (2000).

10. Huang, Y., Unni, A. K., Thadani, A. N. & Rawal, V. H. Hydrogen bonding: single enantiomers from a chiral-alcohol catalyst. *Nature* **424**, 146 (2003).

11. Unni, A. K., Takenaka, N., Yamamoto, H. & Rawal, V. H. Axially chiral biaryl diols catalyze highly enantioselective hetero-Diels − Alder reactions through hydrogen bonding. *J. Am. Chem. Soc.* **127**, 1336–1337 (2005).

12. Jensen, K. H. & Sigman, M. S. Systematically probing the effect of catalyst acidity in a hydrogen-bond-catalyzed enantioselective reaction. *Angew. Chem. Int. Ed.* **46**, 4748–4750 (2007).

13. Boxer, M. B. & Yamamoto, H. 'Super silyl' group for diastereoselective sequential reactions: Access to complex chiral architecture in one pot. *J. Am. Chem. Soc.* **129**, 2762–2763 (2007).

14. Kozmin, S. A. & Rawal, V. H. Preparation and Diels − Alder reactivity of 1-amino-3-siloxy- 1,3-butadienes. *J. Org. Chem.* **62**, 5252–5253 (1997).

15. Kozmin, S. A., He, S. & Rawal, V. H. Preparation of (*E*)-1-dimethylamino-3-tert-butyldimethylsiloxy- 1,3-butadiene. *Org. Synth.* **78**, 152 (2002).

16. Kozmin, S. A., Green, M. T. & Rawal, V. H. On the reactivity of 1-amino-3-siloxy-1,3-dienes: Kinetics investigation and theoretical interpretation. *J. Org. Chem.* **64**, 8045–8047 (1999).

17. Huang, Y. & Rawal, V. H. Hydrogen-bond-promoted hetero-Diels − Alder reactions of unactivated ketones. *J. Am. Chem. Soc.* **124**, 9662–9663 (2002).

18. Smith, III A. B., Wexler, B. A., Tu, C.-Y. & Konopelski, J. P. Stereoelectronic effects in the cationic rearrangements of [4.3.2] propellanes. *J. Am. Chem. Soc.* **107**, 1308–1320 (1985).

19. Danieli, B., Lesma, G., Luzzani, M., Passarella, D. & Silvani, A. Diels-Alder reactions of methyl N-p-methoxybenzensulfonylindole-2-(2-propenoate), a convenient dienophile towards the synthesis of andranginine. *Tetrahedron* **52**, 11291–11296 (1996).

20. Comins, D. L., Zhang, Y.-M. & Zheng, X. Photochemical reactions of chiral 2,3-dihydro-4(1H)- pyridones: asymmetric synthesis of (2)-Perhydrohistrionicotoxin. *Chem. Commun.* 2509–2510 (1998).

21. Adams, A. D., Schlessinger, R. H., Tata, J. R. & Venit, J. J. The structure and kinetic reactivity of a pyrrolidine-derived vinylogous urethane lithium enolate. *J. Org. Chem.* **51**, 3068–3070 (1986).

22. Tsuge, O., Hatta, T., Takahashi, Y., Maeda, H. & Kakehi, A. Reaction of maleimides with new silyloxydienes, 2-amino-4-(trimethylsilyloxy)-1,3-pentadienes. *Heterocycles* **47**, 665–670 (1998).

23. Wang, Z., Li, X. & Huang, Y. Direct α-vinylidenation of aldehydes and subsequent cascade gold and amine catalysts work synergistically. *Angew. Chem. Int. Ed.* **52**, 14219–14223 (2013).

24. Wang, Z., Li, L. & Huang, Y. A general synthesis of ynones from aldehydes via oxidative C–C bond cleavage under aerobic conditions. *J. Am. Chem. Soc.* **136**, 12233–12236 (2014).

25. Tamura, Y., Tsugoshi, T., Mohri, S.-I. & Kita, Y. Michael addition to 1,3-bis(alkoxycarbonyl)allenes: synthesis of heterocyclic compounds having glutaconate structure in the molecules. *J. Org. Chem.* **50**, 1542–1544 (1985).

26. Szeto, J., Sriramurthy, V. & Kwon, O. Phosphine-initiated general base catalysis: Facile access to benzannulated 1,3-diheteroatom five-membered rings via double-Michael reactions of allenes. *Org. Lett.* **13**, 5420–5423 (2011).

27. Sigman, M. S. & Eaton, B. E. Addition of primary amines to conjugated allenyl aldehydes and ketones. *Tetrahedron Lett.* **34**, 5367–5368 (1993).

28. Smith, III A. B. & Adams, C. M. Evolution of dithiane-based strategies for the construction of architecturally complex natural products. *Acc. Chem. Res.* **37**, 365–377 (2004).

29. Wilking, M., Mück-Lichtenfeld, C., Daniliuc, C. G. & Hennecke, U. Enantioselective, desymmetrizing bromolactonization of alkynes. *J. Am. Chem. Soc.* **135**, 8133–8136 (2013).

30. Sasaki, M. & Yudin, A. K. Oxidative cycloamination of olefins with aziridines as a versatile route to saturated nitrogen-containing heterocycles. *J. Am. Chem. Soc.* **125**, 14242–14243 (2003).

31. Dai, W. & Katzenellenbogen, J. A. Stereoselective Z-and E-bromo enol lactonization of alkynoic acids. *J. Org. Chem.* **56**, 6893–6896 (1991).

32. Cheng, Y. A., Chen, T., Tan, C. K., Heng, J. J. & Yeung, Y.-Y. Efficient medium ring size bromolactonization using a sulfur-based zwitterionic organocatalyst. *J. Am. Chem. Soc.* **134**, 16492–16495 (2012).

33. Murai, K. *et al.* Asymmetric bromolactonization catalyzed by a C3-symmetric chiral trisimidazoline. *Angew. Chem. Int. Ed.* **49**, 9174–9177 (2010).

34. Weber, W. P., Felix, R. A. & Willard, A. K. Mass spectral rearrangements. Silyl McLafferty rearrangement. *J. Am. Chem. Soc.* **92**, 1420–1421 (1970).

35. Tius, M. A., Chu, C. C. & Nieves-Colberg, R. An imino Nazarov cyclization. *Tetrahedron Lett.* **42**, 2419–2422 (2001).

36. Haneishi, T., Kitahara, N., Takiguchi, Y., Arai, M. & Sugawara, S. New antibiotics, methylenomycins A and B. I. Producing organism, fermentation and isolation, biological activities and physical and chemical properties. *J. Antibiot.* **27**, 386–392 (1974).

37. Huang, Y.-Z. & Zhou, Q.-L. Nickel-, palladium-, and platinum-catalyzed reactions of perfluoro- and polyfluoroalkyl iodides with tetriary amines. *J. Org. Chem.* **52**, 3552–3558 (1987).

38. Shang, H., Wang, Y., Tian, Y., Feng, J. & Tang, Y. The divergent synthesis of nitrogen heterocycles by Rhodium(II)-catalyzed cycloadditions of 1-sulfonyl 1,2,3-triazoles with 1,3-dienes. *Angew. Chem. Int. Ed.* **53**, 5662–5666 (2014).

39. Kemppainen, E. K., Sahoo, G., Valkonen, A. & Pihko, P. M. Mukaiyama-Michael reactions with acrolein and methacrolein: A catalytic enantioselective synthesis of the C17-C28 fragment of Pectenotoxins. *Org. Lett.* **14**, 1086–1089 (2012).

## Acknowledgements

This work was financially supported by the National Natural Science Foundation of China (21372013) and the Shenzhen Peacock Program (KQTD201103). Y.H. thanks the MOE for the Program for New Century Excellent Talents in University.

## Author contributions

Y.H. conceived and directed the project. X.L. and S.P. performed the experiments. L.L. solved the X-ray structures. Y.H. and X.L. analysed the results. Y.H. wrote the manuscript with the assistance of X.L.

# 4

# Total synthesis of (+)-gelsemine via an organocatalytic Diels–Alder approach

Xiaoming Chen[1,2], Shengguo Duan[1,2], Cheng Tao[1], Hongbin Zhai[1] & Fayang G. Qiu[2]

The structurally complex alkaloid gelsemine was previously thought to have no significant biological activities, but a recent study has shown that it has potent and specific antinociception in chronic pain. While this molecule has attracted significant interests from the synthetic community, an efficient synthetic strategy is still the goal of many synthetic chemists. Here we report the asymmetric total synthesis of (+)-gelsemine, including a highly diastereoselective and enantioselective organocatalytic Diels–Alder reaction, an efficient intramolecular trans-annular aldol condensation furnishing the prolidine ring and establishing the configuration of the C20 quaternary carbon stereochemical centre. The entire gelsemine skeleton was constructed through a late-stage intramolecular $S_N2$ substitution. The enantiomeric excess of this total synthesis is over 99%, and the overall yield is around 5%.

[1] State Key Laboratory of Applied Organic Chemistry, Lanzhou University, Lanzhou 730000, China. [2] Laboratory of Molecular Engineering, and Laboratory of Natural Product Synthesis, Guangzhou Institute of Biomedicine and Health, Chinese Academy of Sciences, 190 Kaiyuan Boulevard, The Science Park of Guangzhou, Guangzhou 510530, China. Correspondence and requests for materials should be addressed to H.Z. (email: zhaih@lzu.edu.cn) or to F.G.Q. (email: qiu_fayang@gibh.ac.cn).

lthough gelsemine was isolated[1] in as early as 1876 from *Gelsemium Sempervirens* Ait., its structure was not determined until 1959 by means of nuclear magnetic resonance (NMR) spectroscopic techniques[2,3] and X-ray crystallographic analysis[4]. This indole alkaloid contains a hexacyclic cage structure and seven contiguous chiral carbon centres (Fig. 1). The complex chemical structures of gelsemine and other members of the alkaloid family[5-8] have attracted considerable attention from synthetic chemists. So far, in addition to the many synthetic efforts[9-37], there are eight total syntheses reported in the literature[38-49] (Fig. 2), two of which are asymmetric[44,48]. Although gelsemine was thought to have no particular biological activities, a recent report indicated that gelsemine exhibited potent and specific antinociception in chronic pain by acting at the three spinal glycine receptors[50]. Besides, gelsemine was nonaddictive, indicating that the mechanism of its action is different from that of morphine. The complex structure and the potential medicinal applications of gelsemine prompted us to initiate a more efficient enantioselective total synthesis.

Herein we wish to report a 12-step, highly enantioselective organocatalytic total synthesis of (+)-gelsemine.

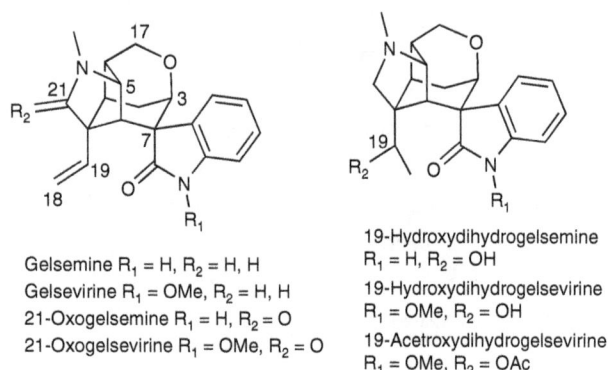

**Figure 1 | The structures of gelsemium alkaloids.** The difference between the members of the gelsemium alkaloids is the presence of the functional groups in the unique carbon skeleton. The major difference appeared in C-19 and C-21.

## Results

**Retrosynthetic analysis.** Gelsemine may be synthesized from intermediate **RS-1** and oxindole via the condensation of the hemiacetal with oxindole followed by an intramolecular $S_N2$ displacement (Fig. 3). Although the condensation may result in four stereoisomers, only two of them may undergo the desired $S_N2$ displacement. The other two isomers, however, may either stay intact or undergo an elimination followed by a Michael addition[51,52] to regenerate the four stereoisomers. This equilibrium is shifted to form the desired product after the intramolecular $S_N2$ displacement, which is irreversible under the reaction conditions (Figs 4 and 5). The $S_N2$ displacement may result in two isomers, one of which is the desired product. Intermediate **RS-1** may be obtained from **RS-2** following a sequence of intramolecular aldol condensation, reduction of the carbonyl group, formation of the sulfonates and then elimination. The intramolecular aldol[53,54] condensation deserves further discussion due to the fact that both the aldehyde and the ketone functionalities may undergo enolization under the reaction conditions, resulting in epimerization of both stereochemical centres attaching the carbonyl groups. Another issue is the direction of the aldol condensation. Since both of the carbonyl groups may be enolized, the aldol condensation from either one may be consequential. However, Cbz is a bulky functional group[55] and it will play a significant role in preventing the aldehyde from being enolized prior to the ketone enolization. In this case, the potential epimerization of the ketone functionality is irrelevant. The third issue is the stereochemistry of the hydroxyl group even if aldol condensation occurs in the desired direction. This difficulty may be overcome when one realizes that the desired product has a more favourable internal hydrogen bond[56,57] than the other isomer. Finally, formation of **RS-3** and its conversion into **RS-2** is straightforward.

**Synthesis of the (+)-gelsemine.** On the basis of the above analysis, the synthetic strategy seemed feasible. If intermediate **3** is made asymmetric, then gelsemine will be made asymmetric. Thus, after a brief literature search[58,59], an asymmetric Diels–Alder reaction was designed and the synthesis began with dihydropyridine **1** (Fig. 6), which may be prepared from 4-methylpyridine in large scale[60].

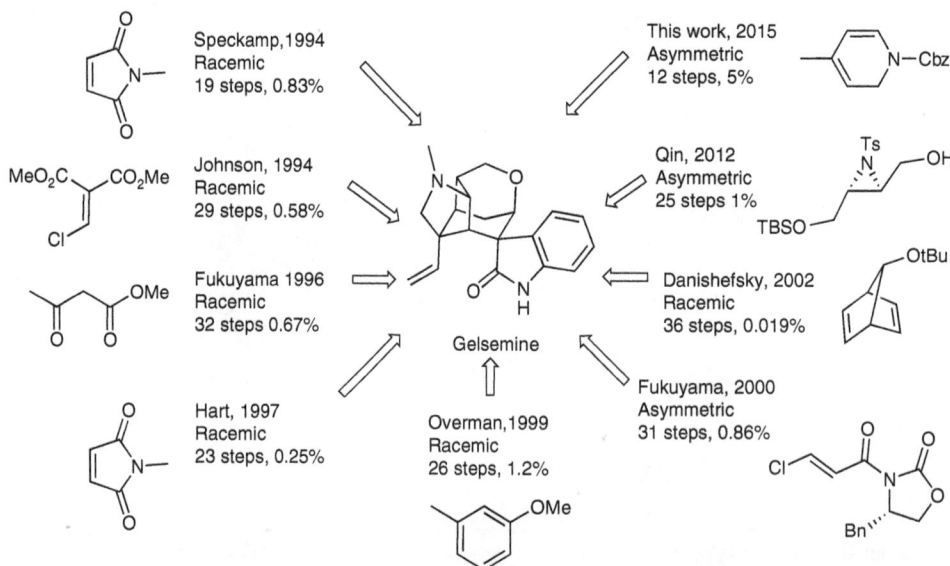

**Figure 2 | Schematic summary of the previous total syntheses of gelsemine.** Among the seven total syntheses completed so far, two of them were asymmetric and the overall yields were around 1%. This molecule has been an active target of total synthesis during the past two decades.

Gratifyingly, the yield of the desired endo product was 47% after reduction of the aldehyde carbonyl group with sodium borohydride, and its enantio excess was determined using chiral high-performance liquid chromatography (HPLC) to be 99.7%, while the exo product was not detected. It was surprising that intermediate **3a** was also produced in 30% yield. Since intermediate **3** was stable under the reaction conditions, **3a** may be a result of the double-bond isomerization of the enal during the catalytic process[61], and the rate of the double-bond isomerization was comparable to that of the Diels–Alder cycloaddition (Fig. 7). Fortunately, **3a** was converted into **3** with DBU (1,8-diazabicycloundec-7-ene) in refluxing toluene in

97% yield, which brought the total yield of the Diels–Alder cycloaddition to 76%. Intermediate **3** was then further selectively reduced to the hemiacetal **4** using Dibal-H at −78 °C in 94% yield. The subsequent Wittig reaction furnished the methyl enol ether, which was directly treated with trimethyl orthoformate and a catalytic amount of p-toluenesulfonic acid to provide intermediate **5** and **5a** (13:1) as a separable mixture in 93% combined yield. Although **5a** may be used as well, it was converted into **5** by treating it with pTSOH in methylene chloride (DCM) and only **5** was used for the next step. After a conventional ozonolysis of intermediate **5** in DCM, the resulting dicarbonyl intermediate was directly treated with sodium methoxide in methanol at 0 °C due to the fact that the dicarbonyl intermediate was unstable for storage. To our delight, the aldol reaction afforded the desired product **6** in 60% combined yield. However, the reaction of **6** with the methanesulfonyl chloride resulted in a complex mixture. Thus, the hydroxyketone intermediate **6** was reduced to diol **7** with sodium borohydride (97%) and the formation of disulfonate **8** with methanesulfonyl chloride was quantitative, the structure of which was confirmed through X-ray crystallographic analysis (Fig. 8). Treatment of intermediate **8** with DBU (1,8-diazabicycloundec-7-ene) in refluxing toluene led to the formation of alkene **9** (85%) and reduction of the Cbz protective group to methyl with lithium aluminium hydride in THF afforded **10** in 86% yield. Subsequent acid hydrolysis of the acetal with aqueous hydrochloric acid in THF provided hemiacetal **11** (96%).

With the key intermediate in hand, we began to test the condensation of **11** with methoxymethyl oxindole and the subsequent S_N2 displacement, another key reaction for

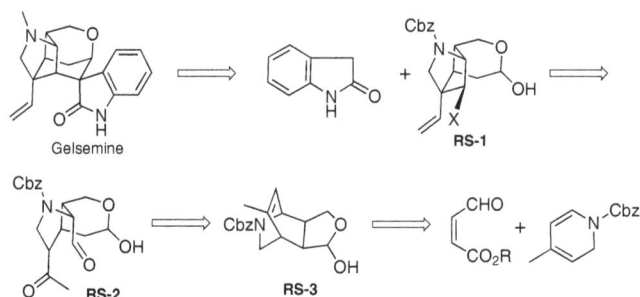

**Figure 3 | Retrosynthetic analysis of gelsemine.** In principle, gelsemine may be constructed from oxindole and intermediate RS-1, where X is a leaving group. After a few transformations, RS-1 may be synthesized from intermediate RS-2, which inturn may be obtained from RS-3 following several reaction steps including ozonolysis. Finally, RS-3 may be synthesized from readily accessible starting materials.

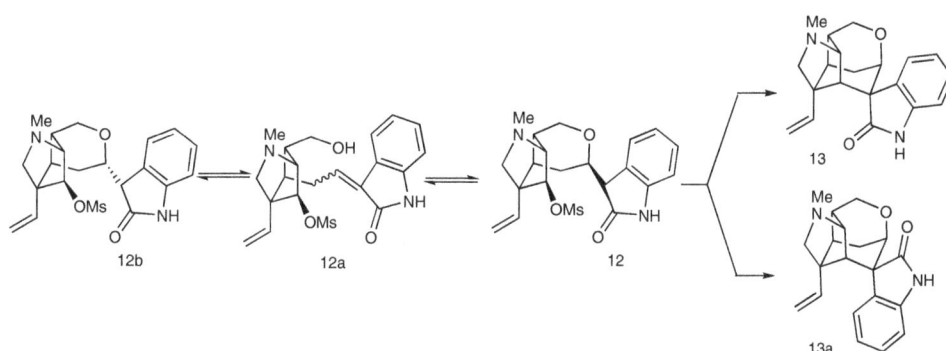

**Figure 4 | Cyclization of intermediate 9 to form the gelsemine framework.** This scheme illustrates the equilibrium between intermediates 12 and 12b via the formation of intermediate 12a. It can be seen that only intermediate 12 can proceed to form the cyclization products 13 and 13a.

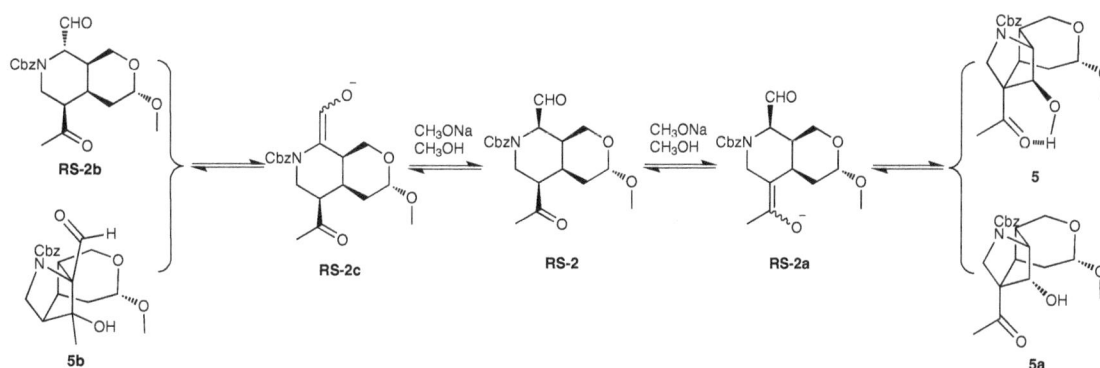

**Figure 5 | The aldol condensation and possible complications.** Enolization of both the aldehyde and the ketone carbonyl groups is possible, while only the cyclization through the ketone carbonyl group enolization can provide the desired product, which is thermodynamically more stable than the other isomer.

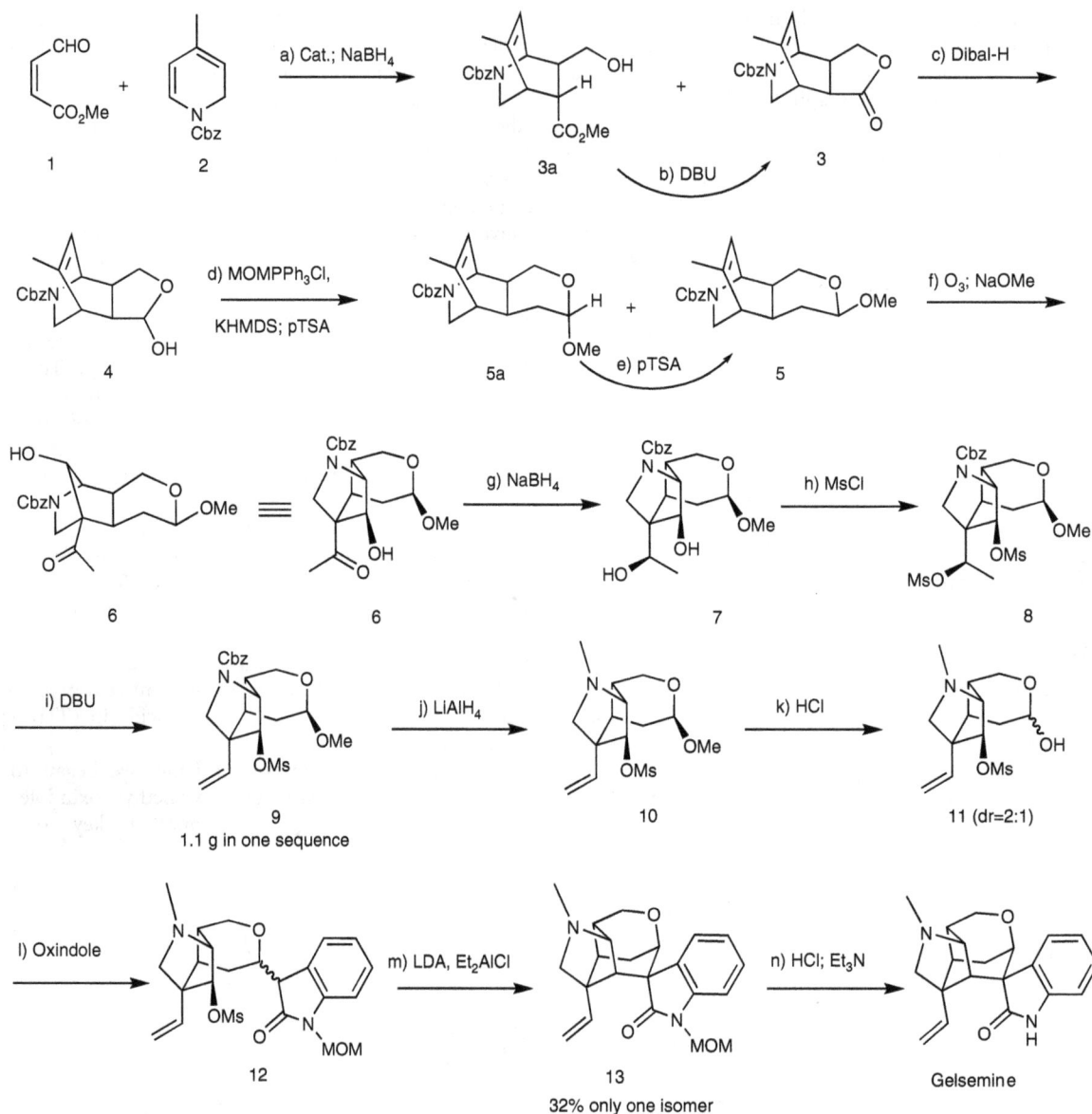

**Figure 6 | The synthesis of ( + )-gelsemine.** Reagents and conditions: (**a**) Cat. (0.1 eq), CH₃CN/H₂O (20:1), − 20 °C, 36 h, then NaBH₄ (1 eq), 0 °C, 30% for **3a**, 47% for **3**; (**b**) DBU, toluene, reflux, 20 h, 97%; (**c**) Dibal-H (1.05 eq), DCM, − 78 °C, 3 h, 90%; (**d**) KHMDS (4.4 eq), MOMPPh₃Cl (4 eq), THF, 0 °C—rt, 3 h, then **4**, 0 °C, 4 h; pTSA (0.1 eq), CH(OMe)₃, DCM, rt, 93% for **5a** and **5** (**5a:5** = 1:13); (**e**) pTSA (0.1 eq), CH(OMe)₃, DCM, rt; (**f**) O₃, DCM, − 78 °C, 30 min; NaOCH₃ (0.3 eq), CH₃OH, 0 °C, 24 h, 60%; (**g**) NaBH₄ (1.1 eq), CH₃OH, 0 °C, 30 min, 93%; (**h**) MsCl (3 eq), DMAP (3 eq), Et₃N (5 eq), DCM, 0 °C, quantitative; (**i**) DBU, toluene, reflux, 24 h, 85%; (**j**) LiAlH₄ (1.2 eq), THF, 0 °C, 10 h, 86%; (**k**) 6 M HCl, THF, H₂O, 3 h, 96%; (**l**) piperidine, 1-MOM-oxindole (1.5 eq), CH₃OH, reflux, 86%; (**m**) LDA (1.2 eq), Et₂AlCl (5 eq), toluene, 32%; (**n**) 6 M HCl, THF, 50 °C, 24 h; Et₃N, CH₃OH, 55 °C, 24 h, 70%. DBU, 1,8-diazabicycloundec-7-ene; Dibal-H, diisobutyl aluminium hydride; KHMDS, potassium hexamethyldisilazane; pTSA, p-toluenesulfonic acid; DCM, dichloromethane; MsCl, methanesulfonyl chloride; DMAP, 4-dimethylaminopyridine; LDA, lithium diisopropylamide; rt, room temperature.

the synthesis of gelsemine. As expected, the condensation of intermediate **11** with oxindole in refluxing methanol and a catalytic amount of piperidine afforded the desired product **12** (85%) as an inseparable mixture of all four possible isomers. The seemed straightforward intramolecular S$_N$2 substitution reaction turned out to be problematic. Many reaction conditions were tested (NaH/THF; NaOCH₃/CH₃OH; KO$^t$Bu/THF; KO$^t$Bu/THF/Bu$^t$OH; LDA/THF; CsF/DMF⁶²; LiHMDS/THF; LiHMDS/HMPA/THF; LiHMDS/LiCl/THF, LiHMDS/ZnCl₂/THF; LiHMDS/DMSO; LDA/Et₂AlCl/THF; LiHMDS/Me₂AlCl/toluene; LiHMDS/Me₂AlCl/THF; NaHMDS/Me₂AlCl/THF, NaH/DMF) but all turned into a complex product mixture. However, when intermediate **12** was treated with LDA and then diethylaluminum chloride in toluene at 90 °C, the reaction

furnished the desired product in 32% yield as a single isomer. Finally, acid hydrolysis of the methyl group from the methoxymethyl protective group and removal of the resulting hydroxymethyl with triethylamine converted **13** into ( + )-gelsemine in 70% combined yield. The synthetic material is identical to the natural product in terms of carbon and proton NMR spectra and optical rotation (see Supplementary Fig. 15).

**Discussion**

The total synthesis of ( + )-gelsemine is completed in a highly enantioselective manner from readily accessible starting materials. This synthesis features an enantioselective organocatalytic Diels–Alder reaction, a formidable intramolecular aldol

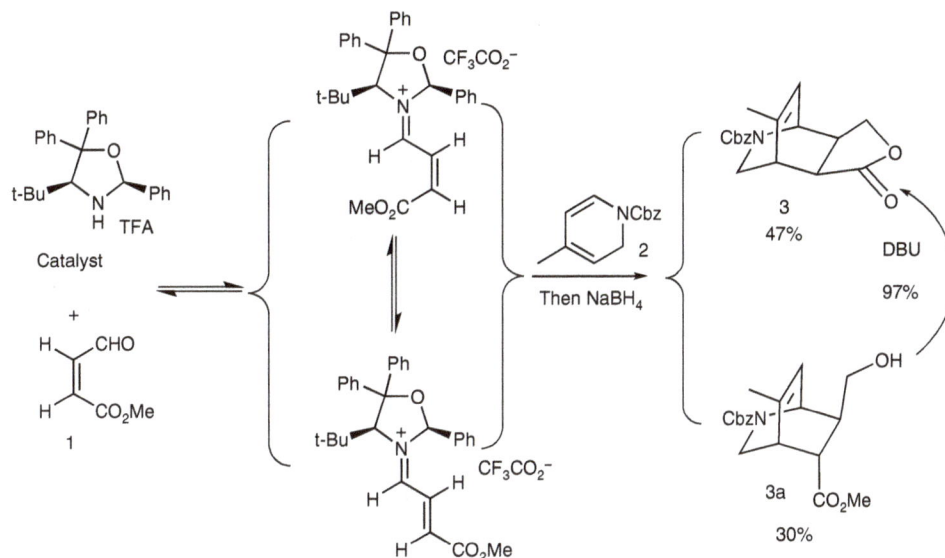

**Figure 7 | Intermediates leading to the formation of 3 and 3a.** The reaction consequence indicates that the carbon–carbon double-bond isomerization of the iminium salt occurred at a rate comparable to that of the cycloaddition.

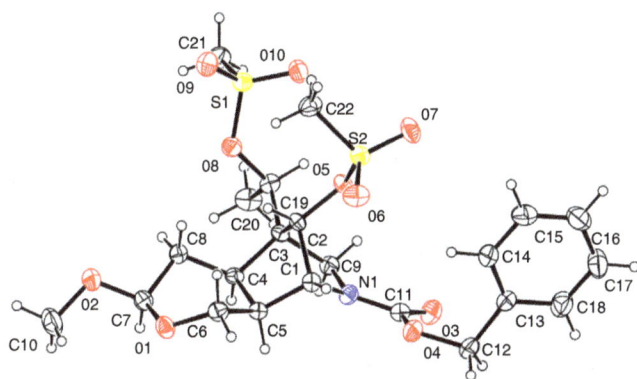

**Figure 8 | The X-ray crystallographic structure of intermediate 8.**
ORTEPs are included in the Supporting Information as a separate file. CCDC 1056043 contains the supplementary crystallographic data.

cyclization and a challenging intramolecular $S_N2$ displacement. The combination of all these features resulted in exceptional overall synthetic efficiency: the enantio excess is over 99%, and the total yield is about 5%.

## Methods
**General.** All reagents were reagent grade and used without purification, unless otherwise noted. All reactions involving air- or moisture-sensitive reagents or intermediates were performed under an inert atmosphere of argon in glassware that was oven dried. Reaction temperatures referred to the temperature of the cooling/heating bath. Chromatography was performed using forced flow (flash chromatography) of the indicated solvent system on 230-400 mesh silica gel (Silicycle flash F60), unless otherwise noted. $^1$H NMR and $^{13}$C NMR spectra were recorded on a Bruker AV-400 or 500 MHz spectrometer. Chemical shifts were referenced to the deuterated solvent (for example, for CDCl$_3$, $\delta = 7.27$ p.p.m. and 77.0 p.p.m. for $^1$H and $^{13}$C NMR, respectively) and reported in parts per million (p.p.m., $\delta$) relative to tetramethylsilane ($\delta = 0.00$ p.p.m.). Coupling constants ($J$) were reported in Hz and the splitting abbreviations used were: s, singlet; d, doublet; t, triplet; q, quartet; m, multiplet; comp, overlapping multiplets of magnetically non-equivalent protons; br, broad; app, apparent. Reactions were monitored using thin-layer chromatography carried out on 0.25-mm E. Merck silica gel plates (60F-254) using ultraviolet light as the visualizing agent or an ethanolic solution of phosphomolybdic acid, cerium sulfate and heat as developing agents. Optical rotations were measured on a PerkinElmer 341 polarimeter. Enantiomeric ratios were determined by chiral HPLC using a chiralpak AD-H (amylose tris (3,5-dimethylphenylcarbamate) coated on 5-μm silica gel) with hexane and i-PrOH as eluents. Tetrahydrofuran, benzene, toluene and diethyl ether were distilled from Na and diphenylketone. DCM, N,N-diisopropylethylamine and triethylamine were

distilled from calcium hydride, while methanol was distilled from dry magnesium turnings immediately before use.

For $^1$H and $^{13}$C NMR spectra of compounds, see Supplementary Figs 1–14. For the comparisons of $^1$H spectra of the natural and synthetic gelsemine, see Supplementary Fig. 15. For the HPLC of 3, see Supplementary Fig. 16. For the experimental procedures and spectroscopic and physical data of compounds and the crystallographic data of compound **8**, see Supplementary Methods.

## References
1. Sonnenschein, F. L. Ueber einige bestandtheile von gelsemium sempervirens. *Ber. Dtsch. Chem. Ges.* **9,** 1182–1186 (1876).
2. Conroy, H. & Chakrabarti, J. K. NMR spectra of gelsemine derivatives. The structure and biogenesis of the alkaloid gelsemine. *Tetrahedron Lett.* **1,** 6–13 (1959).
3. Schun, Y. & Cordell, G. A. Studies on the NMR spectroscopic properties of gelsemine - revisions and refinements. *J. Nat. Prod.* **48,** 969–971 (1985).
4. Lovell, F. M., Pepinsky, R. & Wilson, A. J. C. X-ray analysis of the structure of gelsemine hydrohalides. *Tetrahedron Lett.* **1,** 1–5 (1959).
5. Schun, Y., Cordell, G. A. & Garland, M. 21-oxogelsevirine, a new alkaloid from *Gelsemium rankinii. J. Nat. Prod.* **49,** 483–487 (1986).
6. Schun, Y. & Cordell, A. G. Rankinidine, a new indole alkaloid from *Gelsemium rankinii. J. Nat. Prod.* **49,** 806–808 (1986).
7. Ponglux, D. *et al.* Studies on the indole alkaloids of *gelsemium* elegans (Thailand): structure elucidation and proposal of biogenetic route. *Tetrahedron* **44,** 5075–5094 (1988).
8. Lin, L. Z., Schun, Y., Cordell, A. G., Ni, C.-Z. & Clardy, J. Three oxindole alkaloids from *Gelsemium* species. *Phytochemistry* **30,** 679–683 (1991).
9. Autrey, R. L. & Tahk, F. C. The synthesis and stereochemistry of some isatylideneacetic acid derivatives. *Tetrahedron* **23,** 901–917 (1967).
10. Autrey, R. L. & Tahk, F. C. Oxindoles-II: the products of some Michael additions to isatylideneacetic esters and cinnamyl derivatives. *Tetrahedron* **24,** 3337–3345 (1968).
11. Johnson, R. S., Lovett, T. O. & Stevens, T. S. The alkaloids of *Gelsemium sempervirens.* Part IV. Derivatives of pyridine, isoquinoline, and indol-2(3H)-one as possible initial materials for synthesis of gelsemine. *J. Chem. Soc. C,* 796–800 (1970).
12. Fleming, I., Loreto, M. A., Michael, J. P. & Wallace, I. H. M. Two new stereochemically complementary oxindole synthesis. *Tetrahedron Lett.* **23,** 2053–2056 (1982).
13. Fleming, I., Loreto, M. A., Wallace, I. H. M. & Michael, J. P. Two new oxindole syntheses. *J. Chem. Soc. Perkin Trans. 1* 349–359 (1986).
14. Stork, G., Krafft, M. E. & Biller, S. A. An approach to gelsemine. *Tetrahedron Lett.* **28,** 1035–1038 (1987).
15. Vijn, R. J., Hiemstra, H., Kok, J. J., Knotter, M. & Speckamp, W. N. Synthetic studies towards gelsemine, I. The importance of the antiperiplanar effect in the highly regioselective reduction of non-symmetrical cis-hexahydrophthalimides. *Tetrahedron* **43,** 5019–5030 (1987).
16. Abelman, M. M., Oh, T. & Overman, L. E. Intramolecular alkene arylations for rapid assembly of polycyclic systems containing quaternary centers. A new

synthesis of spirooxindoles and other fused and bridged ring systems. *J. Org. Chem.* **52**, 4130–4133 (1987).

17. Clarke, C. *et al.* An approach to the synthesis of gelsemine: the intramolecular reaction of an allylsilane with an acyliminium ion for the synthesis of one of the quaternary centres. *Tetrahedron* **44**, 3931–3934 (1988).

18. Hiemstra, H., Vijn, R. J. & Speckamp, W. N. Synthetic studies toward gelsemine. 2. Preparation of the tetracyclic skeletal part by way of a highly stereospecific intramolecular reaction of a silyl enol ether with an N-acyliminium ion. *J. Org. Chem.* **53**, 3882–3884 (1988).

19. Earley, W. G., Jacobsen, E. J., Meier, G. P., Oh, T. & Overman, L. E. Synthesis studies directed toward gelsemine. A new synthesis of highly functionalized *cis*-hydroisoquinolines. *Tetrahedron Lett.* **29**, 3781–3784 (1988).

20. Earley, W. G., Oh, T. & Overman, L. E. Synthesis studies directed toward gelsemine. Preparation of an advanced pentacyclic intermediate. *Tetrahedron Lett.* **29**, 3785–3788 (1988).

21. Choi, J. K. *et al.* α-acylamino radical cyclizations: application to the synthesis of a tetracyclic substructure of gelsemine. *J. Org. Chem.* **54**, 279–290 (1989).

22. Fleming, I., Moses, R. C., Tercel, M. & Ziv, J. A new oxindole synthesis. *J. Chem. Soc. Perkin Trans.* 1 617–626 (1991).

23. Hart, D. J. & Wu, S. C. Intramolecular addition of aryl radicals to vinylogous urethanes: studies toward preparation of the oxindole portion of gelsemine. *Tetrahedron Lett.* **32**, 4099–4102 (1991).

24. Koot, W.-J., Hiemstra, H. & Speckamp, W. N. (R)-1-acetyl-5-isopropoxy-3-pyrrolin-2-one: a versatile chiral dienophile from (S)-malic acid. *J. Org. Chem.* **57**, 1059–1061 (1992).

25. Madin, A. & Overman, L. E. Controlling stereoselection in intramolecular heck reactions by tailoring the palladium catalyst. *Tetrahedron Lett.* **33**, 4859–4862 (1992).

26. Overman, L. E. & Sharp, M. J. Reaction of Na₂Fe(CO)₄ with an unsaturated aziridinium ion. Unprecedented rearrangement of an alkyltetracarbonylferrate intermediate. *J. Org. Chem.* **57**, 1035–1038 (1992).

27. Hart, D. J. & Wu, S. C. Gelsemine model studies: alkoxymethylations of decalones and indoles. *Heterocycles* **35**, 135–138 (1993).

28. Takayama, H., Seki, N., Kitajima, M., Aimi, N. & Sakai, S.-I. Application of Tungstate-catalyzed oxidation to the conversion of oxindoles into the corresponding Na-methoxyoxindoles in the gelsemium alkaloid synthesis. *Nat. Prod. Lett.* **2**, 271–276 (1993).

29. Johnson, A. P., Luke, R. W. A., Steele, R. W. & Boa, A. N. Synthesis of 1-acylamino-1-(trimethylsiloxy)alkanes by trimethylsilyl trifluoromethanesulfonate-catalysed addition of bis(trimethylsilyl)amides to aldehydes. *J. Chem. Soc. Perkin Trans.* 1 883–893 (1996).

30. Ng, F., Chiu, P. & Danishefsky, S. J. Toward a potential total synthesis of gelsemine: a regioselective hydroboration directed by a remote olefin. *Tetrahedron Lett.* **39**, 767–770 (1998).

31. Sung, M. J., Lee, C.-W. & Cha, J. K. Ti(II)-mediated cyclization of ω-vinylimides. A stereoselective approach to gelsemine. *Synlett* 561–562 (1999).

32. Avent, A. G., Byrne, P. W. & Pankett, C. S. A photochemical approach to the gelsemine skeleton. *Org. Lett.* **1**, 2073–2075 (1999).

33. Dijkink, J., Cintrat, J.-C., Speckamp, W. N. & Hiemstra, H. Enantioselective synthesis of a key tricyclic intermediate *en route* to (+)-gelsemine. *Tetrahedron Lett.* **40**, 5919–5922 (1999).

34. Pearson, A. J. & Wang, X. A convenient one-pot procedure to afford bicyclic molecules by stereospecific iron carbonyl mediated [6+2] ene-type cyclization: a possible approach to gelsemine. *J. Am. Chem. Soc.* **125**, 13326–13327 (2003).

35. Grecian, S. & Aube, J. Double conjugate addition of a nitropropionate ester to a quinone monoketal: Synthesis of an advanced intermediate to (±)-gelsemine. *Org. Lett.* **9**, 3153–3156 (2007).

36. Tchabanenko, K., Simpkins, N. S. & Male, L. A concise approach to a gelsemine core structure using an oxygen to carbon bridge swapping strategy. *Org. Lett.* **10**, 4747–4750 (2008).

37. Liu, C.-T. & Yu, Q.-S. Biomimetic synthesis of koumine. *Acta Chim. Sinica* **45**, 359–364 (1987).

38. Sheikh, Z., Steel, R., Tasker, A. S. & Johnson, A. P. A total synthesis of gelsemine: synthesis of a key tetracyclic intermediate. *J. Chem. Soc. Chem. Commun.* 763–764 (1994).

39. Dutton, J. K., Steel, R. W., Tasker, A. S., Popsavin, V. & Johnson, A. P. A total synthesis of gelsemine: oxindole spiroannelation. *J. Chem. Soc. Chem. Commun.* 765–766 (1994).

40. Newcombe, N. J., Ya, F., Vijn, R. J., Hiemstra, H. & Speckamp, W. N. The total synthesis of (±)-gelsemine. *J. Chem. Soc. Chem. Commun.* 767–768 (1994).

41. Atarashi, S. *et al.* Free radical cyclizations in alkaloid total synthesis: (±)-21-Oxogelsemine and (±)-gelsemine. *J. Am. Chem. Soc.* **119**, 6226–6241 (1997).

42. Fukuyama, T. & Liu, G. Stereocontrolled total synthesis of (±)-gelsemine. *J. Am. Chem. Soc.* **118**, 7426–7427 (1996).

43. Madin, A. *et al.* Total synthesis of (±)-gelsemine. *Angew. Chem. Int. Ed.* **38**, 2934–2936 (1999).

44. Yokoshima, S., Tokuyama, H. & Fukuyama, T. Enantioselective total synthesis of (+)-gelsemine: determination of its absolute configuration. *Angew. Chem. Int. Ed.* **39**, 4073–4075 (2000).

45. Ng, F. W., Lin, H. & Danishefsky, S. J. Explorations in organic chemistry leading to the total synthesis of (±)-gelsemine. *J. Am. Chem. Soc.* **124**, 9812–9824 (2002).

46. Earley, W. C. *et al.* Aza-Cope rearrangement-Mannich cyclizations for the formation of complex tricyclic amines: stereocontrolled total synthesis of (±)-gelsemine. *J. Am. Chem. Soc.* **127**, 18046–18053 (2005).

47. Madin, A. *et al.* Use of the intramolecular Heck reaction for forming congested quaternary carbon stereocenters: stereocontrolled total synthesis of (±)-gelsemine. *J. Am. Chem. Soc.* **127**, 18054–18065 (2005).

48. Zhou, X., Xiao, T., Iwama, Y. & Qin, Y. Biomimetic total synthesis of (+)-gelsemine. *Angew. Chem. Int. Ed.* **51**, 4909–4912 (2012).

49. Lin, H. & Danishefsky, S. J. Gelsemine: a thought-provoking target for total synthesis. *Angew. Chem. Int. Ed.* **42**, 36–51 (2003).

50. Zhang, J.-Y., Gong, N., Huang, J.-L., Guo, L.-C. & Wang, Y.-X. Gelsemine, a principal alkaloid from *Gelsemium sempervirens* Ait., exhibits potent and specific antinociception in chronic pain by acting at spinal α³ glycine receptors. *Pain* **154**, 2452–2462 (2013).

51. Nising, C. F. & Bräse, S. The oxa-Michael reaction: from recent developments to applications in natural product synthesis. *Chem. Soc. Rev.* **37**, 1218–1228 (2008).

52. Nasir, N. M., Ermanis, K. & Clarke, P. A. Strategies for the construction of tetrahydropyran rings in the synthesis of natural products. *Org. Biomol. Chem.* **12**, 3323–3335 (2014).

53. Guillena, G., Nájera, C. & Ramón, D. J. Enantioselective direct aldol reaction: the blossoming of modern organocatalysis. *Tetrahedron Asymmetry* **18**, 2249–2293 (2007).

54. Xiao, Y.-C. & Chen, Y.-C. Intramolecular reactions. *Comprehensive Enantioselective Organocatalysis* **3**, 1069–1090 (2013).

55. Maria da, C. F. O., Leonardo, S. S. & Ronaldo, A. P. Diastereoselection of the addition of silyloxyfurans to five-, six- and seven-membered *N*-acyliminium ions. *Tetrahedron Lett.* **42**, 6995–6997 (2001).

56. Kolonko, J. K. & Reich, H. J. Stabilization of ketone and aldehyde enols by formation of hydrogen bonds to phosphazene enolates and their aldol products. *J. Am. Chem. Soc.* **130**, 9668–9669 (2008).

57. Denmark, S. E. & Henke, B. R. Investigations on transition-state geometry in the aldol condensation. *J. Am. Chem. Soc.* **113**, 2177–2194 (1991).

58. Nakano, H. *et al.* A novel chiral oxazolidine organocatalyst for the synthesis of an oseltamivir intermediate using a highly enantioselective Diels-Alder reaction of 1,2-dihydropyridine. *Chem. Commun.* **46**, 4827–4829 (2010).

59. Kohari, Y. *et al.* Enantioselective Diels-Alder reaction of 1,2-dihydropyridines with aldehydes using β-amino alcohol organocatalyst. *J. Org. Chem.* **79**, 9500–9511 (2014).

60. Bayly, A. R., White, A. J. P. & Spivey, A. C. Design and synthesis of a prototype scaffold for five-residue α-helix mimetics. *Eur. J. Org. Chem.* **25**, 5566–5569 (2013).

61. Kraus, G. A. & Kim, J. Tandem Diels-Alder/ene reactions. *Org. Lett.* **6**, 3115–3117 (2004).

62. Sato, T. & Otera, J. CsF in organic synthesis. Malonic ester synthesis revisited for stereoselective carbon-carbon bond formation. *J. Org. Chem.* **60**, 2627–2629 (1995).

## Acknowledgements

H.Z. thanks the National Basic Research Program of China (973 Program: 2010CB833200), NSFC (21172100; 21272105; 21290183), Program for Changjiang Scholars and Innovative Research Team in University (PCSIRT: IRT1138), FRFCU (lzujbky-2013-ct02) and '111' Program of MOE for financial support.

## Author contributions

F.G.Q. and H.Z. conceived the synthetic design and directed the project. X.C., S.D. and C.T. conducted the experimental work and data analysis. F.G.Q. wrote the manuscript.

# 5

# The key role of the scaffold on the efficiency of dendrimer nanodrugs

Anne-Marie Caminade[1,2], Séverine Fruchon[3,4], Cédric-Olivier Turrin[1,2], Mary Poupot[5,6], Armelle Ouali[1,2], Alexandrine Maraval[1,2], Matteo Garzoni[7], Marek Maly[8], Victor Furer[9], Valeri Kovalenko[10], Jean-Pierre Majoral[1,2], Giovanni M. Pavan[7] & Rémy Poupot[3,4]

Dendrimers are well-defined macromolecules whose highly branched structure is reminiscent of many natural structures, such as trees, dendritic cells, neurons or the networks of kidneys and lungs. Nature has privileged such branched structures for increasing the efficiency of exchanges with the external medium; thus, the whole structure is of pivotal importance for these natural networks. On the contrary, it is generally believed that the properties of dendrimers are essentially related to their terminal groups, and that the internal structure plays the minor role of an 'innocent' scaffold. Here we show that such an assertion is misleading, using convergent information from biological data (human monocytes activation) and all-atom molecular dynamics simulations on seven families of dendrimers (13 compounds) that we have synthesized, possessing identical terminal groups, but different internal structures. This work demonstrates that the scaffold of nanodrugs strongly influences their properties, somewhat reminiscent of the backbone of proteins.

[1] Laboratoire de Chimie de Coordination du CNRS, UPR 8241, 205 route de Narbonne, BP 44099, 31077 Toulouse Cedex 4, France. [2] Université de Toulouse, UPS, INP, LCC, F-31077 Toulouse, France. [3] Centre de Physiopathologie de Toulouse Purpan, F-31300 Toulouse, France. [4] INSERM, U1043; CNRS, U5282; Université de Toulouse, UPS, Toulouse, France. [5] Centre de Recherche en Cancérologie de Toulouse, F-31300 Toulouse, France. [6] INSERM, U1037; CNRS, U5294; Université de Toulouse, UPS, Toulouse, France. [7] Department of Innovative Technologies, University of Applied Sciences and Arts of Southern Switzerland, Galleria 2, 6928 Manno, Switzerland. [8] Faculty of Science, J.E. Purkinje University, Ceske mladeze 8, 400 96 Ústí nad Labem, Czech Republic. [9] Kazan State Architect and Civil Engineering University, Zelenaya 1, Kazan 420043, Russia. [10] A.E. Arbuzov Institute of Organic and Physical Chemistry of Kazan Scientific Center of Russian Academy of Science, Arbuzov Str., 8, Kazan 420088, Russia. Correspondence and requests for materials should be addressed to A.-M.C. (email: anne-marie.caminade@lcc-toulouse.fr) or to G.M.P. (email: giovanni.pavan@supsi.ch) or to R.P. (email: remy.poupot@inserm.fr).

The large number of potential applications of dendrimers[1] generates each year a tremendous amount of work, often connected to their biological properties. Emphasis is generally put on the modification of the terminal groups and of their number (related to the generation, that is, the number of layers), in connection with the multivalency effect that is the most important property recognized for dendrimers[2-5]. The possibility to design molecules with controlled multivalency is particularly important for biological applications[6], polyvalent interactions being ubiquitous in many biological systems[7]. Only very few publications have experimentally reported so far the influence of the internal structure of dendrimers on their properties, even if among the five critical nanoscale design parameters recently proposed by Tomalia[8] (size, shape, surface chemistry, flexibility and architecture), at least three of them are related to the internal structure. Comparison between PAMAM (polyamidoamine)[9] and PPI (polypropyleneimine)[10] dendrimers has emphasized the difference of the length of branches as the most important characteristics for their use as sensor[11], and for obtaining nanoparticles[12]. Comparison of the physical properties have shown important differences between PAMAM and poly(L-lysine) dendrimers[13], whereas rigid branches of dendrimers with azobenzene core induce significant differences for the isomerization, compared with less rigid branches[14]. In biology, a few examples have compared the efficiency of specific dendrimers with that of PAMAM dendrimers, with particular emphasis on transfection experiments[15]. However, there is no example to date of a study assessing the influence of a large number of dendritic scaffolds on the biological properties *per se*, and getting insights on the reasons for the differences.

Immunogenicity of dendrimers has been investigated for years, and these studies show absence or only weak immunogenicity of these molecules[16]. On the other hand, it is known that phagocytes of the immune system (that is, monocytes and macrophages that are immune white cells playing multiple roles in the immune system[17]) engulf nanoparticles, which in some cases leads to their activation[18]. Moreover, due to their involvement in many different diseases, monocytes are relevant targets to promote curative immunomodulation[19,20]. Some of us have already shown that a first-generation poly(phosphorhydrazone) dendrimer ended by azabis(phosphonic acid) groups has unprecedented biological properties. This compound is able to modulate *in vitro* the response of the human immune system, in particular, by inducing the multiplication of natural killer cells[21,22], activating monocytes[23] through an anti-inflammatory pathway[24]. The efficacy of this molecule has been proven *in vivo* in a mouse model of experimental arthritis relevant to human rheumatoid arthritis[25]. In this model, there is a constitutive inflammatory activation of monocytes/macrophages that is responsible for the onset and the development of the pathology. We have shown that this particular azabis(phosphonic acid)-ended dendrimer targets monocytes/macrophages and inhibits the main physiopathological features of the disease—systemic inflammation, cartilage degradation and bone resorption. The potential of this nanodrug candidate against rheumatoid arthritis has been highlighted[26,27]. This preliminary work has demonstrated that the $N(CH_2P(O)(OH)(ONa))_2$ pincer is the most active part within the structure. Variation on the structure of the pincer strongly decreases the biological activity[21], whereas the replacement of phosphonic acids by carboxylic acids or sulfonic acids precludes any activity[28]. Furthermore, the azabisphosphonic pincer has to be linked to the first-generation poly(phosphorhydrazone) dendrimer (12 terminal groups) through the nitrogen atom. These poly(phosphorhydrazone) dendrimers are still very active with a lower number of such terminal functions (8 or 10), but become poorly active with 6, 4 or 2 terminal functions, and the monomer is non-active at all[29], emphasizing the fact that these dendrimers are not drug carriers[30], but drugs by themselves. An increased number of terminal functions (16, 24 (generation 2)[21] or 30) has also a detrimental influence on the efficiency[29].

Such types of terminal groups appear appealing for studying and rationalizing the influence of the nature of the scaffold on the properties of dendritic nanodrugs. Therefore, we describe the grafting of azabisphosphonic acids as terminal groups (4 to 12 functions) to a series of dendrimers having different internal structures. Seven different families of dendrimers (13 compounds) having identical terminal groups (azabisphosphonic derivatives), but different internal structures (PAMAM, PPI, poly(carbosilane)[31], poly(L-Lysine)[32] and three different types of phosphorus-containing dendrimers) are synthesized. Their efficiency for the activation of human monocytes is described. To identify the reasons of the original and surprising biological results obtained, the modelling of the structures of the dendrimers in aqueous solution by means of all-atom molecular dynamics (MD) simulations is carried out for obtaining high-resolution (atomistic) details of the configuration they assume in the real environment (solvated state).

## Results

**Syntheses of the dendrimers**. The seven different families of dendrimers (13 compounds) ended by azabisphosphonic groups that we have synthesized and tested are shown in Figs 1 and 2. Owing to the different terminal groups of the dendrimers before their functionalization by the azabis(phosphonic acid) groups, we have used two different linkers and developed different synthetic strategies.

The first linker is tyramine, which affords $OC_6H_4CH_2CH_2N(CH_2P(O)(OH)(ONa))_2$ terminal groups. In addition to the first-generation poly(phosphorhydrazone) dendrimer **1-G$_1$** (12 terminal functions) already synthesized[21], this linker has been used for another type of poly(phosphorhydrazone) dendrimer having internal branches extended by an arylether linkage, **2-G$_1$** (12 terminal functions), as well as for poly(thiophosphate)[33] dendrimers **3-G$_1$** (6 terminal functions) and **3-G$_2$** (12 terminal functions) and for a poly(carbosilane) dendrimer **4-G$_1$** (8 terminal functions; Fig. 1).

The second type of linker is obtained from the $NH_2$ terminal groups of the initial dendrimers via an amide linkage $(NHC(O)(CH_2)_xN(CH_2P(O)(OH)(ONa))_2$ with $x = 1$ or 3; Fig. 2). These terminal groups have been grafted also to the surface of the first-generation poly(phosphorhydrazone) dendrimer[34] via the tyramine function, to afford dendrimers **5a-G$_1$** ($x = 1$) and **5b-G$_1$** ($x = 3$) (12 terminal functions for both). The other types of dendrimers that we have synthesized with this second type of linker are different generations of PPI dendrimers, **6a-G$_1$** ($x = 1$) and **6b-G$_1$** ($x = 3$; 4 terminal functions for both) and **6b-G$_2$** ($x = 3$; 8 terminal functions); different generations of PAMAM dendrimers, **7a-G$_1$** ($x = 1$; 4 terminal functions) and **7b-G$_2$** ($x = 3$; 8 terminal functions); and the poly(L-lysine) dendrimer, **8a-G$_2$** ($x = 1$; 8 terminal functions). In all cases, the azabisphosphonic terminal groups are grafted to the dendrimers in the form of the corresponding methyl ester phosphonates, to carry out the reaction in organic solvents in which both reagents are soluble.

As shown in Fig. 3, we have used different strategies for grafting the azabisphosphonates to the various types of dendrimers. The first step of the synthesis of dendrimer **2-G$_1$** is the nucleophilic substitution on the $P(S)Cl_2$ terminal groups by the phenol moieties of the functionalized tyramine **9** in basic conditions; this method is identical to that used for the synthesis

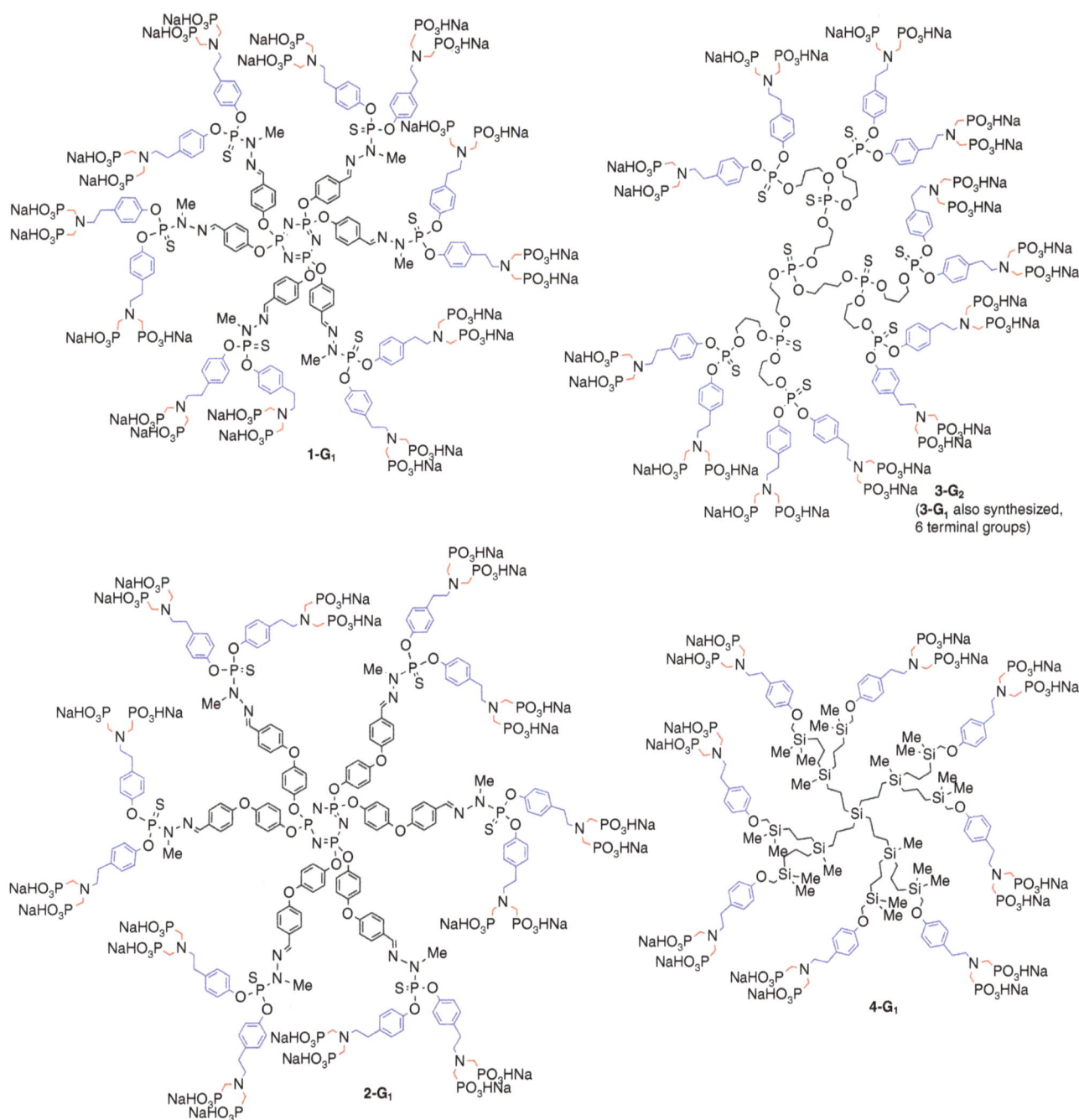

**Figure 1 | Chemical structure of dendrimers 1-G$_1$–4-G$_1$.** The azabisphosphonic salts are in red, the linkers in blue and variable internal structures in black.

of 1-G$_1$ (ref. 21). The same phenol 9 was used for the nucleophilic substitution on the CH$_2$I terminal groups of the carbosilane dendrimer, to finally afford 4-G$_1$. The phosphane bisphosphonate 10 was used for phosphitylation of the alcohol end groups of the thiophosphate dendrimers, followed by oxidation of P$^{III}$ with sulphur, finally affording 3-G$_1$ or 3-G$_2$, depending on the generation of the starting dendrimer. The other types of tyramine derivatives 11a,b (a: CH$_2$ spacer, b: CH$_2$CH$_2$CH$_2$ spacer) were also used for nucleophilic substitutions on P(S)Cl$_2$ terminal groups, to obtain dendrimers 5a-G$_1$ and 5b-G$_1$. Compounds 11a and 11b were obtained by reaction of the carboxylic acids 12a and 12b with the NH$_2$ function of tyramine via peptide coupling. In the case of dendrimers ended by NH$_2$ groups (PPI, PAMAM, polylysine), this peptide coupling was carried out directly on the dendrimers with the derivatives 12a

and 12b, affording finally the series of dendrimers 6a-G$_1$, 6b-G$_1$, 6b-G$_2$ (PPI-type), 7a-G$_1$, 7b-G$_2$ (PAMAM-type) and 8a-G$_2$ (Poly-L-lysine type).

The terminal methyl ester phosphonate groups were converted to phosphonic acid salts by reaction with bromotrimethylsilane, MeOH and then NaOH in water. Remarkably, this process induced the cleavage of the terminal phosphonic ester groups, without destroying the internal structure of the dendrimers, even in the case of dendrimers 3-G$_1$ and 3-G$_2$, which have other types of alkyl phosphonates in their internal structure.

**Monocyte activation by dendrimers.** On stimulation, monocytes/macrophages undergo morphological changes, that is, increase of their size and granularity[35], indicative of an activated

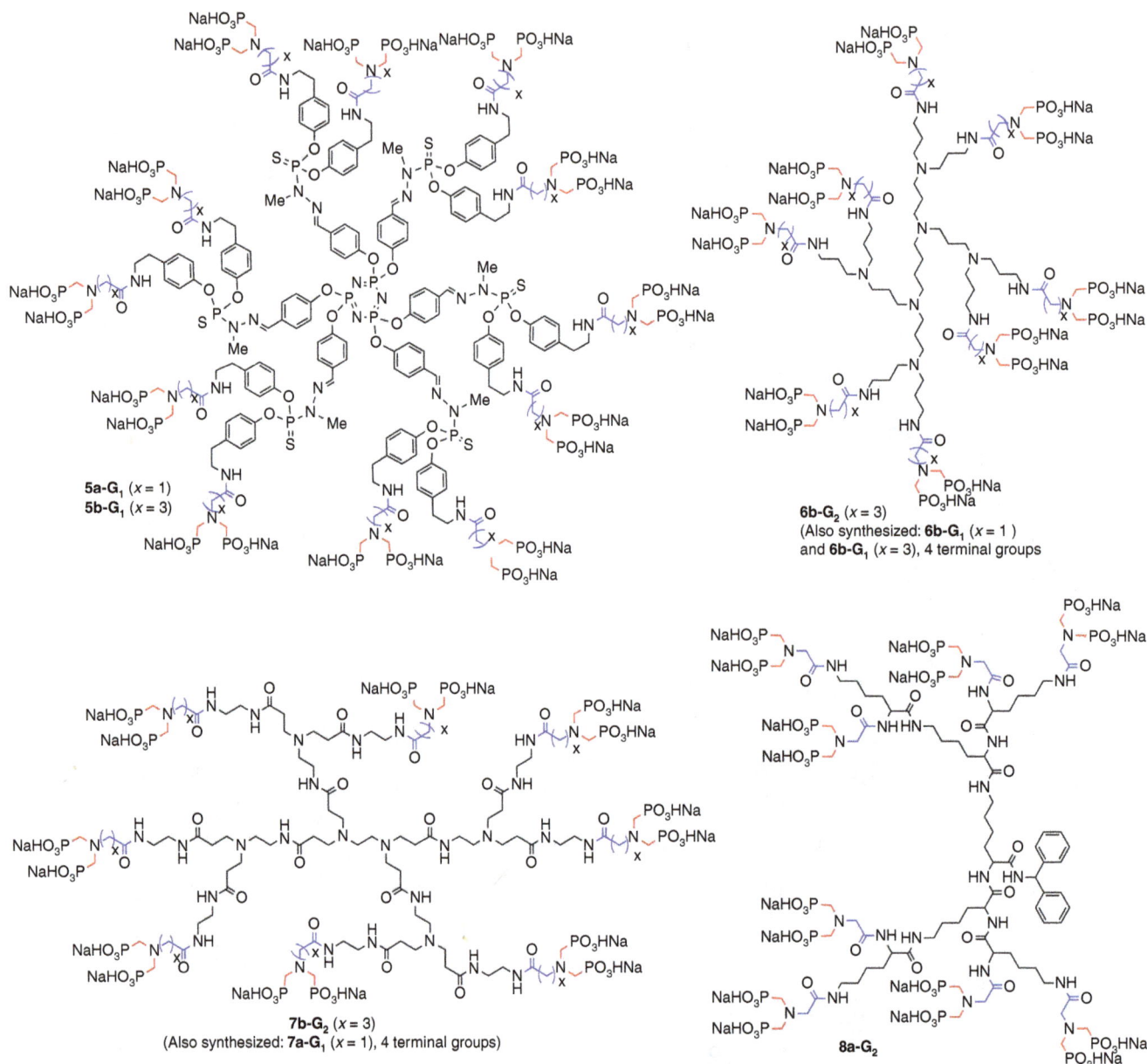

**Figure 2 | Chemical structure of dendrimers 5a,b-G₁–8a-G₁.** The azabisphosphonic salts are in red, the linkers in blue and variable internal structures in black.

status. In particular, we have shown that phosphorus-containing anti-inflammatory dendrimers induce these morphological changes of monocytes[23]. Therefore, we have screened the biological properties of the series of dendrimers synthesized for this study by measuring the morphological features of human monocytes (Figs 4 and 5; the negative control, without any dendrimer, is given first in Fig. 5). These changes appear within a few days in *in vitro* cultures of monocytes, and can be quantified by flow cytometry. These tests were conducted at three different concentrations of dendrimer (20, 2 and 0.2 µM). The more monocytes are activated by the dendrimers, the more their size and granularity are increased. So far, dendrimer $1\text{-}G_1$ appears as the most active among all the dendrimers we have tested[23,25,29]. Thus, this molecule is the standard indicator of the activation of human monocytes in this study. As the dendrimers are dissolved in pure water, the negative control of the tests consists in treating the monocytes with the accurate volume of water. The qualitative results of monocyte activation are shown in Fig. 4 for the active dendrimers and in Fig. 5 for the non-active dendrimers, with

indication of the efficiency score, from + + + for the most active to 0 for the least active (non-active) dendrimers. Dendrimer $1\text{-}G_1$ is the only one displaying strong activity at 2 µM, and which is still active at 0.2 µM. It appears also that dendrimers $2\text{-}G_1$, $3\text{-}G_n$, $4\text{-}G_1$ and $5a,b\text{-}G_1$ display good activity (average for $3\text{-}G_1$) (Fig. 4), whereas dendrimers $6a,b\text{-}G_n$, $7a,b\text{-}G_n$ ($n = 1$, 2) and $8a\text{-}G_2$ are not active at all (Fig. 5). Dendrimers $5a\text{-}G_1$ and $5b\text{-}G_1$ only differ by the length of the linker between the azabisphosphonate terminations and the proximal branching points ($x = 1$ versus $x = 3$, Fig. 2). The bioactivity of these molecules appears to be equivalent. The same observation is made with dendrimers $6a\text{-}G_1$ and $6b\text{-}G_1$, which are both inactive.

**All-atom MD simulations of the dendrimers in explicit solvent.** Molecular modelling was used to understand the striking differences observed in the biological properties. To gain molecular-level information about the dendrimers in the biological conditions, all-atom MD simulations of the 13

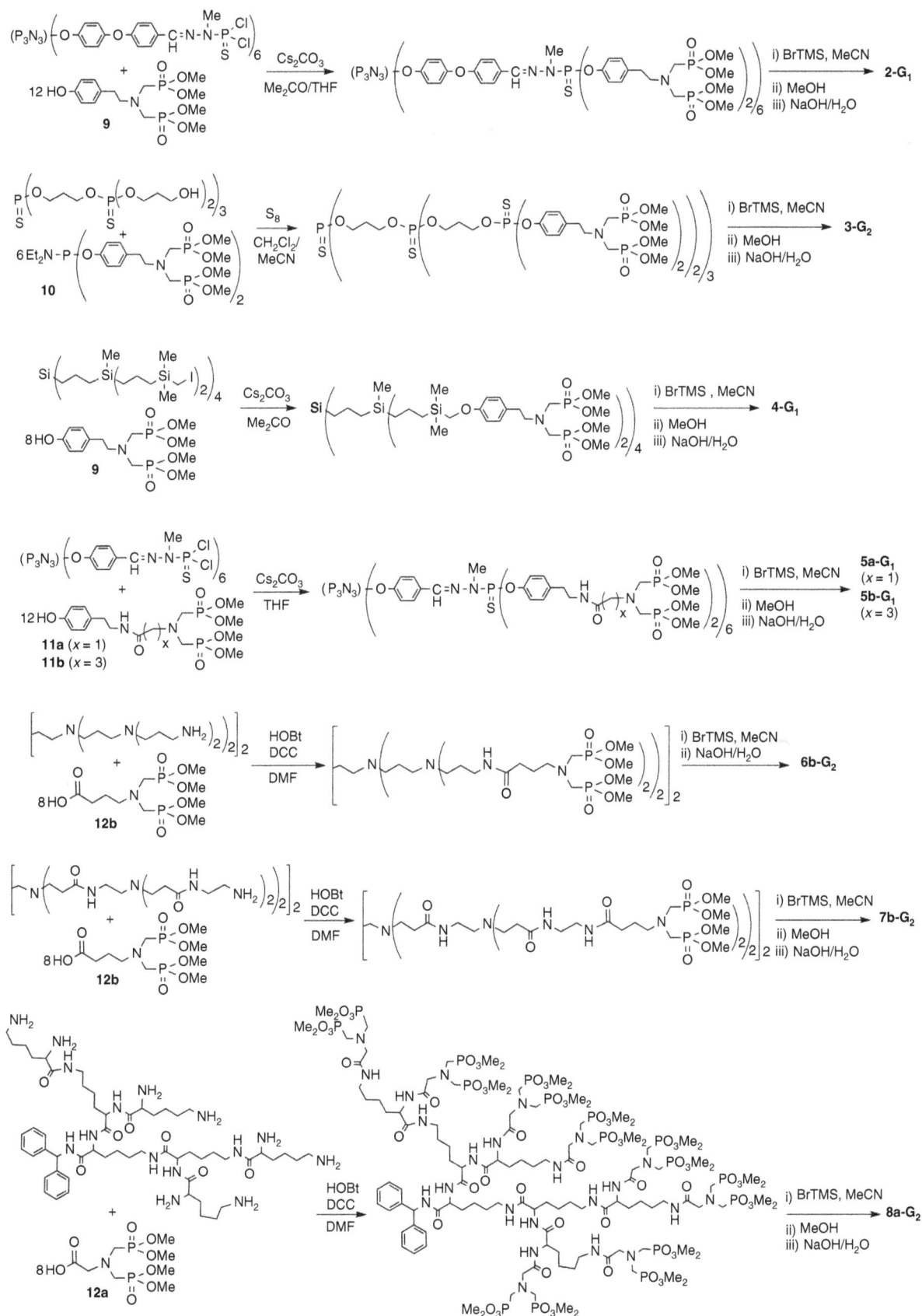

**Figure 3 | The different methods of synthesis of the dendrimers.** Dendrimer **3-G₁** is synthesized as **3-G₂**; dendrimers **6a-G₁** and **6b-G₁** as **6b-G₂**; and dendrimers **7a-G₁** as **7b-G₂** (HOBt: hydroxybenzotriazole, DCC: N,N'-dicyclohexylcarbodiimide, BrTMS: bromotrimethylsilane).

dendrimers in solution were carried out at 37 °C in presence of explicit water molecules and NaCl (150 mM). Each molecular system was equilibrated during 200 ns of MD simulation (Supplementary Fig. 1). Different data were extracted from the equilibrated phase MD trajectories. Figure 6a reports the equilibrated size data (that is, the radius of gyration, $R_g$) for the

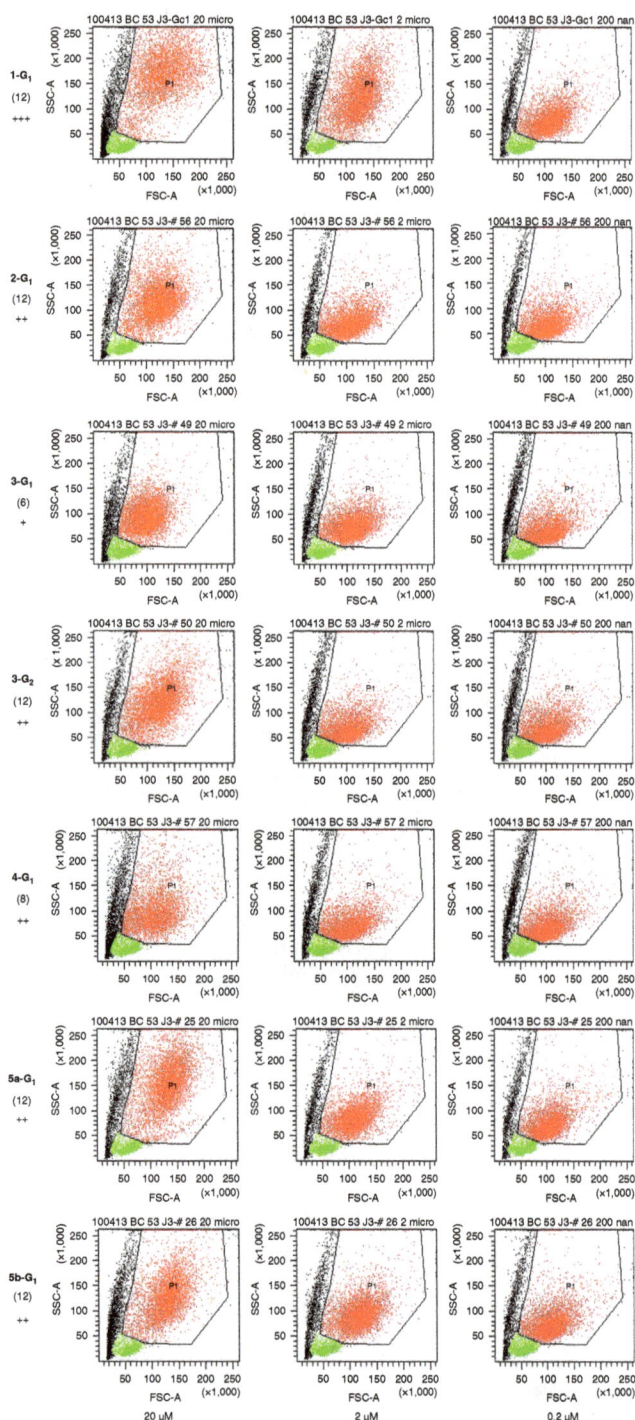

**Figure 4 | Activation of human monocytes by the series of dendrimers 1-G₁, 2-G₂, 3-G₁, 3-G₂, 4-G₁, 5a-G₁ and 5b-G₁.** The bioactivity of the dendrimers is analysed by flow cytometry. Each dot in the plots is indicative of morphological change (size—the Forward Scatter (FSC) parameter on the x axis—and granularity—the Side Scatter (SSC) parameter on the y axis) undergone by purified monocytes in the presence of the different dendrimers at 20, 2, and 0.2 µM (left, middle, and right graphs respectively). Red points are monocytes (gated in the polygon), green points are remaining lymphocytes after purification, black points are died or dying cells. For each dendrimer, the number of terminal functions is indicated in parentheses. The score attributed to each dendrimer appears in red on the left, from 0 (no activation) to + + + (the highest activity, attributed to **1-G₁**). Data are from one representative experiment out of six.

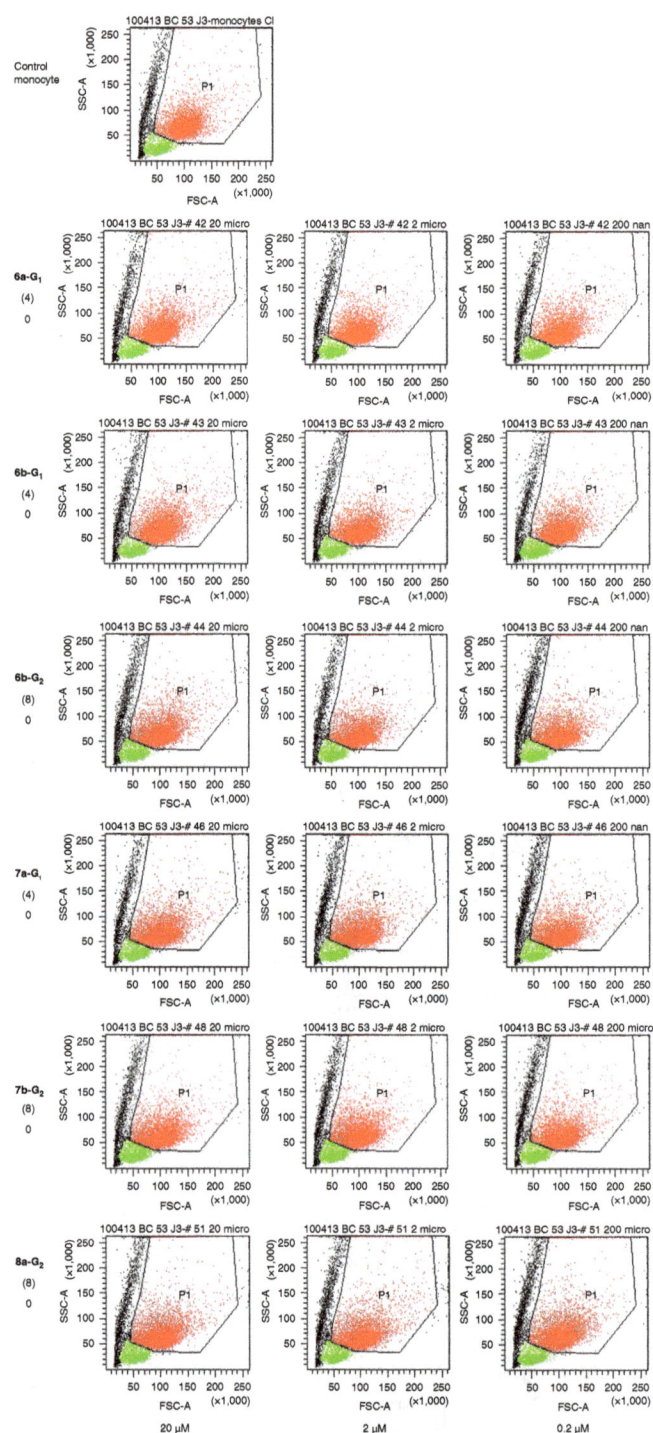

**Figure 5 | Activation of human monocytes by the series of dendrimers 6a-G₁, 6b-G₁, 6b-G₂, 7a-G₁, 7b-G₂ and 8a-G₂.** The bioactivity of the dendrimers is analysed by flow cytometry. Each dot in the plots is indicative of morphological change (size—the forward scatter (FSC) parameter on the x axis—and granularity—the side scatter (SSC) parameter on the y axis) undergone by purified monocytes in the presence of the different dendrimers at 20, 2, and 0.2 µM (left, middle and right graphs, respectively). Red points are monocytes (gated in the polygon), green points are remaining lymphocytes after purification, black points are died or dying cells. For each dendrimer, the number of terminal functions is indicated in parentheses. The score attributed to each dendrimer appears in red on the left, 0 means no activation. The negative control, without any dendrimer, is given first. Data are from one representative experiment out of six.

dendrimers in solution. Comparison of the $R_g$ data with the number of terminal groups indicates that neither the size (generation) of the dendrimers (we used generations 1 and 2), nor the number of terminal functions are exclusively important criteria for the biological activity of each molecule. For instance, the activity of dendrimer 4-$G_1$ (8 terminal functions) is marked $++$, as that of dendrimers 2-$G_1$, 3-$G_1$ and 5b-$G_1$ (12 terminal functions); nevertheless, dendrimers 6b-$G_2$, 7b-$G_2$ and 8a-$G_2$ also have 8 terminal functions, but no activation properties towards monocytes.

The equilibrated snapshots taken from the MD simulations reported in Fig. 6b clearly show that the three-dimensional (3D)-geometrical arrangements of these dendrimers in the solvent are very different. Several information can be extracted from the MD simulations. Analysis of the principal moments of inertia ($I_x$, $I_y$ and $I_z$), of the aspect ratio and of anisotropy of the equilibrated dendrimers highlight that some of them assume spherical-like rather than elongated shape in the real environment (see Supplementary Fig. 2). However, comparison of these data with the biological activity (Figs 4, 5 and 6a) demonstrates that even the overall shape of the dendrimers is not a unique discriminant parameter for their activity. The same is true for the dendrimers solvent-accessible surface area (see Supplementary Fig. 2).

However, deeper structural analysis reveals other important differences related to the location of the active surface groups in the dendrimers structure. Interestingly, as emphasized by the circle drawn around the equilibrated dendrimers in Fig. 6b, dendrimers 3-$G_1$, 6a,b-$G_1$, 6b-$G_2$, 7a-$G_1$, 7b-$G_2$ and 8a-$G_2$ are much more symmetrical than the other ones. The azabisphosphonic terminal functions are spread all over the molecular surface (sphere) for these non-active dendrimers, while, at the equilibrium, the biologically active dendrimers 1-$G_1$, 2-$G_1$, 3-$G_2$, 4-$G_1$ and 5a,b-$G_1$ appear as directional molecules, as in these cases the azabisphosphonic groups are gathered in half-sphere (see also Supplementary

Movies 1 and 2: 1 MD 3D structure of 1-$G_1$, and 2 MD 3D structure of 7b-$G_2$). Thus, the solvated state of the different dendrimers—namely, the conformation assumed in the 'real' environment in terms of localization and density of azabisphosphonic functions on the dendrimers surface—is morphologically very different between active and non-active dendrimers. Since the surface functionalization is identical among all dendrimers, this effect is intimately related to the different internal structure.

Further analysis on the terminal groups' location quantified these structural differences. Plots in Fig. 7a display the radial distribution functions—$g(r)$—of the azabisphosphonate terminal groups (END) calculated with respect to the core unit (CEN) of the dendrimers (Fig. 7a: END to CEN) and respect to each other (END to END: Fig. 7c). The $g(r)$ curves provide indication on the relative probability to find the terminal functions at a certain distance from the centre of the dendrimer or from each other (data are averaged over the last 50 ns of the equilibrated MD trajectories). The positions of the $g(r)$ maximum peaks are particularly interesting. Figure 7b,d shows the normalized peaks of the $g(r)$ curves—only the topmost 10%—revealing the average (most probable) distance for the azabisphosphonic terminal groups (END) respect to the dendrimers central unit (CEN) or respect to each other. Plots in Fig. 7a,b show that, in general, the surface groups are displayed at a larger distance from the dendrimer centre for the active compounds than in the case of the inactive ones ($\sim 1.5 R_g$ versus $\sim R_g$). Moreover, $g(r)$ maximum peaks related to the distance between the different surface groups (Fig. 7d) show that the azabisphosphonate terminal functions are more densely packed in the active dendrimers (the distance between the terminal branching points (N atom) is $\sim 0.5 R_g$) than in the case of the inactive dendrimers ($\sim R_g$). These data provide a picture where active dendrimers look like directional molecules with all surface functions gathered together into 'clusters', far from the dendrimers core unit (CEN), while non-active dendrimers assume a more symmetric configuration in salt water.

**a**

| Dendrimer | 1-$G_1$ | 2-$G_1$ | 3-$G_1$ | 3-$G_2$ | 4-$G_1$ | 5a-$G_1$ | 5b-$G_1$ | 6a-$G_1$ | 6b-$G_1$ | 6b-$G_2$ | 7a-$G_1$ | 7b-$G_2$ | 8a-$G_2$ |
|---|---|---|---|---|---|---|---|---|---|---|---|---|---|
| $R_g$ (Å) | 10.9 | 14.3 | 9.0 | 10.0 | 9.7 | 13.7 | 10.3 | 7.1 | 7.4 | 8.5 | 7.2 | 8.7 | 8.4 |
| END groups | 12 | 12 | 6 | 12 | 8 | 12 | 12 | 4 | 4 | 8 | 4 | 8 | 8 |
| Efficiency | +++ | ++ | + | ++ | ++ | ++ | ++ | 0 | 0 | 0 | 0 | 0 | 0 |

**b**

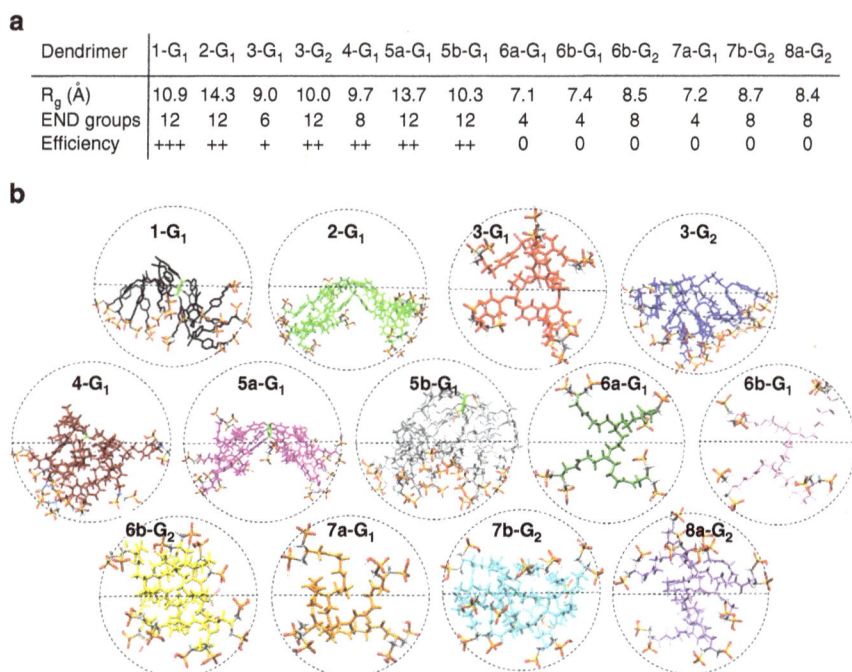

**Figure 6 | Equilibrated configurations of the 13 dendrimers and their size, obtained from the MD simulations.** (**a**) Radius of gyration ($R_g$) of the different dendrimers extracted from the equilibrated phase of the MD simulations in solution, number of azabisphosphonic surface groups (END) and biological efficiency score. Shape analysis: (**b**) MD equilibrated snapshots of the thirteen dendrimers displaying their shape. Dotted circles are added around the dendrimers to emphasize differences in the displacement of the azabisphosphonic functions around the dendrimers surface (symmetrical or directional molecules).

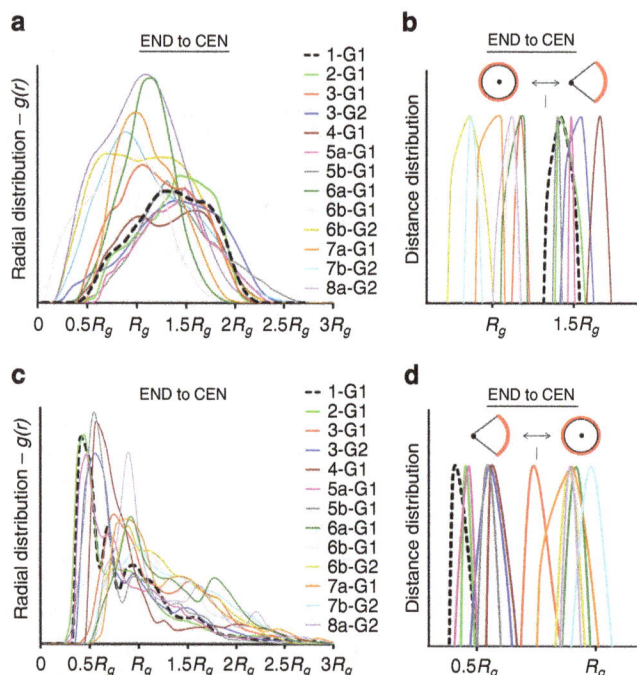

**Figure 7 | Radial distribution functions of the dendrimer terminal groups.**
(a) Radial distribution functions—$g(r)$—of the terminal groups (END) with respect to the dendrimer core unit (CEN); (b) normalized peaks of the $g(r)$ curves (only the topmost 10%) revealing the average (most probable) END to CEN distances in the dendrimers; (c) radial distribution functions, $g(r)$, of the terminal groups with respect to each other; (d) normalized peaks of the $g(r)$ curves of c (only the topmost 10%) revealing the average (most probable) END to END distances. In all graphs, the distance (x axis) is expressed in $R_g$ units to allow comparison between different size dendrimers, and the data corresponding to dendrimer **1-$G_1$** is given in dotted black lines.

To further characterize the solvated state of the dendrimers, the solvation energy ($G_{sol}$)—namely, the solute–solvent interaction energy, or the energy necessary to drag the dendrimers out from water—was also extracted from the equilibrated phase MD simulations (the last 100 ns) and used as a descriptor of molecular hydrophilicity[36]. In general, the higher (the more negative) the $G_{sol}$, the more favourable the interactions of the dendrimer with water, which is a signal of overall hydrophilicity[36]. The $G_{sol}$ data were further normalized for the MW of each dendrimer for comparison between different size, generation and structure of dendrimers (Fig. 8c)[36]. Additional information on the hydration of the dendrimers' interior (water penetration into the scaffold) was also extracted from the MD simulations (the last 100 ns) corroborating this analysis (see Supplementary Fig. 3).

## Discussion

As evidenced from Figs 4 and 5, the various dendrimers induced very different results for the activation of monocytes (measured by the increase of size and granularity), despite their identical terminal groups. It is also clear from Fig. 6a that the number of terminal functions is not exclusively an important criterion. As the striking differences in the properties of all these dendrimers do not seem related to the terminal groups, we hypothesized that they could be related to the internal structure. At first glance, from the chemical structure point of view, the dendrimers could be divided into two families—those having aromatic groups in their structure and those essentially constituted of alkyl linkages. All dendrimers having aromatics in their structure (**1-$G_1$**, **2-$G_1$**,

**5b-$G_1$**) are indeed active, but several dendrimers composed of alkyl linkages (**3-$G_1$**, **3-$G_2$**, **4-$G_1$**) are also active.

In view of these puzzling biological results, theoretical calculations have been carried out to try to rationalize them. We have employed all-atom MD simulations for the 13 different dendrimers immerged in a solvation box containing explicit water molecules and salt ions to gain a molecular-level detailed description of these macromolecules in the real environment.

The radii of gyration ($R_g$) of the equilibrated dendrimers (Fig. 6a) demonstrate that, similar to the number of terminal groups, size is not a discriminant factor controlling molecular activity. One reasonable hypothesis was that dendrimers' activity is somehow related to the shape or configuration assumed by the dendrimers in solution. However, the aspect ratio and the anisotropy parameters for the equilibrated dendrimers extracted from the MD simulations show that even the overall molecular shape is not a key parameter controlling activity (see Supplementary Fig. 2). Nevertheless, the equilibrated conformations assumed by the dendrimers in solution present other interesting differences. In particular, the structure of the active dendrimers is segregated in salt water, with all the hydrophilic terminal functions compacted on one side and the hydrophobic scaffold exposed to the external media (Fig. 6b). On the contrary, the non-active dendrimers have a more symmetrical structure with terminal groups displayed all around the dendrimer surface. Thus, our MD simulations divide the dendrimers into two categories identified by either a directional or a spherical configuration. The radial distribution functions $g(r)$ data of the azabisphosphonic terminal groups (END) provide quantification for this observation (Fig. 7). In particular, the high $g(r)$ peaks at short END–END distance (Fig. 7d) show that the terminal functions are more densely packed in the biologically active dendrimers, in particular for the most active **1-$G_1$**. In general, if the azabisphosphonic terminal groups packing density (position of the END–END $g(r)$ peak) is taken as a score of molecular directionality, Fig. 7c,d show that the latter is in remarkable trend with biological activity.

To better quantify the structure–activity relationships, we have assessed the biological property of the set of dendrimers towards the activation of human monocytes, as there are major players in many different diseases[19,20,25] The biological activity of the molecules has been quantified by flow cytometry. Data shown in Figs 4 and 5 indicate that the different length of the linker ($x = 1$ or $x = 3$) has negligible effect within the same series of dendrimers (**5a-$G_1$/5b-$G_1$** and **6a-$G_1$/6b-$G_1$**). A similar comment can be done also on the results from MD, both in terms of 3D geometrical arrangement of the molecules and of quantification of the structural differences (Figs 6 and 7). Therefore, we discarded the three molecules **5a-$G_1$**, **6a-$G_1$** and **7a-$G_1$** from the initial set of 13. Figure 8a displays the biological activity of the 10 remaining dendrimers, as a function of the number of terminal groups. As taken alone the cell size parameter is poorly discriminating between the dendrimers, we have quantified and used the relative granularity of monocytes (at the highest concentration of dendrimers, that is, 20 μM; Fig. 8b): the more granulous the monocytes, the more active the dendrimer.

For better understanding the key parameters of the dendrimers structure for the biological activity, several other information have been extracted from the MD data. The $G_{sol}$ energies extracted from the equilibrated phase (the last 100 ns) MD simulations of the dendrimers in solution are representative of the level of hydrophilicity/hydrophobicity of their solvated state. Figure 8c shows that also $G_{sol}$ data are in good trend with the molecular activity of the dendrimers (Fig. 8a), which suggests that both hydrophilicity/hydrophobicity (Fig. 8c), and molecular directionality (Fig. 8d) are key discriminant parameters for

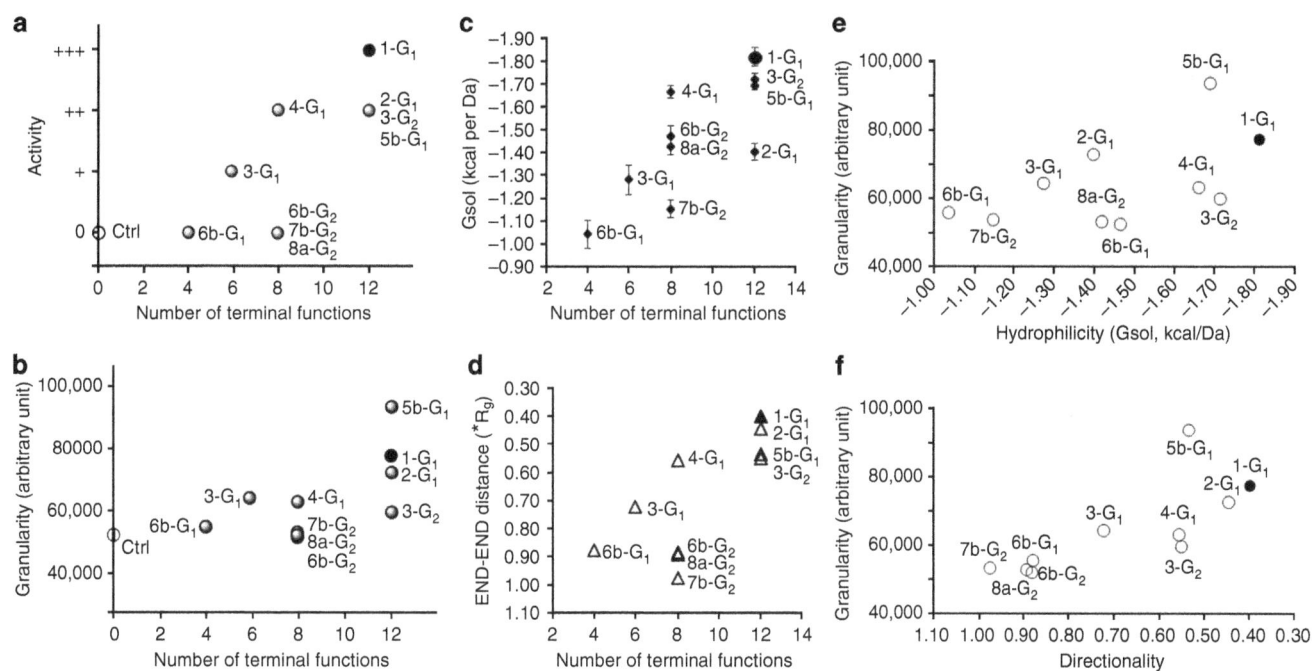

**Figure 8 | Comparison of biological activity and data from MD simulations for 10 dendrimers.** (**a**) The number of terminal groups of dendrimers versus the biological efficiency. (**b**) Quantification of relative granularity (arbitrary units, means from three donors) of the monocytes treated with the dendrimers and of negative control monocytes. (**c**) $G_{sol}$ values from MD (relative hydrophilicity). (**d**) Calculated distance between terminal groups (in $R_g$ units), as an indication of molecular directionality. (**e,f**) Comparison between biological data and MD data: granularity versus hydrophilicity (**e**) and granularity versus directionality (**f**). The black dots correspond in all cases to the lead compound **1-G₁**.

biological activity (Fig. 8a). In fact, the most active dendrimers, with particular emphasis on dendrimer **1-G₁**, possess the most favourable solvation energy ($G_{sol}$) and the highest directionality scores (the smaller the END–END distance, the higher molecular directionality), which also points towards a direct correlation between dendrimers multivalency and activity. In fact, at the molecular level, a higher density of active functions can impact biological efficiency because it intuitively amplifies cooperativity and favours multivalent interactions. As last, we put in direct correlation biological activity (granularity) with hydrophilicity ($G_{sol}$) and molecular directionality. The results are shown in Fig. 8e,f, respectively, showing remarkable trends. Indeed, looking at these dendrimers in solution, these analyses suggest that the dense concentration of active functions in regions of the dendrimers surface and the level of hydrophilicity/hydrophobicity of the structure are crucial to molecular activity. This is the case also of biological molecules like proteins, where the features of the surface, presence of charged or active patches, and overall and local hydrophobicity and hydrophilicity levels produce remarkable effects through a delicate network of multiscale interactions and molecular recognition.

In conclusion, this work carried out on seven different families of dendrimers clearly demonstrates that the internal structure of dendrimers cannot be anymore regarded as an 'innocent' support for active functions, but plays a crucial role, especially when considering biological properties. Indeed, the geometry of the dendrimers with identical terminal groups may be very different intrinsically, and the differences can be amplified in a real environment such as a water solution or biological materials (like blood, biological barriers and cell membranes), depending on the nature of the scaffold. Structural differences induce changes in shape and in the distribution of the terminal active functions of the dendrimers, which are responsible for the biological activity. Regarding monocyte activation, the most active dendrimers are those in which the surface functions are gathered on one side

(directional molecules), a suitable orientation to maximize multivalent interactions with cells. Definitely, this work shows how changing one single parameter, even in the internal moieties, may totally modify the properties of dendrimers. Therefore, the implementation of extensive MD simulation studies of dendrimers appears as a pivotal asset when designing new bio-oriented dendritic devices or to optimize existing dendritic systems. Such finding exceeds the field of dendrimers and embraces those more general of macromolecules and nanostructures. The direct relationships traced here between hydrophobicity/hydrophilicity and activity, and between activity and surface characteristics (that is, directionality in the localization of functional groups) recall the main features of biological macromolecules such as proteins. Our results are reminiscent to what is known about the interplay between natural networks and their surroundings.

## Methods

**General information about the synthesis of dendrimers.** All manipulations are performed under argon using standard Schlenk techniques. Commercial samples were used as received from Aldrich (PAMAM dendrimers, DAB dendrimers, Lysine dendrimer and all other chemicals). The following compounds have been prepared according to published procedures: tyramine azabisphosphonate **9** (ref. 21) and dendrimer **1-G₁** (ref. 21), Salamonczyk's thiophosphite dendrimers[33], chlorine-ended carbosilane dendrimers[37] and phosphorhydrazone dendrimers ended by 12 chlorine atoms[38]. All solvents were dried and distilled according to routine procedure before use. Thin-layer chromatography was carried out on Merck Kieselgel 60F254 precoated silicagel plates. Preparative chromatography was performed on Merck Kieselgel. $^1$H, $^{13}$C, $^{31}$P NMR, HMQC and HMBC measurements were performed on Bruker AC200, AM250 and AV300 and AMX400. All dendrimers have been characterized by $^1$H, $^{13}$C{$^1$H} and $^{31}$P{$^1$H} NMR, and two-dimensional NMR spectra when necessary. All details of the synthesis and the NMR data for all compounds are given in the Supplementary Information. The experimental information for the synthesis of all dendrimers is also given in the Supplementary Methods for chemistry. The synthesis of dendrimer **5a-G₁** is given below as a typical example.

**Typical example for the synthesis of dendrimers.** Synthesis of **5a-G₁**, *first step*: 0.017 mmol of generation 1 phosphorus-containing dendrimer (12 Cl terminations)[38] are placed in solution in 3 ml of dry tetrahydrofuran. $Cs_2CO_3$

(5.04 mmol), then 0.23 mmol of the tyramine azabisphosphonate **11a** ($x = 1$) in solution in 3 ml of dry tetrahydrofuran are successively added to this solution. The mixture is stirred overnight at room temperature and then filtered on celite. The reaction medium is evaporated under reduced pressure then the dry residue is dissolved in a minimum volume of dichloromethane. The product is then precipitated in a large volume of ether. This operation is repeated three times to eliminate the slight excess of starting tyramine derivative. The dendrimer with methylphosphonate terminations is obtained as an off-white powder (88% yield). *Second step*: 0.015 mmol of dendrimer with methylphosphonate terminations are placed in solution under an inert atmosphere in 3 ml of distilled acetonitrile. The solution is taken to 0 °C then 48 equivalents of BrTMS (0.73 mmol) are added dropwise under argon. The mixture is stirred for 30 min at 0 °C, then overnight at room temperature and finally evaporated to dryness under reduced pressure. The residue thus obtained is treated with methanol (2 × 15 ml), then evaporated to dryness and washed with dry ether (20 ml) to afford the dendrimer with phosphonic acid terminations (63% yield). *Third step*: the sodium salt is obtained by reaction of 24 equivalents of NaOH solution at 0.1955 N on one equivalent of dendrimer with phosphonic acid terminations, to produce, after stirring 1 h at room temperature and freeze drying, the expected dendrimer **5a-G₁** (70% yield).

**Purification and activation of human monocytes.** Preparation of PBMC (peripheral blood mononuclear cells) from healthy volunteers and subsequent purification of monocytes are performed as already described[24]. Purity of the monocytes (> 95%) is assessed by flow cytometry on a LSR-II device (BD Biosciences, San Diego, CA, USA).

For activation cultures, purified monocytes are cultured in multi-well plates for 72 h at $10^6$ cells ml$^{-1}$ of complete RPMI 1640 medium (Roswell Park Memorial Institute medium), that is, supplemented with 10% of heat-inactivated fetal calf serum, 1 mM sodium pyruvate, 2 mM L-glutamine (all from Invitrogen Corporation, Paisley, UK), penicillin and streptomycin, both at 100 U ml$^{-1}$ (Cambrex Bio Science). Dendrimers are added at the beginning of the cultures at 20, 2 and 0.2 µM.

**Flow cytometry analyses.** At the end of the culture, monocytes are washed with PBS with 5% fetal calf serum and their morphology (size and granularity) is analysed by flow cytometry (LSR-II device from BD Biosciences) and quantified using FACS-Diva software (BD Biosciences). The most activated monocytes are the most granular, and biggest they appear. A score (from 0 to + + +) is attributed to each dendrimer on these morphological criteria. A score of 0 is given to dendrimers that do not induce morphological changes (monocytes with the same morphology than control, non-activated, monocytes). A score of + + + is given to dendrimer **1-G₁**, the most active one which induces dramatic morphological changes at 20 and 2 µM, still detectable at 0.2 µM. A score of + + is given to dendrimers whose effect is comparable to the one of dendrimer **1-G₁** at 20 µM but with no or weak detectable effect at 2 µM. A score of + is given to dendrimers which has a detectable effect, although weak, at 20 µM.

**All-atom MD simulations.** The MD simulation work is conducted by using the AMBER 12 software[39]. The molecular models for all dendrimers are created and parametrized according to a validated procedure for the simulation of dendrimers in aqueous solution (see Supplementary Table 1 for details)[36,40,41]. The force field parameters for the **4-G₁** carbosilane dendrimer are obtained as previously reported[42]. All the dendrimer models are immerged in a periodic box containing explicit TIP3P water molecules[43] and 150 mM of NaCl to reproduce the experimental conditions. All systems are simulated for 200 ns in NPT conditions at 37 °C. Analyses of the structural features and of the solvation energies ($G_{sol}$) of the dendrimers are performed on the equilibrated phase MD trajectories (Supplementary Fig. 1). Details on the computational procedures are given in the Supplementary Methods and Supplementary Movies 1 and 2 obtained from the MD simulations showing the 3D structures of **1-G₁** (MD 3-D structure of **1-G₁**) and **7b-G₂** (MD 3D structure of **7b-G₂**) as representative cases of directional and symmetrical molecules.

## References

1. Caminade, A. M., Turrin, C. O., Laurent, R., Ouali, A. & Delavaux-Nicot, B. *Dendrimers: Towards Catalytic, Material and Biomedical Uses* (John Wiley & Sons, 2011).
2. Page, D., Zanini, D. & Roy, R. Macromolecular recognition: Effect of multivalency in the inhibition of binding of yeast mannan to concanavalin A and pea lectins by mannosylated dendrimers. *Bioorg. Med. Chem.* **4**, 1949–1961 (1996).
3. Pavan, G. M., Danani, A., Pricl, S. & Smith, D. K. Modeling the multivalent recognition between dendritic molecules and DNA: understanding how ligand 'sacrifice' and screening can enhance binding. *J. Am. Chem. Soc.* **131**, 9686–9694 (2009).
4. Pavan, G. M. Modeling the interaction between dendrimers and nucleic acids—a molecular perspective through hierarchical scales. *ChemMedChem* **9**, 2623–2631 (2014).
5. Darbre, T. & Reymond, J. L. Peptide dendrimers as artificial enzymes, receptors, and drug-delivery agents. *Acc. Chem. Res.* **39**, 925–934 (2006).
6. Lee, C. C., MacKay, J. A., Fréchet, J. M. J. & Szoka, F. C. Designing dendrimers for biological applications. *Nat. Biotechnol.* **23**, 1517–1526 (2005).
7. Mammen, M., Choi, S. K. & Whitesides, G. M. Polyvalent interactions in biological systems: implications for design and use of multivalent ligands and inhibitors. *Angew. Chem. Int. Ed.* **37**, 2754–2794 (1998).
8. Tomalia, D. A. Dendrons/dendrimers: quantisized, nano-element like building blocks for soft-soft and soft-hard nano-compound synthesis. *Soft Matter* **6**, 456–472 (2010).
9. Tomalia, D. A. et al. A new class of polymers—Starburst dendritic macromolecules. *Polym. J.* **17**, 117–132 (1985).
10. De Brabander-van den Berg, E. M. M. & Meijer, E. W. Poly(Propylene Imine) dendrimers—large-scale synthesis by heterogeneously catalyzed hydrogenations. *Angew. Chem. Int. Ed.* **32**, 1308–1311 (1993).
11. Mynar, J. L., Lowery, T. J., Wemmer, D. E., Pines, A. & Fréchet, J. M. J. Xenon biosensor amplification via dendrimer-cage supramolecular constructs. *J. Am. Chem. Soc.* **128**, 6334–6335 (2006).
12. Juttukonda, V. et al. Facile synthesis of tin oxide nanoparticles stabilized by dendritic polymers. *J. Am. Chem. Soc.* **128**, 420–421 (2006).
13. Tomalia, D. A., Hall, M. & Hedstrand, D. M. Starburst dendrimers.3. The importance of branch junction symmetry in the development of topological shell molecules. *J. Am. Chem. Soc.* **109**, 1601–1603 (1987).
14. Liao, X. L., Stellacci, F. & McGrath, D. V. Photoswitchable flexible and shape-persistent dendrimers: comparison of the interplay between a photochromic azobenzene core and dendrimer structure. *J. Am. Chem. Soc.* **126**, 2181–2185 (2004).
15. Merkel, O. M. et al. Triazine dendrimers as nonviral gene delivery systems: effects of molecular structure on biological activity. *Bioconjugate Chem.* **20**, 1799–1806 (2009).
16. Jain, K., Kesharwani, P., Gupta, U. & Jain, N. K. Dendrimer toxicity: let's meet the challenge. *Int. J. Pharm.* **394**, 122–142 (2010).
17. Ginhoux, F. & Jung, S. Monocytes and macrophages: developmental pathways and tissue homeostasis. *Nat. Rev. Immunol.* **14**, 392–404 (2014).
18. Zolnik, B. S., Gonzalez-Fernandez, A., Sadrieh, N. & Dobrovolskaia, M. A. Minireview: nanoparticles and immune system. *Endocrinology* **151**, 458–465 (2010).
19. Murray, P. J. & Wynn, T. A. Protective and pathogenic functions of macrophage subsets. *Nat. Rev. Immunol.* **11**, 723–737 (2011).
20. Sica, A. & Mantovani, A. Macrophage plasticity and polarization: *in vivo* veritas. *J. Clin. Invest.* **122**, 787–795 (2012).
21. Griffe, L. et al. Multiplication of human natural killer cells by nanosized phosphonate-capped dendrimers. *Angew. Chem. Int. Ed.* **46**, 2523–2526 (2007).
22. Portevin, D. et al. Regulatory activity of azabisphosphonate-capped dendrimers on human CD4$^+$ T cell proliferation for ex-vivo expansion of NK cells from PBMCs and immunotherapy. *J. Transl. Med.* **7**, 82 (2009).
23. Poupot, M. et al. Design of phosphorylated dendritic architectures to promote human monocyte activation. *FASEB J.* **20**, 2339–2351 (2006).
24. Fruchon, S. et al. Anti-inflammatory and immuno-suppressive activation of human monocytes by a bio-active dendrimer. *J. Leukocyte Biol.* **85**, 553–562 (2009).
25. Hayder, M. et al. Phosphorus-based dendrimer as nanotherapeutics targeting both inflammation and osteoclastogenesis in experimental arthritis. *Sci. Transl. Med.* **3**, 81ra35 (2011).
26. Leah, E. Experimental arthritis: dendrimer drug mends monocytes. *Nat. Rev. Rheumatol.* **7**, 376 (2011).
27. Lou, K.-J. Dendrimer throws a blanket on RA. *SciBX* **4**, doi:10.1038/scibx.2011.561 (2011).
28. Rolland, O. et al. Efficient synthesis of phosphorus-containing dendrimers capped with isosteric functions of amino-bis(methylene) phosphonic acids. *Tetrahedron Lett.* **50**, 2078–2082 (2009).
29. Rolland, O. et al. Tailored control and optimisation of the number of phosphonic acid termini on phosphorus-containing dendrimers for the *ex-vivo* activation of human monocytes. *Chem. Eur. J.* **14**, 4836–4850 (2008).
30. Wang, Y., Guo, R., Cao, X., Shen, M. & Shi, X. Encapsulation of 2-methoxyestradiol within multifunctional poly(amidoamine) dendrimers for targeted cancer therapy. *Biomaterials* **32**, 3322–3329 (2011).
31. Garber, S. B., Kingsbury, J. S., Gray, B. L. & Hoveyda, A. H. Efficient and recyclable monomeric and dendritic Ru-based metathesis catalysts. *J. Am. Chem. Soc.* **122**, 8168–8179 (2000).
32. Denkewalter, R. G., Kolc, J. & Lukasavage, W. J. Macromolecular highly branched homogeneous compound based on lysine units. US patent 4,289,872 (1981).
33. Salamonczyk, G. M., Kuznikowski, M. & Skowronska, A. A divergent synthesis of thiophosphate-based dendrimers. *Tetrahedron Lett.* **41**, 1643–1645 (2000).
34. Launay, N., Caminade, A. M., Lahana, R. & Majoral, J. P. A general synthetic strategy for neutral phosphorus-containing dendrimers. *Angew. Chem. Int. Ed.* **33**, 1589–1592 (1994).
35. Shafer, L. L., McNulty, J. A. & Young, M. R. Brain activation of monocyte lineage cells: brain-derived soluble factors differentially regulate BV2 microglia and peripheral macrophage immune functions. *Neuroimmunomodulation* **10**, 283–294 (2002).

36. Pavan, G. M., Barducci, A., Albertazzi, L. & Parrinello, M. Combining metadynamics simulation and experiments to characterize dendrimers in solution. *Soft Matter* **9,** 2593–2597 (2013).

37. de Groot, D., Reek, J. N. H., Kamer, P. C. J. & van Leeuwen, P. W. N. M. Palladium complexes of phosphane-functionalised carbosilane dendrimers as catalysts in a continuous-flow membrane reactor. *Eur. J. Org. Chem.* **2002,** 1085–1095 (2002).

38. Launay, N., Caminade, A. M. & Majoral, J. P. Synthesis of bowl-shaped dendrimers from generation 1 to generation 8. *J. Organomet. Chem.* **529,** 51–58 (1997).

39. Case, D. A. *et al. AMBER 12* (University of California, 2012).

40. Garzoni, M., Okuro, K., Ishii, N., Aida, T. & Pavan, G. M. Structure and shape effects of molecular glue on supramolecular tubulin assemblies. *ACS Nano* **8,** 904–914 (2014).

41. Simanek, E. E., Enciso, A. E. & Pavan, G. M. Computational design principles for the discovery of bioactive dendrimers: [s]-triazines and other examples. *Exp. Opin. Drug Disc.* **8,** 1057–1069 (2013).

42. Fuentes-Paniagua, E. *et al.* Carbosilane cationic dendrimers synthesized by thiol–ene click chemistry and their use as antibacterial agents. *RSC Adv.* **4,** 1256–1265 (2014).

43. Jorgensen, W. L., Chandrasekhar, J., Madura, J. D., Impey, R. W. & Klein, M. L. Comparison of simple potential functions for simulating liquid water. *J. Chem. Phys.* **79,** 926–935 (1983).

## Acknowledgements

This work was financially supported by Rhodia (grant to S.F. and A.O.), CNRS (Centre National de la Recherche Scientifique), INSERM (Institut National de la Santé et de la Recherche Médicale), UPS (Université Paul Sabatier) and FRM (Fondation pour la Recherche Médicale, grant DCM20111223039).

## Author contributions

A.-M.C. conceived the idea of the comparison between different types of dendrimers, supervised all the chemical work and wrote most of this paper. C.-O.T. was the day-to-day supervisor of the chemical work. S.F. performed most of the biological experiments under the direct guidance of M.P. A.O. synthesized dendrimers 2-G$_1$, 3-Gn ($n = 1, 2$) and 4-G$_1$. A.M. synthesized dendrimers 5a,b-G$_1$, 6a,b-G$_n$ ($n = 1, 2$), 7a-G$_1$, 7b-G$_2$ and 8a-G$_2$ from commercial sources. G.M.P. is responsible for the modelling part, which has been performed with the help of M.G., M.M., V.F. and V.K., and wrote the simulation part of this paper. J.-P.M. was helpful in discussing the chemical part and obtained the grant for A.O. R.P. supervised all the biological work and wrote the biological part of this paper.

# 6

# Targeting bacteria via iminoboronate chemistry of amine-presenting lipids

Anupam Bandyopadhyay[1], Kelly A. McCarthy[1], Michael A. Kelly[1] & Jianmin Gao[1]

Synthetic molecules that target specific lipids serve as powerful tools for understanding membrane biology and may also enable new applications in biotechnology and medicine. For example, selective recognition of bacterial lipids may give rise to novel antibiotics, as well as diagnostic methods for bacterial infection. Currently known lipid-binding molecules primarily rely on noncovalent interactions to achieve lipid selectivity. Here we show that targeted recognition of lipids can be realized by selectively modifying the lipid of interest via covalent bond formation. Specifically, we report an unnatural amino acid that preferentially labels amine-presenting lipids via iminoboronate formation under physiological conditions. By targeting phosphatidylethanolamine and lysylphosphatidylglycerol, the two lipids enriched on bacterial cell surfaces, the iminoboronate chemistry allows potent labelling of Gram-positive bacteria even in the presence of 10% serum, while bypassing mammalian cells and Gram-negative bacteria. The covalent strategy for lipid recognition should be extendable to other important membrane lipids.

[1] Department of Chemistry, Merkert Chemistry Center, Boston College, 2609 Beacon Street, Chestnut Hill, Massachuetts 02467, USA. Correspondence and requests for materials should be addressed to J.G. (email: jianmin.gao@bc.edu).

I t is increasingly clear that membrane lipids do not merely provide a physical barrier for a cell; instead they play active roles in regulating numerous processes in cell physiology and disease[1]. To support the diverse functions of a membrane, the composite lipids, while maintaining the common feature of amphiphilicity, do vary in their chemical structures to give a complex lipidome (Fig. 1a)[2,3]. The lipid composition of a membrane has significant ramifications in biology. For example, it is well known that the plasma membranes of bacterial and mammalian cells display distinct compositions of lipids: while a mammalian cell membrane primarily consists of phosphatidylcholine (PC) and sphingomyelin (SM), bacterial cells display highly enriched phosphatidyletahnolamine (PE) and phosphatidylglycerol (PG)[4,5]. In addition, some bacterial species present a lysine-modified PG (Lys-PG, Fig. 1a) in high percentages as a resistance mechanism to cationic antibiotics[6].

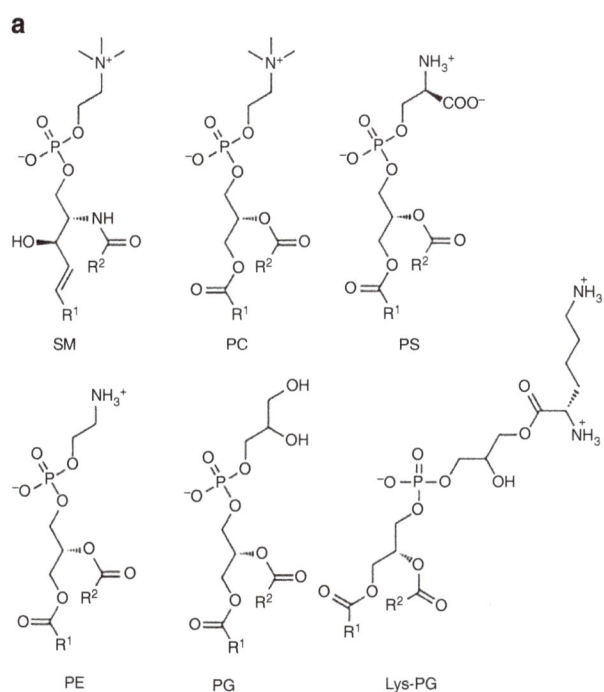

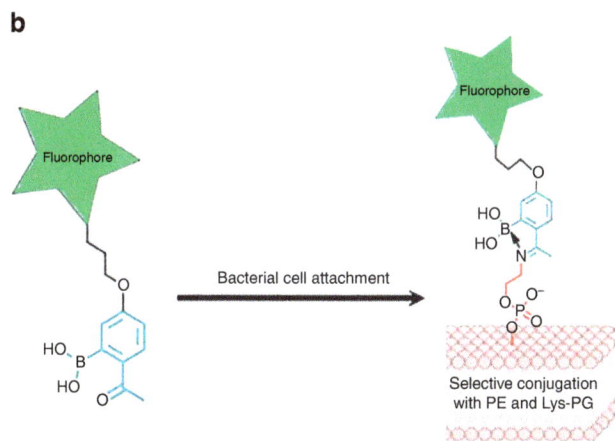

Synthetic molecules that specifically target bacterial lipids may give rise to new imaging methods of bacterial infection, as well as novel solutions to the antibiotic-resistance problem. The critical importance of lipids also manifests in the subcellular distribution of certain lipids in mammalian cells, a change of which may alter the homeostasis of important signalling proteins[7,8]. To further elucidate the diverse roles of membrane lipids, it is highly desirable to have molecular probes that specifically target a lipid of interest as well. Currently known lipid-targeting agents, which are primarily lipid-binding proteins and their synthetic mimetics, achieve lipid recognition by employing networks of *noncovalent* interactions, such as hydrogen bonds and salt bridges[9,10]. It remains to be seen whether membrane lipids can be selectively recognized by *covalently* targeting their unique chemical structure and reactivity with synthetic molecules.

In this contribution, we report the design and synthesis of an unnatural amino acid that selectively conjugates with amine-presenting lipids via formation of iminoboronates. By targeting the membrane lipids enriched in bacterial cells, namely PE and Lys-PG, the iminoboronate chemistry allows highly selective labelling of bacteria over mammalian cells.

## Results

**Design and synthesis of AB1.** The two major bacterial lipids, PE and Lys-PG, differ from their mammalian counterparts (PC and SM) by the presence of primary amino groups. We postulated that these nucleophilic amines could be captured by a 2-acetylphenylboronic acid (2-APBA) motif to form an iminoboronate (Fig. 1b). Although theoretically possible, amines in biology milieu only forms a Schiff base with simple ketones at high concentrations[11]. For example, the association constant of acetone and glycine was reported to be $3.3 \times 10^{-3}\,M^{-1}$. Usually, the imine formation is trapped with a reduction step for biological applications[12]. With the *ortho* boronic acid group serving as an electron trap, the 2-APBA motif conjugates with an amine much more readily to give an iminoboronate[13-17]. Importantly, the reaction proceeds under physiological conditions and in a reversible manner. Furthermore, an iminoboronate conjugate can exchange with other amines to allow for thermodynamic control of the final iminoboronate formation (Supplementary Fig. 1)[15]. These features make the iminoboronate chemistry particularly suitable for facilitating molecular recognition in biological systems.

To test our hypothesis, we have designed and synthesized a novel unnatural amino acid (AB1, Fig. 2) that presents a 2-APBA motif as its side chain. We envisioned that the amino-acid scaffold should allow the 2-APBA motif to be readily conjugated to fluorescent labels or other functional peptides. The synthetic route of AB1 is summarized in Fig. 2. Briefly, with 2′,4′-dihydroxy acetophenone 1 as the starting material, regioselective alkylation of the 4′-OH followed by triflate protection of the 2′-OH yielded 3 with an overall 81% yield. By taking advantage of the powerful thiol-ene chemistry[18], compound 3 was conjugated to two cysteine derivatives, respectively, to give the protected amino acids 4 and 7 in high yields. The key transformation of our synthesis is the Miyaura borylation[19], which converts the triflate to the Bpin moiety. In our hands, rigorous control of temperature was critical to the success of the borylation step: the reaction did not initiate below 95 °C and prolonged heating at higher temperatures caused the complete loss of the Bpin moiety to give the protodeboronated product, a protected AB2 (ref. 20). With optimized conditions, the Bpin moiety was introduced with 70–80% yield. Fortuitously, with the boronic acid moiety eliminated, AB2 served as a perfect negative control for AB1 in the following membrane-binding studies.

**Figure 1 | Covalent recognition of membrane lipids. (a)** Structures of the major membrane lipids from mamallian (sphingomyelin (SM) and phosphatidylcholine (PC)) and bacterial (phosphatidylethanolamine (PE), phosphatidylglycerol (PG), lysylphosphatidylglycerol (Lys-PG)) cells. PE and phosphatidylserine (PS) exist in mammalian cells as minority lipids. **(b)** Illustration of the iminoboronate chemistry for targeting PE on bacterial cell surfaces.

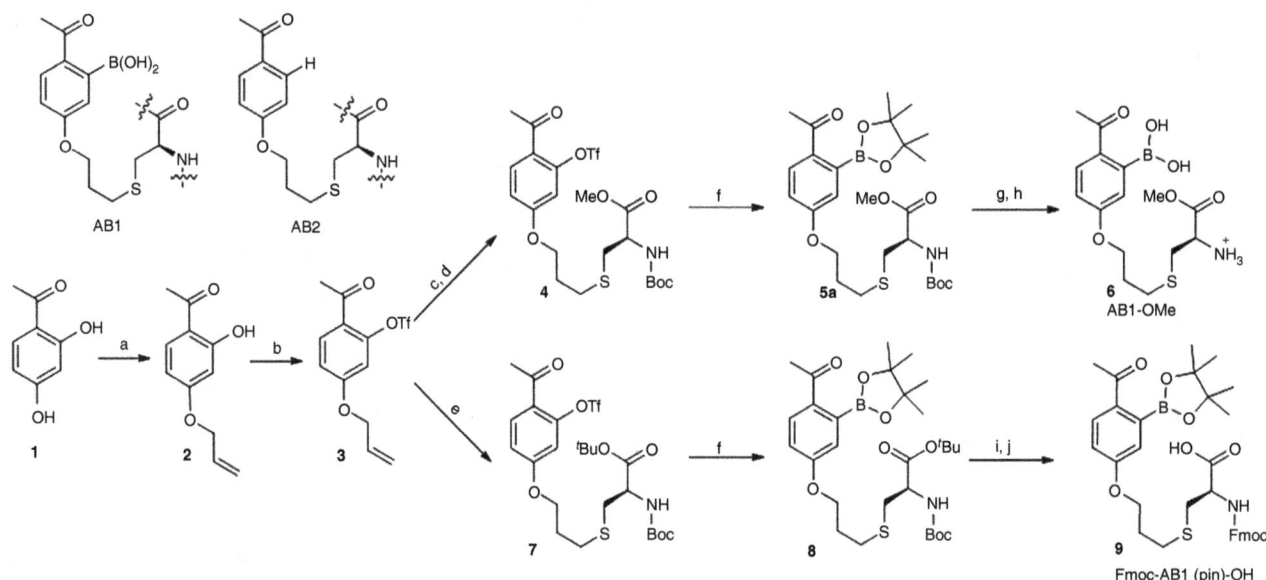

**Figure 2 | Synthesis of AB1 and its derivatives.** (a) Allyl bromide, $K_2CO_3$, NaI, acetone, 81%. (b) $(CF_3SO_2)_2O$, $Et_3N$, DCM, 95%. (c) Cys-OMe, DMPA, MeOH, ~365 nm ultraviolet irradiation. (d) Boc anhydride, $Na_2CO_3$, THF/$H_2O$, 80% over two steps. (e) Boc-Cys-OtBu, DMPA, MeOH, ~365 nm ultraviolet irradiation, 75%. (f) Pd(dppf)$Cl_2$/dppf, $B_2Pin_2$, KOAc, dioxane, ~70–80%. (g) 40% TFA in DCM. (h) diethanolamine, 1 N HCl, 74% over two steps. (i) 60% TFA in DCM. (j) Fmoc-OSu, $Na_2CO_3$, THF/$H_2O$, 81% over two steps. DCM, dichloromethane; DMPA, 2,2-dimethoxy-2-phenylacetophenone; THF, tetrahydrofuran; TFA, trifluoroacetic acid.

**AB1 selectively conjugates with PE and Lys-PG.** The use of cysteine methyl ester (Cys-OMe) in the thiol-ene coupling step yielded the AB1 methyl ester (AB1-OMe, Fig. 2), which can be readily labelled with amine-reactive fluorophores. To assess the binding propensity towards different lipids, a fluorescein isothiocyanate-labelled AB1 methyl ester (Fl-AB1-OMe) was tested against lipid vesicles of varied composition. Specifically, 100nm-sized vesicles were prepared with PC alone or with 40% guest lipids including PE, PS, PG and Lys-PG. The fluorescence anisotropy values of Fl-AB1-OMe were recorded with increasing concentrations of lipids and the data are summarized in Fig. 3a. Interestingly, significant anisotropy increases were observed only with vesicles that present PE and Lys-PG, with other vesicle compositions eliciting marginal changes of anisotropy. Specifically, the presence of PG or PS did not induce more AB1 binding than PC-alone, showcasing the unique reactivity of PE and Lys-PG towards AB1. The lack of PS labelling by AB1 is perhaps surprising given that PS does display an amino group. This is presumably because the amino group of PS, in comparison to that of PE, is sterically more challenging for iminoboronate formation. This observation is consistent with a recent report, in which 2-APBA was found to preferentially react with lysine side chains over the main chain amino group[17]. Importantly, in contrast to Fl-AB1-OMe, Fl-AB2-OMe did not show significant association with the PC/PE or PC/Lys-PG vesicles (Fig. 3b), highlighting the importance of the boronic acid moiety in AB1 binding into vesicles.

To further validate the binding mechanism, we directly characterized the postulated iminoboronate conjugate of Lys-PG. Briefly, the PC/Lys-PG vesicles were treated with 2-APBA, the 'warhead' structure of AB1. Then the mixture was lyophilized, redissolved in $CDCl_3$/$CD_3OD$ (2:1) and subjected to $^{11}$B-NMR and mass spectrometry analysis. The $^{11}$B-NMR spectrum of the treated lipids displays a peak around 13 p.p.m. as expected for iminoboronate structures (Fig. 3c). The mass-spec data clearly present the molecular ions that correspond to the 2-APBA adduct of Lys-PG (Fig. 3d). Further, mass-spec analysis also reveals the iminoboronate conjugate of Lys-PG and an AB1-presenting peptide (Supplementary Fig. 2). These data consistently support

the iminoboronate mechanism for the association of AB1 with lipid membranes.

The iminoboronate mechanism predicts that the iminoboronate formation between AB1 and lipids can be inhibited by the presence lysine and lysine-presenting proteins. Indeed, lysine and bovine serum albumin (BSA) were found to disrupt the association of AB1 and PC/PE vesicle with an $IC_{50}$ of ~0.3 mM and 5 μM, respectively (Supplementary Fig. 3). It is interesting to note that BSA at 5 μM gives ~0.3 mM in lysine concentration given that BSA has a total of 59 lysine residues. In a later section, we will present strategies that minimize the protein interference of AB1-labelling lipids.

**AB1 selectively labels Gram-positive bacteria.** Encouraged by the model membrane studies, we sought to investigate the potential of AB1 in staining bacterial cells. Three strains of bacteria, including *B. subtilis* (American Type Culture Collection (ATCC) 663), *S. aureus* (ATCC 6538) and *E. coli* (BL 21), were selected as the initial set, which are known to have PE and/or Lys-PG as the major lipids of their plasma membranes[4–6]. The bacterial cells were stained with an Alexa Fluor 488 (AF488)-labelled AB1-OMe, which was chosen for cell studies because of the superior brightness and stability of the fluorophore. At concentrations below 1 μM, little fluorescence staining of the cells was observed with AF488-AB1-OMe. With higher concentrations, a quick washing procedure was included to minimize background fluorescence, after which the samples were immediately examined under an epi-fluorescence microscope (Fig. 4a). With wash, AF488-AB1-OMe effectively stained the two Gram-positive bacteria (*B. subtilis* and *S. aureus*) at ≥100 μM concentrations. In sharp contrast, the Gram-negative *E. coli* showed no fluorescence staining at all. As a negative control, AF488-AB2-OMe failed to stain any of the bacterial strains under the same conditions (Supplementary Fig. 4), showcasing the critical importance of the boronic acid moiety for bacteria labelling by AB1. The labelling can be inhibited by the addition of lysine (Supplementary Fig. 5) or BSA (Supplementary Fig. 6), lending further support to the iminoboronate mechanism of

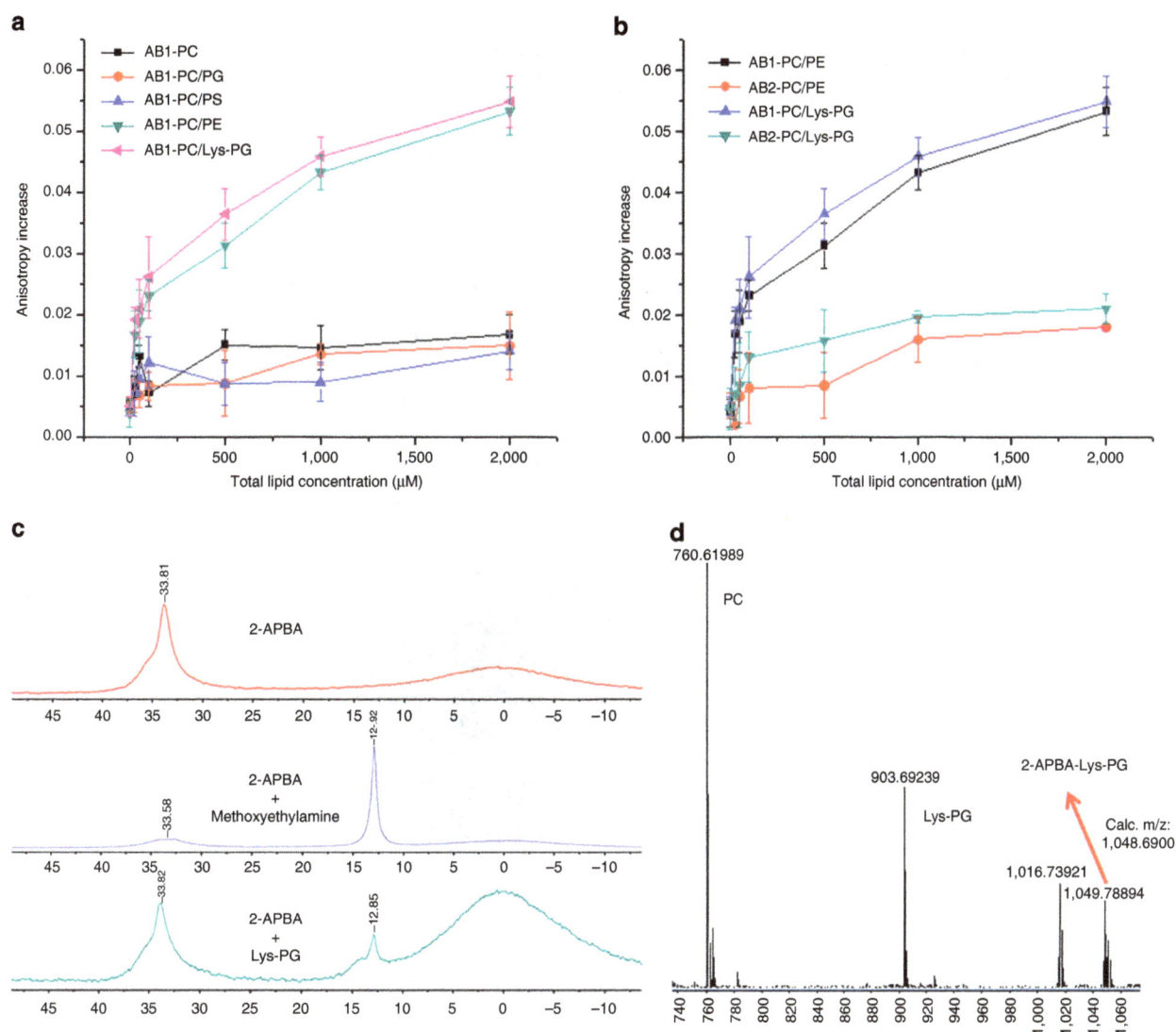

**Figure 3 | Iminoboronate formation on synthetic vesicles.** (**a**) Binding curves of Fl-AB1-OMe to lipid vesicles highlighting its selectivity for PE and Lys-PG. (**b**) Comparison of Fl-AB1-OMe and Fl-AB2-OMe for lipid binding showcasing the critical importance of the boronic acid moiety in vesicle association. All data points were measured with triplicate samples, from which error bars (s.e.m.) were generated. (**c**) [11]B-NMR spectra of 2-APBA (2-acetylphenylboronic acid) and its conjugates with methoxyethylamine and Lys-PG. The peaks around 13 p.p.m. correspond to the iminoboronates and the broad peaks around 0 p.p.m. originate from the NMR tube. (**d**) Mass-spec analysis of the iminoboronate conjugation of 2-APBA to PC/Lys-PG vesicles. The specific lipids used are POPC (1-palmitoyl-2-oleoylphosphatidylcholine) and Lys-DOPG (lysyl 1,2-dioleoylphosphatidylglycerol). 2-APBA-Lys-PG denotes the iminoboronate conjugate of 2-APBA and Lys-DOPG.

conjugation. To gain more mechanistic insights, we analyzed the lipid extract of the *S. aureus* cells treated with 2-APBA: the [11]B-NMR spectrum clearly revealed the characteristic peak ($\sim 13$ p.p.m.) for iminoboronates (Supplementary Fig. 7), although our trials with mass-spec failed to identify the expected conjugates directly.

The failure of AB1 to stain *E. coli* is consistent with the fact that the outer membrane of *E. coli* does not have PE or Lys-PG. It further indicates that AF488-AB1-OMe is unable to permeate through the outer membrane to reach the plasma membrane, where PE does exist. Consistent with the membrane impermeability of AF488-AB1-OMe, we found that the fluorescence staining of *S. aureus* could be rapidly and completely washed away with a pH 5.0 buffer (Supplementary Fig. 8). The failure of *E. coli* staining suggests that AB1 does not label cell surface proteins under the experimental conditions, the exact mechanism of which remains to be further investigated. One possible explanation is that certain features of the membrane, such as

local membrane curvature[21], create kinetic traps for AB1. Supporting this hypothesis, we found that, after the initial washing step, the cell-bound AB1 molecules dissociate from the cell very slowly at neutral pH (Supplementary Fig. 8). Another possibility is that the number of surface proteins might be significantly smaller than that of lipids; consequently labelled surface proteins afford negligible fluorescence in comparison to labelled lipids.

We further assessed the selectivity of AB1 for bacteria over mammalian cells. Excitingly, when a co-culture of Jurkat lymphocytes and *S. aureus* cells was treated with AF488-AB1-OMe and analysed under a confocal microscope, strong fluorescence staining was observed for *S. aureus* cells, whereas the Jurkat cells were minimally labelled (Fig. 4b). Similar to the result of *E. coli* staining, the lack of Jurkat cell staining also indicates that AF488-AB1-OMe does not conjugate with cell surface proteins under our experimental conditions. Further, as we learned from the bacterial staining experiments, AF488-AB1-

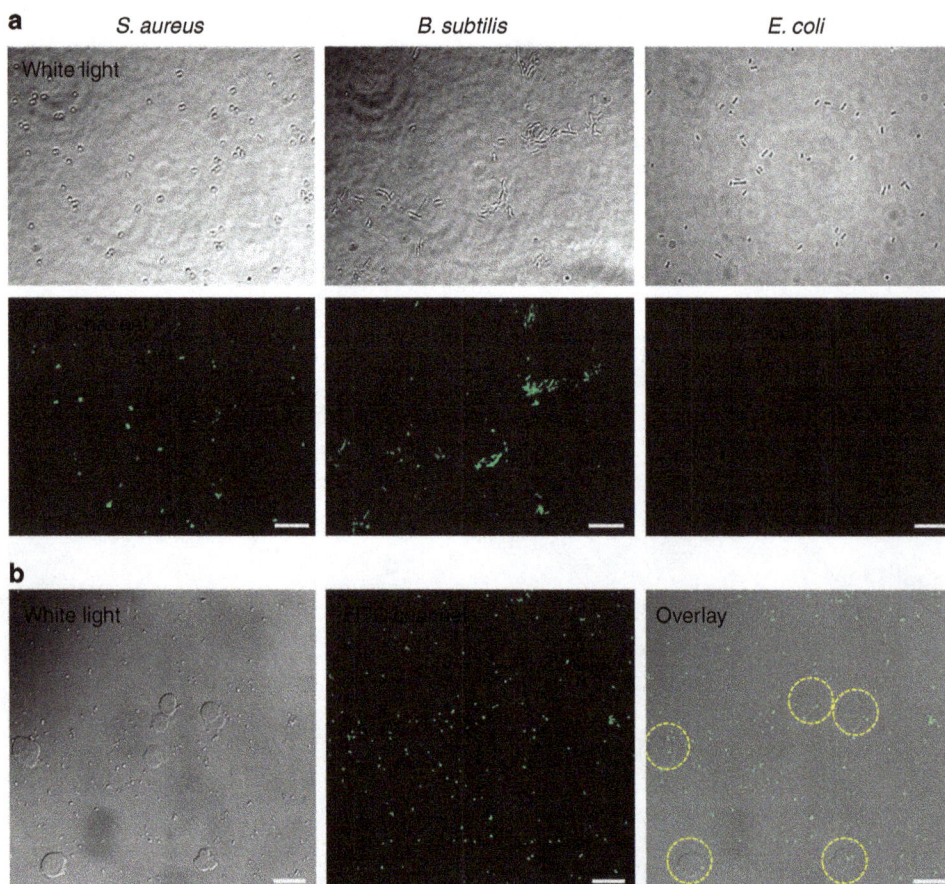

**Figure 4 | Assessing the selectivity of AB1 for various cell types. (a)** Microscopic images of three bacterial strains stained with 200 µM AF488-AB1-OMe (scale bar, 10 µm), showing that the AB1 derivative readily labels Gram-positive bacteria, but not Gram-negatives. **(b)** Confocal microscopic images of a mixed cell culture consisting of *S. aureus* and Jurkat lymphocytes stained with 100 µM AF488-AB1-OMe (scale bar, 25 µm). The Jurkat cells are highlighted by the yellow circles on the overlay image. These results collectively demonstrate the superb selectivity of AB1 for Gram-positive bacteria. FITC, fluorescein isothiocyanate.

OMe is membrane impermeable, which precludes labelling of intracellular targets. Finally and importantly, there are few AB1-reactive lipids on the outer surface of Jurkat cells: a mammalian cell does not unusually produce Lys-PG. Although PE can account for up to 20% of the total lipids of a mammalian cell[1], it is primarily confined to the cytosolic leaflet and therefore not available at the cell surface either[7].

**AB1 synergizes with cationic peptides for potent bacteria labelling.** Despite the remarkable selectivity for Gram-positive bacteria, simple AB1 derivatives like AF488-AB1-OMe suffer from the high concentrations needed to achieve effective bacteria labelling. Furthermore, dictated by the mechanism of iminoboronate formation, AB1 derivatives are expected to react with lysine and lysine residues of various proteins[17], which in turn inhibit the association of AB1 with membranes. For example, BSA was found to inhibit the bacterial cell labelling by AF488-AB1-OMe with an apparent $IC_{50}$ of $\sim 1.5$ mg ml$^{-1}$ ($\sim 22$ µM). At 10 mg ml$^{-1}$ concentration, BSA resulted in $\sim 90\%$ reduction of the fluorescence staining of the AF488-AB1-OMe-treated *S. aureus* cells (Supplementary Fig. 6). We surmised that these problems could be resolved by conjugating AB1 to a directing functionality to bacterial cells. Towards this end, we have synthesized AB1 in its properly protected form (Fmoc-AB1(pin)-OH, Fig. 2) for solid-phase peptide synthesis, which should allow facile conjugation of AB1 to a variety of peptides or peptidomimetics that can serve as bacteria-directing motifs. Given that bacterial

cells are known to be enriched with negatively charged lipids, such as PG and cardiolipin, we thought to employ cationic peptides to direct AB1 to bacterial cell surfaces[22,23]. A small group of peptides were synthesized to incorporate AB1 as the C-terminal residue (Fig. 5a). The bacteria-targeting elements we tested include single cationic residues Lys and Arg, as well as a polycationic peptide Hlys[24,25] with the sequence of RYWVAWRNR. Hlys was reported to give a minimal inhibitory concentration of 24 µM against *S. aureus*, yet minimal haemolytic activity[24]. In addition, we chose Hlys because of its small size and the absence of lysine residues, which could in principle form an intramolecular iminoboronate with AB1. Nevertheless, we did not see intramolecular iminoboronate formation with the peptide K-AB1 (Supplementary Fig. 9). This lack of intramolecular conjugation is possibly due to the potential steric constraint that results from the fact that lysine and AB1 are contiguous in sequence. An analogous observation was recently reported for a cysteine-mediated macrocyclization with adjacent residues[26]. A control peptide (G-AB1) was also synthesized that incorporates a glycine instead of cationic motifs. All peptides were synthesized with an N-terminal cysteine so that they can be easily labelled with AF488-C5-maleimide (Fig. 5a).

The fluorescently labelled peptides were first assessed via flow cytometry analysis of the *S. aureus* cells stained with peptides at varied concentrations (Fig. 5b and Supplementary Fig. 10). The control peptide G-AB1 was only able to give a small fluorescence increase even at concentrations up to 100 µM. This is consistent

**a**

| G-AB1 | Ac-C*G(AB1)-NH$_2$ |
| K-AB1 | Ac-C*K(AB1)-NH$_2$ |
| R-AB1 | Ac-C*R(AB1)-NH$_2$ |
| Hlys-AB1 | Ac-C*RYWVAWRNRG(AB1)-NH$_2$ |
| Hlys | Ac-C*RYWVAWRNR-NH$_2$ |

**b**

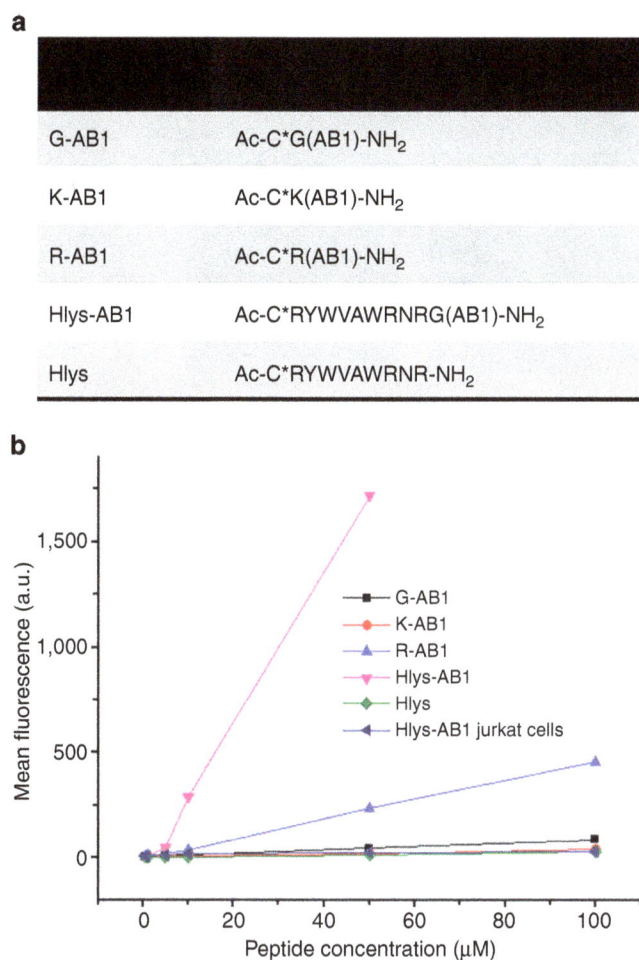

**Figure 5 | Synergizing covalent and noncovalent interactions for bacteria targeting.** (**a**) Sequences of AB1-presenting peptides, where C* represents a cysteine labelled with AF488-C5-maleimide. (**b**) Concentration profiles of *S. aureus* cell staining by the AB1-presenting peptides. All samples were prepared with the washing step right before analysis. The data for Jurkat cell staining with Hlys-AB1 were included to show its superb bacterial selectivity. The flow cytometry experiments were performed twice, which gave consistent results. One set of the data is presented herein.

appearance of dying or dead bacterial cells, which may stain with AB1 differently.

With much improved potency, we assessed Hlys-AB1 for bacteria labelling at nanomolar concentrations, with which the washing procedure is no longer necessary. The microscopic images show that, without wash, Hlys-AB1 effectively stained *S. aureus* cells at concentrations of 100 nM or higher (Fig. 6 and Supplementary Fig. 12). The confocal images revealed the cell envelope localization of the Hlys-AB1 (Supplementary Fig. 13), as expected for its membrane-targeting mechanism. Including the washing step in sample preparation resulted in approximately sevenfold reduction of the fluorescence staining of the cells (Supplementary Fig. 14). This is perhaps not surprising considering the reversible nature of the iminoboronate chemistry. Importantly, Hlys alone did not label the *S. aureus* cells under the same conditions (Fig. 6a), highlighting the critical importance of the AB1 moiety for the Hlys-AB1 staining of bacterial cells. Excitingly, Hlys-AB1 remained highly selective for Gram-positive bacteria under the no-wash conditions: the peptide failed to afford any fluorescence staining for the Gram-negative *E. coli* (Fig. 6b), as well as the Jurkat lymphocytes (Fig. 6e).

With the design of Hlys-AB1, the antimicrobial peptide Hlys is expected to selectively bind bacterial cell membranes and direct AB1 to covalently label PE or Lys-PG on bacterial cells. We hypothesized that this synergistic mechanism would not only improve the potency for bacteria labelling, but also minimize the protein interference of the iminoboronate chemistry. To prove this hypothesis, we assessed the inhibitory effect of fetal bovine serum (FBS) on the bacterial staining of Hlys-AB1. *S. aureus* cells were treated with Hlys-AB1 in presence of FBS at varied concentrations and the samples were analyzed with fluorescence microscopy and flow cytometry (Fig. 6d,f). The results show that, even with 10% FBS, submicromolar concentrations of Hlys-AB1 readily allowed the visualization of *S. aureus* cells under a fluorescence microscope (Fig. 6d), although reduced brightness was observed in comparison to the cells treated without FBS. Flow cytometry analysis yielded consistent results with microscopy: the presence of 10% FBS elicited ~30% reduction of the median fluorescence of the stained *S. aureus* cells (Fig. 6f). Again we attribute the much reduced protein interference of Hlys-AB1 to the synergy of covalent (AB1) and noncovalent (cationic peptide) mechanisms for bacterial cell targeting. The high potency and bacterial selectivity makes Hlys-AB1 potentially useful for targeting bacteria in blood serum or further in living organisms.

## Discussion

To summarize, we have demonstrated that the two major membrane lipids of bacterial cells, namely PE and Lys-PG, can be selectively targeted by synthetic molecules that induce formation of iminoboronate structures. Specifically, we have synthesized an unnatural amino acid (dubbed AB1) that displays a 2-APBA motif and can therefore conjugate with primary amines to form iminoboronates. By targeting the differential abundance and accessibility of Lys-PG and PE on cell surfaces, a fluorophore-labelled AB1 effectively stains Gram-positive bacteria (*B. subtilis* and *S. aureus*), bypassing Gram-negative bacteria and mammalian cells. Conjugating AB1 to cationic peptides greatly enhances its potency for bacteria labelling and importantly minimizes the interference of serum proteins to its bacterial association. Specifically, a hybrid peptide Hlys-AB1 was found to label *S. aureus* cells at nanomolar concentrations even in the presence of 10% FBS.

Nature primarily employs noncovalent mechanisms, such as hydrogen bonding, to achieve specific molecular recognition. Covalent chemistry has been largely avoided in targeting

with the fact that high concentrations of AF488-AB1-OMe are needed to achieve effective staining of bacterial cells. Conjugating AB1 to a lysine (K-AB1) did not improve, and perhaps even compromised AB1's association with *S. aureus*. In contrast, conjugation to an arginine (R-AB1) significantly enhanced the cell labelling. This contrasting results for K-AB1 and R-AB1 can be rationalized by the fact that an arginine side chain can afford stronger interaction with phospholipids than a lysine[27]. The AB1 conjugate with the polycationic peptide Hlys (Hlys-AB1) afforded a dramatic improvement of its potency for bacterial cell staining. For example, at 50 μM, Hlys-AB1 afforded a mean fluorescence intensity 8 times higher than that of R-AB1 and over 40 times better than G-AB1. In contrast, Hlys alone did not afford any fluorescence staining of *S. aureus* cells (Fig. 5b), again highlighting the critical importance of the AB1 moiety that covalently conjugates with the membrane lipids of bacterial cells. Importantly, Jurkat cells were not stained by Hlys-AB1 even with the highest concentration tested (Fig. 5b). Finally, we note that under our experimental conditions Hlys-AB1 caused marginal reduction of the viability of the bacterial cells (Supplementary Fig. 11), indicating the Hlys-enhanced staining is not due to the

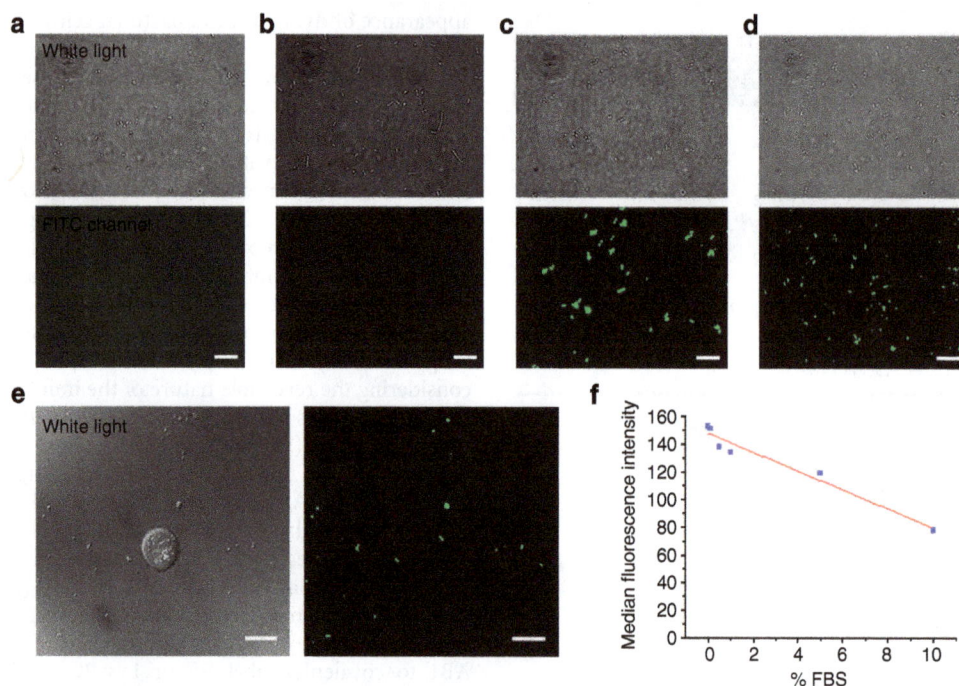

**Figure 6 | Bacterial labelling with submicromolar concentrations of Hlys-AB1.** (**a**) *S. aureus* cells treated with 0.5 μM Hlys as a negative control. (**b**) *E. coli* treated with 0.5 μM Hlys-AB1. (**c**) *S. aureus* cells stained with 0.5 μM Hlys-AB1. (**d**) *S. aureus* cells stained with 0.5 μM Hlys-AB1 in the presence of 10% FBS. (**e**) Confocal images of a *S. aureus* and Jurkat cell mixture stained with 0.5 μM Hlys-AB1. (**f**) FBS inhibition of *S. aureus* cell staining by Hlys-AB1 analysed via flow cytometry. 0.2 μM Hlys-AB1 was used for this experiment. The median fluorescence (*y*-axis) appears to give a linear relationship against the percentage of FBS (*x*-axis). The flow cytometry experiments were performed twice, which gave consistent results. One set of the data is presented. FITC, fluorescein isothiocyanate.

biomolecules of interest as irreversibility could result in modification of unintended targets and consequently toxicity[28,29]. However, reversible covalent chemistry circumvents this problem and should be able to complement the noncovalent mechanisms for molecular recognition. Among the limited number of examples, Wells and co-workers reported a strategy for protein ligand discovery that utilizes reversible disulfide chemistry to target reactive cysteines[30]. More recently, a group of nitrile modified acrylamides was reported by the Taunton group that reacts with cysteines of a protein kinase in a rapidly reversible manner[31]. In addition to targeting thiols, various boronic acid-presenting structures have been developed to target certain carbohydrates via reversible boronic ester formation[32,33]. Our work presented here, together with some recent publications by others[16,17], expands the reversible covalent chemistry toolbox for targeting biological amines. Although the iminoboronate chemistry was previously shown to label purified proteins[17] and aminosugars[16], the selectivity over other biomolecules has not been addressed. In comparison, our work here clearly demonstrates the applicability of the iminoboronate chemistry in complex biological systems (for example, bacteria labelling in the presence of blood serum).

It is highly desirable, yet challenging to differentiate various membrane lipids. A recent report[34] describes the covalent modification of mammalian aminophospholipids (PE and PS) on cell surfaces with an amine reactive reagent named sulfo-NHS-biotin, which allows the capture and quantification of externalized aminophospholipids. The nonselective reactivity of this reagent towards amines precludes its use in complex biological milieu. With the goal of better understanding lipid biology, a number of chemically modified lipids have been developed to display bioorthogonal reacting groups[35–38]. Once incorporated into a membrane, lipids as such can be selectively

labelled to reveal their subcellular distribution and homeostasis behaviour. However, these synthetic lipid probes are not known to afford specificity for bacterial cells. A recent report by Dumont *et al.* describes the metabolic incorporation of an azide-modified sugar into lipopolysaccharide[39], which enables fluorescence labelling of Gram-negative bacteria without genetic modification. Our work differs from these previous reports because the AB1 derivatives selectively target natural endogenous lipids of Gram-positive bacteria. For the purpose of bacterial detection, this contribution complements the elegant work by Dumont and co-workers, which is limited to selected Gram-negative bacteria.

The results of *in vitro* characterization presented here clearly demonstrate the superb bacteria selectivity, as well as the minimal serum interference, of AB1 and derivatives. Ongoing research in our lab seeks to further improve the potency of the AB1 derivatives for fast and more efficient labelling of bacterial cells, as well as to improve their stability towards proteolytic degradation. Our future research will evaluate the potential of optimized AB1 derivatives for biomedical applications, such as detecting bacteria in blood samples or imaging bacterial infection in animal models[40]. Finally, we submit that the covalent strategy for molecular recognition should be extendable to other important lipids of biological membranes. Research towards this end is also currently underway.

## Methods
**Materials and instrumentation.** Chemical reagents for small molecule and peptide synthesis were purchased from various vendors and used as received. The phospholipids were purchased from Avanti Polar Lipids (Alabaster, Al). PBS buffer, DMEM/high-glucose media, RPMI 1640 media and penicillin/streptomycin were purchased from Thermal Scientific. The Gram-positive bacteria (*B. subtilis* (ATCC 663) and *S. aureus* (ATCC 6538)) were purchased from Microbiologics as lyophilized cell pellet. *E. coli* (BL 21) was a gift from the lab of Professor Mary F.

Roberts at Boston College. NMR data of the small molecules were collected on a VNMRS 500 MHz NMR spectrometer. Mass spectrometry data were generated by using an Agilent 6230 LC TOF mass spectrometer. Peptide synthesis was carried out on a Tribute peptide synthesizer from Protein Technologies. The fluorescence anisotropy experiments were performed by using a SpectraMax M5 plate reader. Fluorescence images were taken on a Zeiss Axio Observer A1 inverted microscope. Confocal images were taken on the Leica SP5 confocal fluorescence microscope housed in the Biology Department of Boston College. Flow cytometry analyses were carried out on a BD FACSAria cell sorter also housed in the Biology Department of Boston College.

**Synthesis.** Details of the amino acid and peptide synthesis are provided as Supplementary Methods. Also presented in Supplementary Figs 15–25 are the NMR spectra of novel compounds, exemplary high-performance liquid chromatography traces of the fluorophore-labelled AB1 derivatives, and in Supplementary Table 1 are mass-spec data of the fluorophore-labelled amino acids and peptides.

**Binding assays with lipid vesicles.** Liposomes were prepared by dissolving and mixing the desired phospholipids in chloroform. After evaporating chloroform, the residue was suspended in 50 mM phosphate buffer, pH 7.4. The lipid suspensions were treated through 10 cycles of freeze-and-thaw process, and extruded 11 times through a membrane with pore size of 100 nm. The concentrations of liposome stocks were characterized via the Stewart Assay[41]. The size distribution of each vesicle sample was characterized with a dynamic light scattering instrument (DynaproTM NanoStar, Wyatt Technology Corp.). The diameter of all vesicles were found to fall into the narrow range of 100–110 nm. Lipid vesicles at varied concentrations (25, 50, 100, 500, 1,000, 2,000 μM total lipids) were incubated with 0.5 μM of Fl-AB1/2-OMe for 40 min in a phosphate buffer (50 mM Na•Pi, pH = 7.4). Then the fluorescence anisotropy values of each sample were recorded. To correct for the interference of light scattering, the lipid binding data of fluorescein isothiocyanate-alaninamide were used for blank subtraction. All samples were measured in triplicates and the data were averaged to generate the binding curves.

**NMR and mass spectrometry characterization of iminoboronates.** The PC/Lys-PG (3:2) vesicles (200 μl, 2 mM total lipids) were incubated with 2-APBA (200 μl, 10 mM) and Ac-R-AB1-amide (200 μl, 2 mM), respectively, for 40 min. Then the mixtures were lyophilized and dissolved in $CDCl_3$:$CD_3OD$ (2:1; 600 μl). The iminoboronate formation was confirmed by [11]B NMR and mass spectrometry. All the [11]B-NMR experiments were carried out with $BF_3$ as an external standard, the chemical shift of which was set at 0 p.p.m. $BF_3$ was not used as an internal standard because of its acidic nature, which might disrupt the iminoboronate conjugates.

**Bacterial cell culture and staining.** Bacterial staining experiments were performed against three strains: B. subtilis (ATCC 663), S. aureus (ATCC 6538) and E. coli (BL21). For each strain, bacterial cells from a single colony were grown overnight in LB broth at 37 °C with agitation. An aliquot was taken and diluted (1:50 for E. coli, 1:20 for B. subtilis, 1:200 for S. aureus) in fresh broth and cultured for another ∼ 3 h until the cells reached the mid-logarithmic phase ($OD_{600}$ ∼ 0.5). Then the bacterial cell culture was diluted ten times and used immediately for small molecule labelling. For a typical labelling experiment, 100 μl of the diluted bacterial cell culture was spun down at 7,000 r.p.m. in a centrifuge tube (1.5 ml). The cells were washed once with 100 μl phosphate buffer (50 mM Na•Pi, pH = 7.4), and then mixed with 100 μl solution of an AB1 derivative at desired concentrations. After 40 min incubation, the samples with low AB1 concentrations (≤1 μM) were directly analyzed. The samples with higher AB1 concentrations were subjected to a washing procedure: the cells were spun down at 7,000 r.p.m. and the supernatant was discarded. Then the cells were washed twice with the phosphate buffer (100 μl, 2 min incubation), after which the spun-down cells were re-suspended in 50 μl of the phosphate buffer for analysis.

**Mammalian cell culture and staining.** Jurkat cells were grown and maintained in RPMI 1640 media with 10% FBS and 1% penicillin/streptomycin at 37 °C, 5% $CO_2$ and passed for less than 50 generations. The cell viability and density was checked and counted daily by using 0.2 μM trypan blue as a viability testing dye on a haemocytometer. Before staining with a small molecule, the cells were cultured to a density of 1.5–2.0 × 10^6 cells per ml in a Corning cell culture flask (with vent cap). Small-molecule staining was carried out by a similar protocol as used for the bacterial cells except the speed of centrifugation (Jurkat cells were spun down at 200 r.c.f. (relative centrifugal force)). Samples with high AB1 concentrations (>1 μM) were washed right before analysis.

**Co-culture preparation.** For labelling with AF488-AB1-OMe, 500 μl of Jurkat (1.5 ∼ 2.0 × 10^6 cells per ml) cells and 100 μl of S. aureus (2-3 × 10^8 cells per ml) were separately stained with 100 μM AF488-AB1-OMe in centrifuge tubes for 40 min. Then the Jurkat and S. aureus cells were spun down at 200 r.c.f. and

7,000 r.p.m., respectively. The supernatants were discarded and the cells were further washed twice with 100 μl of the phosphate buffer. Finally, the Jurkat and S. aureus cells were re-suspended in 50 μl of phosphate buffer and mixed together for imaging study. For the labelling experiment with Hlys-AB1, 500 μl of Jurkat (1.5 ∼ 2.0 × 10^6 cells per ml) cells and 100 μl of S. aureus (2-3 × 10^8 cells per ml) were mixed. The mixture was incubated with 0.5 μM Hlys-AB1 for 40 min and then immediately subjected to microscopy analysis.

**Microscopic analysis of AB1-stained cells.** For epi-fluorescence microscopy, 5 μl of the bacterial cell suspension was dropped on a glass slide (Fisherfinest premium, 75 × 25 × 1 mm³). A coverslip (Fisherbrand, 22 × 22 × 0.15 mm³) was pressed down on the cell droplet to give a single layer of cells on the glass slides. White light and fluorescence images were taken on a Zeiss Axio Observer A1 inverted microscope equipped with a filter cube (488 nm excitation, 515–520 nm emission) suitable for detection of AF488 fluorescence. A Plan-NeoFluar × 100 oil objective from Zeiss was used to visualize the bacterial cells. All images were captured with the exposure time of 300 ms for AF488-labelled AB1 derivatives. All fluorescence images were processed following a fixed protocol with the software Fiji ImageJ[42]. For confocal analysis, 5 μl of cells were placed on a glass slide and a 22 × 22 × 1.5 Fisherbrand microscope cover glass was placed on top. Images were taken on a Leica SP5 confocal fluorescence microscope with filters that allowed detection of AF488 (488 nm excitation, 496–564 nm emission). A × 63 oil objective was used with an Argon laser at 10% laser power. Gain was adjusted to between 900 HV and 1,100 HV with an offset of − 0.5%. The images were captured with the software LAS 2.6 and then processed with Fiji ImageJ[42].

**Flow cytometry analysis of AB1 stained cells.** The samples were prepared and stained following the same protocol described for microscopy. The cells stained with sub-micromolar concentrations of Hlys-AB1 were analyzed without wash, while all other samples were subjected to the wash procedure right before analysis. The samples were analyzed on a BD FACSAria cell sorter (BD Biosciences). Data analysis was performed with FlowJo (Tree Star, Inc.), from which the median fluorescence intensities of the stained cells were extracted and plotted against AB1 concentration. For the protein inhibition experiments, the cell samples were prepared in the presence of BSA or FBS at desired concentrations before the addition of the AB1 compounds. The median fluorescence intensity of these cell samples was extracted and plotted against BSA or FBS concentration.

## References

1. van Meer, G., Voelker, D. R. & Feigenson, G. W. Membrane lipids: where they are and how they behave. Nat. Rev. Mol. Cell Biol. 9, 112–124 (2008).
2. Shevchenko, A. & Simons, K. Lipidomics: coming to grips with lipid diversity. Nat. Rev. Mol. Cell Biol. 11, 593–598 (2010).
3. Wenk, M. R. Lipidomics: new tools and applications. Cell 143, 888–895 (2010).
4. Epand, R. F., Schmitt, M. A., Gellman, S. H. & Epand, R. M. Role of membrane lipids in the mechanism of bacterial species selective toxicity by two alpha/beta-antimicrobial peptides. Biochim. Biophys. Acta 1758, 1343–1350 (2006).
5. Epand, R. F., Savage, P. B. & Epand, R. M. Bacterial lipid composition and the antimicrobial efficacy of cationic steroid compounds (Ceragenins). Biochim. Biophys. Acta 1768, 2500–2509 (2007).
6. Roy, H. Tuning the properties of the bacterial membrane with aminoacylated phosphatidylglycerol. IUBMB Life 61, 940–953 (2009).
7. Balasubramanian, K. & Schroit, A. J. Aminophospholipid asymmetry: a matter of life and death. Annu. Rev. Physiol. 65, 701–734 (2003).
8. Yeung, T. et al. Membrane phosphatidylserine regulates surface charge and protein localization. Science 319, 210–213 (2008).
9. Lemmon, M. A. Membrane recognition by phospholipid-binding domains. Nat. Rev. Mol. Cell Biol. 9, 99–111 (2008).
10. Gao, J. & Zheng, H. Illuminating the lipidome to advance biomedical research: peptide-based probes of membrane lipids. Future Med. Chem. 5, 947–959 (2013).
11. Crugeiras, J., Rios, A., Riveiros, E., Amyes, T. L. & Richard, J. P. Glycine enolates: the effect of formation of iminium ions to simple ketones on alpha-amino carbon acidity and a comparison with pyridoxal iminium ions. J. Am. Chem. Soc. 130, 2041–2050 (2008).
12. McFarland, J. M. & Francis, M. B. Reductive alkylation of proteins using iridium catalyzed transfer hydrogenation. J. Am. Chem. Soc. 127, 13490–13491 (2005).
13. Arnal-Herault, C. et al. Functional G-quartet macroscopic membrane films. Angew. Chem. Int. Ed. 46, 8409–8413 (2007).
14. Hutin, M., Bernardinelli, G. & Nitschke, J. R. An iminoboronate construction set for subcomponent self-assembly. Chemistry 14, 4585–4593 (2008).
15. Galbraith, E. et al. Dynamic covalent self-assembled macrocycles prepared from 2-formyl-aryl-boronic acids and 1,2-amino alcohols. N. J. Chem. 33, 181–185 (2009).
16. Gutierrez-Moreno, N. J., Medrano, F. & Yatsimirsky, A. K. Schiff base formation and recognition of amino sugars, aminoglycosides and biological

polyamines by 2-formyl phenylboronic acid in aqueous solution. *Org. Biomol. Chem.* **10**, 6960–6972 (2012).

17. Cal, P. M. *et al.* Iminoboronates: a new strategy for reversible protein modification. *J. Am. Chem. Soc.* **134**, 10299–10305 (2012).

18. Hoyle, C. E. & Bowman, C. N. Thiol-ene click chemistry. *Angew. Chem. Int. Ed.* **49**, 1540–1573 (2010).

19. Ishiyama, T., Itoh, Y., Kitano, T. & Miyaura, N. Synthesis of arylboronates via the palladium(O)-catalyzed cross-coupling reaction of tetra(alkoxo)diborons with aryl triflates. *Tetrahedron Lett.* **38**, 3447–3450 (1997).

20. Lozada, J., Liu, Z. & Perrin, D. M. Base-promoted protodeboronation of 2,6-disubstituted arylboronic acids. *J. Org. Chem.* **79**, 5365–5368 (2014).

21. McMahon, H. T. & Gallop, J. L. Membrane curvature and mechanisms of dynamic cell membrane remodelling. *Nature* **438**, 590–596 (2005).

22. Hancock, R. E. & Diamond, G. The role of cationic antimicrobial peptides in innate host defences. *Trends Microbiol.* **8**, 402–410 (2000).

23. Tew, G. N., Scott, R. W., Klein, M. L. & Degrado, W. F. *De novo* design of antimicrobial polymers, foldamers, and small molecules: from discovery to practical applications. *Acc. Chem. Res.* **43**, 30–39 (2010).

24. Gonzalez, R., Albericio, F., Cascone, O. & Iannucci, N. B. Improved antimicrobial activity of h-lysozyme (107-115) by rational Ala substitution. *J. Pept. Sci.* **16**, 424–429 (2010).

25. Iannucci, N. B., Curto, L. M., Albericio, F., Cascone, O. & Delfino, J. M. Structure-activity relationship analysis of a novel antimicrobial peptide derived from the 107-115 h-lysozyme fragment. *Biopolymers* **100**, 279–279 (2013).

26. Bionda, N., Cryan, A. L. & Fasan, R. Bioinspired strategy for the ribosomal synthesis of thioether-bridged macrocyclic peptides in bacteria. *ACS Chem. Biol.* **9**, 2008–2013 (2014).

27. Tang, M., Waring, A. J., Lehrer, R. I. & Hong, M. Effects of guanidinium-phosphate hydrogen bonding on the membrane-bound structure and activity of an arginine-rich membrane peptide from solid-state NMR spectroscopy. *Angew. Chem. Int. Ed.* **47**, 3202–3205 (2008).

28. Singh, J., Petter, R. C., Baillie, T. A. & Whitty, A. The resurgence of covalent drugs. *Nat. Rev. Drug. Discov.* **10**, 307–317 (2011).

29. Johnson, D. S., Weerapana, E. & Cravatt, B. F. Strategies for discovering and derisking covalent, irreversible enzyme inhibitors. *Future Med. Chem* **2**, 949–964 (2010).

30. Erlanson, D. A. *et al.* Site-directed ligand discovery. *Proc. Natl Acad. Sci. USA* **97**, 9367–9372 (2000).

31. Serafimova, I. M. *et al.* Reversible targeting of noncatalytic cysteines with chemically tuned electrophiles. *Nat. Chem. Biol.* **8**, 471–476 (2012).

32. James, T. D., Sandanayake, K. R. A. S. & Shinkai, S. Saccharide sensing with molecular receptors based on boronic acid. *Angew. Chem. Int. Ed.* **35**, 1910–1922 (1996).

33. Dai, C. F. *et al.* Carbohydrate biomarker recognition using synthetic lectin mimics. *Pure Appl. Chem.* **84**, 2479–2498 (2012).

34. Thomas, C. P. *et al.* Identification and quantification of aminophospholipid molecular species on the surface of apoptotic and activated cells. *Nat. Protoc.* **9**, 51–63 (2014).

35. Neef, A. B. & Schultz, C. Selective fluorescence labeling of lipids in living cells. *Angew. Chem. Int. Ed.* **48**, 1498–1500 (2009).

36. Best, M. D., Rowland, M. M. & Bostic, H. E. Exploiting bioorthogonal chemistry to elucidate protein-lipid binding interactions and other biological roles of phospholipids. *Acc. Chem. Res.* **44**, 686–698 (2011).

37. Yang, J., Seckute, J., Cole, C. M. & Devaraj, N. K. Live-cell imaging of cyclopropene tags with fluorogenic tetrazine cycloadditions. *Angew. Chem. Int. Ed.* **51**, 7476–7479 (2012).

38. Erdmann, R. S. *et al.* Super-resolution imaging of the Golgi in live cells with a bioorthogonal ceramide probe. *Angew. Chem. Int. Ed.* **53**, 10242–10246 (2014).

39. Dumont, A., Malleron, A., Awwad, M., Dukan, S. & Vauzeilles, B. Click-mediated labeling of bacterial membranes through metabolic modification of the lipopolysaccharide inner core. *Angew. Chem. Int. Ed.* **51**, 3143–3146 (2012).

40. Panizzi, P. *et al. In vivo* detection of Staphylococcus aureus endocarditis by targeting pathogen-specific prothrombin activation. *Nature Med.* **17**, 1142–1153 (2011).

41. Stewart, J. C. M. Colorimetric Determination of Phospholipids with Ammonium Ferrothiocyanate. *Anal. Biochem.* **104**, 10–14 (1980).

42. Schindelin, J. *et al.* Fiji: an open-source platform for biological-image analysis. *Nat. Methods* **9**, 676–682 (2012).

## Acknowledgements

We gratefully acknowledge the financial support provided by the Boston College, the National Science Foundation (CHE1112188), and the National Institute of General Medical Sciences (R01GM102735). We also thank Dr Bret Judson for his help on fluorescence microscopy and Dr Patrick Autissier for his help on the flow cytometry experiments.

## Author Contributions

J.G. and A.B. conceived the project, analyzed the data and wrote the manuscript; A.B. performed the majority of the experiments; K.A.M. assisted in the synthesis of the AB1 derivatives; M.A.K. performed the confocal microscopy work with Jurkat cells.

# 7

# Macroscopic ordering of helical pores for arraying guest molecules noncentrosymmetrically

Chunji Li[1,*], Joonil Cho[2,*], Kuniyo Yamada[2], Daisuke Hashizume[2], Fumito Araoka[2,3], Hideo Takezoe[3], Takuzo Aida[1,2] & Yasuhiro Ishida[2,4]

Helical nanostructures have attracted continuous attention, not only as media for chiral recognition and synthesis, but also as motifs for studying intriguing physical phenomena that never occur in centrosymmetric systems. To improve the quality of signals from these phenomena, which is a key issue for their further exploration, the most straightforward is the macroscopic orientation of helices. Here as a versatile scaffold to rationally construct this hardly accessible structure, we report a polymer framework with helical pores that unidirectionally orient over a large area ($\sim 10\,cm^2$). The framework, prepared by crosslinking a supramolecular liquid crystal preorganized in a magnetic field, is chemically robust, functionalized with carboxyl groups and capable of incorporating various basic or cationic guest molecules. When a nonlinear optical chromophore is incorporated in the framework, the resultant complex displays a markedly efficient nonlinear optical output, owing to the coherence of signals ensured by the macroscopically oriented helical structure.

[1]Department of Chemistry and Biotechnology, School of Engineering, The University of Tokyo, Hongo, Bunkyo, Tokyo 113-8656, Japan. [2]RIKEN Center for Emergent Matter Science, 2-1 Hirosawa, Wako, Saitama 351-0198, Japan. [3]Department of Organic and Polymeric Materials, Tokyo Institute of Technology, 2-12-1-S8-42 O-okayama, Meguro, Tokyo 152-8552, Japan. [4]PRESTO, Japan Science and Technology Agency, 4-1-8 Honcho, Kawaguchi, Saitama 332-0012, Japan. * These authors contributed equally to this work. Correspondence and requests for materials should be addressed to Y.I. (email: y-ishida@riken.jp).

When appropriately functionalized molecules are arranged in one-handed helices, they often display intriguing physical phenomena that never occur in centrosymmetric structures, as represented by second harmonic generation (SHG)[1-4] and piezoelectricity[5-8]. These helices would also give a clue to pursue unexplored predictions, including molecular solenoid effects[9,10]. To explore these phenomena further, a key issue is to improve the quality of their output signals. For this purpose, the most straightforward approach is to orient the helices macroscopically so that mutual cancellation of signals is prevented[3-6,8]. Although such hierarchical structures, realizing one-handed helicity and macroscopic orientation at the same time, are rarely obtained, their rational formation would become possible by using a framework with macroscopically oriented helical pores[11-17]. If this framework serves as a scaffold for arraying various molecules in a macroscopically oriented helical structure by simple host–guest complexation, it should facilitate the exploration of the aforementioned phenomena. However, such a framework has not been developed so far. Even in the case of achiral pores, their macroscopic orientation remains a general challenge[18-22]. For the present purpose, individual pores must be helical and capable of precisely positioning the incorporated molecules, which makes this challenge even greater.

With the aim of developing such frameworks, we focused on an advanced version of molecularly imprinted polymers[23] prepared by the crosslinking of liquid crystals (LCs), which have been pioneered by Gin et al.[24] and are now regarded as a new class of solid-state hosts[25-32]. When a multicomponent LC composed of a polymerizable frame unit and a non-polymerizable template

unit (for example, Fig. 1, i) is in situ crosslinked, the frame units are converted into a polymer framework, while the template units are noncovalently captured in the polymer framework and therefore are exchangeable with other molecules. Owing to the well-controlled structure, thus obtained polymer frameworks exhibit unique functions in conversion[24], binding[25-30] and transport[31,32] of guest molecules. As we recently demonstrated, chiral pores are rationally constructed by using chiral template units[27,28]. Although previous polymer frameworks were mostly prepared from randomly oriented LCs except for a few examples[31,32], we noted that LCs are dynamic and potentially orientable macroscopically by application of an external stimulus, such as a force or a field[31-40]. Among these stimuli, the use of a magnetic field has the advantages of being capable of application in a non-destructive and non-contact manner[20,32-34].

Here we report an unprecedented type of polymer framework with macroscopically oriented helical pores, prepared by in situ crosslinking of a supramolecular LC preorganized in a magnetic field. This achievement results from our unexpected finding that a chiral liquid crystalline salt we recently developed[27,28] meets all requirements as the precursor of such framework, that is, multicomponent nature, polymerizability[24-32], orientability[18-22,31-40] and helicity with controlled handedness[11-17]. Before this work, these features have been achieved separately, but never simultaneously. The resultant polymer framework serves as a versatile scaffold for arraying various molecules in a macroscopically oriented helical structure, thereby offering useful motifs for the exploration of the physical phenomena particular to noncentrosymmetric systems.

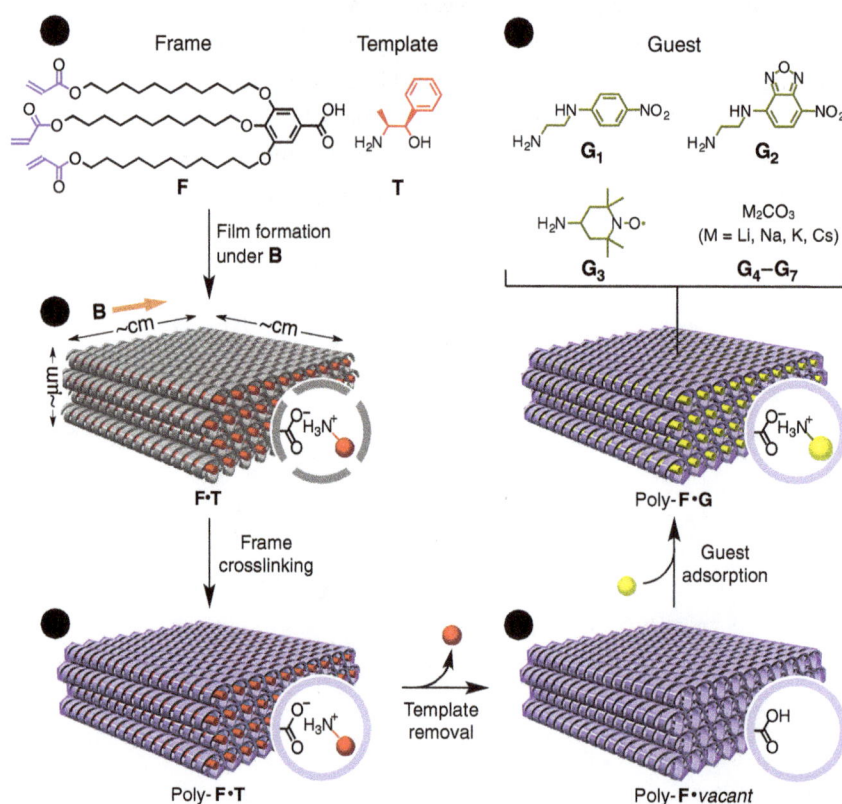

**Figure 1 | Macroscopically oriented polymer framework with helical pores by *in situ* crosslinking of a magnetically preorganized LC salt.** (i) Molecular structure of the frame (polymerizable carboxylic acid, **F**) and template (enantiopure amine, **T**) units that self-assemble into a columnar LC salt. (ii) Processing of the salt (**F•T**) into a macroscopically oriented LC film in a magnetic field. (iii) *In situ* crosslinking of the LC film of **F•T** by radical polymerization to give a polymerized film consisting of poly-**F•T**. (iv) Desorption of **T** from the polymerized film of poly-**F•T** to give a guest-free film of poly-**F•*vacant***. (v) Adsorption of guests **G₁-G₇** by the guest-free film of poly-**F•*vacant*** to give a guest-exchanged film of poly-**F•G**. B, Magnetic field applied during the LC film-preparation process.

## Results

### Synthesis of the macroscopically oriented polymer framework.

The polymer framework is prepared with a supramolecular columnar LC material recently reported by our group (Fig. 1, i)[27,28]. This consists of a polymerizable carboxylic acid that contains three flexible chains (frame, F)[24] and an enantiopure amine (template, T). On mixing in equimolar amounts, these components form a salt (F•T) that exhibits a stable columnar LC mesophase. In attempt to form a macroscopically oriented structure for the LC salt F•T, we began with well-established methods, such as thermal annealing and drop casting. When F•T was slowly cooled from an isotropic molten state at 130 to 20 °C, small, randomly oriented LC domains ($\sim$10 μm) were formed, as confirmed by polarized optical microscopy (POM; Fig. 2a). On the other hand, drop casting of its dichloromethane solution on a glass substrate produced relatively large LC domains, with a size of the order of millimetres and a characteristic crosshatched texture (Fig. 2b). Needless to say, these domains had no orientational regularity at the macroscopic level.

Interestingly, however, the application of a strong magnetic field during the drop-casting process resulted in perfectly controlled orientation of the LC domains over an $\sim$10-cm$^2$-size scale (Fig. 1, ii). For example, F•T (24.9 mg, 25 μmol) in dichloromethane (500 μl) was cast on a glass substrate (2.5 × 7.5 cm) and slowly concentrated to dryness over 2 h at 20 °C in the presence of a 10-T magnetic field oriented parallel to the substrate plane. As shown in Fig. 2c, the resultant LC film exhibited rhombic patterns of multiple LC domains that aligned along the applied magnetic field to form a continuous two-dimensional (2D) array. Over the entire region of the material, these domains were oriented in the same direction (Supplementary Fig. 1). Because of the high viscosity of LC salt F•T after perfect drying, the macroscopically oriented film did not undergo structural relaxation, even when the applied magnetic field was turned off. This sluggish relaxation suggests that the present magnetic orientation took place in an intermediate state during the solvent evaporation, where residual solvent lowered the viscosity. In contrast, when a 10-T magnetic field was directed perpendicular to the substrate plane, essentially no effect was brought on the orientation of the LC domains (Fig. 2d). These observations indicate that only the horizontal (in-plane) component of the magnetic field vector contributes to this magnetic orientation, as discussed below.

For covalent fixation of the magnetically oriented structure, the LC film of F•T was then subjected to *in situ* crosslinking polymerization (Fig. 1, iii). To polymerize the acryloyl groups in F, we chose a γ-ray irradiation method[41] that can be operated without a radical initiator and is a promising method for preserving an organized structure, as we recently disclosed[28]. Otherwise, doping with a radical initiator, even in a small amount (for example, 1 wt% of 2-hydroxy-2-methyl-1-propiophen-1-one), markedly weakened the structural order of the LC film. On irradiating the LC film with γ-ray (6.25 kGy h$^{-1}$) at 20 °C for 16 h, crosslinking polymerization proceeded quantitatively to convert the viscous fluidic F•T into an insoluble and non-meltable solid (Supplementary Figs 2 and 3) consisting of the salt of polymerized F (poly-F) and T (denoted hereafter as poly-F•T). This polymer film was flexible and freestanding, so that it was easily peeled off from the glass substrate (Fig. 3a).

### Structural analysis of the framework.

Having obtained the macroscopically oriented polymer framework in hand, we investigated its structure at various scales from a macroscopic through the mesoscopic to the molecular. As observed in POM (Fig. 3b) or even by the naked eye (Fig. 3a), the crosshatched texture was extended over the entire region of the polymer film, suggesting that the crosslinking and the peel-off processes had no effect on the macroscopic orientation. When the polymer film was rotated in an in-plane manner, its POM showed a contrast every 45°, giving a dark image when the light-polarization angle with respect to the applied magnetic field was 0° or 90° (Supplementary Fig. 4). The polarized infrared absorption of the film also showed an apparent dependency on the polarization angle, in that the peaks at 1,369 cm$^{-1}$ ($-CO_2^-$) and 3,221 cm$^{-1}$ ($-N^+H_3$) became maximal at the polarization angle of 0° and 90°, respectively (Fig. 3c). From these observations, it is obvious that columnar aggregates of F and T, which are afforded by salt-pair formation between the $-CO_2H$ and $-NH_2$ groups, lie along the glass surface and are oriented perpendicular or parallel to the applied magnetic field.

To examine the structure of the polymer framework in more detail, we performed an X-ray diffraction analysis with a synchrotron radiation source. As shown in Fig. 4 (top), the polymer film was exposed to a beam of X-rays from the directions perpendicular (through view) and in-plane (edge and end views) to the film surface to obtain three-dimensional (3D) structural information[42]. In the in-plane exposure, the X-ray beam was directed parallel (edge view) or perpendicular (end view) to the direction of the magnetic field applied during the preparation of the LC film. In the wide-angle region of the 2D X-ray diffraction images, two types of diffractions characteristic of columnar LCs were observed: a diffuse halo (*d*-spacing, 4.4 Å) due to the loose packing of the aliphatic chains (Fig. 4a–c, i) and a pair of obscure spots (*d*-spacing, 3.6 Å) attributable to the stacking of the π-conjugated systems (Fig. 4a,b, ii). Because the diffractions from the π-stacking appeared only in the equatorial region of the through and edge views, the π-conjugated systems in F•T are probably stacked in the direction perpendicular to the magnetic field.

To our surprise, a number of sharp spots appeared at regular intervals in the small-angle region (Fig. 4a–c, iii), unlike usual

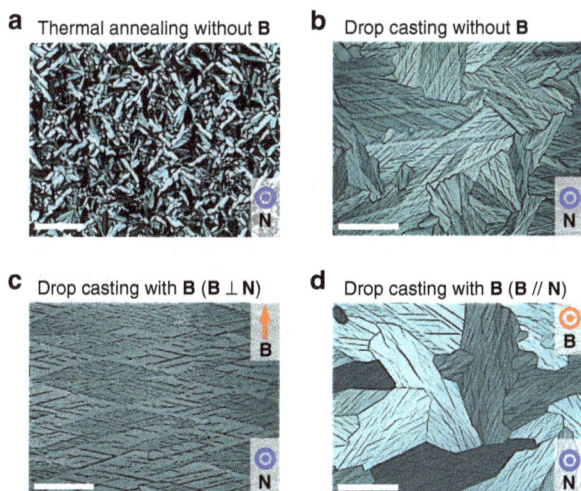

**Figure 2 | POM images of films of F•T under crossed Nicols. (a)** LC film ($\sim$10 μm thick) of F•T processed by slow cooling (−5 °C min$^{-1}$) from an isotropic melt at 130 to 20 °C in the absence of a magnetic field. Scale bar, 100 μm. **(b–d)** LC films ($\sim$10 μm thick) of F•T processed by drop casting of a dichloromethane solution on a glass substrate at 20 °C: in the absence **(b)** and presence of a 10-T magnetic field oriented parallel **(c)** and perpendicular **(d)** to the substrate plane. Scale bar, 500 μm. **B**, Magnetic field applied during the LC film-preparation process. **N**, Normal vector of the film surface.

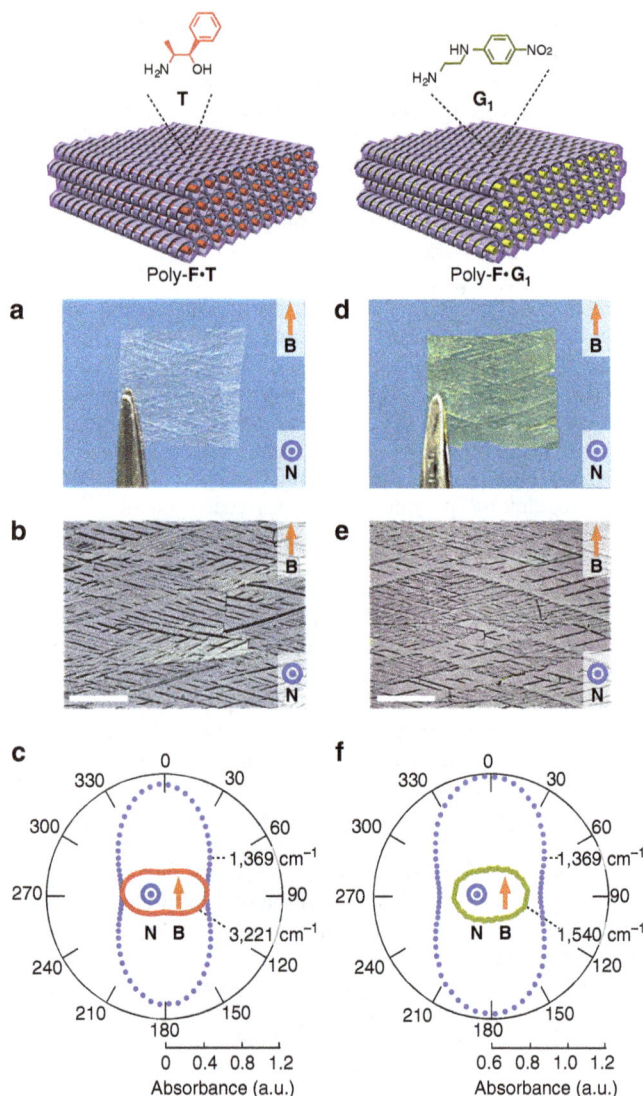

**Figure 3 | Properties of films with macroscopically oriented structures.** (**a–c**) Film of poly-F•T (~10 μm thick) prepared by *in situ* crosslinking of the LC film of F•T with a magnetically oriented structure. (**d–f**) Film of poly-F•G$_1$ (~10 μm thick) prepared by the guest exchange of the film of poly-F•T with a magnetically oriented structure. Pictures (**a,d**), POM images under crossed Nicols (**b,e**) and polar plots of the infrared absorption as a function of the light-polarization angle (**c,f**). Scale bar, 500 μm. **B**, Magnetic field applied during the LC film-preparation process. **N**, Normal vector of the film surface.

fluctuation because of the flexible, polydomain nature of the film, as exemplified by the radially broaden spots in Fig. 4c. In addition, the presence of amorphous regions was suggested by a broad scattering at $q = \sim 0.04\ \text{Å}^{-1}$, where the crystallinity degree of the film was estimated to 82% (Supplementary Fig. 5). Such amorphous regions would be non-negligible particularly at domain boundaries. To remove the effects of polydomain nature, a monodomain clump trimmed from the film was used instead, where much narrower spots were observed (Supplementary Figs 8 and 9). The crystallinity degree was also estimated to be notably higher (92%; Supplementary Fig. 10) but not perfect, probably because of the dynamic nature of the precursor LC and disordering during the *in situ* polymerization.

Using the data of the monodomain clump, the space group of poly-F•T was deduced as follows (Supplementary Methods). After determining the lattice parameters, all indexed reflections were integrated. By calculating the internal residual factor from the reflection intensity and considering symmetry, the Laue group was determined to be 6/*mmm*. Taking account of the systematic extinction rule (Supplementary Fig. 8b), the space group was deduced to be $P6_122$ or its enantiomorph $P6_522$, which contain right- and left-handed sixfold screw axes, respectively. These space groups also have a twofold screw axis that is perpendicular to the sixfold one, indicating that the whole system is intrinsically apolar and that the sixfold helix forms an antiparallel duplex with another homochiral helix. By assuming that the density of poly-F•T is $\sim 1.0\ \text{g cm}^{-3}$, we estimated the $Z$-value of the lattice to be 24, indicating the presence of two crystallographically independent pairs of F•T. In relation to this, the $c$ axis lattice parameter (42.6 Å) is exactly 12 times as large as the observed π-stacking distance (3.6 Å), indicating that one helical pitch consists of 12 stacking units. Taking account of all the parameters thus obtained, we proposed the structural model shown in Fig. 5, where two salt pairs of F and T aggregate to form a bow-tie-shaped unit (Fig. 5c). These stack helically with one another (stacking distance, 3.6 Å; rotation angle, 30°) to form a sixfold helix (Fig. 5b). The helical columns lie along the substrate plane and are oriented perpendicular to the applied magnetic field (Fig. 5a).

For further insight into the assembled structure of F•T, thermal behaviour of an analogous salt F•T′ (T′, C2-stereo-inverted analogue of T) was investigated (Supplementary Fig. 11). As we previously reported, conformational preferences of T and T′ are quite different from each other, where T tends to adopt a more flatten conformation (=larger dihedral angle $\theta$ in Supplementary Fig. 11a) due to the steric hindrance between the phenyl and methyl groups[45]. Such a flatten conformation would facilitate the stacking of the salt pairs of F and T, as suggested by the diffraction due to π-stacking (Fig. 4a,b ii). Indeed, the analogous salt F•T′ cannot form an assembled structure, exhibiting only an isotropic phase (Supplementary Fig. 11b). The X-ray crystal structure of another analogous salt F′•T (F′, analogue of F lacking long alkyl chains) is in consistent with the above hypothesis (Supplementary Fig. 12 and Supplementary Data 1). Thus, the salt pairs of F′ and T, adopting a flatten shape, self-assemble via π-stacking and hydrogen-bonding interactions. In the crystal structure of F′•T, the salt pairs assemble into a one-dimensional (1D) array with lateral offsets of the π-conjugated systems, while in the LC structure of F•T, the salt pairs assemble in a helical array, most likely because of the steric hindrance of the long alkyl chains in F.

The structural model in Fig. 5 accords with the general tendency of π-conjugated systems to orient parallel to an applied magnetic field, and explains the mechanism underlying the present magnetic orientation. Given that an applied magnetic field orients the π-conjugated systems in individual molecules of F and T parallel to the field, it would generate a torque that

polymer materials. The end view (Fig. 4c) exhibited a pattern of sixfold symmetry, indicating that the columnar objects (diameter, 31.5 Å) are packed in a hexagonal manner (Supplementary Fig. 5c). The patterns in the through (Fig. 4a) and edge views (Fig. 4b) correspond to rectangular lattices, suggesting that the columnar objects have a regular periodicity (distance, 42.6 Å) along the column axis (Supplementary Fig. 5a,b), most likely originating from the helical pitch[43,44]. All the reflections could be unambiguously indexed (Fig. 4a–c, iii) by assuming the presence of a 3D hexagonal columnar lattice ($a = 31.5$ Å and $c = 42.6$ Å). Transmission electron microscopy (TEM) images and the Fourier transform also suggested the hexagonal packing of columnar objects (Supplementary Fig. 7).

The 2D X-ray diffraction images, measured for a relatively large film specimen (3 × 3 mm, ~20 μm thick) to evaluate its macroscopic structural order, were affected by the angular

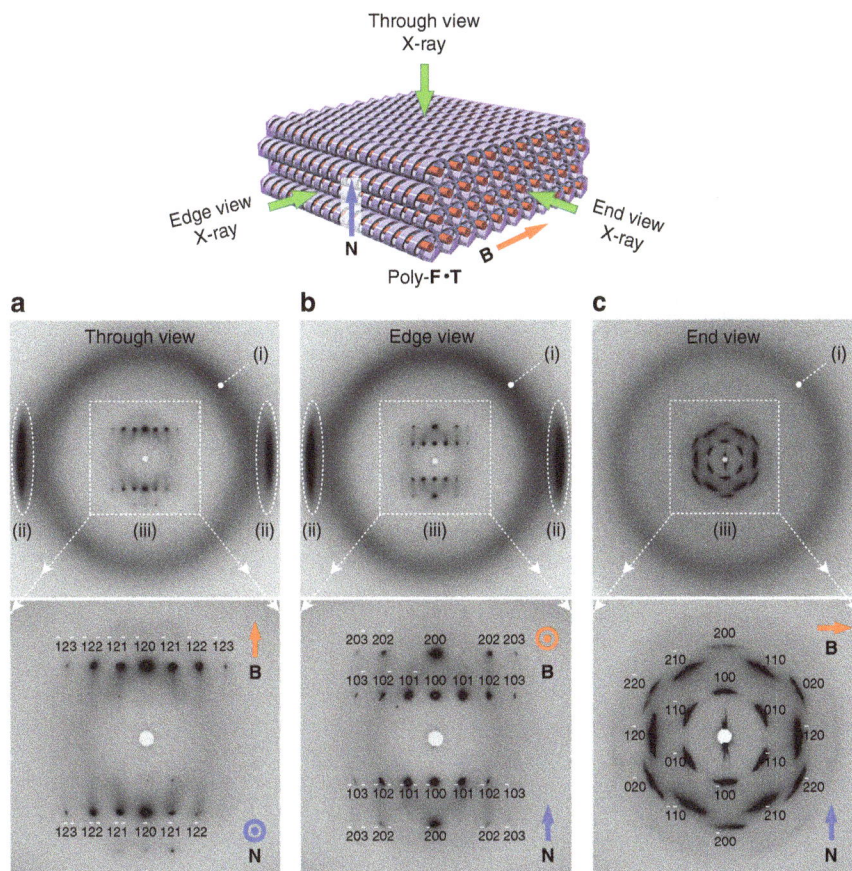

**Figure 4 | 2D X-ray diffraction images of a film of poly-F•T with a macroscopically oriented structure.** A square-shaped film (3 × 3 mm, ∼20 μm thick) was used. (**a**) Through view image (X-ray∥**N**, X-ray⊥**B**). (**b**) Edge view image (X-ray⊥**N**, X-ray∥**B**). (**c**) End view image (X-ray⊥**N**, X-ray⊥**B**). (i), (ii) and (iii) highlight signals due to aliphatic chain packing, π-stacking and 3D lattice, respectively. **B**, Magnetic field applied during the LC film-preparation process. **N**, Normal vector of the film surface.

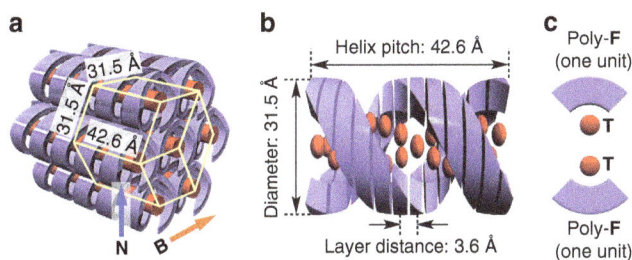

**Figure 5 | Schematic of the structure of poly-F•T.** (**a**) Hexagonally packed helical columns (diameter, 31.5 Å) with structural periodicity along the column axis (distance, 42.6 Å). (**b**) Supramolecular duplex helices formed by the stacking of the repeating unit (shown in **c**). (**c**) Repeating unit consisting of two salt pairs of poly-**F** and **T**. **B**, Magnetic field applied during the LC film-preparation process. **N**, Normal vector of the film surface.

orients their columnar aggregates perpendicular to the magnetic field. Simultaneously, the columns are induced to lie along the surface of the film, most likely because of the interaction of the columns' aliphatic side chains with the air interface during the drop-casting process. When the magnetic field is applied along the substrate plane, the magnetic and surface effects cooperate to produce a unidirectional orientation[34]. This hypothesis is consistent with the observation that a magnetic field applied perpendicular to the substrate plane had no effect on the structural ordering (Fig. 2d), where both the magnetic and surface effects only defined the columns to orient within the substrate

plane and did not restrict their in-plane rotation. We also confirmed that the present magnetic orientation had no effect on the microstructure of the polymer framework; as shown by its 1D SAXS profiles, a film of poly-**F**•**T** with a randomly oriented structure, prepared in the absence of magnet, had the same lattice as that of a magnetically structured film (Supplementary Fig. 6).

**Use of the framework as a solid-state host.** By immersing a film of poly-**F**•**T** in acidified ethanol at 20 °C for 10 h, **T** can be quantitatively desorbed from the framework of poly-**F** (Fig. 1, iv), owing to the lack of covalent interactions between **T** and poly-**F** (Supplementary Fig. 13 and Supplementary Table 2). The resultant film, which consisted exclusively of guest-free polymer framework (poly-**F**•*vacant*), showed infrared absorption characteristic of a free carboxyl acid (–$CO_2H$), while the absorption associated with the carboxylate ion (–$CO_2^-$) disappeared (Supplementary Fig. 15b). Through the desorption of **T**, the macroscopic orientation of the polymer framework was well preserved (Supplementary Fig. 16b). However, the hexagonal columns shrunk in diameter from 31.5 to 30.0 Å (Supplementary Fig. 17c) and lost the periodicity of helical pitch (Supplementary Fig. 17a,b), revealing the flexible nature of the polymer framework[27].

Because poly-**F**•*vacant* contains hollow pores featuring many –$CO_2H$ groups, the guest-free film readily incorporates various basic or cationic guest molecules **G₁**–**G₇** (Fig. 1, v). For example, when a film of poly-**F**•*vacant* was immersed at 20 °C for 8 h in a methanolic solution of an amine bearing a *p*-nitroaniline moiety (**G₁**)[46], the film adsorbed ∼0.9 equivalents of **G₁** with respect to the $-CO_2H$ group content of the film (Supplementary Fig. 14 and Supplementary Table 2). The resultant yellow film (Fig. 3d) retained its macroscopic orientation, as confirmed by POM (Fig. 3e). Changes in infrared absorption indicate that this guest binding is driven by salt-pair formation (Supplementary Fig. 15c). Amine $G_2$ bearing a fluorescent moiety and amine $G_3$ bearing a stabilized radical as well as alkali metal ions ($G_4$–$G_7$) were also incorporated as guests in the polymer framework through salt-pair formation with retention of the macroscopic orientation (Supplementary Figs 15 and 16).

As a representative example of a guest-exchanged film, that of poly-$F•G_1$ was investigated in more detail. In polarized infrared spectroscopy, its absorption at $1,369\ cm^{-1}$ ($-CO_2^-$) and $1,540\ cm^{-1}$ ($-NO_2$) was clearly dependent on the light-polarization angle (Fig. 3f), indicating that the molecular units of $F$ and $G_1$ are anisotropically positioned in the macroscopically oriented helical pores. After incorporation of $G_1$, the polymer framework retained its hexagonal columnar packing, and, moreover, recovered the periodicity of helical pitch. In fact, the 2D X-ray diffraction patterns of poly-$F•G_1$ (Supplementary Fig. 18) were quite similar to those of poly-$F•T$ (Fig. 4) for all three views. In addition, lattice parameters of poly-$F•G_1$ (column diameter $= 31.4\ \text{Å}$, helical pitch $= 42.8\ \text{Å}$) were essentially identical to those of poly-$F•T$ (column diameter $= 31.5\ \text{Å}$, helical pitch $= 42.6\ \text{Å}$), despite the different molecular shapes of $G_1$ and $T$. Such consistency in lattice parameters suggests that a fundamental skeleton of the helical pores might be preserved in poly-$F$ even after the helical pitch disordering because of the desorption of $T$ (Supplementary Fig. 17). It is probable that molecules of $G_1$ filled the spaces originally occupied by $T$ molecules, thereby reducing the structural strain in the polymer framework and recovering the helical pitch[27]. From these observations, we also deduced that the molecules of $G_1$ in the polymer framework were present in a helical arrangement, as in the case of $T$. In this case, a chiral structure emerged from achiral components ($F$ and $G_1$) because of the chiral template effect of $T$ (ref. 47).

Owing to their push–pull-substituted aromatic systems, $p$-nitroaniline derivatives such as $G_1$ are known to exhibit the second-order nonlinear optical (NLO) properties, when they are arrayed in a noncentrosymmetric ($=$ polar, chiral or both) arrangement[46,48]. Because the molecules of $G_1$ confined in poly-$F$ are likely to align in helical arrays, as described above, they potentially show second-order NLO properties[1–4]. Although there have been reported several host systems that induce the NLO output of chromophore guests, most of them contain the contribution of polar structures[46,49]. Contrary to them, our polymer framework with a chiral and apolar space group ($P6_122$ or $P6_522$) would afford NLO output genuinely because of chirality, which is suitable for pursuing nonlinear phenomena in chiral architectures. Furthermore, the macroscopic orientation of poly-$F$ would be expected to have positive effects on the NLO output because the NLO signals generated from the ordered structure should be less prone to mutual cancellation and therefore coherent.

To confirm this possibility, we measured the SHG of the film of poly-$F•G_1$ (Fig. 6a and Supplementary Fig. 19). We used circularly polarized light (CPL) with right- or left-handedness as a fundamental beam, so that we could determine, independently of any effect of birefringence, whether the origin of the SHG was the chiral arrangement induced by the polymer framework or whether it arose from coincidental polar ordering[50]. To detect the SHG circular dichroism clearly, the incident angle was set to $45°$, and the s- and p-polarized contents of the SHG output were separately monitored (Methods)[50]. When right-handed CPL (wavelength, 800 nm) was irradiated to a 5-$\mu$m-thick film of poly-$F•G_1$ with a magnetically oriented structure (Fig. 6a, red), strong SHG output was observed for both

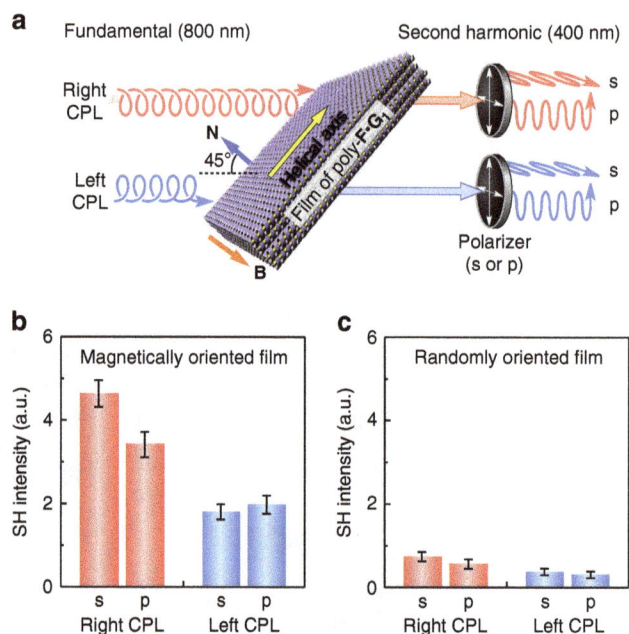

Figure 6 | SHG circular dichroism measurement of films of poly-$F•G_1$. (a) Optical set-up (for details, see Methods and Supplementary Fig. 19). As a fundamental beam, right- (red) or left-handed (blue) CPL (800 nm wavelength) was incident at $45°$ to a film of poly-$F•G_1$ with magnetically oriented structure. The film was set so that the direction of the magnetic field that was applied during the LC film-formation process was perpendicular to the incident plane, that is, the helical axes of the columnar objects were parallel to the incident plane. Generated frequency-doubled light (400 nm wavelength) was detected from the transmitted direction by using an s- or p-analyser. (b,c) SHG output power for a film of poly-$F•G_1$ ($\sim$5 $\mu$m thick) with a magnetically oriented structure (b) and for an analogous film ($\sim$5 $\mu$m thick) with a randomly oriented structure (c). Each error bar represents the s.d. of 100 replicate measurements. B, Magnetic field applied during the LC film-preparation process. N, Normal vector of the film surface.

of the s- and p-polarized contents (Fig. 6b, red). On switching the handedness of the CPL from right to left (Fig. 6a, blue), the intensity of the SHG reduced by half (Fig. 6b, blue), proving that the SHG observed here is because of the chiral arrangement of $G_1$. As we had conjectured, the macroscopic orientation of the framework had a significant effect on the SHG output. Indeed, the SHG intensity of the film with magnetically oriented structure (Fig. 6b) was seven times greater than that of an analogous 5-$\mu$m-thick film with a randomly oriented structure (Fig. 6c).

## Discussion

Although many types of ordered porous materials, such as zeolites, porous silicates and metal–organic frameworks, have been developed so far, our new polymer framework (poly-$F$), prepared by *in situ* crosslinking of a chiral supramolecular LC salt ($F•T$) preorganized in a magnetic field (Fig. 1), is the first porous material that realizes helicity with controlled handedness[11–17] and macroscopic orientation[18–22,31–40] at the same time. Our framework, capable of incorporating various basic or cationic guest molecules and restricting their positions, is expected to serve as a 'universal' scaffold for arraying molecules into a helical and macroscopically oriented structure. The resulting arrangement of the guest molecules might be suitable for inducing molecular events to proceed in a directionally controlled and mutually correlated manner, as exemplified by the SHG of a NLO chromophore ($G_1$) demonstrated in this work.

Further potential applications of our framework include CPL-emitting devices, piezoelectric materials, chiral magnets, anisotropic ion-conductive materials or membranes for chiral separation.

## Methods

**General.** POM was performed on a Nikon model Eclipse LV100POL optical polarizing microscope equipped with a Mettler Toledo model FP90 Central Processor connected with a model FP 82HT hot stage. $^{60}$Co γ-ray irradiation was carried out in the Takasaki Advanced Radiation Research Institute of Japan Atomic Energy Agency. Film thicknesses were measured using a Mitutoyo model MDQ-30M micrometer. HPLC analysis was performed on a JASCO model PU-980 intelligent HPLC pump equipped with a model UV-970 intelligent ultraviolet–vis detector. Elemental analysis was performed on a Yanaco CHN CORDER MT-6 elemental analyser. $^{1}$H NMR spectra were measured on a JEOL model NM-Excalibur NM-500 spectrometer operated at 500 MHz. Infrared spectra were measured on a JASCO model FT/IR-4100 Fourier transform infrared spectrometer with a model ATR PRO450-S-attenuated total reflection equipment. X-ray crystallographic study was carried out using a Rigaku AFC-8 diffractometer with graphite monochromated Mo Kα radiation at 300 K.

**Materials.** 3,4,5-Tris(11-acryloyloxyundecyloxy)benzoic acid (**F**)[24], (1R,2R)-pseudonorephedrine (**T′**)[51], 3,5-dimethoxybenzoic acid (**F′**)[52], N-(2-aminoethyl)-p-nitroaniline (**G₁**)[46] and N-(7-nitrobenz-2-oxa-1,3-diazol-4-yl)aminoethylamine (**G₂**)[53] were prepared according to literature methods. Water was obtained from a Millipore model Milli-Q integral water purification system. (1R,2S)-Norephedrine (**T**) was purchased from TCI and purified by distillation. Other reagents were used as received from Kanto (CH₂Cl₂, EtOH, MeOH and HClO₄ (60% aqueous solution)), Sigma-Aldrich (Cs₂CO₃ (**G₇**)), TCI (4-amino-2,2,6,6-tetramethylpiperidine-1-oxyl (**G₃**)) and Wako (HCO₂H, hexane, Li₂CO₃ (**G₄**), K₂CO₃ (**G₆**) and Na₂CO₃ (**G₅**)).

**Synthesis of a film of poly-F•T with magnetically oriented structure.** A solution of salt **F•T** (24.9 mg, 25 μmol) in CH₂Cl₂ (500 μl) was cast on a glass substrate (2.5 × 7.5 cm). The glass substrate was placed inside a pair of loosely closed Petri dishes that were immediately inserted into the bore of a superconducting magnet with its 10-T field oriented parallel to the substrate plane. The cast solution of **F•T** was carefully and slowly concentrated to dryness over 2 h at 20 °C. The resultant material was further dried under reduced pressure (∼1 mm Hg) at 20 °C overnight to remove residual volatiles to give an LC film of **F•T** with a magnetically oriented structure. To achieve in situ polymerization of the acryloyl groups in **F**, the LC film on the glass substrate was placed in a glass tube equipped with a three-way cock. The glass tube was evacuated by means of a rotary oil pump (∼1 mm Hg) at 20 °C for 1 h and was purged with argon. The film in the glass tube was then exposed to $^{60}$Co γ-ray (6.25 kGy h$^{-1}$) at 20 °C for 16 h so that the LC film was converted into a film of poly-**F•T** with a magnetically oriented structure. The film was immersed in CH₂Cl₂ at 20 °C for a few minutes and then detached from the glass substrate. By using a micrometre gauge, the thickness of the film was estimated to be 10–15 μm. The thickness of the film could be tuned by adjusting the concentration of the **F•T** solution and the drop-casting area in the LC film-preparation process. Thicknesses of film samples used for the measurements are summarized in Supplementary Table 1.

**Synthesis of a film of poly-F•T with randomly oriented structure.** Salt **F•T** was sandwiched between two glass plates and warmed to 130 °C so that the salt turned into an isotropic melt. It was then allowed to cool to 20 °C at 5.0 °C min$^{-1}$ and then left to stand at 20 °C overnight to give an LC film with a randomly oriented structure. For in situ polymerization of the acryloyl groups in **F**, the LC film sandwiched between the two glass plates was subjected to $^{60}$Co γ-ray irradiation, as described above, to give a film of poly-**F•T** with a randomly oriented structure.

**Desorption of T from a film of poly-F•T.** A film of poly-**F•T** (20.1 mg, containing 20.2 μmol of **T** and the –CO₂H groups) was immersed in a 7.5 M solution of HCO₂H in EtOH (50 ml) at 20 °C for 10 h. The resultant mixture was separated into the film and the supernatant. The film was then washed with EtOH (5 ml) and dried under reduced pressure (∼1 mm Hg) to give a film of poly-**F•vacant**. For the quantification of desorbed **T**, the supernatant and the washings were combined and concentrated to dryness, dissolved in a pH 2.0 aqueous solution of HClO₄ (4.00 ml) containing L-tyrosine (3.0 mM) as an internal standard, and subjected to HPLC. The concentration of **T** was calculated from the data of authentic solutions of **T** (Supplementary Fig. 13). Column, Daicel CROWNPAK CR (+) (4.6 × 153 mm); eluent, pH 2.0 aqueous HClO₄; temperature, 20 °C; flow rate, 0.60 ml min$^{-1}$; injection volume, 5.0 μl; detection, UV absorption at 200 nm; elution time, 8.9 (L-tyrosine) and 13.3 min (**T**).

**Adsorption of G₁–G₇ by films of poly-F•vacant.** A film of poly-**F•vacant** (10.0 mg, containing 11.8 μmol of the –CO₂H groups) was immersed in a 2.5 mM

solution of **G₁** in MeOH (5 ml) at 20 °C for 8 h. From the resultant mixture, the film was separated, washed with MeOH (2.5 ml) and dried under reduced pressure (∼1 mm Hg) to afford a film of poly-**F•G₁**. For the quantification of the uptake of **G₁**, the film of poly-**F•G₁** was immersed in a 7.5-M solution of HCO₂H in EtOH (25 ml) at 20 °C for 10 h. From the resultant mixture, the supernatant was separated, while the film was washed with EtOH (2.5 ml). The supernatant and the washings were combined and concentrated to dryness, dissolved in a pH 2.0 aqueous solution of HClO₄ (2.00 ml) containing L-tyrosine (3.0 mM) as an internal standard and subjected to HPLC analysis. The concentration of **G₁** was calculated from the data of authentic solutions of **G₁** (Supplementary Fig. 14). Column, Daicel CROWNPAK CR (+) (4.6 × 153 mm); eluent, pH 2.0 aqueous HClO₄; temperature, 20 °C; flow rate, 1.00 ml min$^{-1}$; injection volume, 5.0 μl; detection, ultraviolet absorption at 228 nm; elution time, 5.4 (L-tyrosine) and 39.3 min (**G₁**).

The other amino guests (**G₂** and **G₃**) were adsorbed by similar procedures. Alkali metal ion guests (**G₄–G₇**) were adsorbed from a mixture of MeOH and water (90:10, v/v) at 70 °C for 8 h.

**Synchrotron 2D X-ray diffraction measurement at SPring-8 BL45XU.** 2D X-ray diffraction measurement of the polydomain films and powders of poly-**F•T** and poly-**F•G₁** was carried out at BL45XU in SPring-8 (Hyogo, Japan)[54]. Diffraction data were collected using a Rigaku imaging plate area detector model R-AXIS IV + +. The incident X-ray beam (1.00 Å wavelength) was monochromated by a diamond (1 1 1) double-crystal monochromator. The sample-to-detector distance was 0.40 m. Scattering vector q and position of an incident X-ray beam on the detector were calibrated using several orders of layer reflections from silver behenate (d = 58.380 Å). Film samples were mounted on nylon fibres (φ = 0.2 mm) and exposed at 20 °C to an X-ray beam for 30 (edge and end views) or 300 s (through view). Powder samples were placed into 1.5-mm-φ glass capillaries and exposed at 20 °C to an X-ray beam for 50 s.

**Synchrotron 2D X-ray diffraction measurement at SPring-8 BL26B2.** 2D X-ray diffraction measurement for the determination of the space group of poly-**F•T** was carried out at BL26B2 in SPring-8 (ref. 55). Diffraction data were collected using a MarMosaic225 detector. The incident X-ray beam (1.0000 Å wavelength) was monochromated by a Si (1 1 1) double-crystal monochromator. The sample-to-detector distance was 500 mm. A monodomain clump (0.7 × 0.6 × 0.2 mm) trimmed from a film of poly-**F•T** was held on a MicroMounts (MiTeGen) that was attached to a brass pin (Supplementary Fig. 8a), mounted on a goniometer head, and exposed to an X-ray beam at 27 °C. A total of 18 frames of data were collected with an oscillation range of 10° and an exposure time of 10 s for each frame (Supplementary Fig. 8b). The total oscillation angle was 180°.

**TEM measurement.** TEM measurement was performed on a JEOL model JEM-2100F/SP operated at 200-kV accelerating voltage. Films of poly-**F•T** were embedded in EPON812 (TAAB Laboratories Equipment) and sectioned at 20 °C with a diamond knife mounted on a Leica Ultracut UCT ultramicrotome. The resultant sections (∼60 nm thick) were floated on water, retrieved on TEM grids, stained in vapour of a 0.5% aqueous solution of RuO₄ (TAAB Laboratories Equipment) at 20 °C for 10 min and subjected to TEM observation.

**SHG circular dichroism measurement.** SHG circular dichroism measurement[50] was performed with planar freestanding films of poly-**F•G₁**. For simplicity, films thinner than the coherent length (5 μm thick) were used. As the fundamental beam, right- and left-handed CPLs were created by using 800-nm light of a titanium:sapphire laser (averaged power, 200 mW; duration, 200 fs; repetition, 80 MHz). The laser beam was successively passed through a polarizer and a quarter-wave plate, where the rotation angle of the quarter-wave plate with respect to the polarizer was set to be 45° or 225° for the right-handed CPL and 135° or 315° for the left-handed CPL. Thus, created CPL was guided into a sample with 45° incidence. In the case of film with a magnetically oriented structure, the film was set so that the direction of the magnetic field that was applied during the LC film-formation process was perpendicular to the incident plane, that is, the helical axes of the columnar objects were parallel to the incident plane. SHG signals were detected in the transmission direction by a Hamamatsu model H7421-40 cooled photomultiplier tube after successively passing through an infrared-cut filter, an analyser (s or p), and a 400-nm band pass filter. The signals were recorded by a photon-counting system C8855-01 for a window width of 1 s. For optical scheme, see Supplementary Fig. 19.

## References

1. Evans, O. R. & Lin, W. Crystal engineering of NLO materials based on metal–organic coordination networks. *Acc. Chem. Res.* **35**, 511–522 (2002).
2. Verbiest, T. & Persoons, A. In *Materials-Chirality: Vol. 24 of Topics in Stereochemistry.* (eds Green, M. M., Nolte, R. J. M. & Meijer, E. W.) Chapter 9 (Wiley, 2003).
3. Verbiest, T. et al. Strong enhancement of nonlinear optical properties through supramolecular chirality. *Science* **282**, 913–915 (1998).

4.  Gonella, G. *et al.* Control of the orientational order and nonlinear optical response of the "push-pull" chromophore RuPZn via specific incorporation into densely packed monolayer ensembles of an amphiphilic 4-helix bundle peptide: second harmonic generation at high chromophore densities. *J. Am. Chem. Soc.* **132**, 9693–9700 (2010).

5.  Eckert, T., Finkelmann, H., Keck, M., Lehmann, W. & Kremer, F. Piezoelectricity of mechanically oriented $S_C$*-elastomers. *Macromol. Rapid Commun.* **17**, 767–773 (1996).

6.  Jaworek, T., Neher, D., Wegner, G., Wieringa, R. H. & Schouten, A. J. Electromechanical properties of an ultrathin layer of directionally aligned helical polypeptides. *Science* **279**, 57–60 (1998).

7.  Haussuhl, S., Bohaty, L. & Becker, P. Piezoelectric and elastic properties of the nonlinear optical material bismuth triborate, $BiB_3O_6$. *Appl. Phys. A* **82**, 495–502 (2006).

8.  Lee, B. Y. *et al.* Virus-based piezoelectric energy generation. *Nat. Nanotechnol.* **7**, 351–356 (2012).

9.  Tagami, K., Tsukada, M., Wada, Y., Iwasaki, T. & Nishide, H. Electronic transport of benzothiophene-based chiral molecular solenoids studied by theoretical simulations. *J. Chem. Phys.* **119**, 7491–7497 (2003).

10. Goh, M., Matsushita, S. & Akagi, K. From helical polyacetylene to helical graphite: synthesis in the chiral nematic liquid crystal field and morphology-retaining carbonization. *Chem. Soc. Rev.* **39**, 2466–2476 (2010).

11. Miyata, M., Tohnai, N. & Hisaki, I. Crystalline host–guest assemblies of steroidal and related molecules: diversity, hierarchy, and supramolecular chirality. *Acc. Chem. Res.* **40**, 694–702 (2007).

12. Che, S. *et al.* Synthesis and characterization of chiral mesoporous silica. *Nature* **429**, 281–284 (2004).

13. Yu, J. & Xu, R. Chiral zeolitic materials: structural insights and synthetic challenges. *J. Mater. Chem.* **18**, 4021–4030 (2008).

14. Bradshaw, D., Claridge, J. B., Cussen, E. J., Prior, T. J. & Rosseinsky, M. J. Design, chirality, and flexibility in nanoporous molecule-based materials. *Acc. Chem. Res.* **38**, 273–282 (2005).

15. Percec, V. *et al.* Principles of self-assembly of helical pores from dendritic dipeptides. *Proc. Natl Acad. Sci. USA* **103**, 2518–2523 (2006).

16. Miyagawa, T., Yamamoto, M., Muraki, R., Onouchi, H. & Yashima, E. Supramolecular helical assembly of an achiral cyanine dye in an induced helical amphiphilic poly(phenylacetylene) interior in water. *J. Am. Chem. Soc.* **129**, 3676–3682 (2007).

17. Gan, Q. *et al.* Helix-rod host-guest complexes with shuttling rates much faster than disassembly. *Science* **331**, 1172–1175 (2011).

18. Feng, S. & Bein, T. Growth of oriented molecular sieve crystals on organophosphonate films. *Nature* **368**, 834–836 (1994).

19. Lee, J. S., Lee, Y.-J., Tae, E. L., Park, Y. S. & Yoon, K. B. Synthesis of zeolite as ordered multicrystal arrays. *Science* **301**, 818–821 (2003).

20. Tolbert, S. H., Firouzi, A., Stucky, G. D. & Chmelka, B. F. Magnetic field alignment of ordered silicate-surfactant composites and mesoporous silica. *Science* **278**, 264–268 (1997).

21. Melosh, N. A., Davidson, P., Feng, P., Pine, D. J. & Chmelka, B. F. Macroscopic shear alignment of bulk transparent mesostructured silica. *J. Am. Chem. Soc.* **123**, 1240–1241 (2001).

22. Miyata, H. *et al.* Remarkable birefringence in a $TiO_2$-$SiO_2$ composite film with an aligned mesoporous structure. *J. Am. Chem. Soc.* **133**, 13539–13544 (2011).

23. Wulff, G. Enzyme-like catalysis by molecularly imprinted polymers. *Chem. Rev.* **102**, 1–27 (2002).

24. Smith, R. C., Fischer, W. M. & Gin, D. L. Ordered poly(*p*-phenylenevinylene) matrix nanocomposites via lyotropic liquid-crystalline monomers. *J. Am. Chem. Soc.* **119**, 4092–4093 (1997).

25. Lee, H.-K. *et al.* Synthesis of a nanoporous polymer with hexagonal channels from supramolecular discotic liquid crystals. *Angew. Chem. Int. Ed.* **40**, 2669–2671 (2001).

26. Courty, S., Tajbakhsh, A. R. & Terentjev, E. M. Stereo-selective swelling of imprinted cholesteric networks. *Phys. Rev. Lett.* **91**, 085503 (2003).

27. Ishida, Y., Amano, S., Iwahashi, N. & Saigo, K. Switching of structural order in a cross-linked polymer triggered by the desorption/adsorption of guest molecules. *J. Am. Chem. Soc.* **128**, 13068–13069 (2006).

28. Amano, S., Ishida, Y. & Saigo, K. Solid-state hosts by the template polymerization of columnar liquid crystals: locked supramolecular architectures around chiral 2-amino alcohols. *Chem. Eur. J.* **13**, 5186–5196 (2007).

29. Broer, D. J., Bastiaansen, C. M. W., Debije, M. G. & Schenning, A. P. H. J. Functional organic materials based on polymerized liquid-crystal monomers: supramolecular hydrogen-bonded systems. *Angew. Chem. Int. Ed.* **51**, 7102–7109 (2012).

30. van Kuringen, H. P. C., Eikelboom, G. M., Shishmanova, I. K., Broer, D. J. & Schenning, A. P. H. J. Responsive nanoporous smectic liquid crystal polymer networks as efficient and selective adsorbents. *Adv. Funct. Mater.* **24**, 5045–5051 (2014).

31. Yoshio, M. *et al.* One-dimensional ion-conductive polymer films: alignment and fixation of ionic channels formed by self-organization of polymerizable columnar liquid crystals. *J. Am. Chem. Soc.* **128**, 5570–5577 (2006).

32. Feng, X. *et al.* Scalable fabrication of polymer membranes with vertically aligned 1 nm pores by magnetic field directed self-assembly. *ACS Nano* **8**, 11977–11986 (2014).

33. Shklyarevskiy, I. O. *et al.* High anisotropy of the field-effect transistor mobility in magnetically aligned discotic liquid-crystalline semiconductors. *J. Am. Chem. Soc.* **127**, 16233–16237 (2005).

34. Kim, H.-S. *et al.* Uniaxially oriented, highly ordered, large area columnar superstructures of discotic supramolecules using magnetic field and surface interactions. *Adv. Mater.* **20**, 1105–1109 (2008).

35. Van Winkle, D. H. & Clark, N. A. Freely suspended strands of tilted columnar liquid-crystal phases: one-dimensional nematics with orientational jumps. *Phys. Rev. Lett.* **48**, 1407–1410 (1982).

36. Mas-Torrent, M., den Boer, D., Durkut, M., Hadley, P. & Schenning, A. P. H. J. Field effect transistors based on poly(3-hexylthiophene) at different length scales. *Nanotechnology* **15**, S265–S269 (2004).

37. Gibbons, W. M., Shannon, P. J., Sun, S. T. & Swetlin, B. J. Surface-mediated alignment of nematic liquid crystals with polarized laser light. *Nature* **351**, 49–50 (1991).

38. van de Craats, A. M. *et al.* Meso-epitaxial solution-growth of self-organizing discotic liquid-crystalline materials. *Adv. Mater.* **15**, 495–499 (2003).

39. Liu, C.-Y. & Bard, A. J. In-situ regrowth and purification by zone melting of organic single-crystal thin films yielding significantly enhanced optoelectronic properties. *Chem. Mater.* **12**, 2352–2362 (2000).

40. Tracz, A. *et al.* Uniaxial alignment of the columnar super-structure of a hexa (alkyl) hexa-*peri*-hexabenzocoronene on untreated glass by simple solution processing. *J. Am. Chem. Soc.* **125**, 1682–1683 (2003).

41. Hohn, W. & Tieke, B. γ-ray polymerization of mesogenic diacrylates in the crystalline, liquid crystalline and liquid state – a kinetic study. *Macromol. Chem. Phys.* **197**, 821–831 (1996).

42. Wang, M.-D., Nakanishi, E. & Hibi, S. Effect of molecular weight on rolled high density polyethylene: 1. Structure, morphology and anisotropic mechanical behavior. *Polymer* **34**, 2783–2791 (1993).

43. Pisula, W., Feng, X. & Müllen, K. Tuning the columnar organization of discotic polycyclic aromatic hydrocarbons. *Adv. Mater.* **22**, 3634–3649 (2010).

44. Percec, V. *et al.* Self-assembly of dendronized Perylene bisimides into complex helical columns. *J. Am. Chem. Soc.* **133**, 12197–12219 (2011).

45. Ishida, Y. *et al.* Tunable chiral reaction media based on two-component liquid crystals: regio-, diastereo-, and enantiocontrolled photodimerization of anthracenecarboxylic acids. *J. Am. Chem. Soc.* **132**, 17435–17446 (2010).

46. Prakash, M. J. & Radhakrishnan, T. P. Second harmonic generation from a homologous series of molecular crystals: impact of supramolecular interactions. *Chem. Mater.* **18**, 2943–2949 (2006).

47. Choi, S. S., Morris, S. M., Huck, W. T. S. & Coles, H. J. Simultaneous red–green–blue reflection and wavelength tuning from an achiral liquid crystal and a polymer template. *Adv. Mater.* **22**, 53–56 (2010).

48. Verbiest, T., Houbrechts, S., Kauranen, M., Clays, K. & Persoons, A. Second-order nonlinear optical materials: recent advances in chromophore design. *J. Mater. Chem.* **7**, 2175–2189 (1997).

49. Caro, J., Marlow, F. & Wübbenhorst, M. Chromophore–zeolite composites: the organizing role of molecular sieves. *Adv. Mater.* **6**, 413–416 (1994).

50. Araoka, F. *et al.* Twist-grain-boundary structure in the B4 phase of a bent-core molecular system identified by second harmonic generation circular dichroism measurement. *Phys. Rev. Lett.* **94**, 137801 (2005).

51. Groeper, J. A., Hitchcock, S. R. & Ferrence, G. M. A scalable and expedient method of preparing diastereomerically and enantiomerically enriched pseudonorephedrine from norephedrine. *Tetrahedron* **17**, 2884–2889 (2006).

52. Song, Y. M. *et al.* Synthesis of novel azo-resveratrol, azo-oxyresveratrol and their derivatives as potent tyrosinase inhibitors. *Bioorg. Med. Chem. Lett.* **22**, 7451–7455 (2012).

53. Cotté, A., Bader, B., Kuhlmann, J. & Waldmann, H. Synthesis of the N-terminal lipohexapeptide of human $G_{\alpha O}$-protein and fluorescent-labeled analogues for biological studies. *Chem. Eur. J.* **5**, 922–936 (1999).

54. Fujisawa, T. *et al.* Small-angle X-ray scattering station at the SPring-8 RIKEN beamline. *J. Appl. Crystallogr.* **33**, 797–800 (2000).

55. Ueno, G. *et al.* RIKEN structural genomics beamlines at the SPring-8; high throughput protein crystallography with automated beamline operation. *J. Struct. Funct. Genomics* **7**, 15–22 (2006).

## Acknowledgements

C.L. thanks the Japanese Government Scholarship (MEXT, Japan) and the RIKEN Junior Research Associate Fellowship. This work was supported by ImPACT Program of Council for Science, Technology and Innovation (Cabinet Office, Government of Japan) and JSPS Grant-in-Aid for Scientific Research (B), Grant Number 15H03820. The synchrotron X-ray diffraction experiments were performed at BL45XU and BL26B1 in SPring-8 with the approval of RIKEN SPring-8 Center (proposal 20140073). We thank

T. Tose for the synthesis of **F**, M. Takata, T. Hikima and G. Ueno for their support in the X-ray diffraction measurements at SPring-8, T. Kikitsu and D. Inoue for the transmission electron microscopy measurement and S. Amano and K. Saigo for the X-ray crystallographic study.

## Author contributions

Y.I. conceived the project; C.L., J.C. and Y.I. designed, performed and analysed experiments; C.L., J.C. and K.Y. synthesized compounds; D.H. and Y.I. measured and analysed the X-ray diffraction data; F.A. and H.T. measured the SHG; T.A. and Y.I. co-wrote the paper.

# Chemical reaction mechanisms in solution from brute force computational Arrhenius plots

Masoud Kazemi[1] & Johan Åqvist[1]

Decomposition of activation free energies of chemical reactions, into enthalpic and entropic components, can provide invaluable signatures of mechanistic pathways both in solution and in enzymes. Owing to the large number of degrees of freedom involved in such condensed-phase reactions, the extensive configurational sampling needed for reliable entropy estimates is still beyond the scope of quantum chemical calculations. Here we show, for the hydrolytic deamination of cytidine and dihydrocytidine in water, how direct computer simulations of the temperature dependence of free energy profiles can be used to extract very accurate thermodynamic activation parameters. The simulations are based on empirical valence bond models, and we demonstrate that the energetics obtained is insensitive to whether these are calibrated by quantum mechanical calculations or experimental data. The thermodynamic activation parameters are in remarkable agreement with experiment results and allow discrimination among alternative mechanisms, as well as rationalization of their different activation enthalpies and entropies.

[1] Department of Cell and Molecular Biology, Uppsala University, Biomedical Center, Box 596, SE-751 24 Uppsala, Sweden. Correspondence and requests for materials should be addressed to J.Å. (email: aqvist@xray.bmc.uu.se).

The existence of an entropy of activation for chemical reactions is inherent in transition state (TS) theory, where the activated complex is assumed to be in thermodynamic equilibrium with the reactants. For solution reactions where the transmission coefficient can be assumed to be close to unity, this entropy of activation is typically obtained from experimental Arrhenius plots of the logarithm of the rate against inverse temperature. If activation entropies could be reliably predicted theoretically, then such calculations would be very useful for distinguishing between alternative TS structures of similar energy[1]. However, due to the huge number of degrees of freedom involved in solution reaction dynamics, the extensive configurational sampling required to rigorously obtain activation entropies is presently beyond the scope of quantum chemical calculations. While ideal gas rigid-rotor and harmonic oscillator approximations, in combination with parametrized continuum solvent models, are useful for obtaining thermally corrected activation free energy estimates from quantum mechanical calculations, they do not usually provide sufficiently accurate descriptions of entropic effects. Here we explore a combined method where TS structures and energies are obtained from density functional theory (DFT) calculations, which can then be used to parametrize empirical valence bond (EVB) models[2,3] that allow very extensive all-atom sampling of the reacting system in aqueous solution. This method is used to obtain computational Arrhenius plots for the hydrolytic deamination of cytidine and dihydrocytidine, thereby allowing for direct comparisons with experimental thermodynamic activation parameters.

The spontaneous deamination reaction of cytidine to uridine is of major interest due to its importance for genome instabilities, as cytosine is known to be the nucleic acid base that is most susceptible to hydrolytic deamination[4]. The actual reaction mechanism and energetics of the uncatalysed deamination via attack of a water molecule is also highly relevant for assessing the catalytic power of the enzyme cytidine deaminase. This enzyme, which produces uridine and ammonia from cytidine, has been taken as a prototypic example of an enzyme that achieves its catalytic effect primarily by reducing the loss in activation entropy[5,6]. That is, the spontaneous deamination reaction in aqueous solution has been shown to proceed with a large entropy loss of $T\Delta S^{\ddagger} = -8.3\,\text{kcal mol}^{-1}$ at room temperature. The corresponding value for the enzyme-catalysed reaction is $T\Delta S^{\ddagger} = +0.9\,\text{kcal mol}^{-1}$, obtained from the temperature dependence of $k_{\text{cat}}$[5]. Since the entropy effect on substrate binding ($T\Delta S^{\ddagger} = -7.6\,\text{kcal mol}^{-1}$, derived from $K_{\text{m}}$) closely matches the difference in activation entropy this may appear as an example of Jencks' so-called 'Circe effect'[7], which is by many enzymologists considered to be the main explanation of enzyme catalysis. This hypothesis focuses on the substrate configurational entropy, and essentially states that if it is significantly reduced on binding, that could eliminate the entropy loss associated with reaching the TS.

The overall activation free energy for the uncatalysed deamination of cytidine is $30.4\,\text{kcal mol}^{-1}$ at room temperature[5], corresponding to a rate of about $3 \times 10^{-10}\,\text{s}^{-1}$, which is close to the observed rate for spontaneous cytosine deamination per site of single-stranded DNA[8]. Several theoretical studies have addressed the reaction mechanism of cytidine deamination and proposed that formation of a tetrahedral intermediate is the rate-limiting step of the reaction[9–12]. Almatarneh et al. examined the gas-phase reaction of cytosine with water by quantum mechanical calculations, which showed very large energy barriers that can hardly be relevant for the solution reaction[9]. They also explored the reaction pathway for deamination by hydroxide ion, which resulted in a negatively charged cytosine as the reactant state, lying some $65\,\text{kcal mol}^{-1}$ below the $\text{Cyt} + \text{H}_2\text{O} + \text{OH}^-$ starting point[10]. Despite the fact that the predicted activation barrier from the $\text{Cyt}^-$ reactant state was close to the experimentally observed value, this reactant state would not be accessible for the reaction at physiological pH. For the attack by neutral water, Matsubara et al. obtained similar results from their density functional calculations, and also found that an auxiliary water molecule could reduce the potential energy barrier to about $40\,\text{kcal mol}^{-1}$ in the gas phase[11]. Similar results were obtained in a recent work[12], which also reported calculations with continuum solvent models. There, the free energy barriers in solution were predicted to be about $35\,\text{kcal mol}^{-1}$ for cytosine and $31–33\,\text{kcal mol}^{-1}$ for 5,6-dihydrocytosine.

Here we analyse the mechanism and energetics of cytidine and 5,6-dihydrocytidine deamination in water using M06-2X/6-311++G(d,p) DFT calculations[13] with the SMD continuum solvent model[14], which together with experimental data[5,6] serve as an input for extensive EVB simulations[2,3]. The key point with using the latter method is that it can be unambiguously parametrized for different mechanistic pathways and then used for extensive molecular dynamics (MD) sampling and free energy calculations. This allows us for the first time to accurately obtain the temperature dependence of the activation free energy for a solution reaction directly from computer simulations, and thereby decompose the activation barrier into its enthalpic and entropic components. The thermodynamic decomposition of the free energy barrier is not only in excellent agreement with experimental results, but it also allows us to determine the operational mechanism of cytidine deamination. It thus turns out that, both for cytidine and 5,6-dihydrocytidine, the only pathway compatible with all experimental activation parameters is concerted and involves three water molecules in an eight-membered TS.

## Results

**Stepwise mechanisms.** It is clear from the DFT calculations with the SMD solvent model that the deamination of cytidine at neutral pH occurs via the formation of a transient tetrahedral intermediate, resulting from water attack and protonation of the N3 nitrogen. This intermediate is then subsequently deaminated to yield uridine and ammonia (Fig. 1). Formation of the tetrahedral intermediate could go either via a stepwise or a concerted pathway. In the former case, cytidine is protonated at N3 by a water molecule and the resulting hydroxide ion then attacks C4. The energy profile for this two-step pathway was calculated with and without one or two auxiliary water molecules, which do not directly participate in this half of the reaction, but rather act as screening waters. The reaction free energy for the proton transfer (PT) step is predicted to be 15.9 and $14.4\,\text{kcal mol}^{-1}$ with two and three water molecules, respectively (Fig. 1a, Supplementary Table 1). Both of these values are very close to the experimental reaction free energy of $\Delta G^0 = 15.1\,\text{kcal mol}^{-1}$ (at 298 K) that can be estimated from the $pK_a$ values of water (15.7) and 1-methylcytosine (4.6)[15]. The fact that the three-water case appears more favourable is likely due to improved solvation energy of the protonated cytidine-hydroxide ion complex[16]. It can also be noted that the fact that our calculated PT reaction free energies agree well with the corresponding estimate from bulk $pK_a$ values indicate that the entropic cost of bringing the resulting hydroxide ion and protonated cytidine into contact, from a 1 M standard state, is counterbalanced by the favourable electrostatic interaction.

The second step involves nucleophilic attack by hydroxide on the C4 carbon, and the TS for this process was also optimized with and without screening waters (Fig. 1a). Similar to the protonation step, the three-water system gives the lowest energy

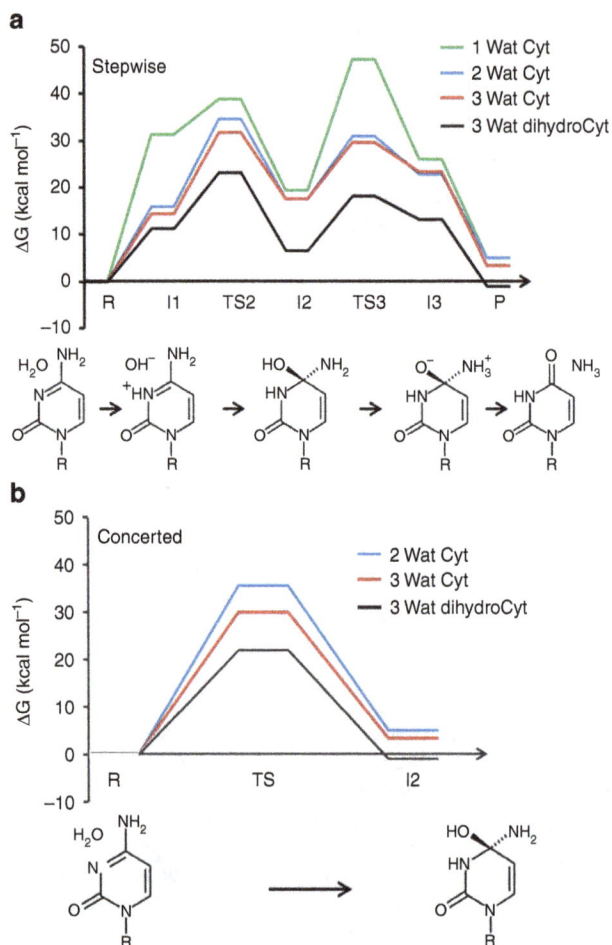

**Figure 1 | Energetics of hydrolytic deamination of cytidine and 5,6-dihydrocytidine.** Calculated free energy profiles (kcal mol$^{-1}$) in aqueous solution for the stepwise (**a**) and concerted (**b**) reaction pathways with the two substrates at the M06-2X/6-311++G**(SMD) level of theory, in the presence of one or two additional explicit water molecules. These do not directly participate in the first half of the stepwise reactions, but mediate PT from the nucleophile to the N3 nitrogen in the concerted cases. The valence bond structures used in subsequent EVB simulations are also shown.

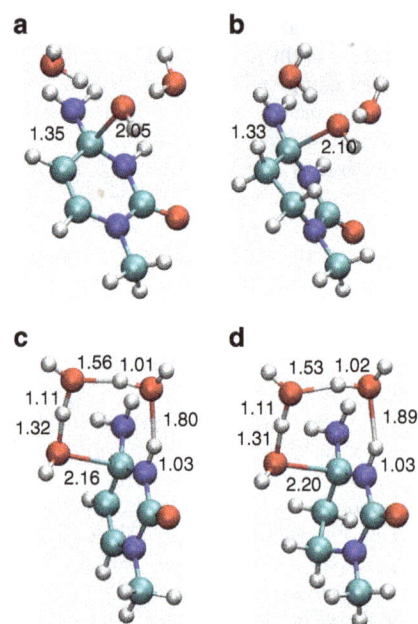

**Figure 2 | TSs for different possible reaction mechanisms.** Optimized rate-limiting transition structures for the stepwise deamination pathways of cytidine (**a**) and 5,6-dihydrocytidine (**b**) and for the corresponding concerted pathways (**c,d**), respectively (the two substrates are capped by methyl groups at N1). The lowest-energy TSs with a total of three explicit water molecules are shown and relevant bond distances are given in Å.

estimate (17.4 kcal mol$^{-1}$) for nucleophilic attack at C4, while the two-water case gives a somewhat higher barrier. Overall, the calculations with three water molecules yield a rate-limiting barrier of 31.8 kcal mol$^{-1}$ for the formation of the tetrahedral intermediate, which is in excellent agreement with the experimentally derived barrier of 30.4 kcal mol$^{-1}$ (ref. 5). The corresponding optimized rate-limiting TS is shown in Fig. 2 and stationary points for the entire reaction are shown in Supplementary Fig. 1. The auxiliary waters thus do not assist directly in any PTs before protonation of the leaving ammonia group in the last TS (Supplementary Fig. 1).

Since the three-water model closely reproduces the available experimental data for cytidine deamination, this model was also used to examine the energetics of the 5,6-dihydrocytidine reaction. The stepwise reaction path resulted in PT to N3 being 11.3 kcal mol$^{-1}$ uphill and a subsequent barrier for the nucleophilic attack of 11.9 kcal mol$^{-1}$ (Fig. 1a, Supplementary Table 1). This yields an overall rate-limiting barrier of 23.2 kcal mol$^{-1}$, which is again in excellent agreement with the experimental barrier of 23.5 kcal mol$^{-1}$ (ref. 6). Furthermore, the optimized TSs for the nucleophilic attack (Fig. 2) are very similar for the two reactants, the C–O bond being marginally longer for

5,6-dihydrocytidine. For both substrates, following the intrinsic reaction coordinate path from TS3 resulted in a local minimum (Supplementary Figs 1 and 2) in which the ammonia group is still bonded to the heterocyclic ring. This intermediate is most stable in the case of 5,6-dihydrocytidine with an activation free energy of 3.3 kcal mol$^{-1}$ for its decomposition (data not shown).

The above results show that the three-water model gives a very good approximation to the activation free energy for both substrates with the stepwise water attack mechanism. Another possible stepwise pathway would be the attack of hydroxide ion on the unprotonated cytidine, which was also examined (data not shown). The predicted activation free energy for such an attack in the presence of two screening waters is 22.8 kcal mol$^{-1}$. This value is, in fact, also in excellent agreement with the corresponding experimental barrier for OH$^{-}$ catalysed deamination of cytidine (24 kcal mol$^{-1}$ at 85 °C and a 55 M standard state, corresponding to the van der Waals contact complex)[17]. However, since the energetic cost of forming a hydroxide ion at pH 7 is about 12 kcal mol$^{-1}$, the overall activation barrier for this type of mechanism appears too high to be compatible with the experimental data. Comparison of the experimental deamination rates at neutral pH and with 1 M KOH also allowed the direct hydroxide mechanism to be ruled out earlier[17].

**Concerted mechanisms.** If the tetrahedral intermediate is formed in a concerted reaction, the protonation at N3 and nucleophilic attack at C4 occur essentially simultaneously. With just a single water molecule (four-membered TS), such a process would involve considerable strain, as was shown by Matsubara et al.[11] who obtained a barrier of 58.6 kcal mol$^{-1}$ in vacuum. These authors, however, also demonstrated a significant barrier reduction down to 39.6 kcal mol$^{-1}$ when a second water participates in the PT chain (six-membered TS). At the M06-2X/6-311++G(d,p)/SMD level of theory, including continuum solvation, we obtain an activation free energy of 35.5 kcal mol$^{-1}$ for the same system (Fig. 1b), with a very similar TS (not shown).

Further relaxation of strain is achieved by a three-water mechanism, corresponding to an eight-membered TS (Fig. 2), for which the activation barrier becomes $29.9\,\text{kcal}\,\text{mol}^{-1}$ (Fig. 1b). Just as for the stepwise mechanism with three waters, this value is extraordinarily close to the experimentally derived value of $30.4\,\text{kcal}\,\text{mol}^{-1}$. The same goes for the concerted reaction with 5,6-dihydrocytidine, where the predicted barrier for the eight-membered TS (Fig. 2) is $22.0\,\text{kcal}\,\text{mol}^{-1}$, while the experimental value is $23.5\,\text{kcal}\,\text{mol}^{-1}$ (ref. 6). It can be noted here that the solution free energy barriers predicted in ref. 12 with three water molecules ($\sim35\,\text{kcal}\,\text{mol}^{-1}$ for cytosine and 31–$33\,\text{kcal}\,\text{mol}^{-1}$ for 5,6-dihydrocytosine), at the B3LYP/6-31G(d,p) level of theory with PCM and SMD, correspond to a different type of TS (six-membered) where the third water molecule is dangling rather than participating in an eight-membered TS.

Taken together, our results show that the stepwise and concerted mechanisms have very similar activation energies and that the formation of the tetrahedral intermediate is indeed rate limiting. This also holds for both of the examined substrates and the predicted activation free energies are in good agreement with the available experimental data. However, the experimental Arrhenius plots for spontaneous deamination of cytidine and 5,6-dihydrocytidine give additional and highly interesting information with regard to the reaction energetics[5,6]. That is, in the case of cytidine, the entropy contribution ($-T\Delta S^{\ddagger} = 8.3\,\text{kcal}\,\text{mol}^{-1}$) to the activation free energy at $25\,°C$ is about one-third of the corresponding enthalpy contribution ($\Delta H^{\ddagger} = 22.1\,\text{kcal}\,\text{mol}^{-1}$). For 5,6-dihydrocytidine, on the other hand, the corresponding values are $-T\Delta S^{\ddagger} = 10.1$ and $\Delta H^{\ddagger} = 13.4\,\text{kcal}\,\text{mol}^{-1}$. The activation enthalpy is thus considerably smaller than for cytidine and the entropic contribution to the free energy barrier is now almost as large.

**Arrhenius plots from EVB simulations**. To examine to what extent the different alternative deamination mechanisms can explain the above activation enthalpy–entropy partitioning, we turned to MD/EVB simulations of the chemical reactions. It is necessary here to be able to obtain enthalpy and entropy changes for the entire solute–solvent system, to make the connection to the experiment. Without resorting to simplified approximations for the solutes and solvent, as discussed above, the only way is to carry out very extensive configurational sampling of a fully microscopic system. The EVB model[2,3] is ideally suited for this purpose since the reaction surface is parametrized by mixing the key valence bond structures describing the reaction (Fig. 1). Each of these is represented by an analytical force field, which makes the calculations very fast, and allows convergent free energy profiles to be obtained along the relevant reaction paths (for the two reactions considered here, the reported results correspond to over $3\,\mu s$ of MD simulation). The free energy profiles are obtained using a standard umbrella sampling technique[2,3,18]. Since we have reliable and similar free energies for the rate-limiting step of the reaction, both from experiments and the DFT calculations, it is straightforward to calibrate EVB potentials that exactly reproduce the desired activation free energies. By running multiple reaction simulations with such a model at different temperatures, computational Arrhenius plots can be constructed and the activation enthalpies and entropies obtained from the relation

$$\Delta G^{\ddagger}/T = \Delta H^{\ddagger}/T - \Delta S^{\ddagger} \qquad (1)$$

by plotting $\Delta G^{\ddagger}/T$ versus $1/T$.

For calibration of the EVB potential surfaces, we can use either the DFT results or the experimental free energy barriers, since they are very similar. Since the former may be associated with

larger errors, we find it more unbiased to illustrate the present approach by parametrization against the same experimental data for all relevant possible reaction pathways, that is, the stepwise and the two- and three-water concerted variants. The corresponding results from parametrization directly against the DFT data are given in Supplementary Table 2 and yield exactly the same conclusions. The resulting average EVB reaction free energy profiles at 298 K, each based on 15 independent simulations, are shown in Fig. 3 for the stepwise and three-water concerted pathways for both of the substrates. For the stepwise pathways, the reaction free energy of initial PT step was parametrized from $pK_a$ values of 4.6 and 6.6 for 1-methylcytosine and 1-methyl-5,6-dihydrocytosine, respectively[15,19]. The (non-rate-limiting) intervening activation barriers were taken from Eigen's accurate free energy relationships[20] as described elsewhere[21]. The overall activation free energies were set to 30.4 and $23.5\,\text{kcal}\,\text{mol}^{-1}$ for the two compounds, respectively, in accordance with the experimental results[5,6].

The EVB simulations give Arrhenius plots with remarkably good fits to straight lines (Fig. 4), which allows the thermodynamic parameters to be extracted with sufficient accuracy. Focusing on the entropic contribution, the overall activation entropies are thus the sum of the reaction entropy for PT from water to the substrate and the activation entropy for nucleophilic attack at C4. The equilibrium constant for the PT reactions show, as expected, only weak temperature dependence as revealed by the van't Hoff plots of $\Delta G^0/T$ versus $1/T$, which have small slopes (Fig. 4a). Hence, the PT reaction free energy is dominated by the $-T\Delta S^0$ term, which is found to be 16.5 and $17.1\,\text{kcal}\,\text{mol}^{-1}$ at 298 K for cytidine and 5,6-dihydrocytidine, respectively (Table 1). The high entropy contribution for this reaction step is mainly due

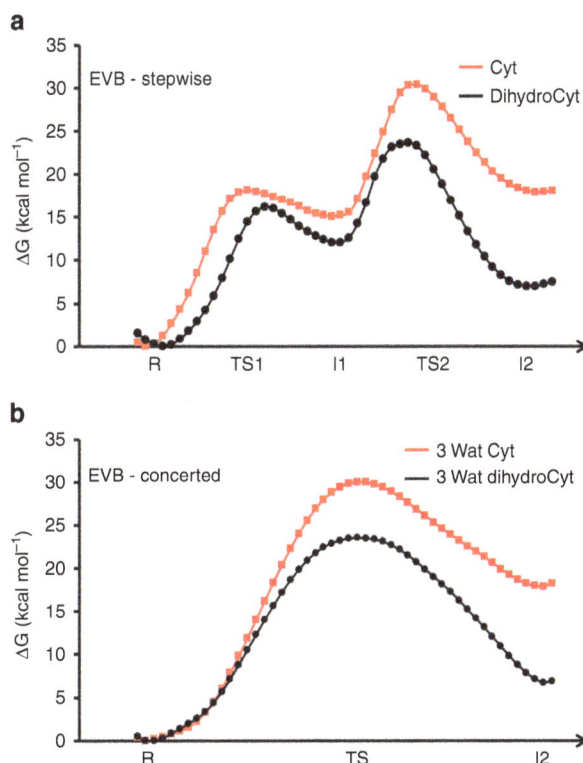

**Figure 3 | Free energy profiles from EVB simulations.** Calculated free energy profiles ($\text{kcal}\,\text{mol}^{-1}$) at 298 K from MD/EVB simulations of the stepwise (**a**) and concerted (**b**) pathways for the rate-limiting part of the cytidine and 5,6-dihydrocytidine deamination reactions. The concerted reaction path corresponds to the three-water reaction with an eight-membered TS.

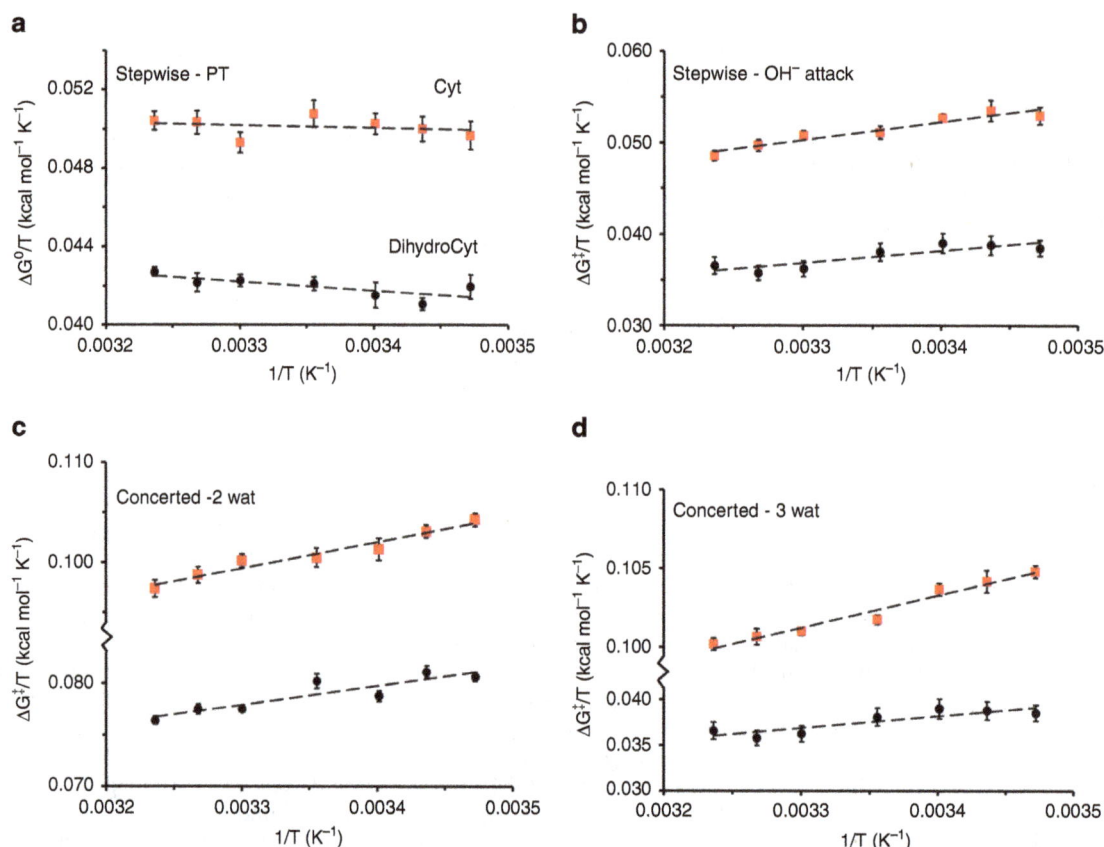

**Figure 4 | Calculated temperature dependence of different reaction mechanisms.** Van't Hoff (**a**) and Arrhenius (**b**) plots from MD/EVB simulations of the stepwise deamination mechanism for cytidine (red squares) and 5,6-dihydrocytidine (black circles). (**c,d**) show the corresponding Arrhenius plots for the concerted pathways involving two or three water molecules, respectively. Error bars, 1 s.e.m. from 15 independent simulations.

**Table 1 | Thermodynamic activation parameters for different mechanisms at 298 K from MD/EVB simulations*.**

| Reaction pathway | $\Delta G^{\ddagger}$ | $\Delta H^{\ddagger}$ | $T\Delta S^{\ddagger}$ | s.e.m.[†] |
|---|---|---|---|---|
| Cyt—proton transfer[‡] | 15.0 | −1.5 | −16.5 | 0.18 |
| Cyt—nucleophilic attack | 15.3 | 19.9 | 4.6 | 0.21 |
| Cyt—stepwise | 30.3 | 18.4 | −11.9 | 0.28 |
| Cyt—2W concerted | 30.3 | 24.2 | −6.1 | 0.24 |
| Cyt—3W concerted | 30.5 | 21.4 | −9.1 | 0.13 |
| Cyt—experimental | 30.4 | 22.1 | −8.3 | |
| dihCyt—proton transfer[‡] | 12.5 | −4.6 | −17.1 | 0.13 |
| dihCyt—nucleophilic attack | 11.1 | 13.6 | 2.5 | 0.28 |
| dihCyt—stepwise | 23.6 | 9.0 | −14.6 | 0.31 |
| dihCyt—2W concerted | 23.5 | 18.6 | −4.9 | 0.14 |
| diCyt—3W concerted | 23.6 | 12.7 | −10.9 | 0.12 |
| dihCyt—experimental | 23.5 | 13.4 | −10.1 | |

*Energies in kcal mol$^{-1}$.
[†]Average s.e.m. for 15 calculated $\Delta G^{\ddagger}$ values at each of the seven temperatures used to construct Arrhenius (and van't Hoff) plots.
[‡]The overall reaction thermodynamic parameters $\Delta G^0$, $\Delta H^0$ and $T\Delta S^0$ are given for the proton transfer step.

to the ordering of water molecules on going from neutral to zwitterionic reaction species. This behaviour is thus comparable to the ionization of acetic acid in pure water, which proceeds with $\Delta G^0 = 6.5$ and $\Delta H^0 = -0.1$ kcal mol$^{-1}$ (ref. 1). The nucleophilic attack on the protonated substrates, on the other hand, shows a positive activation entropy with $T\Delta S^{\ddagger} = 4.6$ and 2.5 kcal mol$^{-1}$ at 298 K for cytidine and 5,6-dihydrocytidine, respectively (Table 1). This favourable contribution can be interpreted such that the

solvation effects again dominate, but now the highly localized charges are partially neutralized and the accompanying increase in solvent entropy dominates over the reduction of the configurational space of the reacting groups in the TS.

The EVB simulations for the stepwise mechanisms thus predict the overall activation entropies corresponding to $T\Delta S^{\ddagger} = -11.9$ and $-14.6$ kcal mol$^{-1}$ (at 298 K) for the two deamination reactions, respectively. Both of these values are about 4 kcal mol$^{-1}$ off from the experimental results, which indicates that the stepwise mechanisms may not properly describe the reactions. The same type of MD/EVB simulations were also carried out for the concerted reaction pathways with both the two- and three-water mechanisms, yielding six- and eight-membered TSs, as discussed above. While the two-water concerted mechanism could probably be excluded already from the DFT results, owing to its significantly higher activation barrier (see above), it is nevertheless instructive to examine its predicted thermodynamic activation parameters. The computed Arrhenius plots for the concerted deamination mechanisms of the two substrates are shown in Fig. 4c,d and the thermodynamic data are summarized in Table 1. There it can be seen that the concerted reaction paths all have significantly less negative activation entropies than the stepwise mechanisms. This basically reflects the avoidance of visiting the zwitterionic state where solvent reorganization imposes a major entropy penalty. It can also be seen that engaging three water molecules (eight-membered TS) in a concerted mechanism, instead of two (six-membered TS), increases the entropic penalty by 3–6 kcal mol$^{-1}$. This energetic cost of ordering an additional water molecule in the TS is, however, counterbalanced by the relieving enthalpic strain in the TS (Table 1).

Comparing the overall thermodynamic data in Table 1, we find that the concerted three-water mechanism is the one that best coincides with the experimental data[5,6]. For this mechanism, we see that both the predicted enthalpy and entropy terms are within 1 kcal mol$^{-1}$ of the corresponding experimentally derived values, and for both substrates, which is quite remarkable. It should again perhaps be emphasized that this conclusion is not dependent on our choice of EVB calibration to experimental free energies, since calibration to the DFT results lead to the same conclusion (Supplementary Table 2). It is also noteworthy that the DFT-SMD results for the activation free energies also, in fact, point to this mechanism being slightly more favourable than the competing ones.

## Discussion

To summarize, we have shown how direct computer simulations of the temperature dependence of free energy profiles for chemical reactions in solution can be used to extract reliable thermodynamic activation parameters. It thus appears that this approach is sufficiently robust for making mechanistic predictions and direct comparisons to experiment. A prerequisite is that extensive configurational sampling can be carried out, which was achieved here with the EVB method, but could eventually perhaps be done by other QM/MM methods. The results are also informative with regard to the origin of different activation entropies for alternative mechanisms and highlight the importance of the solvent in this respect. In fact, the underlying reason for why the activation enthalpy–entropy partitioning becomes very precise, although it was in no way built into the EVB models, is that it is mainly determined by the configurational entropy of the solvent, which is correctly captured by the MD sampling. Hence, it is likely that the same approach as used here can also be applied to obtain accurate thermodynamic parameters, via computational van't Hoff plots, for solvation and ligand-binding processes[22–24].

For the case of spontaneous cytidine deamination, the simulations clearly predict that a concerted eight-membered TS mechanism is at play. A comparison with the enzyme cytidine deaminase, where the same reaction occurs essentially without any entropy loss, further suggests that the origin of this effect may be that hydroxide ion attack dominates the observed activation entropy in that case. Such an explanation would thus be rather different from the view that 'freezing' substrate motion on binding[7] is at the heart of favourable enzyme activation entropies.

## Methods

**Quantum mechanical calculations.** The different molecular systems explored by DFT calculations consisted of the cytosine and 5,6-hydrocytosine bases, capped by a methyl group at N1, and one, two or three water molecules participating in the hydrolytic reaction. Geometry optimization of all systems was carried out with the hybrid M06-2X hybrid functional[13] and the 6-311 + + G(d,p) basis set, using an ultrafine numerical integration grid. The TS structures were validated by frequency calculations at the same level of theory and basis set to confirm stationary points. To verify that the correct minima are connected, intrinsic reaction coordinate calculations were performed for the TSs in both directions. The stepwise mechanisms were optimized with the SMD solvation model[14] (Supplementary Table 5). The concerted pathways did not yield convergence with SMD and were therefore optimized in the gas phase, with solvation energies calculated at the gas-phase geometries added as corrections to the free energy profiles (Supplementary Table 6). In contrast to the stepwise mechanisms, the charge separation for the concerted pathways is not very significant and, hence, the gas-phase geometries should provide good approximations for evaluating solvation effects[25]. The Gaussian 09 programme[26] was used for all DFT calculations. All reported DFT energies are free energies obtained from the standard gas-phase thermochemical corrections[26] in Gaussian 09 plus the SMD solvation free energies. The reactant reference points were the corresponding complexes with water molecules. For 5,6-dihydrocytidine, two different initial conformations were considered in the optimizations, with either C5 or C6 out of plane. In both cases, the final structure converged to the conformation with C6 out of plane and this was used throughout the subsequent calculations.

**EVB simulations.** EVB/MD simulations were performed with the programme Q[27] utilizing spherical boundary conditions, where the full cytidine and 5,6-dihydrocytidine nucleosides were immersed in a TIP3P water droplet of 20 Å radius. The OPLS-AA force field[28] was used for parametrization of the different valence bond structures via the ffld_server utility in maestro (version 9.2, Schrödinger, LLC, New York, NY, 2011). The non-bonded parameters used in the calculations are given in Supplementary Table 4. Water molecules close to the sphere boundary were subjected to radial and polarization restraints according to the SCAAS model[27,29] and the cytidine ring was restrained to the centre of the sphere with a weak force constant of 0.5 kcal mol$^{-1}$Å$^{-2}$ applied to the C4 atom. Note that, although the difference between the Gibbs' and Helmholtz' free energies is vanishingly small for a solution reaction at normal temperatures and pressures, the SCAAS model formally corresponds more closely to the former as the volume is not strictly constant, but subject to harmonic restraints. The MD simulations were carried out with a 1-fs time step without any nonbonded interaction cutoffs applied to the reacting groups. For water–water interactions (excluding those participating in the reaction), a direct cutoff of 12 Å was applied together with the local reaction field method[30], which gives an accurate representation of long-range electrostatics.

The valence bond structures used to represent the different reaction pathways are shown in Fig. 1. The stepwise mechanism was simulated via consecutive transformation $R \rightarrow I1 \rightarrow I2$, while the concerted pathway was represented by the direct transformation $R \rightarrow I2$. The ground-state EVB free energy profiles $\Delta G(X_n)$ were calculated as described elsewhere[18] from

$$\Delta G(X_n) = \sum_{m \supset \Delta \varepsilon = X_n} w_m \left[ \Delta G(\vec{\lambda}_m) \right.$$
$$\left. - RT \ln < \exp\left\{ - \left[ E_g(X_n) - \varepsilon_m(X_n) \right] / RT \right\} >_m \right] / \sum_{m \supset \Delta \varepsilon = X_n} w_m \qquad (2)$$

where the discretized reaction coordinate, $X_n = \Delta \varepsilon = \varepsilon_i - \varepsilon_j$, is the energy gap between the initial and final diabatic surfaces of the given reaction step. The MD average $\langle \rangle_m$ is evaluated on a mapping potential surface $\varepsilon_m$, given by $\varepsilon_m = \lambda_i^m \varepsilon_i + \lambda_j^m \varepsilon_j$, where the mapping vector, $\vec{\lambda}_m = (\lambda_i^m, \lambda_j^m)$, defines a linear combination between the end-point potentials and changes between the values (1,0) and (0,1). $\Delta G(\vec{\lambda}_m)$ is the free energy on this mapping potential and is obtained from Zwanzig's exponential formula[31]. $E_g(X_n)$ is the EVB ground-state energy that is obtained from mixing the diabatic states, via the off-diagonal Hamiltonian matrix elements $H_{ij}$, and solving the corresponding secular equation[2,3]. Finally, different mapping vectors contribute to a given reaction coordinate interval $X_n$ and are weighted proportionally to the total contribution in that interval $(w_m / \sum w_m)$. It may be noted that a major advantage with using the energy gap reaction coordinate together with MD sampling along a linear combination of the end-point potentials is that the system itself is allowed to choose a path of least action, as opposed to imposing geometric constraints to define the reaction path.

The EVB free energy profiles for each reaction step were calculated using 101 different values of $\vec{\lambda}_m$, with 50 ps of MD sampling at each $\vec{\lambda}_m$. Every such simulation was also repeated 15 times with different initial velocities, yielding about 76 ns of simulation time for each reaction step at each temperature. These simulations were then carried out at seven temperatures from 288 to 309 K to obtain reliable Arrhenius plot of $\Delta G^\ddagger / T$ versus $1/T$, for extracting the activation enthalpy and entropy. The stepwise and concerted EVB models for cytidine and 5,6-dihydrocytidine were parametrized at 298 K to either experimental or DFT results by adjusting the relevant $H_{ij}$ parameters and gas-phase energy shifts[2,3,18] (Supplementary Table 3). The barriers for the non-rate-limiting initial PT in the stepwise mechanisms were taken from accurate experimental linear free energy relationships[20,21], as this barrier was difficult to locate in the DFT optimizations (because it is low) and may also be underestimated by that method. Therefore, we consider the experimental estimates[20,21] to be the most reliable.

## References

1. Schaleger, L. L. & Long, F. A. Entropies of activation and mechanisms of reactions in solution. *Adv. Phys. Org. Chem.* **1**, 1–33 (1963).
2. Warshel, A. *Computer Modeling of Chemical Reactions in Enzymes and Solutions* (John Whiley & Sons, 1991).
3. Åqvist, J. & Warshel, A. Simulation of enzyme reactions using valence bond force fields and other hybrid quantum/classical approaches. *Chem. Rev.* **93**, 2523–2544 (1993).
4. Lindahl, T. Instability and decay of the primary structure of DNA. *Nature* **362**, 709–715 (1993).
5. Snider, M. J., Gaunitz, S., Ridgway, C., Short, S. A. & Wolfenden, R. Temperature effects on the catalytic efficiency, rate enhancement, and transition state affinity of cytidine deaminase, and the thermodynamic consequences for catalysis of removing a substrate "anchor". *Biochemistry* **39**, 9746–9753 (2000).
6. Snider, M. J., Lazarevic, D. & Wolfenden, R. Catalysis by entropic effects: the action of cytidine deaminase on 5,6-dihydrocytidine. *Biochemistry* **41**, 3925–3930 (2002).

7. Jencks, W. P. Binding energy, specificity, and enzyme catalysis: the Circe effect. *Adv. Enzymol. Relat. Areas Mol. Biol.* **43,** 219–410 (1975).

8. Frederico, L. A., Kunkel, T. A. & Shaw, B. R. A sensitive genetic assay for the detection of cytosine deamination: determination of rate constants and the activation energy. *Biochemistry* **29,** 2532–2537 (1990).

9. Almatarneh, M. H., Flinn, C. G., Poirier, R. A. & Sokalski, W. A. Computational study of the deamination reaction of cytosine with $H_2O$ and $OH^-$. *J. Phys. Chem. A* **110,** 8227–8234 (2006).

10. Almatarneh, M. H., Flinn, C. G. & Poirier, R. A. Mechanisms for the deamination reaction of cytosine with $H_2O/OH^-$ and $2H_2O/OH^-$: a computational study. *J. Chem. Inf. Model.* **48,** 831–843 (2008).

11. Matsubara, T., Ishikura, M. & Aida, M. A quantum chemical study of the catalysis for cytidine deaminase: contribution of the extra water molecule. *J. Chem. Inf. Model.* **46,** 1276–1285 (2006).

12. Uddin, K. M., Flinn, C. G., Poirier, R. A. & Warburton, P. L. Comparative computational investigation of the reaction mechanism for the hydrolytic deamination of cytosine, cytosine butane dimer and 5,6-saturated cytosine analogues. *Comput. Theor. Chem.* **1027,** 91–102 (2014).

13. Zhao, Y. & Truhlar, D. G. The M06 suite of density functionals for main group thermochemistry, thermochemical kinetics, noncovalent interactions, excited states, and transition elements: two new functionals and systematic testing of four M06-class functionals and 12 other functionals. *Theor. Chem. Acc.* **120,** 215–241 (2008).

14. Marenich, A. V., Cramer, C. J. & Truhlar, D. G. Universal solvation model based on solute electron density and on a continuum model of the solvent defined by the bulk dielectric constant and atomic surface tensions. *J. Phys. Chem. B* **113,** 6378–6396 (2009).

15. Blackburn, G. M., Jarvis, S., Ryder, M. C. & Solan, V. Kinetics and mechanism of reaction of hydroxylamine with cytosine and its derivatives. *J. Chem. Soc. Perkin Trans. I* 370–385 (1975).

16. Pliego, J. R. & Riveros, J. M. The cluster-continuum model for the calculation of the solvation free energy of ionic species. *J. Phys. Chem. A* **105,** 7241–7247 (2001).

17. Frick, L., MacNeela, J. P. & Wolfenden, R. Transition state stabilization by deaminases: rates of nonenzymatic hydrolysis of adenosine and cytosine. *Bioorg. Chem.* **15,** 100–108 (1987).

18. Bjelic, S. & Åqvist, J. Catalysis and linear free energy relationships in aspartic proteases. *Biochemistry* **45,** 7709–7723 (2006).

19. Brown, D. M. & Hewlins, M. J. E. Dihydrocytosine and related compounds. *J. Chem. Soc. (C)* 2050–2055 (1968).

20. Eigen, M. Proton transfer, acid-base catalysis, and enzymatic hydrolysis. Part I: Elementary processes. *Angew. Chem. Int. Ed.* **3,** 1–72 (1964).

21. Åqvist, J. in *Computational Approaches To Biochemical Reactivity.* (eds Naray-Szabo, G. & Warshel, A. ) 341–362 (Kluwer Academic Publishers, 1997).

22. Kollman, P. A. Free energy calculations: applications to chemical and biochemical phenomena. *Chem. Rev.* **93,** 2395–2417 (1993).

23. Mobley, D. L., Bayly, C. I., Cooper, M. D., Shirts, M. R. & Dill, K. A. Small molecule hydration free energies in explicit solvent: an extensive test of fixed-charge atomistic simulations. *J. Chem. Theory Comput.* **5,** 350–358 (2009).

24. Carlsson, J. & Åqvist, J. Absolute hydration entropies of alkali metal ions from molecular dynamics simulations. *J. Phys. Chem. B* **113,** 10255–10260 (2009).

25. Almerindo, G. I. & Pliego, Jr. J. R. *Ab initio* investigation of the kinetics and mechanism of the neutral hydrolysis of formamide in aqueous solution. *J. Braz. Chem. Soc.* **18,** 696–702 (2007).

26. Frisch, M. J. *et al. GAUSSIAN 03, Revision B.03* (Gaussian Inc., 2003).

27. Marelius, J., Kolmodin, K., Feierberg, I. & Åqvist, J. Q: A molecular dynamics program for free energy calculations and empirical valence bond simulations in biomolecular systems. *J. Mol. Graph. Model.* **16,** 213–225 (1998).

28. Jorgensen, W. L., Maxwell, D. S. & Tirado-Rives, J. Development and testing of the OPLS all-atom force field on conformational energetics and properties of organic liquids. *J. Am. Chem. Soc.* **118,** 11225–11236 (1996).

29. King, G. & Warshel, A. A surface constrained all-atom solvent model for effective simulations of polar solutions. *J. Chem. Phys.* **91,** 3647–3661 (1989).

30. Lee, F. S. & Warshel, A. A local reaction field method for fast evaluation of long-range electrostatic interactions in molecular simulations. *J. Chem. Phys.* **97,** 3100–3107 (1992).

31. Zwanzig, R. W. High-temperature equation of state by a perturbation method. I. Nonpolar gases. *J. Chem. Phys.* **22,** 1420–1426 (1954).

## Acknowledgements

Support from the Swedish Research Council (VR), the Knut and Alice Wallenberg Foundation and the Swedish National Infrastructure for Computing (SNIC) is gratefully acknowledged.

## Author contributions

M.K. performed the experiments. J.Å. designed the experiments. M.K. and J.Å. analysed the data and wrote the paper.

# 9

# Iron-catalysed cross-coupling of organolithium compounds with organic halides

Zhenhua Jia[1], Qiang Liu[1], Xiao-Shui Peng[1,2] & Henry N.C. Wong[1,2]

In past decades, catalytic cross-coupling reactions between organic halides and organometallic reagents to construct carbon–carbon bond have achieved a tremendous progress. However, organolithium reagents have rarely been used in cross-coupling reactions, due mainly to their high reactivity. Another limitation of this transformation using organolithium reagents is how to control reactivity with excellent selectivity. Although palladium catalysis has been applied in this field recently, the development of an approach to replace catalytic systems of noble metals with nonprecious metals is currently in high demand. Herein, we report an efficient synthetic protocol involving iron-catalysed cross-coupling reactions employing organolithium compounds as key coupling partners to unite aryl, alkyl and benzyl fragments and also disclose an efficient iron-catalysed release-capture ethylene coupling with isopropyllithium.

[1] Department of Chemistry, State Key Laboratory of Synthetic Chemistry and Centre of Novel Functional Molecules, Chinese University of Hong Kong, Shatin, New Territories, Hong Kong SAR, China. [2] Shenzhen Center of Novel Functional Molecules and Shenzhen Municipal Key Laboratory of Chemical Synthesis of Medicinal Organic Molecules, Shenzhen Research Institute, Chinese University of Hong Kong, No.10, Second Yuexing Road, Shenzhen 518507, China. Correspondence and requests for materials should be addressed to H.N.C.W. (email: hncwong@cuhk.edu.hk).

Transition metal-catalysed cross-coupling has emerged as a highly useful, selective and widely applicable method for synthesizing structurally diverse organic compounds via carbon–carbon bond formation[1,2]. Since the discoveries of cross-coupling reactions, palladium-catalysed cross-coupling with organic halides and organometallic reagents, has dominated this area as an exceptionally powerful approach to assemble C–C bond (Fig. 1a)[3]. Although Murahashi et al. disclosed a palladium-catalysed cross-coupling reaction of alkenyl halides with various organolithium compounds, direct use of organolithium reagents in cross-coupling reactions has been neglected for a long time, mainly due to the high reactivity and low stability of organolithium reagents[4–7]. Recently, Feringa and co-workers developed palladium-based catalytic systems to directly generate C–C bond using organolithium compounds as cross-coupling partners (Fig. 1b)[8–13]. Although palladium-based catalysts typically mediated such reactions, there are increasing concerns about their long-term sustainability in the synthetic community because of its high cost, low natural abundance, environmentally deleterious extraction, toxicity and competition for its use from the automotive and consumer electronics sectors[14]. Therefore, there is a growing interest in replacing palladium-based catalysts with those more Earth-abundant elements. With its low cost, high natural abundance and low toxicity, iron is indeed a particularly appealing alternative, and accordingly, the development of iron-catalysed cross-coupling is undergoing an explosive growth[15–20]. Herein, we develop an iron-catalysed cross-coupling strategy of organolithium reagents with organic halides to form C–C bonds, examples including $C(sp^2)$-$C(sp^3)$ bonds, $C(sp^3)$-$C(sp^3)$ bonds and a rare method to form a novel $C(sp^2)$-$C(sp^3)$ bond via in-situ generation of ethylene from tetrahydrofuran (THF).

## Results

**Serendipity.** Previously, we demonstrated that the rigid tetraphenylene (tetrabenzo[a,c,e,g]cyclooctatetraene) is a structurally and functionally exceptional molecule[21,22]. To improve the efficiency of the coupling step, we proposed to synthesize tetraphenylene derivatives through a one-pot iron-catalysed intramolecular cross-coupling protocol[23]. Although we obtained a trace amount of tetraphenylene, the serendipity is that 2-n-butylbiphenyl was observed (Fig. 2). Therefore, we recognized the potential use of alkyllithium reagents in iron-catalysed cross-coupling reactions.

**Optimization.** Encouraged by the reaction shown in Fig. 2, we attempted to couple 4-methoxybromobenzene (**1a**) with n-BuLi

(**2a**) under the same condition. As expected, the target product p-methoxybutylbenzene (**3a**), together with a trace amount of the isomerized cross-coupling product (**3a'**), were detected by gas chromatography-mass spectrometry (GC–MS), although the homo-coupling product (**4**) is the major product, together with dehalogenated product (**5**; Table 1, entry 1). In the absence of triethylamine (Et$_3$N) and at 22 °C, the ratio of the desired product was almost the same (Table 1, entries 2–3). Then, in the presence of FeCl$_2$ (10 mol%), ligands (20 mol%) and **1a** (0.2 mmol) in THF (1.0 ml), several traditional bidentate ligands with different bite angles (for expanded screening results, see Supplementary Table 1) and monodentate electron-rich phosphine ligands (Fig. 3) were examined through a slow addition of dilute organolithium reagent **2a** (0.3 mmol, 0.35 M) within 1 h at 22 °C using a syringe pump (Table 1, entries 4–11). To our delight, a distinct improvement was displayed by GC–MS. When trimethyl phosphite (L$_5$) was used as a ligand, the desired product was isolated in 68% yield (Table 1, entry 8). Further screening of iron salts revealed that iron(II) chloride gave a promising yield with trimethyl phosphite as the ligand (Table 1, entries 12–15). Recently, Fürstner pioneered the use of iron(III) acetylactonate-catalysed cross-coupling reactions with Grignard reagents[24–27], however, this iron catalyst demonstrated lower reactivity with lithium reagents (Table 1, entry 16). Surprisingly, upon addition of tetramethylethylenediamine (TMEDA, L$_8$) into the solution of iron(III) chloride in THF, the homo-coupling by-product (**4**) was suppressed dramatically (Table 1, entry 17). This catalytic system developed by Nakamura had been employed in cross-coupling reactions with Grignard reagents and arylzinc reagents[28–31]. Therefore, the complex of TMEDA with iron(III) chloride was prepared according to Nakamura's procedure for use in our next stage of optimization (Table 1, entries 18–24). When the cross-coupling reaction was conducted at 0 °C, the generation of the dehalogenated by-product (**5**) was reduced and the expected product was isolated in 85% (0.2 mmol scale; Table 1, entry 19). To further improve this procedure, we also screened the reaction media. A comparison of results obtained in THF revealed that the ratios of desired product were decreased in toluene and diethyl ether (Table 1, entries 20–21). When the catalyst loading was reduced to 3 and 1 mol%, respectively, the overall efficiency was not reduced in an obvious manner (Table 1, entries 22–23).

## $C(sp^2)$-$C(sp^3)$ cross-coupling of aryl halides with alkyllithiums.

To expand the scope of the iron-catalysed reactions, $C(sp^2)$-$C(sp^3)$ cross-coupling of aryl halides with alkyllithium reagents was further investigated. Initially, we compared the

**a** Carbon–Carbon bond formation via catalytic cross-coupling - in literature

$$R^1-X \ + \ R^2-M \xrightarrow[\text{Catalysis}]{\text{Transition metal}} R^1-R^2$$

**b** Pd catalytic cross-coupling of organolithium compounds - recently reported

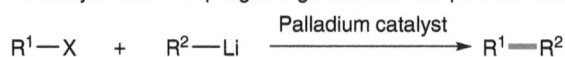

$$R^1-X \ + \ R^2-Li \xrightarrow{\text{Palladium catalyst}} R^1-R^2$$

**c** Iron catalytic cross-coupling of organolithium compounds - remain elusive (this research)

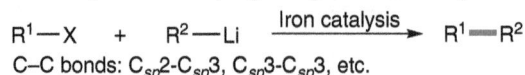

$$R^1-X \ + \ R^2-Li \xrightarrow{\text{Iron catalysis}} R^1-R^2$$

C–C bonds: C$_{sp}$2-C$_{sp}$3, C$_{sp}$3-C$_{sp}$3, etc.

• Earth-abundant metal catalyst (Fe)
• Mild reaction conditions
• Chem- and regioselective

**Figure 1 | Transition metal-catalysed cross-coupling to form carbon-carbon bonds.** (a) C–C bond formation via catalytic cross-coupling. (b) Palladium-catalysed cross-coupling of organolithium compounds. (c) Iron-catalysed cross-coupling of organolithium reagents.

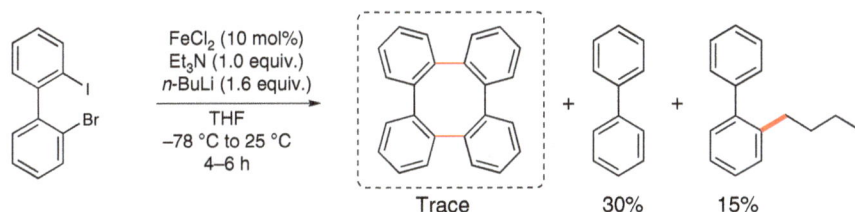

**Figure 2 | Serendipity induced the discovery of iron-catalysed cross-coupling of organolithium compounds.** Dimerization of 2-bromo-2'-iodo-1,1'-biphenyl to synthesize tetraphenylene via a one-pot iron-catalysed intramolecular cross-coupling protocol.

**Table 1 | Selected optimization results for C($sp^2$)-C($sp^3$) cross-coupling of 4-methoxybromobenzene (1a) and n-BuLi (2a)*.**

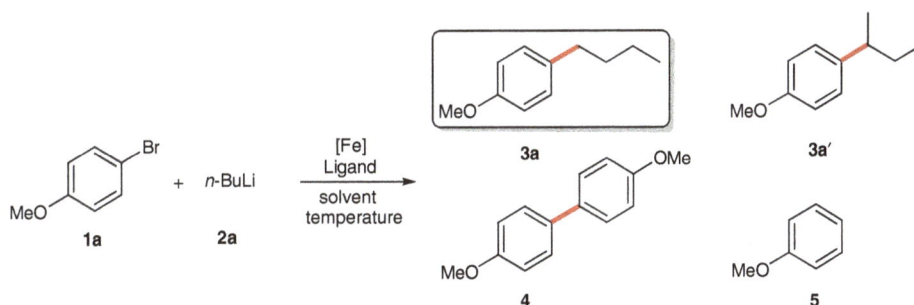

| Entry | [Fe] (mol%) | Ligand (mol%) | Solvent | T (°C) | 3a/3a'/4/5[†] |
|---|---|---|---|---|---|
| 1 | FeCl$_2$ (10) | — | THF | −78 | 20/3/62/15[‡] |
| 2 | FeCl$_2$ (10) | — | THF | 22 | 18/2/45/35[‡] |
| 3 | FeCl$_2$ (10) | — | THF | 22 | 22/1/17/60[§] |
| 4 | FeCl$_2$ (10) | L$_1$ (20) | THF | 22 | 39/0/16/45 |
| 5 | FeCl$_2$ (10) | L$_2$ (20) | THF | 22 | 35/3/19/43 |
| 6 | FeCl$_2$ (10) | L$_3$ (20) | THF | 22 | 30/2/23/45 |
| 7 | FeCl$_2$ (10) | L$_4$ (20) | THF | 22 | 27/3/26/44 |
| 8 | FeCl$_2$ (10) | L$_5$ (20) | THF | 22 | 77/1/12/10 |
| 9 | FeCl$_2$ (10) | L$_6$ (20) | THF | 22 | 44/1/28/27 |
| 10 | FeCl$_2$ (10) | L$_7$ (20) | THF | 22 | 21/0/19/60 |
| 11 | FeCl$_2$ (10) | L$_8$ (20) | THF | 22 | 20/1/17/62 |
| 12 | FeBr$_2$ (10) | L$_5$ (20) | THF | 22 | 47/3/19/31 |
| 13 | Fe(OAc)$_2$ (10) | L$_5$ (20) | THF | 22 | 37/0/28/35 |
| 14 | FeCl$_3$ (10) | L$_5$ (20) | THF | 22 | 55/1/17/27 |
| 15 | FeBr$_3$ (10) | L$_5$ (20) | THF | 22 | 44/2/22/32 |
| 16 | Fe(acac)$_3$ (10) | — | THF | 22 | 11/2/35/52 |
| 17 | FeCl$_3$ (10) | L$_8$ (20) | THF | 22 | 75/0/4/21 |
| 18[‖] | FeCl$_3$ (5) | L$_8$ (7.5) | THF | 22 | 74/0/5/21 |
| 19[‖] | FeCl$_3$ (5) | L$_8$ (7.5) | THF | 0 | 95/0/3/2 |
| 20[‖] | FeCl$_3$ (5) | L$_8$ (7.5) | Tol | 0 | 66/1/27/6 |
| 21[‖] | FeCl$_3$ (5) | L$_8$ (7.5) | Et$_2$O | 0 | 11/0/82/7 |
| 22[‖] | FeCl$_3$ (3) | L$_8$ (4.5) | THF | 0 | 94/0/3/3 |
| 23[‖] | FeCl$_3$ (1) | L$_8$ (1.5) | THF | 0 | 90/0/5/5 |

*Reaction conditions: a solution of n-BuLi (0.30 mmol, 2.4 M solution in hexane diluted with indicated solvent to a final concentration of 0.35 M) was added by a syringe pump in 1h to a solution of **1a** (0.2 mmol), iron catalyst (0.02 mmol for entries 1-17), ligand (0.04 mmol for entries 4-17) in THF (1 ml) at the indicated temperature.
[†]Determined by gas chromatography-mass spectrometry analysis of the crude reaction mixture.
[‡]1 Equiv. Et$_3$N was introduced as an additive.
[§]No ligand and additive were used.
[‖]The complex [(FeCl$_3$)$_2$(TMEDA)$_3$] was used as catalyst.

reactivity of different aryl halides (Table 2, **3a**). 4-Methoxychlorobenzene (5% conversion) was more inert than 4-methoxybromobenzene (**1a**). The target product was generated exclusively, when 4-methoxyiodobenzene was used as a starting substrate. Unexpectedly, 4-methoxyphenyltrifluoromethanesulfonyl triflate (an aryl triflate) decomposed to the corresponding phenol (see note in Table 2). In consideration of their commercial availability, we made use of aryl bromides for further investigation (Table 2, 3**b**-3**r**). It was uncovered that varying the position of the methoxy group on the benzene ring

led to a pronounced effect on the reaction outcome, presumably due to chelation of oxygen with lithium (Table 2, **3b**-**3c**). In the case of bromobenzene, the GC yield was given due to the volatility issue (Table 2, **3d**). Electron-donating and bulky functional groups facilitated cross-coupling reaction without sacrificing the yield of the corresponding products (Table 2, **3e**-**3g**). However, a strongly electron-withdrawing substituent was found to lead to halogen-metal exchange (Table 2, **3h**). Remarkably, a series of alkyllithiums were freshly prepared and were found to be compatible with this protocol, being able to couple with

Selected ligands:

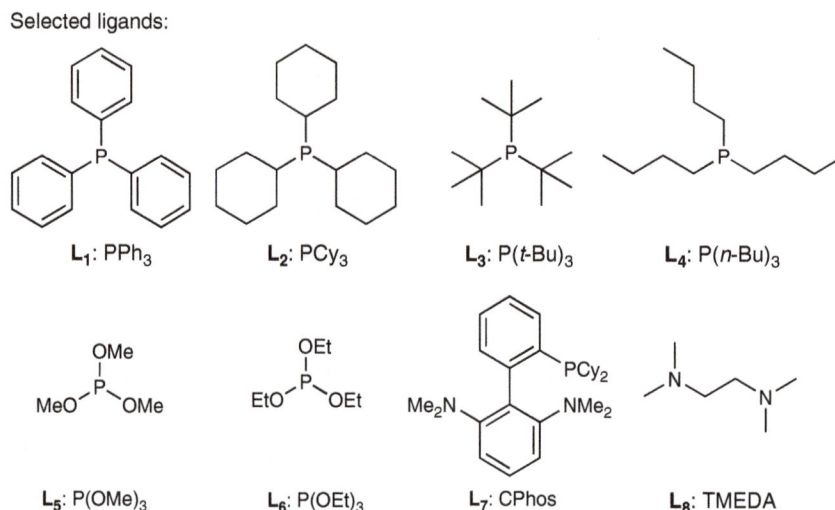

**L₁**: PPh₃     **L₂**: PCy₃     **L₃**: P(*t*-Bu)₃     **L₄**: P(*n*-Bu)₃

**L₅**: P(OMe)₃     **L₆**: P(OEt)₃     **L₇**: CPhos     **L₈**: TMEDA

**Figure 3 | Selected ligands.** Monodentate and bidentate ligands were screened, also see Supplementary Information.

---

**Table 2 | Iron-catalysed cross-coupling of aryl halides with alkyllithium reagents\*.**

$$Ar-X + Li-Alkyl \xrightarrow[\text{THF, 0°C, 2h}]{[(FeCl_3)_2(TMEDA)_3]\ 3\ mol\%} Ar-Alkyl$$

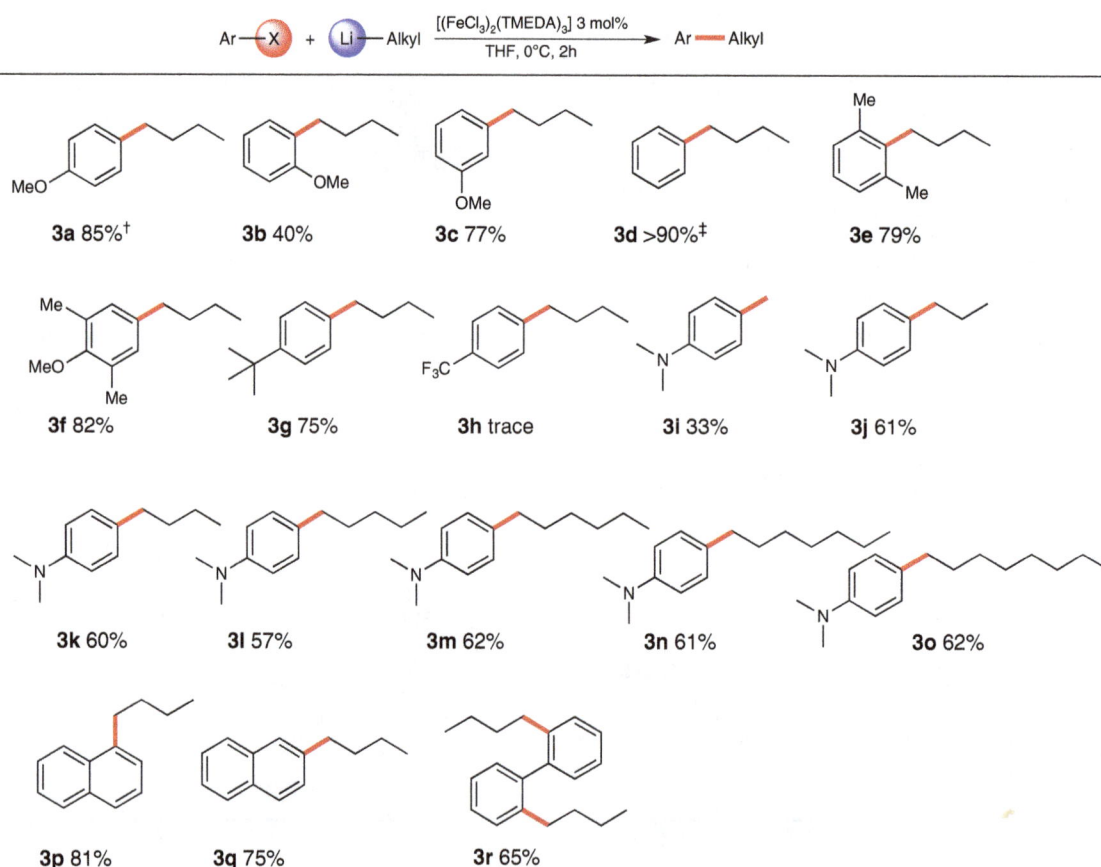

**3a** 85%†    **3b** 40%    **3c** 77%    **3d** >90%‡    **3e** 79%

**3f** 82%    **3g** 75%    **3h** trace    **3i** 33%    **3j** 61%

**3k** 60%    **3l** 57%    **3m** 62%    **3n** 61%    **3o** 62%

**3p** 81%    **3q** 75%    **3r** 65%

\*Reaction conditions: a solution of 0.30 mmol alkyllithium reagents (0.3 mmol) diluted with THF to a final concentration of 0.35 M was added by a syringe pump in 1 h to a THF solution (1.0 ml) of aryl halides (0.2 mmol) and [(FeCl₃)₂(TMEDA)₃] (3 mol%) at 0 °C.
†X = Cl, 5% conversion, X = I, 99% yield of **3a**, X = OTf as leaving group, 83% starting material recovered.
‡Gas chromatography yield.

---

4-bromo-*N,N*-dimethylaniline (Table 2, **3i-3o**). Polyaromatic compounds were found to undergo alkylation in moderate yields (Table 2, **3p-3q**). In addition, a double alkylation product was obtained in 65% yield (Table 2, **3r**).

**Release-capture ethylene coupling with isopropyllithium.** When isopropyllithium, a typical secondary organolithium, was utilized in the iron catalysis system with 4-methoxy-bromobenzene (**1a**), 1-isopentyl-4-methoxybenzene (**3a**ₜₕF) was obtained together with a trace amount of cross-coupling product 1-isopropyl-4-methoxybenzene. After prolonging the reaction time to overnight at 22 °C, the yield of $3a_{THF}$ was optimized up to 71% (Table 3, $3a_{THF}$). To our best knowledge, this is an unusual example of transition metal-catalysed cross-coupling reaction involving freshly prepared ethylene generated by decomposing THF with isopropyllithium. Several aryl bromides were then investigated to explore the substituent effect at various positions of the benzene ring. Possible chelation effect and steric effect were demonstrated when the benzene ortho-position was occupied by a methoxy group or a bulky group (Table 3, $3b_{THF}$ and $3c_{THF}$). It is noted that when $FeCl_2$ with $P(OMe)_3$ was used as the catalyst in place of $[(FeCl_3)_2(TMEDA)_3]$, the yield of $3c_{THF}$ could be improved. Remote substituents could be tolerated, leading to the formation of the corresponding products in 37–77% yield (Table 3, $3d_{THF}$-$3f_{THF}$). Moreover, a naphthyl compound was found to participate efficiently (Table 3, $3g_{THF}$). On the basis of the previously reported reactions[32], we would like to propose a plausible pathway for this release-capture ethylene process. Thus, as shown in Fig. 4, THF is deprotonated at its 2-position by isopropyllithium to form 2-lithioTHF (I). Then, a subsequent intramolecular reverse $[3 + 2]$ cycloaddition of the anion would release ethylene and generate the lithium enolate (II). Finally, the resulting ethylene could be caught in situ to give the doubly homologated lithium product (III). Moreover, further evidence for our proposed pathway was obtained from a relevant deuterium-labelled crossover experiment utilizing deuterated tetrahydrofuran (THF-$d_8$) as solvent. Thus, treatment of 4-methoxybromobenzene (1a) with isopropyllithium in THF-$d_8$, led to the release of ethylene-$d_4$. The expected deuterated product $3a_{THF-d8}$ (Table 3) was obtained in 61% yield.

**Cross-coupling of alkyl bromides with organolithiums**. We next extended the iron catalysis strategy to alkyl bromides with organolithium reagents. Typically, commercially available 1-bromo-3-phenylpropane was assessed with n-BuLi to explore the possibility of $C(sp^3)$-$C(sp^3)$ cross-coupling. Gratifyingly, the reaction proceeded smoothly and the desired product was isolated in 77% yield (Table 4, 3aa). Other organolithium reagents, such as cyclopropyllithium, 9H-fluoren-9-yllithium and (trimethylsilyl)methyllithium, were allowed to couple with 1-bromo-3-phenylpropane to provide the corresponding $C(sp^3)$-$C(sp^3)$ cross-coupling products in good to excellent yields (Table 4, 3ab-3ad). Benzylic compounds, possessing a typical $C(sp^3)$-Br bonds, were also used as coupling partners. As expected, the cross-coupling products with yields ranging from 11 to 71% were generated, when n-BuLi and (trimethylsilyl)methyllithium were used as coupling partners (Table 4, 3ae-3am). Under the same condition, 2-(3-bromopropyl)naphthalene was also alkylated (Table 4, 3an). Subsequently, bromocyclohexane successfully underwent a similar reaction to form the relevant coupling product in 44% yield (Table 4, 3ao).

---

**Table 3 | Iron-catalysed release-capture ethylene coupling with isopropyllithium*.**

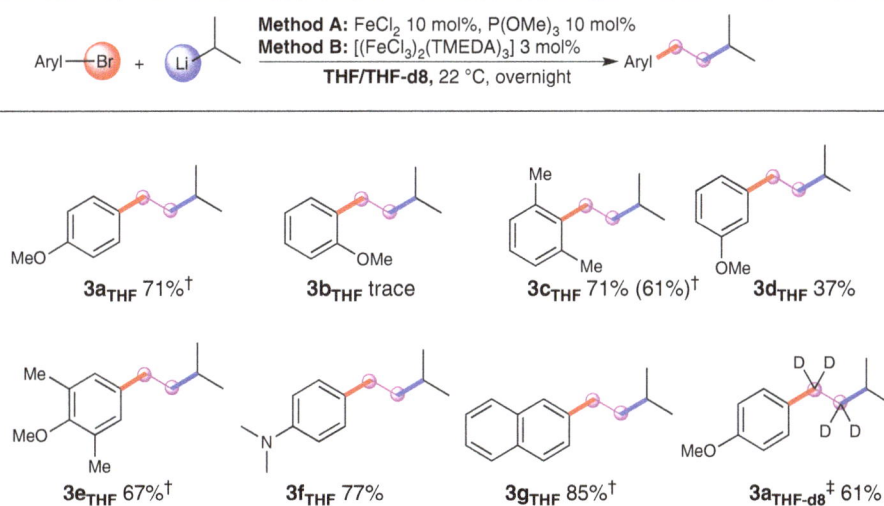

*Reaction conditions: Under − 78 °C, a solution of isopropyllithium (0.50 mmol) was added slowly into a solution of aryl bromides (0.2 mmol) in THF (1 ml). Subsequently, the mixture was added by a syringe pump in 1 h to a solution of $[(FeCl_3)_2(TMEDA)_3]$ (3 mol%) in THF (1.0 ml) at 22 °C.
†Method A: $FeCl_2$ combined with $P(OMe)_3$ as catalyst was used in place of $[(FeCl_3)_2(TMEDA)_3]$.
‡Deuterated $3a_{THF-d8}$ was obtained in 61% when THF-$d_8$ was used in a deuterium-labelled experiment.

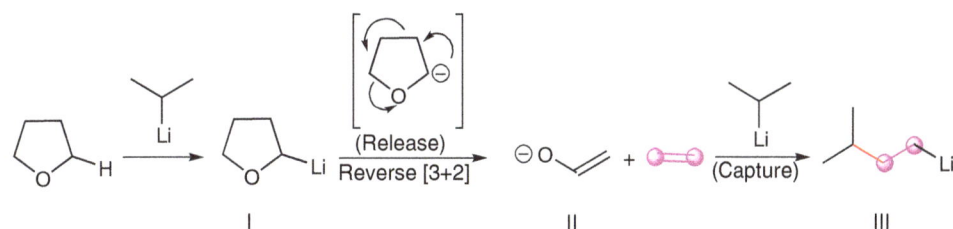

**Figure 4 | Possible pathway of release-capture ethylene.** (**I**) 2-LithioTHF. (**II**) Lithium enolate. (**III**) Doubly homologated of isopropyllithium to generate the isopentyllithium in situ.

## Table 4 | Iron-catalysed cross-coupling of alkyl bromides with organolithium reagents*.

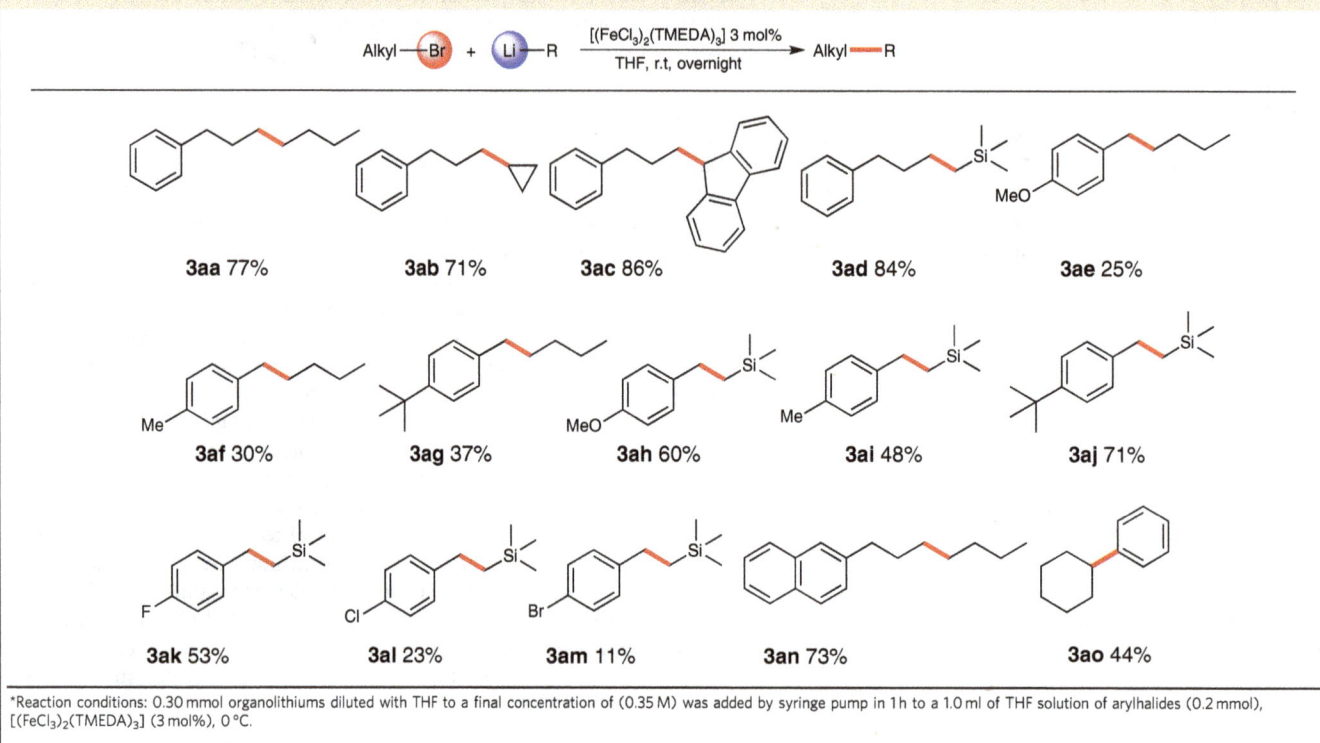

Alkyl—Br + Li—R $\xrightarrow[\text{THF, r.t, overnight}]{[(FeCl_3)_2(TMEDA)_3]\ 3\ mol\%}$ Alkyl—R

| | | | | |
|---|---|---|---|---|
| 3aa 77% | 3ab 71% | 3ac 86% | 3ad 84% | 3ae 25% |
| 3af 30% | 3ag 37% | 3ah 60% | 3ai 48% | 3aj 71% |
| 3ak 53% | 3al 23% | 3am 11% | 3an 73% | 3ao 44% |

*Reaction conditions: 0.30 mmol organolithiums diluted with THF to a final concentration of (0.35 M) was added by syringe pump in 1 h to a 1.0 ml of THF solution of arylhalides (0.2 mmol), [(FeCl₃)₂(TMEDA)₃] (3 mol%), 0 °C.

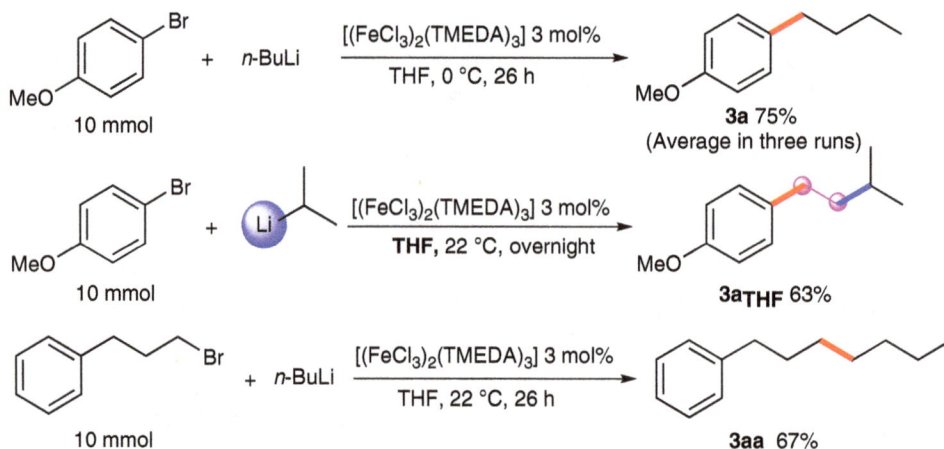

**Figure 5 | Gram scale reactions.** Three model substrates were selected to scale up to 10 mmol scale and the corresponding target products were isolated in satisfied yields.

In summary, we have disclosed iron-catalysed cross-coupling of organolithium compounds to form diverse carbon–carbon bonds efficiently. These results are expected to expand the scope of iron catalysis as well as the use of organolithium reagents. We trust that these reactions would provide milder, cheaper and more environmentally friendly approaches towards cross-coupling products. An extension of this catalytic system to broaden its scope, and to investigate its mechanistic nature is underway in our laboratory.

### Discussion

To provide a support against the involvement of trace amounts of other metal species, such as Pd, Pt, Co and Ni in our iron catalysts that would catalyse C–C bond formation, inductively coupled plasma mass spectrometry was performed on samples of FeCl₃ to detect the trace quantities of these metals (see Supplementary Information for details). Moreover, we conducted experiments to mimic the catalyst system to prove that relevant products were not isolated when the concentration of Co and Ni were as low as those present in the iron salts (see Supplementary Information for details). We also performed preliminary mechanistic analysis of this transformation utilizing several control experiments (see Supplementary Information for details). It was likely that the reaction involved radical species.

Noteworthy, the capability to procure useful product quantities for laboratory and industry usage through scalable routes is emerging as a very essential goal in catalytic reactions today. Therefore, we also confirmed the scalable feasibility of these iron-catalysed reactions, as shown in Fig. 5. As can be seen, several typical scale-up reactions in multi-gram scales provided relevant desired products in satisfied yields.

## Methods

**Iron-catalysed cross-coupling of 4-methoxybromobenzene (1a) and *n*-BuLi (2a).** To an oven-dried vial, equipped with a magnetic stirring bar, was charged with $[(FeCl_3)_2(TMEDA)_3]$ (3.96 mg, 0.006 mmol, 3 mol%) in a glove box, following by the subsequent addition of 4-methoxybromobenzene (1a; 0.2 mmol) and THF (1.0 ml). Then, after the sealed vial with a rubber stopper was taken out from the glove box, the reaction mixture was cooled to 0 °C, *n*-BuLi (2a; 0.30 mmol, 1.6 M or 2.4 M in hexane, diluted with THF to a final concentration of 0.35 M) was added to the mixture using a syringe pump in 1 h. After the addition was completed, the reaction mixture was stirred at 0 °C for 1 h. Then, after quenching with a saturated solution of aqueous $NH_4Cl$, the reaction mixture was extracted with $CH_2Cl_2$ three times. The combined organic solvent was evaporated under reduced pressure to afford the crude product, which was then purified by column chromatography on silica gel or preparative thin-layer chromatography.

## References

1. de Meijere, A., Brase, S. & Oestreich, M. *Metal-Catalyzed Cross-Coupling Reactions and More, Vol. 1, 2 and 3* (Wiley-VCH, 2014).
2. Negishi, E. Magical power of transition metals: past, present, and future. *Angew. Chem. Int. Ed.* **50,** 6738–6764 (2011).
3. Nicolaou, K. C., Bulger, P. G. & Sarlah, D. Palladium-catalyzed cross-coupling reactions in total synthesis. *Angew. Chem. Int. Ed.* **44,** 4442–4489 (2005).
4. Murahashi, S., Yamamura, M., Yanagisawa, K., Mita, N. & Kondo, K. Stereoselective synthesis of alkenes and alkenyl sulfides from alkenyl halides using palladium and ruthenium catalysts. *J. Org. Chem.* **44,** 2408–2417 (1979).
5. Clayden, J. *Organolithiums: Selectivity for Synthesis* (Oxford, 2002).
6. Rappoport, Z. & Marek, I. *The Chemistry of Organolithium Compounds* (Wiley-VHC, 2004).
7. Luisi, R. & Capriati, V. *Lithium Compounds in Organic Synthesis: From Fundamentals to Applications* (Wiley-VCH, 2014).
8. Giannerini, M., Fañanás-Mastral, M. & Feringa, B. L. Direct catalytic cross-coupling of organolithium compounds. *Nat. Chem* **5,** 667–672 (2013).
9. Vila, C., Giannerini, M., Hornillos, V., Fañanás-Mastral, M. & Feringa, B. L. Palladium-catalysed direct cross-coupling of secondary alkyllithium reagents. *Chem. Sci* **5,** 1361–1367 (2014).
10. Vila, C. et al. Palladium-catalysed direct cross-coupling of organolithium reagents with aryl and vinyl triflates. *Chem.-Eur. J* **20,** 13078–13083 (2014).
11. Castello, L. M. et al. Palladium-catalyzed cross-coupling of aryllithium reagents with 2-alkoxy-substituted aryl chlorides: Mild and efficient synthesis of 3,3'-diaryl BINOLs. *Org. Lett.* **17,** 62–65 (2015).
12. Hornillos, V., Giannerini, M., Vila, C., Fañanás-Mastral, M. & Feringa, B. L. Direct catalytic cross-coupling of alkenyllithium compounds. *Chem. Sci* **6,** 1394–1398 (2015).
13. Heijnen, D., Hornillos, V., Corbet, B. P., Giannerini, M. & Feringa, B. L. Palladium-catalyzed C($sp^3$)–C($sp^2$) cross-coupling of (trimethylsilyl) methyllithium with (hetero)aryl halides. *Org. Lett.* **17,** 2262–2265 (2015).
14. Nakamura, E. & Sato, K. Managing the scarcity of chemical elements. *Nat. Mater.* **10,** 158–161 (2011).
15. Plietker, B. *Iron Catalysis in Organic Chemistry* (Wiley-VCH, 2008).
16. Bolm, C., Legros, J., Le Paih, J. & Zani, L. Iron-catalyzed reactions in organic synthesis. *Chem. Rev.* **104,** 6217–6254 (2004).
17. Correa, A., Garcia Mancheno, O. & Bolm, C. Iron-catalysed carbon-heteroatom and heteroatom-heteroatom bond forming processes. *Chem. Soc. Rev.* **37,** 1108–1117 (2008).
18. Sherry, B. D. & Fürstner, A. The promise and challenge of iron-catalyzed cross coupling. *Acc. Chem. Res.* **41,** 1500–1511 (2008).
19. Bauer, I. & Knölker, H.-J. Iron catalysis in organic synthesis. *Chem. Rev.* **115,** 3170–3387 (2015).
20. Bedford, R. B. How Low Does Iron Go? Chasing the active species in Fe-catalyzed cross-coupling reactions. *Acc. Chem. Res.* **48,** 1485–1493 (2015).
21. Han, J.-W., Li, X. & Wong, H. N. C. Our expedition in eight-membered ring compounds: From planar dehydrocyclooctenes to tub-shaped chiral tetraphenylenes. *Chem. Rec.* **15,** 107–131 (2015).
22. Han, J.-W., Chen, J.-X., Li, X., Peng, X.-S. & Wong, H. N. C. Recent developments and applications of chiral tetraphenylenes. *Synlett.* **24,** 2188–2198 (2013).
23. Toummini, D., Ouazzani, F. & Taillefer, M. Iron-catalyzed homocoupling of aryl halides and derivatives in the presence of alkyllithiums. *Org. Lett.* **15,** 4690–4693 (2013).
24. Fürstner, A. & Leitner, A. Iron-catalyzed cross-coupling reactions of alkyl-Grignard reagents with aryl chlorides, tosylates, and triflates. *Angew. Chem. Int. Ed.* **41,** 609–612 (2002).
25. Fürstner, A., Leitner, A., Méndez, M. & Krause, H. Iron-catalyzed cross-coupling reactions. *J. Am. Chem. Soc.* **124,** 13856–13863 (2002).
26. Fürstner, A. & Méndez, M. Iron-catalyzed cross-coupling reactions: Efficient synthesis of 2, 3-allenol derivatives. *Angew. Chem. Int. Ed.* **42,** 5355–5357 (2003).
27. Scheiper, B., Glorius, F., Leitner, A. & Fürstner, A. Catalysis-based enantioselective total synthesis of the macrocyclic spermidine alkaloid isooncinotine. *Proc. Natl Acad. Sci. USA* **101,** 11960–11965 (2004).
28. Nakamura, M., Matsuo, K., Ito, S. & Nakamura, E. Iron-catalyzed cross-coupling of primary and secondary alkyl halides with aryl Grignard reagents. *J. Am. Chem. Soc.* **126,** 3686–3687 (2004).
29. Noda, D., Sunada, Y., Hatakeyama, T., Nakamura, M. & Nagashima, H. Effect of TMEDA on iron-catalyzed coupling reactions of ArMgX with Alkyl halides. *J. Am. Chem. Soc.* **131,** 6078–6079 (2009).
30. Ito, S., Fujiwara, Y.-I., Nakamura, E. & Nakamura, M. Iron-catalyzed cross-coupling of alkyl sulfonates with arylzinc reagents. *Org. Lett.* **11,** 4306–4309 (2009).
31. Bedford, R. B. et al. TMEDA in iron-catalyzed kumada coupling: Amine adduct versus homoleptic "ate" complex formation. *Angew. Chem. Int. Ed.* **53,** 1804–1808 (2014).
32. Bartlett, P. D., Friedman, S. & Stiles, M. The reaction of isopropyllithium and *t*-butyllithium with simple olefins. *J. Am. Chem. Soc.* **75,** 1771–1772 (1953).

## Acknowledgements

This work was supported by a grant to the State Key Laboratory of Synthetic Chemistry from the Innovation and Technology Commission, the National Natural Science Foundation of China/Research Grants Council Joint Research Scheme (N_CUHK451/13), the Research Grants Council of the Hong Kong SAR, China (GRF Project 403012 and CRF projects), the Chinese Academy of Sciences-Croucher Foundation Funding Scheme for Joint Laboratories and the National Natural Science Foundation of China (NSFC no. 21272199). The Shenzhen Science and Technology Innovation Committee for the Municipal Key Laboratory Scheme (ZDSY20130401150914965) and the Shenzhen Basic Research Program (JCYJ20120619151721025, JCYJ20140425184428455) are also gratefully acknowledged.

## Author contributions

Z.J. performed the experiments and wrote the draft of the manuscript, Q.L. helped to perform experiments to assess scope of coupling-partners as well as to help preparing the manuscript. X.-S.P. and H.N.C.W. provided overall supervision. All authors discussed the results and commented on the manuscript.

# 10

# Induction and control of supramolecular chirality by light in self-assembled helical nanostructures

Jisung Kim[1], Jinhee Lee[1], Woo Young Kim[2], Hyungjun Kim[1], Sanghwa Lee[1], Hee Chul Lee[3], Yoon Sup Lee[1], Myungeun Seo[4] & Sang Youl Kim[1]

Evolution of supramolecular chirality from self-assembly of achiral compounds and control over its handedness is closely related to the evolution of life and development of supramolecular materials with desired handedness. Here we report a system where the entire process of induction, control and locking of supramolecular chirality can be manipulated by light. Combination of triphenylamine and diacetylene moieties in the molecular structure allows photoinduced self-assembly of the molecule into helical aggregates in a chlorinated solvent by visible light and covalent fixation of the aggregate via photopolymerization by ultraviolet light, respectively. By using visible circularly polarized light, the supramolecular chirality of the resulting aggregates is selectively and reversibly controlled by its rotational direction, and the desired supramolecular chirality can be arrested by irradiation with ultraviolet circularly polarized light. This methodology opens a route to ward the formation of supramolecular chiral conducting nanostructures from the self-assembly of achiral molecules.

[1] Department of Chemistry, Korea Advanced Institute of Science and Technology (KAIST), 291 Daehak-ro, Yuseong-gu, Daejeon 305-701, Korea.
[2] Department of Mechanical Engineering, KAIST, Daejeon 305-701, Korea. [3] Department of Electronic Engineering, KAIST, Daejeon 305-701, Korea.
[4] Graduate School of Nanoscience and Technology, KAIST, Daejeon 305-701, Korea. Correspondence and requests for materials should be addressed to M.S. (email: seomyungeun@kaist.ac.kr) or to S.Y.K. (email: kimsy@kaist.ac.kr).

Circularly polarized light (CPL) is considered to be a true chiral entity and has been attributed as the origin of the homochirality of biological molecules[1,2], as chiral information encoded in CPL can be transferred to molecules. Irradiation with CPL of a racemic mixture of chiral molecules can selectively enrich one enantiomer up to a small extent, called 'symmetry breaking'[3,4]. It has been speculated that this slight chiral bias might have been amplified via autocatalytic reactions or crystallization processes during the evolutionary process[5], resulting in homochirality. Amplification of the chiral information and macroscopic expression of chirality has been observed with liquid crystals, where the excess enantiomer generated by CPL acts as a chiral dopant and induces a chiral liquid crystalline phase[6–8]. Chiral information in CPL could be also transferred to helical polymers and expressed at a secondary structure level, as the handedness of the helix was controlled by the direction of CPL[9–11].

Supramolecular chirality has been generally induced by the self-assembly of homochiral molecules or a mixture of chiral and achiral molecules, where the molecular chirality is transferred (and amplified as well in case of the mixture) to the handedness of the self-assembled helical structure[12–15]. Only a few papers have described generation of supramolecular chirality by irradiation of azobenzene-containing achiral molecules with CPL, presumably through alignment of the azobenzene chromophores following the direction of CPL[16–18].

In contrast to the preceding research, here we apply CPL to a light-induced self-assembly process so that the CPL irradiation triggers the self-assembly and simultaneously transfers the chiral information to the self-assembled structure. As a key molecular motif for the light-induced self-assembly process, we use a triphenylamine (TPA) moiety that contains three phenyl rings connected to a central nitrogen atom similar to the blades of a propeller. TPA-containing molecules have been shown to self-assemble on light exposure in chlorinated solvents by the formation of triphenylammonium radicals that adopt more planar configurations than the neutral forms and facilitate stacking[19–23]. Simply by employing CPL as a light source in the self-assembly process of a TPA-containing molecule, we show that self-assembled aggregates with supramolecular chirality can be generated, driven by the helical stacking of the TPA moieties. Our data suggest that the chiral bias generated by irradiation of the TPA moiety with CPL is amplified by the self-assembly process, which promotes enrichment of one enantiomer by the enantioselective helical stacking and forms aggregates with

specific handedness. The handedness of the aggregates is exclusively dictated by the rotational direction of the CPL and can even be switched in a totally reversible manner. Furthermore, by incorporating a diacetylene (DA) moiety that undergoes topochemical photopolymerization by ultraviolet (254 nm)[24,25], we demonstrate that the desired handedness can be permanently locked by circularly polarized ultraviolet light (CPUL) irradiation, which knits the self-assembled structure with covalent bonds[26–28]. We believe this to be the first example of light-induced self-assembling systems where the entire process of induction, control and locking of supramolecular chirality can be solely manipulated by CPL.

## Results

**Synthesis and photoinduced self-assembly.** Achiral compound **1** was synthesized by amide coupling reaction between an amine-functionalized TPA molecule (**2**) and a carboxylic acid-functionalized DA molecule (**3**) as shown in Fig. 1.

Although the $^1$H nuclear magnetic resonance (NMR) spectrum of **1** in deuterated dimethyl sulfoxide (DMSO-$d_6$) showed clearly resolved aromatic and amide protons, irradiating with light from a fluorescent lamp for a few minutes on deuterated 1,2-dichloroethane (DCE-$d_4$) solution of **1** turned the colour of the solution from orange to green and its $^1$H NMR spectrum revealed no signal in the 6–11 p.p.m. range where the aromatic and amide protons should appear (Supplementary Fig. 1). This observation, consistent with the literature[20–22,29], indicates that a small number of triphenylammonium radicals were generated by light exposure and triggered stacking of the TPA moieties with the aid of amide hydrogen bonding. Chlorinated solvents seem to play an important role in the radical generation process, as solvent radicals are easily formed by light exposure, and reversibly transfer electrons to the TPA moiety. Formation of the triphenylammonium radicals is believed to facilitate stacking of the radicals, as $\pi$–$\pi$ interaction between the radicals becomes effective as they adopt a more planar configuration and electric interaction of the radical cation–anion pairs further stabilizes the stack. Simulation of a TPA-containing molecule similar to **1** suggests that the energy gain for the neutral precursor by joining the stack is greater than the energy penalty for the required configuration change to a more planar form; thus, eventually self-assembled aggregates containing a small number of triphenylammonium radicals would be generated.

The photoinduced self-assembly of **1** was observed in various chlorinated solvents and the morphology was visualized by field-emission scanning electron microscopy (FE-SEM) after evaporation of the solvent. Figure 2 shows the self-assembled structure of **1** formed in DCE, revealing fibrillar aggregates with several tens

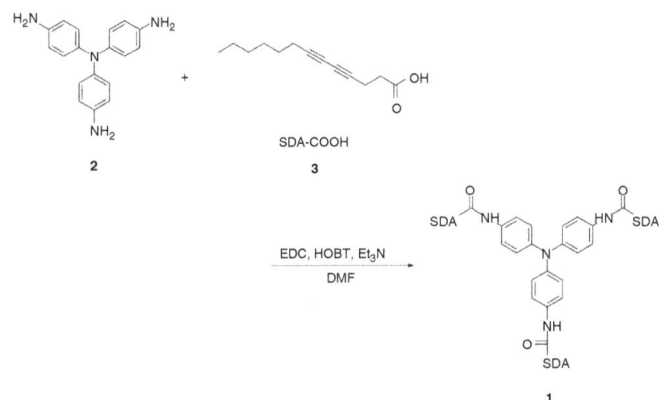

**Figure 1 | Synthesis of the compound 1.** The achiral compound **1** consisting of the TPA moiety in the centre and the diacetylene units at the periphery was synthesized by amide coupling reaction between a TPA-containing molecule **2** and a diacetylene-containing molecule **3**.

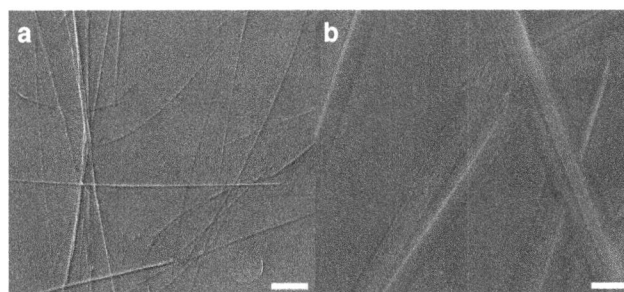

**Figure 2 | Fibrillar aggregates formed by light irradiation.** FE-SEM images of the aggregates of **1** formed in DCE solution (2 mg ml$^{-1}$ concentration) by visible light irradiation. The sample was coated with Au before imaging. (**a**) Low magnification. Scale bar, 10 μm. (**b**) High magnification. Scale bar, 500 nm.

of micrometres in length and an average diameter of 980 ($\pm$ 180) nm. A close look indicated that the fibril was composed of bundles of nanofibres with diameters of *ca.* 23 ($\pm$ 4.9) nm, which is consistent with the transmission electron microscopy images of the aggregates (Supplementary Fig. 2a). For comparison, an FE-SEM image of a sample prepared without light exposure showed no distinct morphology, suggesting stacking of the TPA moieties triggered by light exposure is important in the formation of the one-dimensional self-assembled structure (Supplementary Fig. 2b).

An X-ray diffraction (XRD) pattern of the solid obtained by the photoinduced self-assembly exhibited a single broad peak at $2\theta = 21°$ corresponding to spacing ($d$) of 0.42 nm, which is attributed to the intermolecular distance between stacked TPA moieties (Supplementary Fig. 3a). Appearance of the diffraction peaks in the small-angle regime also indicates periodicity of the aggregates at a nanometre-length scale (Supplementary Fig. 3b). We consider the peak at $2\theta = 3.0°$ in particular corresponds to $d = 2.94$ nm, originating from the distance between the stacks, assuming the intercalation of alkyl groups. Second- and third-order diffractions at $2\theta = 6.0°$ and $9.2°$ were also observed. Assuming a hexagonal geometry, the peak at $2\theta = 5.5°$ and $8.1°$ may be assigned as $\sqrt{3}$ and $\sqrt{7}$ diffractions. We also note that a similar TPA-containing molecule with long alkyl chains without DA moieties has been shown to form self-assembled nanofibres in solution, presumably by alkyl chain packing and amide hydrogen bonding, resulting in physical gelation of the solvent[30]. The existence of amide hydrogen bonding was also observed in our system by attenuated total reflection infrared spectroscopy of **1** after light irradiation and evaporation of solvent (Supplementary Fig. 4)[22,30,31].

The self-assembled structure of **1** was estimated by computation based on density functional theory (DFT). To properly predict the most thermodynamically stable structure, $\omega$B97X-D functional and 6–31G* basis set were used to approximate the electron density and molecular interactions of the self-assembled structure composed of large numbers of atoms[32]. Figure 3 depicts an optimized trimer obtained with a simplified molecular structure, suggesting that **1** self-assembles into helical stacks via hydrogen bonding of the amides (see Methods and Supplementary Fig. 5 for calculation details). The structure resembles the C3 symmetric helical supramolecular structures based on 1,3,5-benzenetricarboxylic acid and others thoroughly studied by Meijer and colleagues[33–35].

**Induction and control of supramolecular chirality by CPL.** When a normal light was used for the radical generation process, the solution after irradiation showed no circular dichroism (CD)

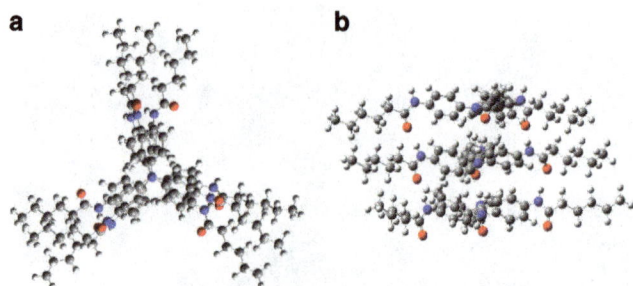

activity, suggesting that it contained a racemic mixture of both left- and right-handed helical aggregates of **1**. However, irradiation of the DCE solution of **1** with *l*-CPL produced intense CD activity with two positive maxima at 222 and 319 nm, and two negative maxima at 240 and 360 nm (Fig. 4a). The intensity increased over irradiation time and became level after 10 min (Fig. 4b).

The use of *r*-CPL produced the exact identical result, except with the opposite sign of CD activity, suggesting that helical aggregates with a desired handedness can be selectively generated by choice of the rotational direction of CPL[36,37]. More interestingly, the supramolecular chirality generated by *l*- or *r*-CPL could be reversibly switched by irradiation with the CPL of the opposite rotational direction. For instance, time-dependent CD activity change was monitored during irradiation of *r*-CPL on the DCE solution of **1**, which had been pre-irradiated by *l*-CPL (Fig. 4c). The initial CD activity generated by *l*-CPL was rapidly suppressed after 10 min of exposure, whereas the CD activity of the opposite sign evolved over time and reached the maximal value after 60 min. The maximal intensity achieved after 60 min was identical to the value that was obtained by irradiation of *r*-CPL on the unexposed solution to light, indicating full inversion of supramolecular chirality. As expected, chirality inversion by *l*-CPL produced the exactly identical result, except with the opposite direction of CD activity change (Fig. 4d). The inversion process could be reversibly repeated without any sign of loss of optical activity (Supplementary Fig. 7), but irradiation with non-polarized light decreased CD activity over time (Supplementary Fig. 8).

Although there has been no literature reporting the generation of supramolecular chirality using the TPA moiety, we suggest that irradiation with CPL induces radical formation at the nitrogen atom of the TPA moiety and also generates preferential chiral arrangements of the phenyl rings. As illustrated by the DFT calculation results, the phenyl rings in the TPA radical would be twisted with respect to the C3 symmetric plane resembling the blades of a propeller. We posit that the interaction of $\pi$ electrons in the phenyl rings with the angular momentum of CPL would determine the preferential twisting direction by photoresolution effect[38] to produce the TPA moiety with a certain enantiomeric form. 'Enantioselective' stacking of the TPA moieties possessing phenyl rings with the identical twisting direction should result in helical aggregates with one handedness. Simulation of CD spectra of the estimated self-assembled structure by DFT calculation suggests that indeed such a stacking structure can generate CD activity, which was qualitatively in agreement consistent with the experimental data, and *l*-CPL and *r*-CPL (as defined from the point of view of the receiver) produce left-handed and right-handed helicity, respectively (Supplementary Fig. 6). Chiral asymmetry would be more amplified, because only the TPA moieties with the same arrangements (that is, same enantiomer) would participate in the self-assembly process and be stabilized by the enthalpic gain. The fact that the evolution of CD activity on CPL irradiation was faster on the unexposed solution than the solution pre-irradiated by CPL suggests the process of supramolecular chirality inversion would include reversible dissociation of terminal molecules of a helical aggregate, chirality inversion at a molecular level and nucleation/growth of the helical aggregate with opposite handedness[39–41].

**Figure 3 | Self-assembled structure of 1 estimated by calculation.** The self-assembled structure of **1** (trimer) was calculated by DFT method. The molecular structure was simplified by substituting the side chains with *n*-pentyl groups. Images were taken from the top (**a**) and the side (**b**) of the stacks.

**Locking of supramolecular chirality by photopolymerization.** As noted above, the generated supramolecular chirality disappears under exposure to non-CPL such as ambient light from a common fluorescent lamp on the ceiling (Supplementary Fig. 8). To permanently lock the supramolecular chirality, we

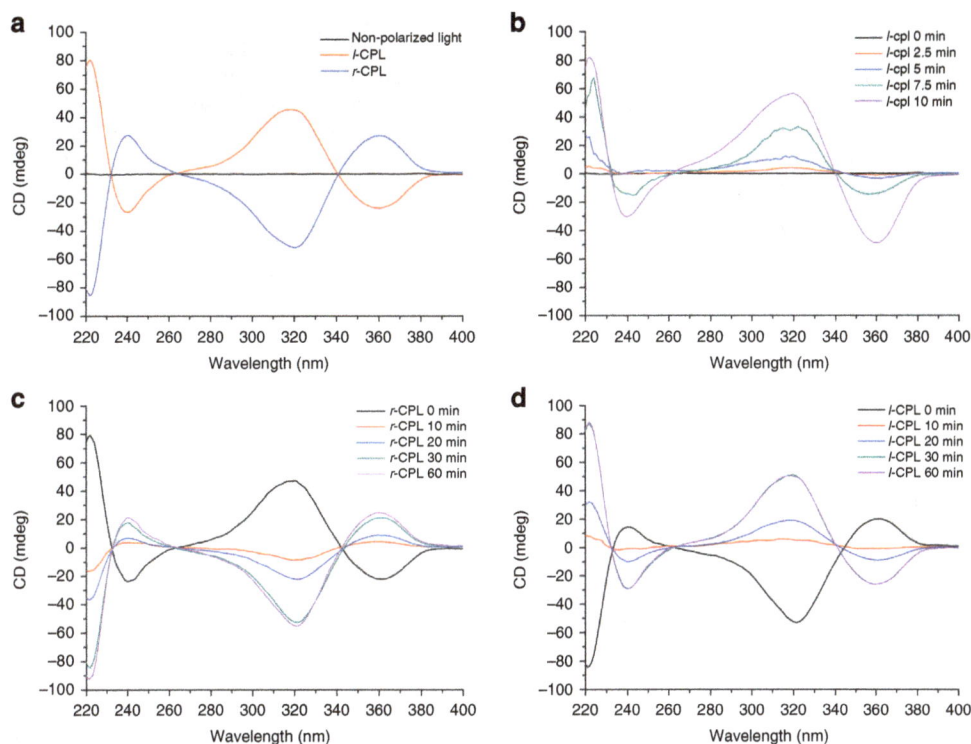

**Figure 4 | Induction and control of supramolecular chirality by CPL.** CD spectra of **1** in DCE solution after irradiation with CPL. (**a**) Spectra after irradiation with non-polarized light (black), l-CPL (red) and r-CPL (blue) for 10 min. (**b**) Time-dependent evolution of CD activity on irradiation with l-CPL. (**c**) Inversion of CD activity by irradiation with r-CPL of the solution that had been pre-exposed to l-CPL. (**d**) Inversion of CD activity by irradiation with l-CPL of the solution that had been pre-exposed to r-CPL.

exploited the photopolymerization of DA moieties, by applying ultraviolet irradiation (254 nm, non-polarized) to the DCE solution of **1**, which had been pre-irradiated with visible light and consisted of self-assembled aggregates. The photopolymerizability of the self-assembled structure was verified by Fourier transform (FT) Raman spectroscopy with an excitation wavelength of 633 nm. The sample obtained by spin coating of the ultraviolet-irradiated solution on glass showed clearly discernable peaks at 1,492 and 2,113 cm$^{-1}$ corresponding to $v(C=C)$ and $v(C\equiv C)$, respectively, of the conjugated DA polymer, indicating successful polymerization of DA (Fig. 5a)[42,43]. The broad background peak was attributed to the inherent fluorescence from the TPA moiety[44,45].

As a comparison, a sample not exposed to ultraviolet radiation exhibited almost no intensity at the characteristic vibration frequencies for the conjugated DA polymer. The result suggests that only preorganized DA moieties in the self-assembled structure can undergo topochemical polymerization, which requires an optimal and long-range arrangement of adjacent DA moieties. Stability of the polymerized assembly was also evaluated by dynamic light scattering analysis. Although the polymerized sample maintained its scattering intensity in DCE as well as in DMSO, the non-polymerized sample showed no scattering intensity in DMSO, indicating dissolution of the assembly into individual molecules (Supplementary Fig. 9).

SEM imaging did not reveal particular changes in the morphology of the aggregates after polymerization (Supplementary Fig. 10). XRD patterns of the polymerized sample were also similar to that of non-polymerized one, but showed more refined peaks in the small-angle regime, suggesting increased periodicity between stacks by polymerization of DA moieties (Supplementary Fig. 3c,d). We envision that lateral ordering of the alkyl chains would be self-corrected by topochemical polymerization to form more regular stacks. In

addition, formation of a new C–C bond per alkyl chain by the polymerization would restrict mobility of the alkyl chains and stiffen the covalently joined stacks, resulting in the more well-ordered packing of the stacks. The diffraction peaks in the wide-angle regime did not show a noticeable difference, suggesting aromatic packing was not affected by polymerization.

However, the use of non-polarized ultraviolet induced racemization of the self-assembled structure resulting in the polymerized assembly with no CD activity. To maintain the desired handedness during the photopolymerization process, we employed CPUL. Irradiation of the DCE solution of **1** with CPL and subsequent irradiation with CPUL of the same rotational direction produced polymerized aggregates with controlled handedness, corresponding to the rotational direction of the polarized light. The CD activity was completely retained, suggesting the helical arrangement was not affected by polymerization.

Moreover, the polymerized aggregates showed remarkable stability against inversion of supramolecular chirality on irradiation with CPL of the opposite rotational direction. In contrast to non-polymerized aggregates that changed their handedness completely after 60 min of CPL irradiation with the opposite rotational direction, the polymerized aggregates showed no sign of CD activity change on irradiation (Fig. 5b,c). The result clearly indicates photopolymerization of the DA moieties by CPUL successfully locked the supramolecular chirality.

We also tested whether the polymerized aggregate with certain handedness could nucleate further stacking of **1**, following its handedness, under non-CPL irradiation. However, we observed that the CD activity as well as the particle size did not noticeably increase after the non-CPL irradiation, indicating that new aggregates were formed with a 50:50 population of both handedness (Supplementary Fig. 11a). It seems that the aggregate end may adopt a slightly different configuration and the enthalpic gain obtained by interaction of **1** and the aggregate end may not be greater than between **1**.

The whole process of induction, control and locking of supramolecular chirality by CPL and CPUL is summarized in Fig. 6. Irradiation with *r*- or *l*-CPL induces self-assembly of **1** in chlorinated solution and control over handedness of the forming helical aggregates. The resulting handedness is totally switchable by irradiating with CPL of the opposite rotational direction. Subsequent irradiation with CPUL of the identical rotational direction polymerizes adjacent DA moieties, to convert the self-assembled aggregate into a covalently joined nano-object, and permanently locks the handedness. To our knowledge, this is the first example of controlling and fixing supramolecular chirality from an achiral compound by light.

**Conductivity measurement.** As it was previously demonstrated that one-dimensional self-assembled structures of TPA derivatives formed by photoirradiation have conductivity along the assembly axis[19,46-48], we also investigated the current–voltage relationship of the self-assembled structures of **1**, to determine the effect of supramolecular chirality and photopolymerization of the DA moieties on conductivity (Supplementary Fig. 12). When the DCE solution of **1** was placed on a gold nanogap of 500 nm with 2 V of continuous potential difference and visible light was turned on, current between the nanogap increased by approximately four orders of magnitude after 100 s giving conductivity of $1.4\,S\,cm^{-1}$, which was comparable to other conducting organic materials. Consistent with the previous result, conductivity was maintained in the solution state under light exposure (that is, as long as radicals were being continuously generated) and could even be reversibly controlled by turning the light on and off (Supplementary Fig. 12b).

Interestingly, the conductivity of the photopolymerized aggregates was similar to the non-polymerized ones, and was also modulated by light irradiation, despite the fact that the covalently joined self-assembled structure could not dissociate under any circumstances (Supplementary Fig. 12c). The result suggests photopolymerization of the DA moieties does not affect the electrical properties of the TPA moieties and the key to conductivity is maintaining stable radicals at the nitrogen atom.

## Discussion

We developed a light-controlled self-assembling system that can evolve and arrest supramolecular chirality following the

information encoded in CPL. Our results suggest that the chiral bias generated by CPL can not only be transferred to a supramolecular level but also considerably amplified when combined with a self-assembly process, which preferentially incorporates one enantiomer into the self-assembled helical structure with specific handedness dictated by the rotational direction of CPL. This methodology provides complete control over the entire evolutionary process of supramolecular chirality using light and will be useful for future applications such as chiral sensors and chiral discrimination.

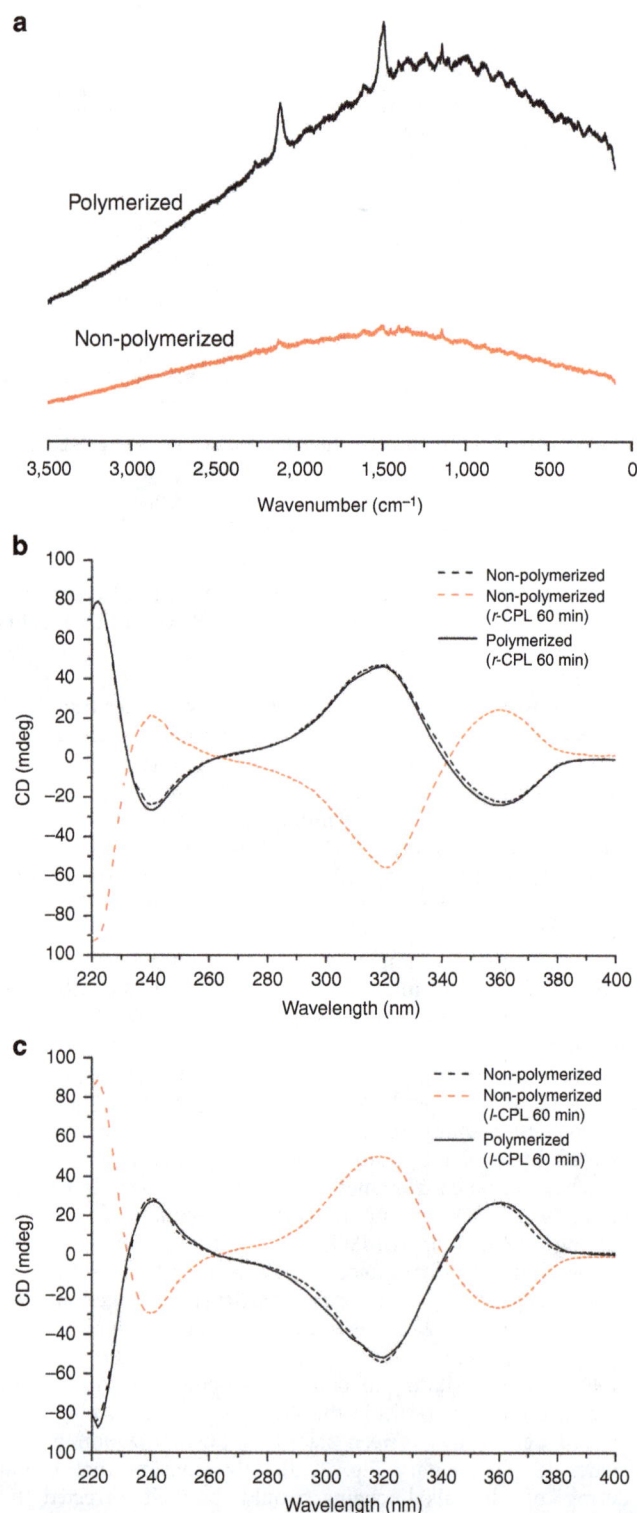

**Figure 5 | Locking of supramolecular chirality by photopolymerization.** (**a**) Fourier-transform Raman (FT-Raman) spectra of spin-coated films of DCE solution of **1** on glass. Film prepared after photopolymerization in solution (black solid line). Film prepared without polymerization (red solid line). (**b,c**) Persistence of CD activity of DCE solution of **1** after photopolymerization by CPUL against inversion by irradiation with CPL of the opposite rotational direction. The polymerized aggregate obtained by sequential irradiation with CPL and CPUL of the identical rotational direction possesses permanent handedness. Notice that the CD spectrum of the polymerized sample (black solid line) does not change its CD activity on irradiation with CPL of the opposite rotation direction for 60 min; the spectrum almost completely overlaps the original CD spectrum of the sample before photopolymerization (black dashed line). For comparison, inversion of CD activity of the non-polymerized sample on irradiation with CPL of the opposite direction was also shown (red dashed line, same data from Fig. 4). (**b**) Spectra of **1** polymerized with *l*-CPUL and then exposed to *r*-CPL (black solid line). Spectra of non-polymerized **1** after irradiation with *l*-CPL (black dashed line) and then exposed to *r*-CPL (red dashed line) are also shown. (**c**) Spectra of **1** obtained with CPL and CPUL of the opposite rotational direction.

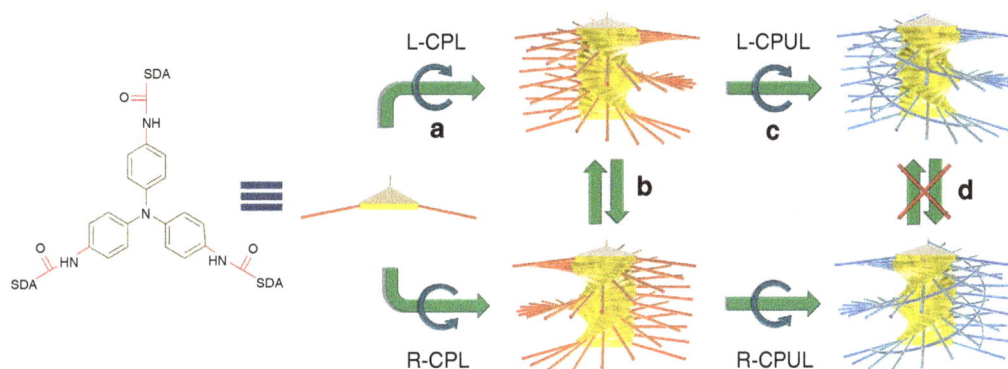

**Figure 6 | Manipulation of supramolecular chirality entirely by light.** The whole process of induction, control and locking of supramolecular chirality by CPL and CPUL from an achiral compound **1** containing TPA (yellow) and DA (red: non-polymerized, blue: polymerized) moieties is schematically depicted. Irradiation with CPL induces self-assembly of **1** with control over handedness (**a**) and the handedness can be reversibly switched by irradiating with the counter CPL (**b**). Photopolymerization of the DA moieties by CPUL irradiation knits the self-assembled structure to produce a covalently joined nano-object (**c**) and the supramolecular chirality is permanently locked, exhibiting no further handedness change on CPL irradiation (**d**).

## Methods

**Materials.** 1-Hydroxybenzotriazole monohydrate (HOBT) (97%) and 1-(3-dimethylaminopropyl)-3-ethylcarbodiimide hydrochloride (EDC) (98%) were purchased from Tokyo Chemical Industry Co. Triethylamine (Et$_3$N) (99.5%) was purchased from Sigma-Aldrich. N,N-Dimethylformamide (DMF) (99.5%) and DCE were purchased from Junsei Chemical Co., Ltd. All chemicals were used as received. Tris(4-aminophenyl)amine (**2**)[49] and trideca-4,6-diynoic acid (**3**)[50] were synthesized according to the literature procedures.

**Characterization.** $^1$H and $^{13}$C NMR spectroscopies were conducted using a Bruker Science Avance 400 MHz NMR spectrometer with the residual solvent signal as an internal reference. High-resolution mass spectroscopy was performed on a Bruker Daltonik microTOF-Q II mass spectrometer using electrospray ionization method. Elemental analysis was conducted using a Thermo Scientific FLASH 2000 series elemental analyser. FE-SEM was performed on a FEI company Nova 230 SEM. Samples were prepared on a silicon wafer by dropping DCE solution of **1** (2 mg ml$^{-1}$) and evaporation of the solvent. The samples were coated with Au before imaging. Transmission electron microscopy was performed on a JEOL Ltd. JEM-2100F 200 kV field-emission transmission electron microscope. Samples were prepared on a 300-mesh carbon film by dropping DCE solution of **1** (2 mg ml$^{-1}$) and evaporation of the solvent. The samples were stained by exposing ruthenium tetroxide vapour before imaging. XRD patterns were obtained with a Rigaku D/MAX-2500 X-ray diffractometer with scan speed of 2° per minute and sampling width of 0.01°. Cu $K_\alpha$ ($\lambda = 0.154$ nm) was used as a light source. Samples for wide-angle XRD were prepared by spin coating of DCE solution of **1** (2 mg ml$^{-1}$) on a glass substrate. For small-angle XRD experiments, solid obtained by evaporation of the DCE solution was ground and loaded on a sample holder using carbon tape. Fourier-transform Raman spectra were obtained on an ARAMIS Horiba Jobin Yvon spectrometer with 633 nm He-Ne laser source. Samples were prepared by spin coating of DCE solution of **1**. Dynamic light scattering measurement was performed on a Brookhaven Instruments Corp. 90Plus/BI-MAS particle size analyser at wavelength of 658 nm with scattering angle of 90 °. Samples were prepared by dilution of DCE solution of **1** (2 mg ml$^{-1}$) into 1/20 concentration (0.1 mg ml$^{-1}$). CD spectra were obtained using Jasco Inc., Jasco-815 Spectropolarimeter equipped with temperature controller. Measurement was conducted with DCE solution of **1** in a 0.2-nm quarts cell at $-10$ °C using scan speed of 200 nm min$^{-1}$ and sampling width of 1 nm. FT infrared spectroscopy was performed by attenuated total reflection infrared spectroscopy with a Bruker Science Platinum-ATR Alpha FT-infrared spectrometer. Samples were prepared by dropping the DCE solution of **1** (2 mg ml$^{-1}$) on a sample window.

**Synthesis of tris(4-trideca-4,6-diynamidophenyl)amine (1).** Tris(4-aminophenyl)amine (0.209 g, 0.720 mmol), trideca-4,6-diynoic acid (0.520 g, 2.520 mmol) and 1-hydroxybenzotriazole monohydrate (0.486 g, 3.60 mmol) were dissolved in 50 ml of N,N-dimethylformamide. The reaction mixture was stirred at 0 °C for 30 min. Next, 1-(3-dimethylaminopropyl)-3-ethylcarbodiimide hydrochloride (0.690 g, 3.60 mmol) and triethylamine (0.364 g, 3.60 mmol) were added into the solution and the reaction mixture was warmed up to room temperature (RT). After stirring for 12 h, the reaction mixture was poured into brine and extracted with tetrahydrofuran. The organic layer was washed with water, dried over anhydrous MgSO$_4$ and condensed in a rotary evaporator, to yield a crude product. Column chromatography on silica gel with tetrahydrofuran/hexane 1:1 as an eluent and recrystallization in ethanol produced pure **1** as a red solid (0.219 g, yield 35.57%). Electrospray ionization–mass spectrometry, m/e 877.50 for $[M + Na]^+$ (Calcd:

854.51). $\delta_H$ (400 MHz, DMSO-$d_6$; p.p.m.) 9.89 (3H, s, NHCO), 7.44 (6H, d, $J = 8.9$ Hz, Ar), 6.87 (6H, d, $J = 8.8$ Hz, Ar), 2.54 (12H, t, $J = 7.2$ Hz, C≡CCH$_2$), 2.23 (6H, t, $J = 6.9$ Hz, COCH$_2$), 1.40 (6H, p, $J = 6.7$ Hz, C≡CCH$_2$CH$_2$), 1.29 (6H, m, CH$_2$CH$_3$), 1.20 (12H, m, CH$_2$CH$_2$CH$_2$CH$_2$), 0.80 (9H, t, $J = 6.6$ Hz, CH$_3$). $\delta_C$ (100 MHz, DMSO-$d_6$; p.p.m.) 168.66, 142.78, 133.95, 123.61, 120.35, 78.38, 76.99, 65.43, 65.24, 34.66, 30.63, 27.85, 27.63, 21.95, 18.26, 14.80, 13.88. Anal. Calcd for C$_{57}$H$_{66}$N$_4$O$_3$: C 80.06, H 7.78, N 6.55; Found: C 80.08, H 7.82, N 6.65.

**Photoinduced self-assembly of 1.** DCE solution of **1** (2 mg ml$^{-1}$) was irradiated with a hand-held lamp having 400–600 nm range of wavelength and 450 nm of maximum intensity with power of 1.5 W at RT. CPL was produced by placing a linear polarizer and quarter-lambda wave plate in front of the light source. The lamp was placed at 3 cm from each sample in every experiment and external light was prevented to leak into experimental samples.

**DFT calculation.** DFT calculation was conducted using Gaussian 09 (ref. 51). The long-range corrected hybrid density functional with damped atom–atom dispersion correction, $\omega$B97X-D, was used[52]. This functional can handle $\pi$-$\pi$ interaction empirically and have been successfully used to model polymer structures with large number of atoms[53–55]. The 6-31G* basis set was used to describe atomic orbitals. For efficient calculation, we simplified the molecular structure by assuming a planar C3 symmetric structure for the TPA moiety and substituting the side chains with n-pentyl groups and then optimized its pentamer structure. A unit positive charge was added to the pentamer and the doublet spin state was calculated to account for the electronic effect of the cationic radical species in the aggregate.

The optimized pentamer structure is shown in Supplementary Fig. 5. The trimer structure shown in Fig. 2 was obtained from the optimized pentamer by removing the terminal molecule, to avoid unrealistic end group contribution and illustrate a proper arrangement in the middle of the stack. In the trimer, the degree of rotation per molecule was 12.1° on average and intermolecular distance between the TPA moieties was estimated to be 0.436 nm, which was comparable to the spacing of 0.42 nm obtained by the XRD analysis. We also note that amide hydrogen bonding between the alkyl chains of the adjacent molecules reinforces the assembly. The intermolecular distance between the terminal methyl carbons was calculated to be an average 0.454 nm, suggesting the feasibility of topochemical polymerization of the adjacent DA groups by light[56–58].

CD spectra were simulated based on the time-dependent DFT method[59]. The trimer geometry was employed to calculate the excitation energy assuming that the geometric change during the electronic excitation would be negligible. The same functional $\omega$B97X-D was used to calculate the electronic excitation energy, but the size of the basis set was reduced to 3-21G* due to the computational cost. Only singlet excited states was considered and the number of roots for the singlet state was 30. The raw Gaussian 09 output file was processed using GaussSum[60], to yield CD spectra.

**Conductivity measurement.** Au nanogap was fabricated as follows. On the thermally grown SiO$_2$/Si wafer cleaned by a mixture of H$_2$SO$_4$ and H$_2$O$_2$ for 10 min, Cr/Au was thermally evaporated at $1 \times 10^{-6}$ torr to form a Cr layer with 10 nm thick and a Au layer with 40 nm thick, respectively. By conventional photolithography, metal lines with width of 44 $\mu$m were patterned. After then, a focused ion beam (FB-2100, Hitachi, Co.) was used to form a 500-nm gap. In light and electromagnetic wave shielded probe station at RT, the fabricated nanogap covered with DCE solution of **1** was placed in a sealed vessel made of quartz.

Visible light (1.5 W) were exposed to samples through the quartz vessel and 2 V of continuous potential difference was held across the nanogap. Photopolymerization of assembled structure of **1** between nanogap was conducted by irradiation of ultraviolet from 8 W hand-held ultraviolet lamp at the distance of 3 cm through the quartz vessel. The current–voltage ($I$–$V$) relationships were obtained by precision semiconductor parameter analyser (HP 4156A, Hewlett Packard Co.).

## References

1. Meinert, C. *et al.* Photonenergy-controlled symmetry breaking with circularly polarized light. *Angew. Chem. Int. Ed.* **53**, 210–214 (2014).
2. Bailey, J. *et al.* Circular polarization in star-formation regions: implications for biomolecular homochirality. *Science* **281**, 672–674 (1998).
3. Feringa, B. L. & van Delden, R. A. Absolute asymmetric synthesis: the origin, control, and amplification of chirality. *Angew. Chem. Int. Ed.* **38**, 3419–3438 (1999).
4. Inoue, Y. Asymmetric photochemical-reactions in solution. *Chem. Rev.* **92**, 741–770 (1992).
5. Blackmond, D. G. The origin of biological homochirality. *Cold Spring Harb. Perspect. Biol.* **2**, a002147 (2010).
6. Huck, N. P. M., Jager, W. F., deLange, B. & Feringa, B. L. Dynamic control and amplification of molecular chirality by circular polarized light. *Science* **273**, 1686–1688 (1996).
7. Tejedor, R. M., Oriol, L., Serrano, J. L., Urena, F. P. & Gonzalez, J. J. L. Photoinduced chiral nematic organization in an achiral glassy nematic azopolymer. *Adv. Funct. Mater.* **17**, 3486–3492 (2007).
8. Burnham, K. S. & Schuster, G. B. Transfer of chirality from circularly polarized light to a bulk material property: propagation of photoresolution by a liquid crystal transition. *J. Am. Chem. Soc.* **121**, 10245–10246 (1999).
9. Muller, M. & Zentel, R. Interplay of chiral side chains and helical main chains in polyisocyanates. *Macromolecules* **29**, 1609–1617 (1996).
10. Li, J., Schuster, G. B., Cheon, K. S., Green, M. M. & Selinger, J. V. Switching a helical polymer between mirror images using circularly polarized light. *J. Am. Chem. Soc.* **122**, 2603–2612 (2000).
11. Yang, G. *et al.* Enantioselective synthesis of helical polydiacetylenes in the visible light region. *Chem. Commun.* **50**, 2338–2340 (2014).
12. Green, M. M. *et al.* A helical polymer with a cooperative response to chiral information. *Science* **268**, 1860–1866 (1995).
13. Green, M. M. *et al.* The macromolecular route to chiral amplification. *Angew. Chem. Int. Ed.* **38**, 3139–3154 (1999).
14. Palmans, A. R. A. & Meijer, E. W. Amplification of chirality in dynamic supramolecular aggregates. *Angew. Chem. Int. Ed.* **46**, 8948–8968 (2007).
15. Palmans, A. R. A., Vekemans, J. A. J. M., Havinga, E. E. & Meijer, E. W. Sergeants-and-soldiers principle in chiral columnar stacks of disc-shaped molecules with c-3 symmetry. *Angew. Chem. Int. Ed.* **36**, 2648–2651 (1997).
16. Choi, S. W. *et al.* Circular-polarization-induced enantiomeric excess in liquid crystals of an achiral, bent-shaped mesogen. *Angew. Chem. Int. Ed.* **45**, 1382–1385 (2006).
17. Nikolova, L. *et al.* Photoinduced circular anisotropy in side-chain azobenzene polyesters. *Opt. Mater.* **8**, 255–258 (1997).
18. Iftime, G., Labarthet, F. L., Natansohn, A. & Rochon, P. Control of chirality of an azobenzene liquid crystalline polymer with circularly polarized light. *J. Am. Chem. Soc.* **122**, 12646–12650 (2000).
19. Faramarzi, V. *et al.* Light-triggered self-construction of supramolecular organic nanowires as metallic interconnects. *Nat. Chem.* **4**, 485–490 (2012).
20. Moulin, E. *et al.* The hierarchical self-assembly of charge nanocarriers: a highly cooperative process promoted by visible light. *Angew. Chem. Int. Ed.* **49**, 6974–6978 (2010).
21. Moulin, E. *et al.* Light-triggered self-assembly of triarylamine-based nanospheres. *Nanoscale* **4**, 6748–6751 (2012).
22. Armao, J. J. *et al.* Healable supramolecular polymers as organic metals. *J. Am. Chem. Soc.* **136**, 11382–11388 (2014).
23. Nyrkova, I. *et al.* Supramolecular self-assembly and radical kinetics in conducting self-replicating nanowires. *ACS Nano* **8**, 10111–10124 (2014).
24. Okada, S., Peng, S., Spevak, W. & Charych, D. Color and chromism of polydiacetylene vesicles. *Acc. Chem. Res.* **31**, 229–239 (1998).
25. Sun, X. M., Chen, T., Huang, S. Q., Li, L. & Peng, H. S. Chromatic polydiacetylene with novel sensitivity. *Chem. Soc. Rev.* **39**, 4244–4257 (2010).
26. Kim, J. H., Seo, M. & Kim, S. Y. Lithographically patterned breath figure of photoresponsive small molecules: dual-patterned honeycomb lines from a combination of bottom-up and top-down lithography. *Adv. Mater.* **21**, 4130–4133 (2009).
27. Hsu, L., Cvetanovich, G. L. & Stupp, S. I. Peptide amphiphile nanofibers with conjugated polydiacetylene backbones in their core. *J. Am. Chem. Soc.* **130**, 3892–3899 (2008).
28. Weiss, J., Jahnke, E., Severin, N., Rabe, J. P. & Frauenrath, H. Consecutive conformational transitions and deaggregation of multiple-helical poly(diacetylene)s. *Nano Lett.* **8**, 1660–1666 (2008).
29. Ishi-i, T. *et al.* Self-assembled triphenylamine-hexaazatriphenylene two-photon absorption dyes. *Tetrahedron* **69**, 29–37 (2013).
30. Yasuda, Y., Takebe, Y., Fukumoto, M., Inada, H. & Shirota, Y. 4,4',4''-tris(stearoylamino)triphenylamine as a novel material for functional molecular gels. *Adv. Mater.* **8**, 740–741 (1996).
31. Kamiyama, T., Yasuda, Y. & Shirota, Y. A novel family of low molecular-weight organic gels. 1,3,5-tris(n-phenyl-n-4-stearoylaminophenylamino)benzene and 4,4',4''-tris(n-phenyl-n-4-stearoylaminophenylamino)triphenylamine/organic solvent systems. *Polym. J.* **31**, 1165–1170 (1999).
32. Teberekidis, V. I. & Sigalas, M. P. Theoretical study of hydrogen bond interactions of felodipine with polyvinylpyrrolidone and polyethyleneglycol. *J. Mol. Struct. (Theochem)* **803**, 29–38 (2007).
33. van Gorp, J. J., Vekemans, J. A. J. M. & Meijer, E. W. C-3-symmetrical supramolecular architectures: Fibers and organic gels from discotic trisamides and trisureas. *J. Am. Chem. Soc.* **124**, 14759–14769 (2002).
34. Smulders, M. M. J., Schenning, A. P. H. J. & Meijer, E. W. Insight into the mechanisms of cooperative self-assembly: The 'sergeants-and-soldiers' principle of chiral and achiral c-3-symmetrical discotic triamides. *J. Am. Chem. Soc.* **130**, 4204–4204 (2008).
35. Gillissen, M. A. J. *et al.* Triple helix formation in amphiphilic discotics: demystifying solvent effects in supramolecular self-assembly. *J. Am. Chem. Soc.* **136**, 336–343 (2014).
36. Jung, J. H., Ono, Y. & Shinkai, S. Sol-gel polycondensation in a cyclohexane-based organogel system in helical silica: Creation of both right- and left-handed silica structures by helical organogel fibers. *Chem. Eur. J.* **6**, 4552–4557 (2000).
37. Kuzyk, A. *et al.* DNA-based self-assembly of chiral plasmonic nanostructures with tailored optical response. *Nature* **483**, 311–314 (2012).
38. Tang, Y. Q. & Cohen, A. E. Optical chirality and its interaction with matter. *Phys. Rev. Lett.* **104**, 163901 (2010).
39. de Greef, T. F. A. & Meijer, E. W. Materials science—supramolecular polymers. *Nature* **453**, 171–173 (2008).
40. Jonkheijm, P., van der Schoot, P., Schenning, A. P. H. J. & Meijer, E. W. Probing the solvent-assisted nucleation pathway in chemical self-assembly. *Science* **313**, 80–83 (2006).
41. Korevaar, P. A. *et al.* Pathway complexity in supramolecular polymerization. *Nature* **481**, 492–496 (2012).
42. Luo, L. *et al.* Poly(diiododiacetylene): Preparation, isolation, and full characterization of a very simple poly(diacetylene). *J. Am. Chem. Soc.* **130**, 7702–7709 (2008).
43. Shchegolikhin, A. N. & Lazareva, O. L. Nir-ft raman image of solid-state polymerization of pts diacetylene. *Spectrochim. Acta A* **53**, 67–79 (1997).
44. Xu, W. J., Chen, H. Z., Shi, M. M., Huang, Y. G. & Wang, M. Poly(triphenylamine) related copolymer noncovalently coated mwcnt nanohybrid: fabrication and observation of enhanced photoconductivity. *Nanotechnology* **17**, 728–733 (2006).
45. Park, K. N., Cho, Y. R., Kim, W., Park, D. W. & Choe, Y. Raman spectra and current-voltage characteristics of 4,4',4''-tris(2-naphthylphenylamino)triphenylamine thin films. *Mol. Cryst. Liq. Cryst.* **498**, 183–192 (2009).
46. Shirota, Y. Organic materials for electronic and optoelectronic devices. *J. Mater. Chem.* **10**, 1–25 (2000).
47. Thelakkat, M. Star-shaped, dendrimeric and polymeric triarylamines as photoconductors and hole transport materials for electro-optical applications. *Macromol. Mater. Eng.* **287**, 442–461 (2002).
48. Shirota, Y. & Kageyama, H. Charge carrier transporting molecular materials and their applications in devices. *Chem. Rev.* **107**, 953–1010 (2007).
49. Juang, T. Y. *et al.* A reactive modifier that enhances the thermal mechanical properties of epoxy resin through the formation of multiple hydrogen-bonded network. *J. Polym. Res.* **18**, 1169–1176 (2011).
50. Nie, X. P. & Wang, G. J. Synthesis and self-assembling properties of diacetylene-containing glycolipids. *J. Org. Chem.* **71**, 4734–4741 (2006).
51. Frisch, M. J. *et al.*Gaussian, Inc., Wallingford, CT (2009).
52. Chai, J. D. & Head-Gordon, M. Long-range corrected hybrid density functionals with damped atom-atom dispersion corrections. *Phys. Chem. Chem. Phys.* **10**, 6615–6620 (2008).
53. Wu, J. S. *et al.* Synthesis and morphology studies of a poly(5,6-difluorobenzo-2,1,3-thiadiazole-4,7-diyl-alt-quaterchalcogenophene) copolymer with 7.3% polymer solar cell efficiency. *Polym. Chem.* **5**, 6472–6479 (2014).
54. Hong, S. *et al.* Non-covalent self-assembly and covalent polymerization co-contribute to polydopamine formation. *Adv. Funct. Mater.* **22**, 4711–4717 (2012).
55. Salzner, U. & Aydin, A. Improved prediction of properties of pi-conjugated oligomers with range-separated hybrid density functionals. *J. Chem. Theory Comput.* **7**, 2568–2583 (2011).
56. Rosenthal, M. *et al.* A diacetylene-containing wedge-shaped compound: synthesis, morphology, and photopolymerization. *Chem. Eur. J.* **19**, 4300–4307 (2013).

57. Masuda, M., Jonkheijm, P., Sijbesma, R. P. & Meijer, E. W. Photoinitiated polymerization of columnar stacks of self-assembled trialkyl-1,3,5-benzenetricarboxamide derivatives. *J. Am. Chem. Soc.* **125,** 15935–15940 (2003).

58. Lauher, J. W., Fowler, F. W. & Goroff, N. S. Single-crystal-to-single-crystal topochemical polymerizations by design. *Acc. Chem. Res.* **41,** 1215–1229 (2008).

59. You, L., Pescitelli, G., Anslyn, E. V. & Di Bari, L. An exciton-coupled circular dichroism protocol for the determination of identity, chirality, and enantiomeric excess of chiral secondary alcohols. *J. Am. Chem. Soc.* **134,** 7117–7125 (2012).

60. O'Boyle, N. M., Tenderholt, A. L. & Langner, K. M. Cclib: A library for package-independent computational chemistry algorithms. *J. Comput. Chem.* **29,** 839–845 (2008).

## Acknowledgements

This work was supported by a National Research Foundation of Korea (NRF) grant funded by MSIP through the NRL (R0A-2008-000-20121-0) and the ERC (R11-2007-050-00000-0) programmes.

## Author contributions

J.K. and J.L. proposed the research. J.K., J.L. and S.L. synthesized and characterized the molecule. J.K., W.Y.K. and H.C.L. measured *I*–*V* relationship. H.K. and Y.L. calculated the model structure. J.K. and M.S. investigated the self-assembly and chirality. J.K. and M.S. prepared the manuscript. S.Y.K. supervised the whole project and revised the manuscript.

# 11

# *De novo* branching cascades for structural and functional diversity in small molecules

Miguel Garcia-Castro[1], Lea Kremer[1,2], Christopher D. Reinkemeier[1], Christian Unkelbach[2], Carsten Strohmann[2], Slava Ziegler[1], Claude Ostermann[3] & Kamal Kumar[1,2]

The limited structural diversity that a compound library represents severely restrains the discovery of bioactive small molecules for medicinal chemistry and chemical biology research, and thus calls for developing new divergent synthetic approaches to structurally diverse and complex scaffolds. Here we present a *de novo* branching cascades approach wherein simple primary substrates follow different cascade reactions to create various distinct molecular frameworks in a scaffold diversity phase. Later, the scaffold elaboration phase introduces further complexity to the scaffolds by creating a number of chiral centres and incorporating new hetero- or carbocyclic rings. Thus, employing *N*-phenyl hydroxylamine, dimethyl acetylenedicarboxylate and allene ester as primary substrates, a compound collection of sixty one molecules representing seventeen different scaffolds is built up that delivers a potent tubulin inhibitor, as well as inhibitors of the Hedgehog signalling pathway. This work highlights the immense potential of cascade reactions to deliver compound libraries enriched in structural and functional diversity.

[1] Max-Planck-Institut für Molekulare Physiologie, Abteilung Chemische Biologie, Otto-Hahn-Strasse 11, 44227 Dortmund, Germany. [2] Technische Universität Dortmund, Fakultät für Chemie und Chemische Biologie, Otto-Hahn-Strasse 6, 44227 Dortmund, Germany. [3] Compound Management and Screening Center (COMAS), Max-Planck-Institut für Molekulare Physiologie, Otto-Hahn-Strasse 11, 44227 Dortmund, Germany. Correspondence and requests for materials should be addressed to K.K. (email: kamal.kumar@mpi-dortmund.mpg.de).

Structural diversity has a profound impact on the performance of a compound collection exposed to biological screenings[1-3]. Probe and drug discovery research, therefore, beseech quality-based compound libraries rich in structural diversity[4,5]. The latter in turn is primarily determined by the number of diverse scaffolds or chemotypes that represent a compound library[6-8]. Consequently, new synthetic challenges have emerged aiming at divergent access to structurally distinct and complex scaffolds[9-11]. Different approaches have been developed to incorporate structural diversity to a compound collection, for instance, in the build-couple-pair strategies[12,13], folding[14] and branching pathways[15-17], structural variations either in the building blocks or the reacting partners of common substrates derive the formation of new scaffolds. Synthesis of natural product scaffold-based[18-21] compound libraries either employs accessible complex natural products or their derivatives for generating new and complex scaffold structures or build up compound libraries around privileged scaffolds[1,22]. However, most of the above mentioned synthesis designs require carefully functionalized substrates, as well as their reacting partners and often deliver structural diversity in a compound library at the cost of tedious multistep synthesis protocol. Nature displays it's amazing ability to assemble a limited pool of simple building blocks into structurally and functionally diverse natural products[23,24]. For instance, terpenes that represent a large class of natural products are formally generated from only one biosynthetic unit[25]. Infact, terpenes and sesquiterpenes represent interesting examples of de novo biosynthesis designs where compound collections are generated in two important phases; the first cyclase phase builds up scaffolds from simple acyclic substrates with the help of cyclase enzymes and the next phase elaborates these scaffolds, for instance, with oxidative modifications, to generate a number of diversely functionalized molecules (Fig. 1a)[25,26].

In contrast to divergent biosynthetic designs, laboratory syntheses targeting diverse scaffolds either choose the substrates already equipped with the ring systems desired in different products or follow multistep and tedious synthetic validations for every target scaffold before producing a library of molecules. Cascade or domino reaction sequences are highly efficient synthetic tools that rapidly build up molecular complexity[27,28]. However, their potential in scaffold diversity synthesis remains underexplored. The 'branching cascades' strategy is a scaffold diversity synthesis approach wherein common precursors follow different cascade or domino reactions and provide structurally distinct scaffolds for library synthesis[29]. Inspired by the de novo biogenesis of secondary metabolites, we envisioned that the de novo branching cascades could be designed to transform simple acyclic substrates into appreciable scaffold diversity (Fig. 1b). In the absence of enzyme weaponry that nature exploits in biogenesis of diverse natural products, the de novo scaffold diversity synthesis with cascade reactions is a formidable synthetic challenge that seeks careful reaction design. Herein we report our efforts towards scaffold diversity synthesis with a de novo branching cascades strategy.

Taking cognizance of the biosynthetic library design, we planned to first establish a scaffold diversity phase (SD phase) by transforming simple acyclic substrates into diverse ring systems via branching cascades. While the same set of simple substrates generates different scaffolds, molecular properties like molecular weight and clogP of the resulting distinct scaffolds remain within a threshold limit and can be controlled by small structural variations in the substrates. Therefore, there is a large scope to further modify the generated scaffolds and their appended functionalities in subsequent scaffold elaboration phase (SE phase) and deliver a compound collection rich in structural diversity and molecular complexity (Fig. 1b). To establish the principle, a de novo cascade reaction design is chosen in which simple acyclic substrates, N-phenylhydroxylamine, acetylenedicarboxylates and allene esters undergo branching cascades reactions under different reaction conditions to provide seven distinct scaffolds in SD phase. Four of these scaffolds are further

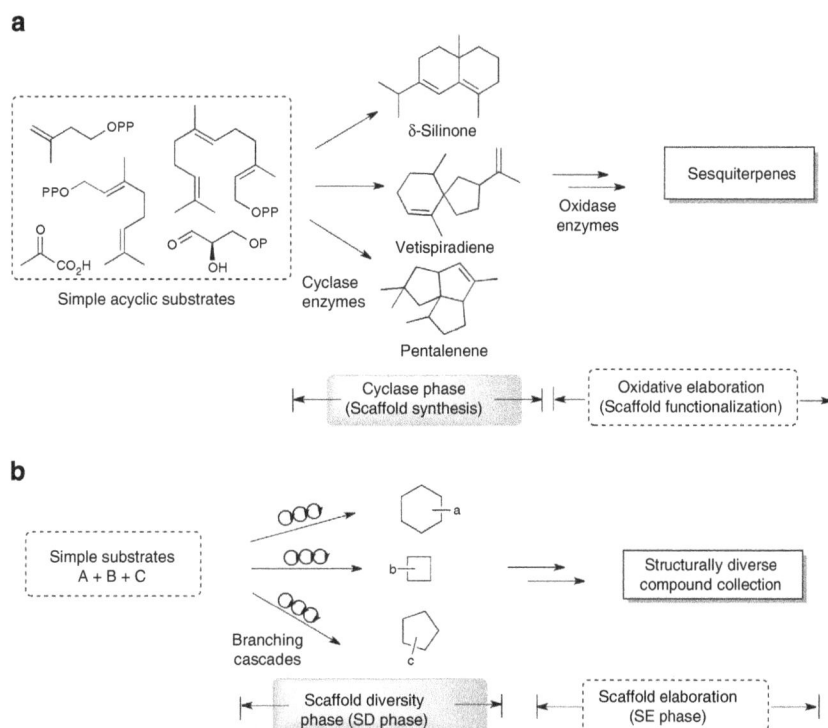

**Figure 1 | Nature's divergent biogenesis strategy inspires synthetic planning to scaffold diversity.** (a) Biogenesis of diverse terpenes from simple substrates; (b) a de novo branching cascades strategy to build a structurally diverse compound collection.

elaborated in SE phase to yield a collection of 61 molecules spread over 17 molecular frameworks. Cell-based screenings identify potent and novel bioactive molecules, representing four different structural classes, as inhibitors of tubulin polymerization or hedgehog signalling, and thereby validate the notion that functional diversity is bequeathed to a compound collection by scaffold diversity.

## Results

**Planning of the *de novo* branching cascades approach.** Targeting a compound collection of diverse azaheterocycles and employing simple substrates lacking any azaring system, we focused on a cascade reaction sequence that could provide a suitable point of divergence, that is, an intermediate that could be transformed into distinct chemotypes (azaheterocycles) by merely modulating the reaction conditions. N-phenyl nitrones (**1**) undergo [3 + 2] cycloadditions with allenic esters (**2**) and the corresponding isoxazolidines (**3**) follow a pericyclic rearrangement leading to azepinones (**5**, Fig. 2)[30–34]. Azepinone **5** beholds different reactive functionalities like a secondary amine and a β-ketoester that can be exploited in branching cascade approach to build distinct azaheterocyclic frameworks. Moreover, similar azepinones are known to rearrange and yield vinyl indoles (**6**)[35]. Cascade reaction sequences, thus, can be designed to explore various reactive functionalities in either the azepinone **5** or vinyl indole **6** to generate new and distinct scaffolds (Fig. 2).

**Scaffold diversity phase.** With this planning, reaction conditions were optimized (Supplementary Table 1) for an *in situ* generation of the nitrone (**10**) by addition of phenyl hydroxylamine (**7**) to dimethyl acetylenedicarboxylate (DMAD) (**8**), and followed by a [3 + 2] cycloaddition with allene ester (**2**), leading to azepinone **11** (*de novo* cascade I, Fig. 3a). DMAD facilitates *in situ* generation of nitrone[36,37] **10** and provides reactive functionalities to build new ring systems in SD phase, as well as to elaborate scaffolds later in SE phase. Figure 3 depicts building of SD phase in a *de novo* branching cascades approach from primary substrates (indicated by blue arrows), as well as from **11** (indicated by red arrows). Formation of scaffolds **13–17** from azepinone **11** not only confirmed its intermediacy in different cascade reaction sequences, but also provided an efficient access to scaffolds **14** and **17**, which were rather difficult to access directly from primary substrates (Fig. 3b).

Azepinone **11** itself is an interesting scaffold rich in $sp^3$ character. Using differently substituted allene esters (**2**), the first cascade reaction sequence (cascade I) under optimized reaction conditions (Supplementary Table 1) delivered highly substituted

azepinones (**11a–o**). Except for the α-substituted allene esters, which provided **11n–o** ($R^4$ = Me and prenyl) with two consecutive quaternary centres in low yields, azepinones **11a–m** were obtained in moderate to high yields (40–76%; Fig. 3b; Supplementary Fig. 1). Interestingly, just changing the reaction condition in cascade I from 80 °C to room temperature provided a different scaffold **12**, that is, a dihydroisoxazole (cascade II; Fig. 3b; Supplementary Fig. 1).

To find suitable reaction conditions that could transform the primary substrates into diverse scaffolds via azepinone as intermediate, **11a** ($R^1$–$R^2$, $R^4$ = H, $R^3$ = Et) was screened against different Lewis and Brønsted acids. While Lewis acids like $BF_3.OEt_2$ or TMSOTf provided only fragmentation products of the azepinone, $AlCl_3$ yielded an inseparable complex mixture of products (Supplementary Table 2). In a successful case (cascade III), catalytic trifluoroacetic acid (TFA) transformed azepinone **11a** into allyl indole (**13a**; Fig. 3b) at high temperature (100 °C). After slight modification in this reaction condition, we were able to synthesize allyl indole **13** directly from primary substrates (cascade IV; Fig. 3b). Allyl indole **13** supports an electron-poor olefin appended to an electron-rich indole, and therefore is a potential building block for higher-order polycyclic indoles. This opportunity was realized by overnight heating of azepinones **11** in toluene with 20 mol% of TFA that would have led to allyl indole (**13**), and followed by treatment with sodium hydroxide (NaOH) at room temperature in dioxane that provided us benzo[*b*]indolizine molecules (**14**) embodying a naturally occurring ring scaffold[38,39] in high yields (cascade V; Fig. 3b). In a separate reaction, treatment of allyl indoles **13** with 1 M NaOH also yielded benzo[*b*]indolizines **14** in excellent yields (80–94%). However, attempts to develop a *de novo* access to **14** directly from primary substrates (**2 + 7 + 8**) did not succeed in this case.

Azepinone **11** contains more than one nucleophilic site in proximity to different carbonyl functions. Therefore, various modes of cyclization and/or ring-distortion reactions can be realized in the presence of a suitable base, leading to different heterocyclic systems. With this hope, we resorted to a reaction screen of azepinone **11a** employing various bases and in different solvents (Supplementary Table 3). The reaction screening revealed various levels of reaction control, that is, effects of different reagents, solvents and temperature in directing substrates to follow a preferred reaction sequence among many possible pathways. For instance, overnight stirring of **11a** with potassium carbonate in dimethylformamide (DMF) at room temperature provided a novel scaffold **15** as the only product in low yield (<15%). Increasing the temperature (100 °C) led to the formation of **15** along with another scaffold **16** in

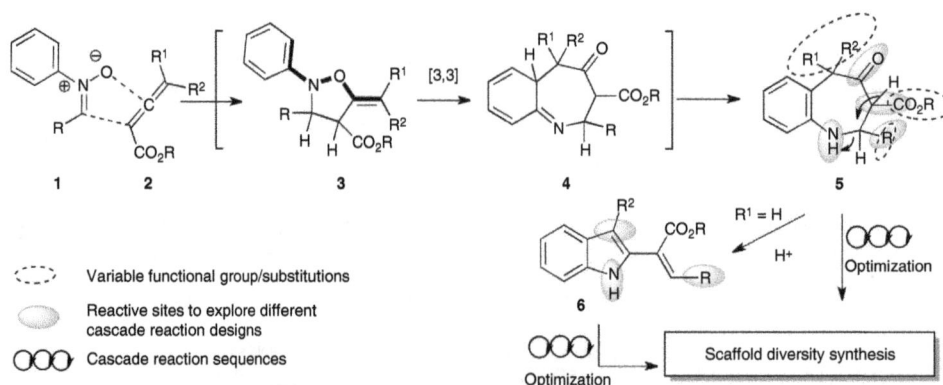

**Figure 2 | A cascade reaction design to explore the *de novo* branching cascades approach.** The key intermediates, azepinone (**5**) and vinyl indole (**6**) that could be transformed into structurally distinct molecular frameworks under cascade reaction conditions.

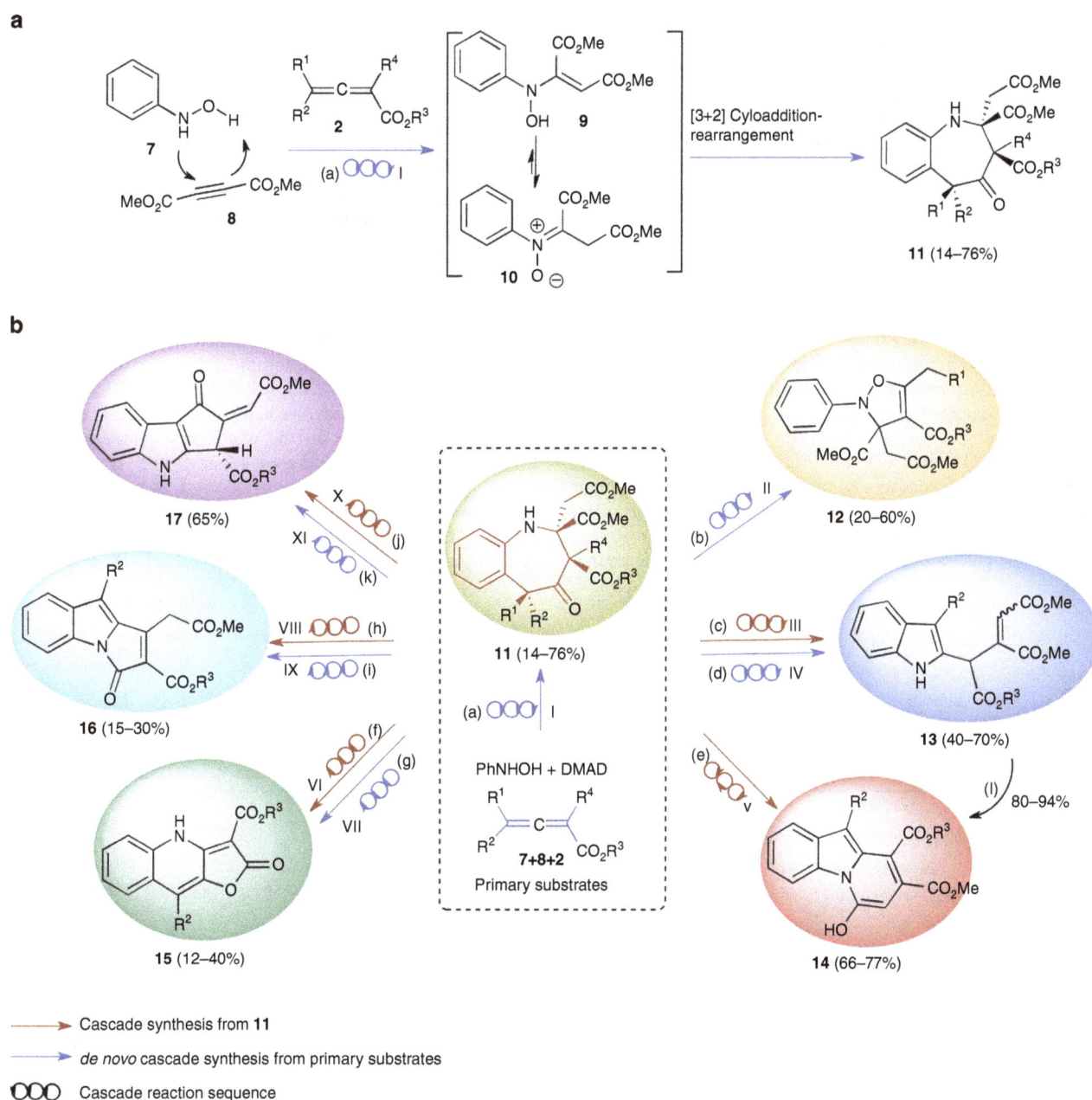

**Figure 3 | Building scaffold diversity phase (SD phase) by *de novo* branching cascades. (a)** Cascade synthesis of azepinone **11** from primary substrates. **(b)** SD phase; *de novo* branching cascades from primary substrates (blue arrows) and branching cascades from azepinone (red arrows) leading to diverse scaffolds, for details see Supplementary Methods. (a) **7** + **8** in MeCN, 10 min, room temperature (RT), then **2**, 80 °C, 8 h, 14–76%; (b) **7** + **8** in MeCN, 10 min, RT, then **2**, RT, 8 h, 20–60%; (c) TFA (20 mol%) in toluene, 100 °C, 6 h, 60–89%; (d) **7** + **8** in MeCN, 10 min, RT, then **2**, 80 °C, 8 h, then TFA (20 mol%) in toluene, 100 °C, 6 h, 40–70%; (e) TFA (20 mol%) in toluene, 100 °C, 6 h, then NaOH(aq) 1 M, 1,4-dioxane, RT, 6 h and neutralization with HCl (3 M) until pH = 6-7, 66–77%; (f) K$_2$CO$_3$, DMF, 100 °C, 6 h, 25–76%; (g) **7** + **8** in DMF, 10 min, RT, then **2**, 80 °C, 8 h, then K$_2$CO$_3$, 100 °C, 6 h, 12–40%; (h) KOAc, EtOH, 60 °C, 6 h, 69% or K$_2$CO$_3$, DMF, 100 °C, 6 h, 15–32%; (i) **7** + **8** in DMF, 10 min, RT, then **2**, 80 °C, 8 h, then K$_2$CO$_3$, 100 °C, 6 h, 15–30%; (j) NaH, DMF, 0 °C, 30 min, 65%; (k) **7** + **8** in DMF, 10 min, RT, then **2**, 80 °C, 8 h, then KOAc, EtOH, 60 °C, 4%; and (l) NaOH(aq) 1 M, 1,4-dioxane, RT, 6 h and neutralization with HCl 3 M until pH = 6-7, 80–94%.

appreciable yields. While some other bases attempted in the reaction screening either led to incomplete reactions or heavy decomposition (Supplementary Table 3), this reaction condition was adapted in *de novo* cascade synthesis from primary substrates delivering substituted scaffolds **15** and **16** in acceptable yields (cascades VII and IX; Fig. 3b; Supplementary Table 3).

To our delight, treatment of azepinone **11a** (($R^1$–$R^2$, $R^4$ = H, $R^3$ = Et) with NaH in DMF at 0 °C (decomposition at room temperature) yielded another novel scaffold **17** in high yield

(65%, cascade X). Compound **17** embodies an indole-fused cyclopentanone supporting an exocyclic olefin (Fig. 3b) that apparently infers C3-indole cyclization to one of the ester function derived from DMAD. Unfortunately, the corresponding *de novo* synthesis of **17** from primary substrates was very low yielding (4% via cascade XI) and could not be improved. Overall, the *de novo* branching cascades successfully delivered five scaffolds (**11**–**13** and **15**–**16**) and two (**14** and **17**) were obtained from azepinone **11**. A compound collection of 46 molecules was thus generated in the SD phase (Supplementary Fig. 1).

The proposed reaction mechanisms for the *de novo* cascades I, II and IV leading to the formation of azepinones 11, isoxazolidines (12) or allyl indoles (13, Fig. 4a), respectively, are supported by various reported cycloaddition reactions between *N*-phenyl nitrones and allenes[30–32]. Notably, *de novo* cascade II to 12 does not pass through azepinone and the initial [3 + 2] cycloadducts 18 (Fig. 4a) preferred energetically favoured olefinic isomerization to 12 over a sigmatropic rearrangement to azepinone 11 that requires higher temperature. Formation of allyl indoles (13) apparently occurs via isomerization of vinyl indoles (20) formed from 11 under optimized reaction conditions (Fig. 4a). Benzo[*b*]indolizines (14) were synthesized under the reaction conditions that first generate an allyl indole 13, which cyclizes on treatment with a base to yield 14 (cascade V; Fig. 4a). Although we also expected the formation of scaffold 21 by *N*-indole cyclization to C2-ester in allyl indoles 13, however, no trace of 21 was detected (Fig. 4a).

Structural features of 17 (formed under basic reaction conditions from 11) were suggestive of a C3-indole cyclization to one of the ester moieties in allyl indole 13. However, treatment of 13 with NaH in a separate reaction did not yield indolocyclopentanone 17 (Fig. 4a). We assume that treatment of 11 with NaH generates a benzylic anion (23) via vinylic aminal

22 that adds to one of the esters regioselectively before aromatizing to yield 17 (Fig. 4a, cascade XI).

Formation of scaffolds 15 and 16 (for X-ray analysis see Supplementary Figs 9–10) was rather unexpected. For the synthesis of 15 (cascade VII; Fig. 4b), we propose that azepinone 11 undergoes a retro-Michael reaction generating an anilide 24 that adds to highly electron-poor Michael acceptor in 24 to give an enolate intermediate 25. Addition of enolate to ethyl ester moiety generates a cyclopropane that opens concomitantly leading to formation of enolate 27 that cyclizes to form tricyclic lactone 28. Electrocyclic ring opening in the latter forms the benzylic vinylogous amide 29 that enolizes under basic reaction conditions and cyclizes to provide the scaffold 15.

Scaffold 21 (expected but not formed) and 16 are structurally similar but support the ester and methyl acetate functions (arising from allene ester and DMAD, respectively) on different positions (Fig. 4a,b). Mechanistically, we assume that addition of enolate in 30 to the methyl ester on the quaternary centre generates the intermediate 31 that undergoes 1,2-acyl shift to yield anionic malenoate 32. Ketone formation followed by concomitant cyclopropane and azepane ring opening forms anilide 33 that condenses with keto moiety to form vinyl indole 34. *N*-indole cyclization in 34 then generates the final adducts 16 (cascade IX; Fig. 4b).

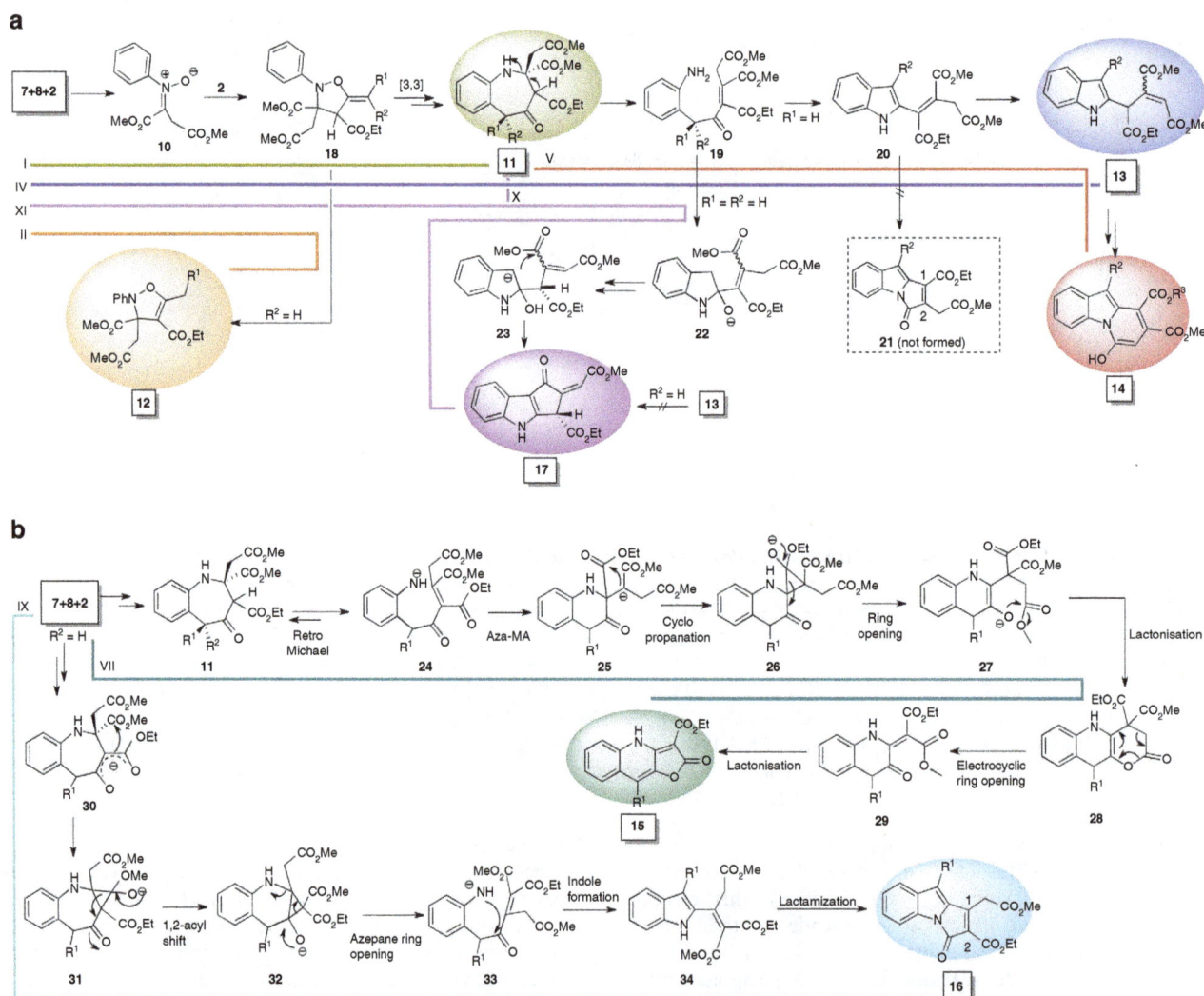

**Figure 4 | Proposed cascade routes to diverse scaffolds in SD phase.** (**a**) Mechanistic proposal for the cascade synthesis of scaffolds **11–14**, **17** and (**b**) for **15–16**. Compound numbers shown in boxes depict isolated molecules. Differently coloured thick bars display different cascade reaction sequences leading to diverse scaffolds in SD phase.

**Scaffold elaboration phase.** The *de novo* cascade reactions were designed to adorn scaffolds in SD phase with functional groups that can be exploited to accomplish further complex scaffolds in terms of embodying ring systems, percentage of $sp^3$ carbons and number of chiral centres, and thus furnishing greater structural diversity to the library. To provide a proof of this principle, we chose four scaffolds generated by *de novo* cascade design, including structurally flat scaffolds (**15–16**) for structural elaboration in SE phase (Fig. 5). Following simple one-step transformations, azepinone **11** delivered further complex benzazepane scaffolds **35–37**. Treatment of **11** with ammonium fluoride yielded azepinamine **35**. To our delight, two different chemo- and diastereoselective reductions of phenacetone in **11** provided hydroxyl azepane **36** supporting three consecutive chiral centres in high yield and a tricyclic benzazepane lactone **37** that was isolated in low yield as single diastereoisomers (Fig. 5).

Birch reduction transformed the novel but flat scaffold **15** into $sp^3$-rich framework **38** in high yields (Supplementary Fig. 1). Another flat but novel scaffold **16** was transformed into six distinct scaffolds **39–44** carrying greater $sp^3$ character and number of chiral centres (Fig. 5). A stereoselective palladium catalysed hydrogenation of **16** yielded single diastereoisomer of 1,2-dihydro-3*H*-pyrrolo[1,2-*a*]indol-3-one (**39**). Employing higher pressure of the hydrogen in this reaction provided, along

with **39**, another $sp^3$-rich scaffold, tetrahydro-3*H*-pyrrolo[1,2-*a*]-indol-3-one (**40**), although as a mixture of diastereomers (dr ~ 5:1, Fig. 5). Oxidation of **16** with oxone-generated scaffold **41** presumably via an intermediary epoxide that opened up under basic condition affording *Z*-**41** as single diastereomer in high yield (Fig. 5, see Supplementary Fig. 53 for the nuclear overhauser effect spectroscopy (NOESY) experiment to confirm relative stereochemistry in **41**).

Scaffold **16** beholds a tetrasubstituted electron-poor olefin that might be explored in different cycloaddition/annulation reactions and could generate two consecutive quaternary centres. However, tetrasubstituted olefins could provide steric resistance to reacting partners. Gratifyingly, scaffold **16** proved to be a nice dipolar-ophile for various dipolar cycloaddition reactions. A dipolar cycloaddition of **16** with an *in situ* generated azomethine ylide from *N*-methoxymethyl-*N*-(trimethylsilylmethyl)benzylamine[40] led to tetracyclic indole scaffold **42** in high yields (71–81%) and with excellent stereoselectivity. Another *in situ* generated dipole, the nitrile iminoester[41], in a low conversion reaction with **16** afforded cycloadduct **43** in a complete regio- and stereoselective manner. Phosphine-catalysed [3 + 2] cycloaddition of **16** with the zwitterion generated from allene ester (**2**, $R^1$–$R^2$, $R^4$ = H, $R^3$ = Et) went smoothly and in a regio- and stereoselective manner to provide adduct **44** in moderate yield (Fig. 5)[42]. Notably, this

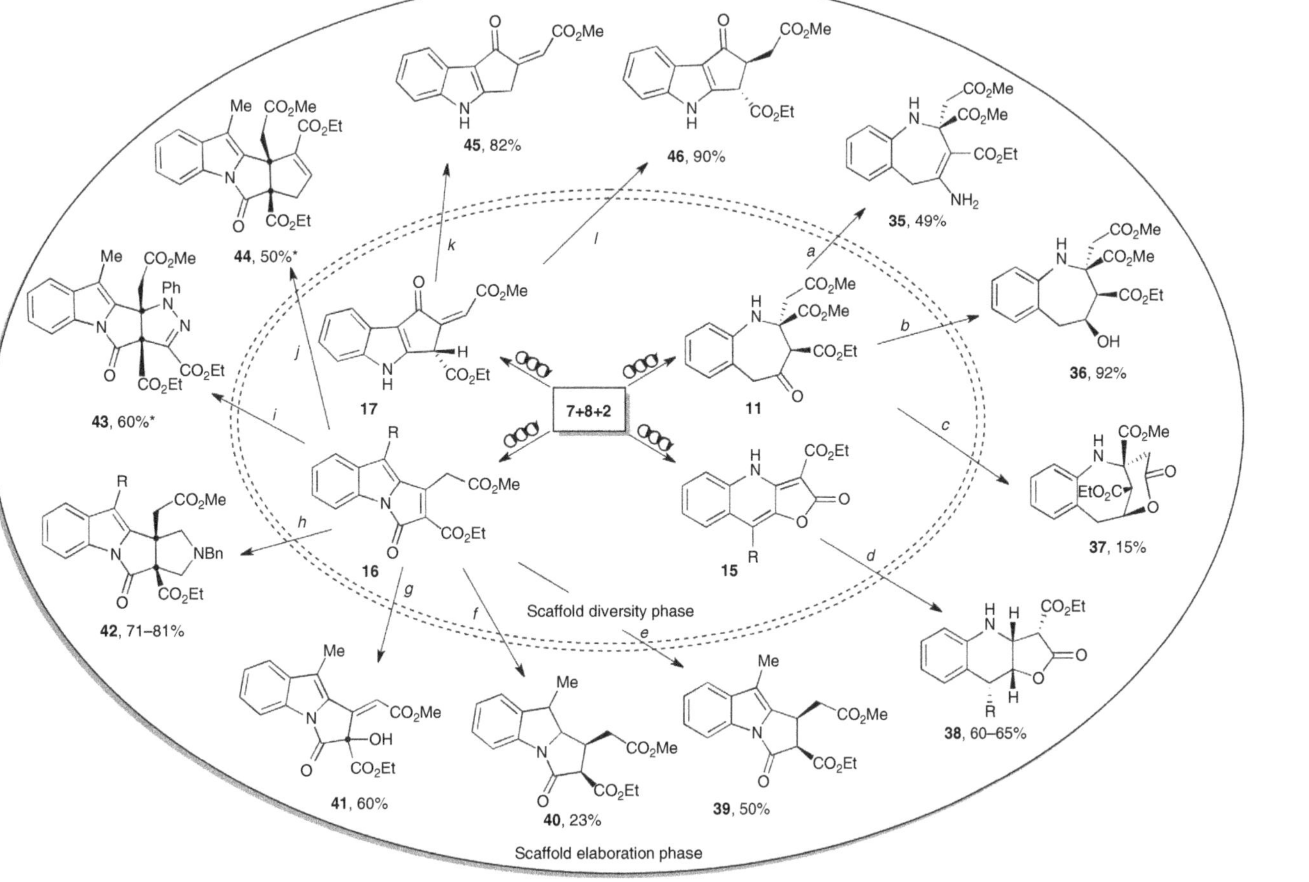

**Figure 5 | SD and SE phases in *de novo* branching cascades strategy.** For details, see Supplementary Methods. (a) NH₄F, MeOH, room temperature (RT), 8 h, 49%; (b) BH₃-morpholine, toluene, RT, 6 h, 92% (single diastereomer); (c) DIBAL-H 1.0 M, DCM, from − 78 °C to 0 °C, 1 h, 15% (single diastereomer); (d) Na, NH₃(liq), THF, − 78 °C, 15 min, 60–65% (dr = 2.3–4.0:1); (e) 1 bar H₂, MeOH, RT, Pd-C, 50% of **39** (single diastereomer); (f) 8 bar H₂, MeOH, RT, Pd-C, 34% of **39** and 23% of **40** (dr = 4.9:1); (g) Oxone, K₂CO₃, MeCN-H₂O, RT, 1 h, 60% (single diastereomer); (h) *N*-Methoxymethyl-*N*-(trimethylsilylmethyl)benzylamine, TFA (1.2 eqv.), toluene, RT, 8 h, 71–81% (single diastereomer for **42a**); (i) Hydrazonoyl chloride, TEA (2.0 eqv.), toluene, 70 °C, 8 h, 60% (*brsm, single diastereomer); (j) allene ester (**2a**), tris(4-methoxyphenyl)-phosphine (40 mol%), toluene, RT, 8 h, 50% (*brsm, single diastereomer); (k) TEA (3.5 eqv.), toluene, reflux, 8 h, 82% (*E:Z* = 1:4); and (l) 1 bar H₂, Pd-C, MeOH, RT, 12 h, 90% (single diastereomer). Dots in **41-44** mark the new quaternary centres generated from scaffold **16**.

phosphine-catalysed transformation is among the rare cases of dipolar annulations of allene-derived zwitterions with tetrasubstituted olefins[43,44]. All cycloadducts formed in above cases, that is, **42–44** behold two consecutive quaternary chiral centres. Elaboration of the fourth scaffold **17** involved a base-mediated selective decarboxylation to yield scaffold **45** and a Pd-catalysed hydrogenation to yield scaffold **46** as a single diastereoisomer in excellent yields (Fig. 5). Functional groups like esters or amines present in the scaffolds from SD and SE phases can be utilized to generate a suitable number of compounds adequately representing each scaffold in the compound library.

**Scaffold diversity and molecular properties analysis.** Scaffold diversity synthesis outlined in Figures 3 and 5 provided a collection of 61 molecules based on 17 distinct molecular frameworks (Fig. 6a; Supplementary Fig. 1). Unlike the conventional multistep synthesis of the complex molecules, each distinct scaffold in SD and SE phases was a one-step product that could be easily purified by silica gel column chromatography providing molecules in sufficient amounts for a wide range of biological screenings (see Supplementary Methods). Some of the cascade

reactions in SD phase were performed at 1–2-g scale providing enough amounts of scaffolds for further elaborations. For instance, *de novo* cascades I, VII and IX provided differently substituted **11**, **15** and **16** in appreciable yields (18–76%) at 1–2-g scale. The fact that no combinatorial synthesis step such as amide synthesis, reductive aminations or coupling reactions etc. were applied to the scaffolds, the potential to build a large and structurally diverse library with this approach is remarkably high.

Although the molecular architectures of the 17 scaffolds in this compound collection are clearly distinctive, the scaffold diversity was quantified by applying similarity metrics to 17 Bemis–Murcko frameworks (Fig. 6a)[45]. To this end, Tanimoto coefficients[46] were generated using extended connectivity fingerprints (ECFP_6, for further details see Supplementary Fig. 2)[47]. Figure 6b clearly depicts that most of the 17 scaffolds are largely distinct from one another (from 1.0 to 0, 1.0 for being the same scaffold). Moreover, the compound collection is represented by ring systems found in natural products, drugs and some unprecedented azaheterocycles, and thus covers biologically relevant as well as novel chemical space.

Substitutions on substrates directly influence the physical properties of library members. In the *de novo* branching cascade approach, simple acyclic and low molecular weight substrates

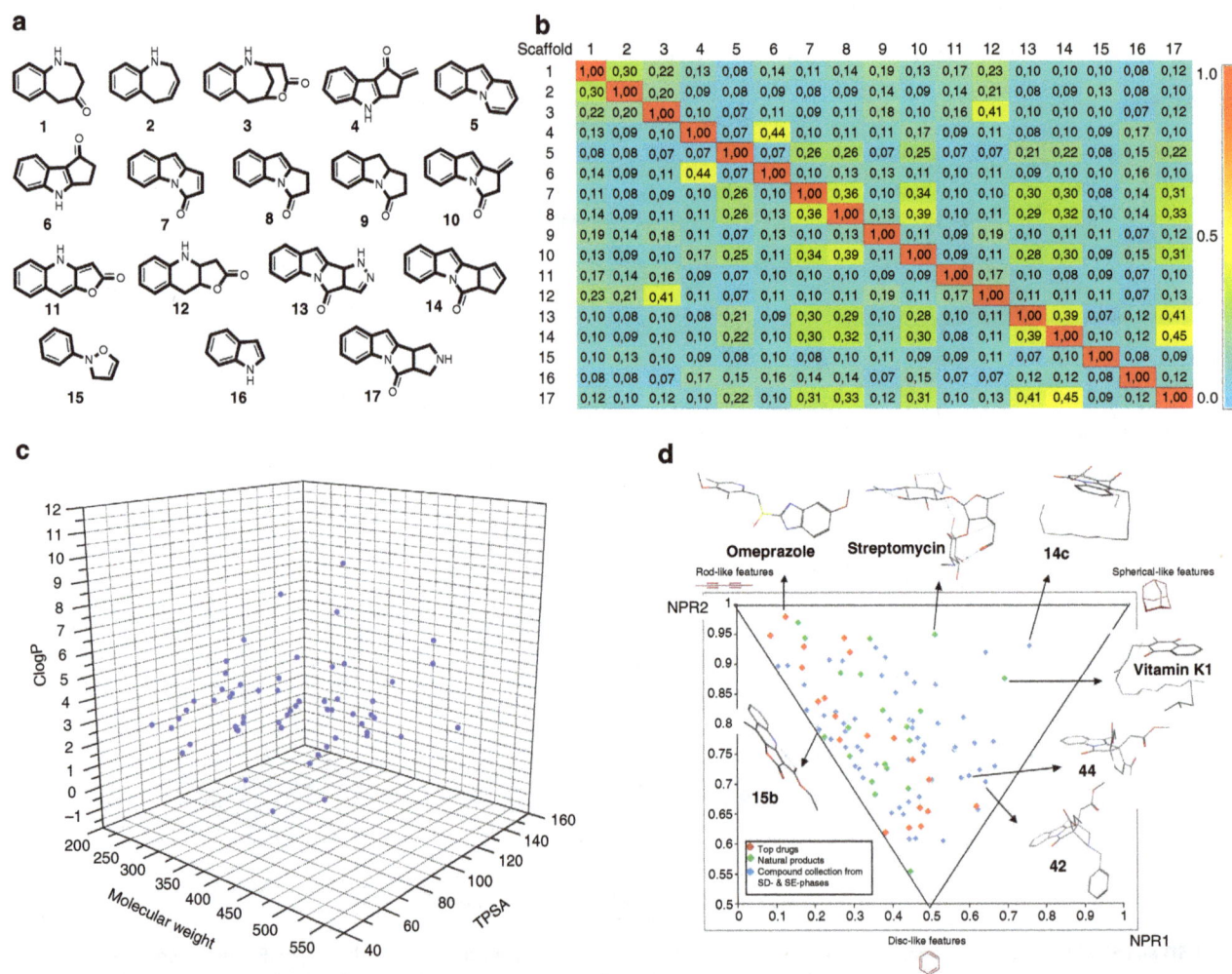

**Figure 6 | Diversity and molecular properties analysis of compounds created in SD and SE phases by *de novo* branching cascades approach.**
(**a**) Structures of 17 diverse scaffolds that make up the core frameworks in the compound collection. (**b**) Relative tanimoto similarity coefficients for 17 scaffolds (1.0 represents perfect similarity and 0.0 totally distinct scaffolds), for a tanimoto-matrix analysis with different connectivity (see Supplementary Fig. 2). (**c**) Molecular properties of the compounds generated in SD and SE phases. (**d**) Principal moment of inertia (PMI) plot. The molecular shape of the compounds generated in SD and SE phases with branching cascade approach (blue squares) and comparison with 20 natural products (green square) and 20 drugs (red squares).

provide scaffold diversity, and thereby keep an inherent check over physical properties of the library members. A major portion of the generated library possesses molecular properties within Lipinski's limits[48] as far as molecular weight, polar surface area and clog$P$ values are concerned. The only molecules that were generated from allene esters supporting lipophilic alkyl chains at $\gamma$-position displayed the expected deviations (Fig. 6c; Supplementary Table 4). In branching cascades approach, some of the reaction sequences were apparently directed by low-energy pathways leading to thermodynamically stable aromatic and flat products. Introduction of greater $sp^3$ character to the library members enhances the chance of a bioactive molecule to specifically interact with a protein target. Therefore, structural elaboration of the flat scaffolds in SE phase is highly significant. We observed that transition from SD phase to SE phase on average reduced the clog$P$ value by 16% and enhanced the fraction of $sp^3$-hybridized carbons (Fsp3) value by >50% (Supplementary Fig. 3)[49], thereby supporting the elaboration of primary scaffolds into more complex frameworks that deliver bioactive molecules (Supplementary Fig. 4).

Principal moment of inertia (PMI) analysis of the lowest-energy conformations of library members, selected natural products and available drugs (Fig. 6d; Supplementary Fig. 5) was performed to compare their three-dimensional shape diversity[50]. The selected natural products and drugs include a number of well-known substances (for example, Taxol, Penicillin and etc.), as well as molecules embodying scaffolds similar to those presented in the SD and SE phases. The library members appeared to cover a large area from the central towards the rod-disc side of the triangle quite in a similar manner to biologically active natural products and drug molecules (Fig. 6d; for details see Supplementary Figs 6 and 7 and Supplementary Tables 5–7).

**Biological evaluation.** Scaffold diversity in a compound collection is expected to bestow the molecules' ability to modulate different biological functions, and thereby sets a platform for medicinal chemistry and probe discovery research projects. To investigate this possibility, the compound collection was screened in two cell-based assays. In the first case, molecules were subjected to a high-content screen that monitors changes in cytoskeleton and DNA in the human cervical carcinoma HeLa cell line[51]. Treatment of cells with compounds at a concentration of 30 µM for 24 h and subsequent staining of DNA, actin filaments and microtubules revealed structurally similar molecules **17** and **45** causing the cells to round up as if they were entering mitosis (Supplementary Figs 11 and 12). The phenotype caused by **17** was similar to that of nocodazole, a known microtubule destabilizer. Mitotic accumulation induced by **17** was further confirmed by a concentration-dependent increase in the percentage of mitotic HeLa cells that were stained for the mitotic marker phospho-histone H3 (see Fig. 7a and Supplementary Fig. 13). Treatment with **17** and **45** for 48 h reduced the viability of HeLa cells with similar half-maximal inhibitory concentrations (IC$_{50}$) of $3.87 \pm 0.01$ µM and $3.86 \pm 0.77$ µM, respectively (Supplementary Fig. 14). Live-cell imaging of HeLa cells treated with 2.5 µM **17** demonstrated that cells were arrested in mitosis for several hours before undergoing apoptosis as detected by membrane blebbing and cell shrinkage (Supplementary Movies 1–3).

A closer look at the influence of **17** (10 µM), which is a more functionalized analogue of **45**, on the cytoskeleton in interphase cells revealed a disorganization of the microtubule network already after 2 h of treatment (Fig. 7b). In contrast to control cells wherein microtubules emerged from the microtubule organizing centre (MTOC) near the nucleus (visible as the most intense tubulin staining) and extend towards the cell periphery,

microtubules in **17** (10 µM)-treated HeLa cells did not converge in the MTOC and their radial organization was distorted. Microtubules are highly dynamic structures that are important for the maintenance of cell shape, inner cellular transport and cell division[52], and remain an attractive target for anticancer treatment[52,53].

The influence of **17** on microtubule dynamics was investigated *in vitro* using porcine brain tubulin. A concentration-dependent inhibition of tubulin polymerization was monitored by means of the increase of 4′,6-diamidino-2-phenylindole (DAPI) fluorescence on binding to microtubules (Fig. 7c)[54]. **17** also inhibited the polymerization of microtubules in HeLa cells after cold treatment (Fig. 7d; Supplementary Fig. 15). Microtubules reversibly disintegrate at low temperatures and their repolymerization can be monitored after rewarming cells to 37 °C. While in dimethylsulphoxide (DMSO)-treated cells, microtubules started to repolymerize 2 min after rewarming, and nearly complete reconstitution of the mictrotubule cytoskeleton was observed after 10 min, in HeLa cells treated with 10 µM solution of **17**, no regrowth of microtubules was detected even 10 min after rewarming. Thus, **17** is a novel microtubule destabilizing small molecule[55].

Among the three well-characterized binding sites in tubulin, tubulin destabilizers bind to either colchicine or vinca alkaloid-binding sites. Binding of **17** to these binding sites was assessed by means of competition experiments. On binding to tubulin, the intrinsic fluorescence of colchicine increases and displacement of colchicine by small molecules leads to the decrease in fluorescence as detected for nocodozole (Supplementary Fig. 16a)[56]. Unfortunately, due to autofluorescence of **17**, it was not possible to determine a putative influence of the compound on the binding of colchicine to tubulin (Supplementary Fig. 16b). However, **17** but not colchicine could displace BODIPY-FL-vinblastine from tubulin in a concentration-dependent manner with a half-maximal effective concentration (EC$_{50}$) of $0.67 \pm 1.51$ µM (Fig. 7e; Supplementary Fig. 17), and thus most likely binds to the vinca site in tubulin.

Molecules from SD and SE phases were also subjected to a screen that monitors modulation of Hedgehog signalling. The Hedgehog pathway plays a fundamental role during animal embryonic and post-embryonic development by regulating proliferation, migration and differentiation[53,57]. In adults, the pathway is silenced and can be reactivated for tissue repair and regeneration[58]. Moreover, the Hedgehog pathway can be involved in tumorigenesis since aberrant Hedgehog signalling is detected in various cancers[53,58]. Therefore, small-molecule modulators of the Hedgehog pathway are highly desired for drug discovery and chemical biology investigations. To find inhibitors of Hedgehog signalling, we employed the pluripotent mesenchymal C3H10T1/2 cells that undergo osteogenic differentiation on activation with Hedgehog ligands or purmorphamine, which in turn is characterized by the expression of alkaline phosphatase (AP) and can be used to monitor Hedgehog signalling[59,60]. Screening of the compound collection identified **11b** (dr ~ 5:2:1), **15g** and **16c** as inhibitors of Hedgehog signalling, which dose-dependently decreased AP activity in C3H10T1/2 cells (without influencing cell viability) with IC$_{50}$ of 0.79, 0.84 and 0.16 µM, respectively (Fig. 8a,b). The high-performance liquid chromatography purified major diastereomer of azepinone **11b** (for purification and structural assignment, see Supplementary Methods) was found to be more potent than the two different isomeric mixtures of **11b** employed in the assay (Supplementary Fig. 18). Binding of lipophilic Hedgehog proteins to the transmembrane protein patched 1 (Ptch-1) triggers a signalling cascade by relieving Ptch-1-induced repression of the transmembrane protein Smoothened (Smo). This results in the activation of glioma-

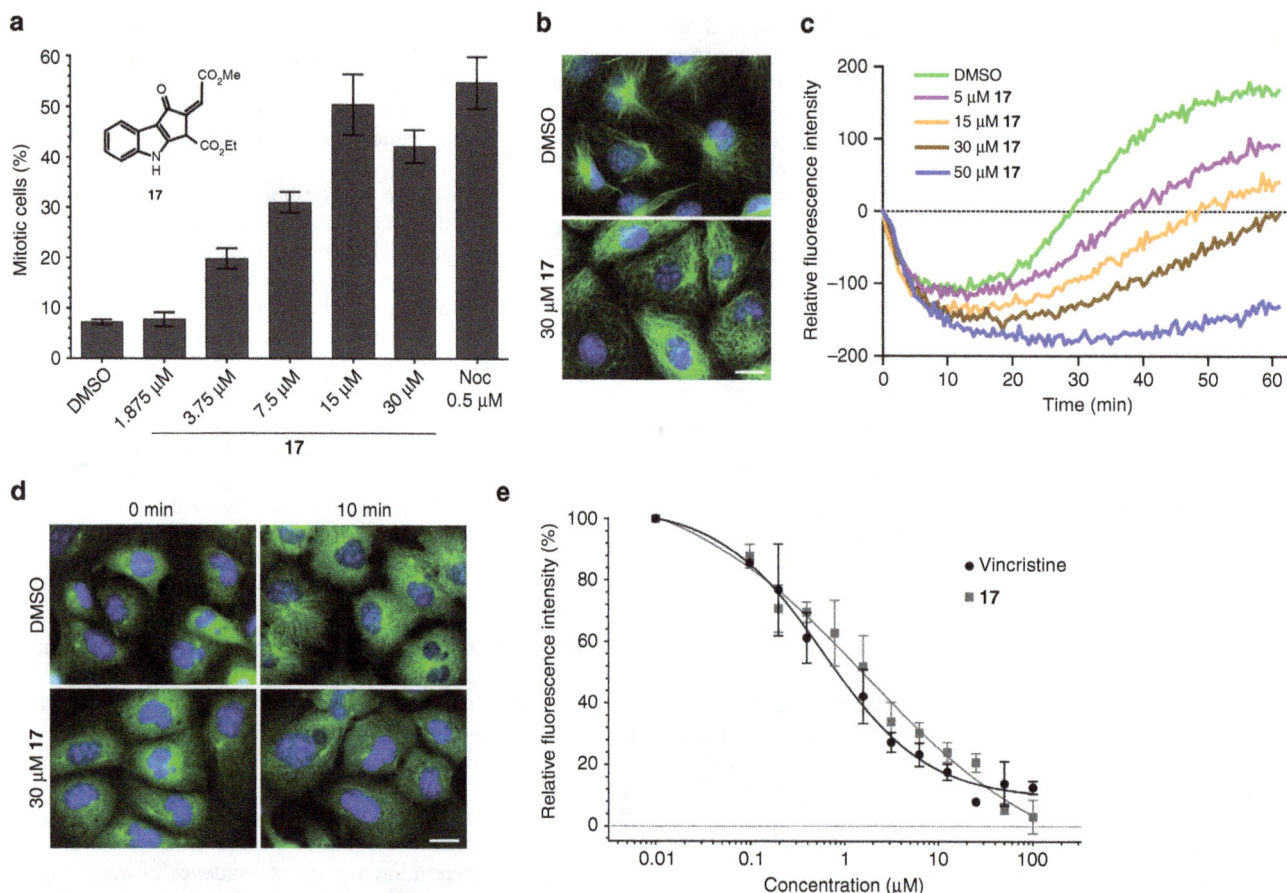

**Figure 7 | Influence of 17 on HeLa cells and microtubules. (a)** 17 induces mitotic arrest. HeLa cells were treated for 24 h with different concentrations of **17** or DMSO and nocodazole (Noc) as controls. Cells were then fixed and stained for DNA, the mitotic marker phospho-histone H3 and tubulin. High-content analysis was performed to determine the percentage of mitotic cells using the MetaMorph software. Data are shown as mean values ($n = 3$) ± s.d. (**b**) **17** impairs the microtubule cytoskeleton in HeLa cells. Cells were treated with 10 µM **17** or DMSO for 2 h. After fixation, cells were stained for tubulin and DNA using an anti-tubulin antibody coupled to FITC (green) and DAPI (blue), respectively. Scale bar, 20 µm. (**c**) **17** inhibits tubulin polymerization *in vitro*. Tubulin polymerization was monitored by means of increase of fluorescence intensity of DAPI on binding to microtubules at ex/em 340/460 nm. Data are representative of three independent experiments. (**d**) **17** inhibits the regrowth of microtubules in HeLa cells. Cells were treated with 10 µM **17** and DMSO for 2 h before incubation on ice for 1 h. Cells were rewarmed at 37 °C for given time intervals and then fixed and stained as described in **b**. Scale bar, 20 µm. Pictures shown are representative of three biological replicates. (**e**) **17** displaces BODIPY-FL-vinblastine from tubulin. Porcine brain tubulin was incubated with BODIPY-FL-vinblastine and different concentrations of **17** or vincristine as a control for 40 min at 25 °C. Fluorescence intensity was then monitored at ex/em 470/514 nm. Decrease in fluorescence indicates competition with BODIPY-FL-vinblastine for binding to the vinca-binding site. Data are shown as mean values ($n = 3$) ± s.d. and were normalized to DMSO.

associated oncogene homologues (Gli)-dependent transcription of Hedgehog pathway-specific target genes[61]. **11b**, **15g** and **16c** suppressed the expression of the Hedgehog target gene Ptch-1 (ref. 62) in NIH/3T3 cells on stimulation with purmorphamine (Fig. 8c). Furthermore, all three molecules inhibited the expression of a Gli-responsive luciferase reporter gene in Shh-LIGHT2 cells[63] (Fig. 8d). Several Hedgehog inhibitors operate by binding to and inhibiting Smo[64]. However, **11b**, **15g** and **16c** failed to displace BODIPY-cyclopamine from Smo, and thus most likely do not bind to this receptor (Fig. 8e; Supplementary Fig. 19). As a result, three new structural classes of Hedgehog inhibitors were discovered providing vital starting points in medicinal chemistry research that targets Hedgehog inhibition-based therapeutics. To obtain an acceptable structure activity relationship (SAR), screening of a larger set of molecules is required. Therefore, a conclusive SAR for the Hedgehog inhibition by molecules based on scaffolds **15** and **16** could not be realized (see Supplementary Fig. 4 for results from the primary screen). However, the results of the primary high-throughput screening for Hedgehog inhibition with

benazepinones (**11**) indicate that the benzylic substitution significantly modulates the bioactivity and prefers methyl and ethyl groups over the bulkier ones (Supplementary Fig. 4). Although **11e** (with an ethyl-β-ketoester and an ethyl group on benzylic carbon) appeared to be active in the primary screening, it displayed an IC$_{50}$ > 10 µM for the inhibition of purmorphamine-induced osteogenesis (data not shown).

## Discussion

Recent analyses of large data sets of synthetic compounds have indicated that a major part of them is presented by a small percentage of scaffolds and the same is true for drug molecules[65]. The redundancy in the scaffolds representing compound libraries that are used in the discovery research severely restrains the biological scope of the small molecules. Synthetic designs leading to significant scaffold diversity are expected to yield highly useful novel small-molecule candidates for drug and probe discovery research. The *de novo* branching cascades strategy imbibes inspiration from biogenesis of natural products, employs simple

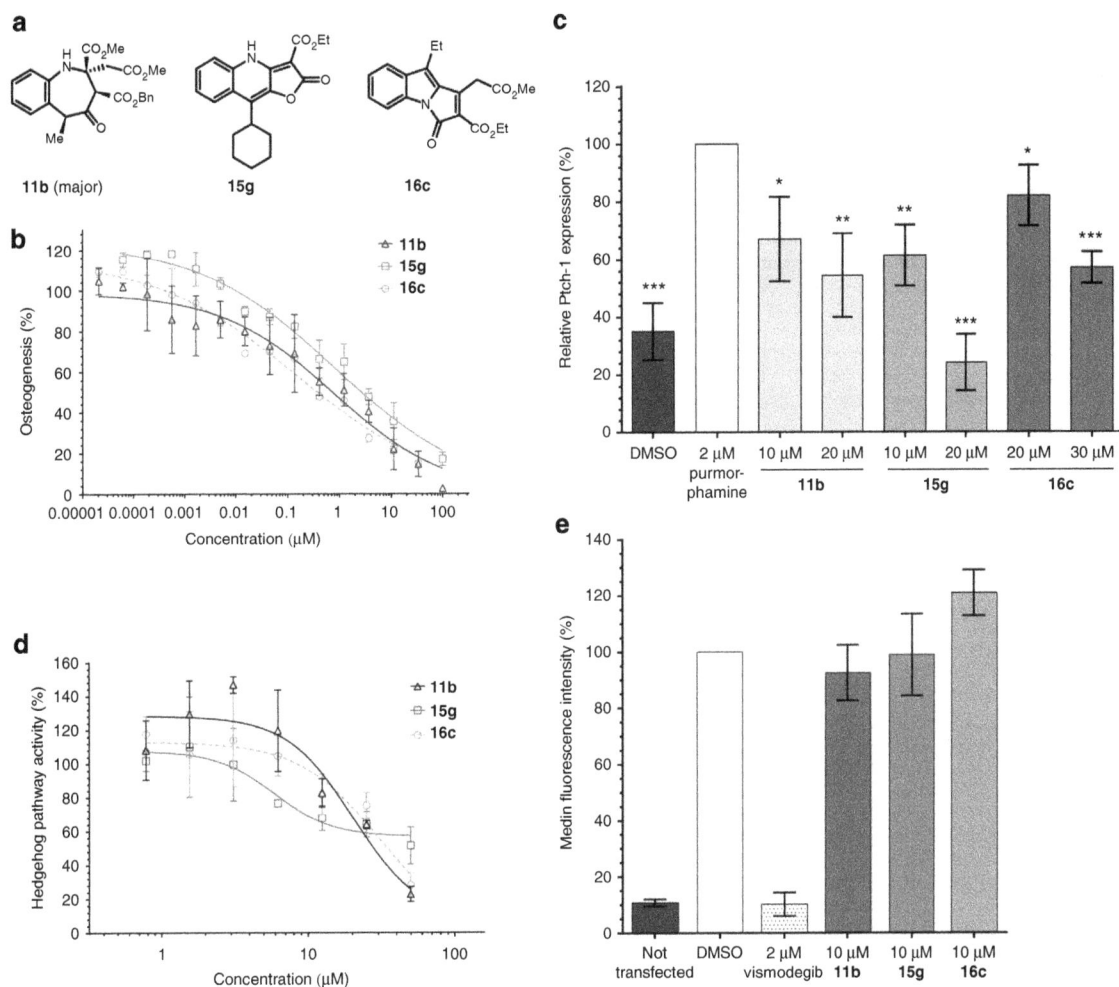

**Figure 8 | Influence of selected compounds on Hedgehog signalling.** (**a**) Chemical structures of **11b** (structure for the major diastereomer is shown), **15g** and **16c**. (**b**) Influence of **11b**, **15g** and **16c** on purmorphamine-induced osteogenesis in C3H10T1/2 cells as determined by the activity of alkaline phosphatase. C3H/10T1/2 cells were treated for 96 h with 1.5 μM purmorphamine and different concentrations of the compounds or DMSO as control. Activity of alkaline phosphatase was determined using a luminescent readout. Nonlinear regression was performed using a four parameter fit. Data are mean values ($n = 3$) ± s.d. and were normalized to purmorphamine-treated cells. (**c**) Influence of **11b**, **15g** and **16c** on the relative expression of Ptch-1. NIH/3T3 cells were incubated with 2 μM purmorphamine and different concentrations of the compounds or DMSO as control for 48 h. Following cDNA preparation, the relative expression levels of Ptch-1 and Gapdh were determined by means of quantitative PCR. Data are mean values ($n = 3$) ± s.d. and were normalized to purmorphamine-treated cells (*$P < 0.05$, **$P < 0.01$ and ***$P < 0.001$). (**d**) Influence of **11b**, **15g** and **16c** on Gli-mediated reporter gene expression. Shh-LIGHT2 cells were treated with 4 μM purmorphamine and different concentrations of the compounds or DMSO as control for 48 h. Luciferase activity was determined as a measure of Hedgehog pathway activity. Data are mean values ($n = 3$) ± s.d. and normalized to cells treated with purmorphamine. (**e**) Influence of the compounds on the binding of BODIPY-cyclopamine to Smo. HEK293T cells were transfected with a Smo-expression construct. Two days later, cells were treated with the compounds or DMSO as control in the presence of 5 nM BODIPY-cyclopamine for 5 h. Cells were then subjected to flow cytometric analysis to detect Smo-bound BODIPY-cyclopamine. The graph shows the median BODIPY-cyclopamine fluorescence intensity on treatment with the compounds. Data are mean values ($n = 3$) ± s.d.

substrates in the reaction designs to generate suitably functionalized scaffolds via cascade reactions and leads to a compound collection rich in scaffold diversity. With just three simple substrates, that is, phenylhydroxylamine, DMAD and allene ester, a collection of 61 molecules represented by 17 scaffolds was generated without employing any combinatorial synthesis step. The structurally diverse compound collection in turn delivered functionally diverse small molecules as potent inhibitors of the tubulin cytoskeleton or the Hedgehog signalling pathway, thus paving the way for their further biological applications.

We believe that endeavours to develop divergent access to novel chemical space get more exciting and challenging when driven alongside by chemists' desire to explore novel chemical reactivity. In fact, many reactive intermediates reported in various cascade or domino reactions including multi-component reactions might be explored in scaffold diversity synthesis and in unravelling new chemical transformations of broader synthetic applications. This work highlights on the one hand, the immense potential of cascade reactions in building structural diversity and molecular complexity from simple substrates and on the other hand validates the notion that functional diversity of a compound collection is a direct consequence of its scaffold diversity.

## Methods

**Chemical synthesis.** Compounds were synthesized according to the procedures specified in Supplementary Methods. X-ray crystallographic data and images are reported in Supplementary Figs 8–10 and Supplementary Tables 8–20. For [1]H, [13]C and two-dimensional nuclear magnetic resonance spectra of compounds see Supplementary Figs 20–59).

**Scaffold diversity assesment.** Two different heat maps were generated to assess the scaffold similarity. The calculation was performed using the software PipelinePilot 9.0.2 from the company Accelrys. Calculation based on the ECFP_4 and ECFP_6 (extended connectivity feature-based fingerprint on four and six bonds, respectively; Supplementary Fig. 2). Molecular properties were calculated using ChemBioDraw 12.0 software (Supplementary Table 4). Representative values of Fsp3 and clogP for SD phase and SE phase molecules are shown in Supplementary Fig. 3.

**PMI calculations.** We compared the molecular shape diversity of our library with established reference sets of 20 top-selling brand name drugs and 20 diverse natural products (Supplementary Figs 6 and 7).

PMI were calculated using Molecular Operating Environment, MOE software package, after minimization of energy of each molecule using a MMFF94x force field with the generalized Born solvation model; eps = r, cutoff [8,10] and gradient = 0.1 RMS Kcal $mol^{-1}A^{-2}$. The PMI and related calculations are performed in units of daltons (AMU) and angstroms. The stochastic conformational search algorithm in the MOE software package was used to generate three-dimensional conformers for each compound. Sampling and minimization parameters were implemented as follows: stochastic search limit: 7; refinement conformation limit: 300; stochastic search failure limit: 100; stochastic search iteration limit: 1,000; energy minimization iteration limit: 200; and energy minimization gradient test: 0.01; only the conformer with the lowest energy was retained for PMI calculations in each conformational sampling run (Supplementary Fig. 5; Supplementary Tables 5–7).

Normalized PMI ratios (I1/I3 and I2/I3) of these conformers were obtained from MOE and then plotted on a triangular graph, with the coordinates (0,1), (0.5,0.5) and (1,1) representing a perfect rod, disc and sphere, respectively, based on the report in ref. 50.

**Phenotypic screen.** HeLa cells were obtained from DSMZ GmbH, Germany and were seeded in black clear bottom 96-well microtiter plates. After incubation overnight, cells were treated with the compounds for 24 h at 30 μM. Cells were fixed with 3.7% formaldehyde in Tris-buffered saline (TBS) and permeabilized with 0.1% Triton X-100 in TBS for 15 min each before blocking using 2% bovine serum albumin (BSA) in TBS/0.1% Tween-20 (TBS-T). Cells were then stained for actin, tubulin and DNA with phalloidin coupled to tetramethylrhodamine, anti-α-tubulin antibody coupled to fluorescein isothiocyanate (FITC) and DAPI, as well as anti-phospho-histone H3 (phospho S10) coupled to AlexaFluor594 (in case of quantification of mitotic cells). Image acquisition was performed on an automated microscope Axiovert M200 (Carl Zeiss, Germany) at 20 × magnification using MetaMorph 7.7.8.0 software (Molecular Devices, USA).

**Immunocytochemistry.** After seeding and treatment with compounds, cells were fixed with 3.7% formaldehyde and permeabilized with 0.1% Triton X-100 in TBS. Samples were then blocked with 2% BSA in TBS-T before staining for tubulin and DNA with an anti-α-tubulin antibody coupled to FITC and DAPI, respectively. Axiovert Observer Z1 or Axiovert M200 microscopes (Carl Zeiss) were used for image acquisition.

**Fluorescence-based tubulin polymerization assay.** 10 μM Porcine α/β-tubulin (>99% pure, cytoskeleton, USA, in 80 mM Na-PIPES pH 6.9, 1 mM $MgCl_2$, 1 mM EGTA and 0.88 mM Na-glutamate) was dissolved in general tubulin buffer containing 16.67% tubulin glycerol buffer, 1 mM GTP and 0.01 mg/ml DAPI on ice. The compound or DMSO were added and fluorescence was measured at 37 °C using the Infinite® M200 plate reader (Tecan, Austria) with excitation/emission wavelengths of 340/460 nm.

**Microtubule regrowth assay.** HeLa cells were seeded on cover slips and incubated overnight followed by treatment with 10 μM **17** or DMSO as a control for 2 h. Depolymerization of microtubules was achieved by cold treatment at 4 °C for 1 h. Afterwards, the microtubule cytoskeleton was allowed to repolymerize by placing cells to 37 °C. Cells were fixed in 3.7% formaldehyde in TBS at given intervals before or after rewarming. Microtubules were visualized with an anti-α-tubulin antibody coupled to FITC. DAPI was used to stain DNA. Axiovert Observer microscope Z1 (Carl Zeiss) was used for image acquisition.

**Colchicine competition assay.** A 1:2 dilution series of the compound or nocodazole as a control was prepared on ice using a master mix containing 5 mM tubulin (dissolved in general tubulin buffer), 1 mM GTP, 50 μM colchicine, 16.9% v/v tubulin glycerol buffer and 76.6% v/v TR-FRET buffer. After incubation for 40 min at 25 °C, fluorescence intensity was measured in black 96-well plates at ex/em 365/435 nm using the Infinite M200 plate reader (Tecan). Blank values were subtracted from all sample values. Values were normalized to the DMSO control.

**Vinblastine competition assay.** A 1:2 dilution series of the compound or vincristine as a control was prepared on ice using a master mix containing 5 mM tubulin (dissolved in general tubulin buffer), 1 mM GTP, 5 μM BODIPY-FL-

vinblastine (Invitrogen, Germany), 16.9% v/v tubulin glycerol buffer and 76.6% v/v TR-FRET buffer. After incubation for 40 min at 25 °C, fluorescence intensity was measured in black 96-well plates at ex/em 470/514 nm using the Infinite M200 plate reader (Tecan). Blank values were subtracted from all sample values. Values were normalized to the DMSO control.

**Osteogenesis.** C3H10T1/2 cells were obtained from ATCC, USA. Eight hundred C3H/10T1/2 cells were seeded per well in white 384-well plates. On the next day, cells were treated with 1.5 μM purmorphamine and different concentrations of the compounds or DMSO as a control. After 96 h, the luminogenic AP substrate CDP-Star (Roche) was added to the wells to detect AP activity. One hour after addition of CDP-Star, luminescence was measured on an Infinite M200 plate reader (Tecan). Nonlinear regression was performed using a four parameter fit (GraphPad Prism 6, GraphPad Software, La Jolla, California, USA).

**Quantitative PCR.** NIH/3T3 cells were obtained from DSMZ GmbH and were seeded in 24-well plates ($2 \times 10^4$ cells per well). After incubation overnight, cells were treated with 2 μM purmorphamine and the compounds or DMSO as a control for 48 h. Complementary DNA (cDNA) was prepared using the FastLane Cell cDNA Kit (Qiagen) following the manufacturer's instructions. The relative messenger RNA amount of the Hedgehog target gene Ptch-1 and the housekeeping gene Gapdh (glyceraldehyde-3-phosphate dehydrogenase) was assessed using the QuantiFast SYBR Green PCR Kit (Qiagen) and the following primers: 5′-CAGTG CCAGCCTCGTC-3′ and 5′-CAATCTCCACTTTG-CCACTG-3′ for Gapdh; and 5′-CTCTGGAGCAGATTTCCAAGG-3′ and 5′-TGCCGCAGTTCTTTTGA ATG-3′ for Ptch-1 (ref. 62) The SYBR Green signal was detected with an iQ5 Real-Time PCR Detection System (Bio-Rad, Germany). Expression levels of Ptch-1 were normalized to Gapdh and were related to the expression level of purmorphamine-treated cells. Significance was determined using the unpaired $t$-test using the GraphPad Prism 6 software (San Diego, USA). Differences were considered statistically significant at $P < 0.05$, confidence interval: 95%.

**Gli-mediated reporter gene assay.** For detection of the Gli-mediated reporter gene expression, the reporter cell line Shh-LIGHT2 was employed. Shh-LIGHT2 cells are NIH/3T3 cells, stably transfected with a Gli-responsive firefly luciferase reporter plasmid and a pRL-TK vector for constitutive expression of Renilla luciferase[61,63]. Shh-LIGHT2 cells ($3.0 \times 10^4$) were seeded per well in 96-well plates. After incubation overnight, cells were treated with 4 μM purmorphamine and the compounds or DMSO as a control for 48 h. Luciferase expression and activity were detected by means of the Dual-Luciferase Reporter Assay System (Promega) using the Infinite M200 plate reader (Tecan). Nonlinear regression was performed using a four parameter fit (GraphPad Prism 6, GraphPad Software).

**Smoothened-binding assay.** Flow cytometric analysis of BODIPY-cyclopamine-labelled cells was performed as described in ref. 66. Briefly, $2.5 \times 10^5$ HEK293T cells were seeded per well in 6-well plates. After incubation overnight, the cells were transfected with a Smo-expression construct (pGEN-mSmo, a gift from Philip Beachy, Addgene no. 37673)[63] using Fugene HD (Promega). Two days after transfection, cells were treated with the compounds or DMSO in DMEM containing 0.5% FBS and 5 nM BODIPY-cyclopamine (Carbosynth Limited) for 5 h at 37 °C. Cells were then detached and diluted in DMEM containing 0.5% FBS before centrifugation at 129 RCF for 5 min at room temperature. Cells were washed twice in ice-cold PBS and were finally collected by centrifugation at 129 RCF for 5 min at 4 °C. Cells were resuspended in ice-cold PBS and subjected to flow cytometric analysis employing the BD LSR II Flow Cytometer (laser line: 488 nm, emission filter: 530/30) to detect BODIPY. Data were analysed with the FlowJo software, version 7.6.5 (Tree Star Inc., USA) and the Flowing software, version 2.5.1 (by Perttu Terho, University of Turku, Finland/ Turku Bioimaging).

# References

1. Shelat, A. A. & Guy, R. K. Scaffold composition and biological relevance of screening libraries. *Nat. Chem. Biol.* **3,** 442–446 (2007).
2. Dandapani, S. & Marcaurelle, L. A. Accessing new chemical space for 'undruggable' targets. *Nat. Chem. Biol.* **6,** 861–863 (2010).
3. Ibbeson, B. M. *et al.* Diversity-oriented synthesis as a tool for identifying new modulators of mitosis. *Nat. Commun.* **5,** 3155 (2014).
4. Villar, H. O. & Hansen, M. R. Design of chemical libraries for screening. *Expert Opin. Drug Discov.* **4,** 1215–1220 (2009).
5. Burke, M. D., Berger, E. M. & Schreiber, S. L. Generating diverse skeletons of small molecules combinatorially. *Science* **302,** 613–618 (2003).
6. Lee, M. L. & Schneider, G. Scaffold architecture and pharmacophoric properties of natural products and trade drugs: application in the design of natural product-based combinatorial libraries. *J. Comb. Chem.* **3,** 284–289 (2001).
7. Renner, S. *et al.* Bioactivity-guided mapping and navigation of chemical space. *Nat. Chem. Biol.* **5,** 585–592 (2009).
8. Ertl, P., Jelfs, S., Muhlbacher, J., Schuffenhauer, A. & Selzer, P. Quest for the rings. In silico exploration of ring universe to identify novel bioactive heteroaromatic scaffolds. *J. Med. Chem.* **49,** 4568–4573 (2006).

9. Tan, D. S. Diversity-oriented synthesis: exploring the intersections between chemistry and biology. *Nat. Chem. Biol.* **1**, 74–84 (2005).

10. Kumar, K. & Waldmann, H. Synthesis of natural product inspired compound collections. *Angew. Chem. Int. Ed. Engl.* **48**, 3224–3242 (2009).

11. Burke, M. D. & Schreiber, S. L. A planning strategy for diversity-oriented synthesis. *Angew. Chem. Int. Ed. Engl.* **43**, 46–58 (2004).

12. Marcaurelle, L. A. *et al.* An aldol-based build/couple/pair strategy for the synthesis of medium- and large-sized rings: discovery of macrocyclic histone deacetylase inhibitors. *J. Am. Chem. Soc.* **132**, 16962–16976 (2010).

13. Morton, D., Leach, S., Cordier, C., Warriner, S. & Nelson, A. Synthesis of natural-product-like molecules with over eighty distinct scaffolds. *Angew. Chem. Int. Ed. Engl.* **48**, 104–109 (2009).

14. Oguri, H. & Schreiber, S. L. Skeletal diversity via a folding pathway: Synthesis of indole alkaloid-like skeletons. *Org. Lett.* **7**, 47–50 (2005).

15. Kwon, O., Park, S. B. & Schreiber, S. L. Skeletal diversity via a branched pathway: Efficient synthesis of 29 400 discrete, polycyclic compounds and their arraying into stock solutions. *J. Am. Chem. Soc.* **124**, 13402–13404 (2002).

16. Kumagai, N., Muncipinto, G. & Schreiber, S. L. Short synthesis of skeletally and stereochemically diverse small molecules by coupling Petasis condensation reactions to cyclization reactions. *Angew. Chem. Int. Ed. Engl.* **45**, 3635–3638 (2006).

17. Robbins, D. *et al.* Synthesis of natural-product-like scaffolds in unprecedented efficiency via a 12-fold branching pathway. *Chem. Sci.* **2**, 2232–2235 (2011).

18. Morrison, K. C. & Hergenrother, P. J. Natural products as starting points for the synthesis of complex and diverse compounds. *Nat. Prod. Rep.* **31**, 6–14 (2014).

19. Huigens, R. W. A ring-distortion strategy to construct stereochemically complex and structurally diverse compounds from natural products. *Nat. Chem.* **5**, 195–202 (2013).

20. Beckmann, H. S. G. *et al.* A strategy for the diversity-oriented synthesis of macrocyclic scaffolds using multidimensional coupling. *Nat. Chem.* **5**, 861–867 (2013).

21. Cordier, C., Morton, D., Murrison, S., Nelson, A. & O'Leary-Steele, C. Natural products as an inspiration in the diversity-oriented synthesis of bioactive compound libraries. *Nat. Prod. Rep.* **25**, 719–737 (2008).

22. Newman, D. J. & Cragg, G. M. Natural product scaffolds as leads to drugs. *Future Med. Chem.* **1**, 1415–1427 (2009).

23. Firn, R. D. & Jones, C. G. Natural products—a simple model to explain chemical diversity. *Nat. Prod. Rep.* **20**, 382–391 (2003).

24. Fischbach, M. A. & Clardy, J. One pathway, many products. *Nat. Chem. Biol.* **3**, 353–355 (2007).

25. Maimone, T. J. & Baran, P. S. Modern synthetic efforts toward biologically active terpenes. *Nat. Chem. Biol.* **3**, 396–407 (2007).

26. Razzak, M. & De Brabander, J. K. Lessons and revelations from biomimetic syntheses. *Nat. Chem. Biol.* **7**, 865–875 (2011).

27. Tietze, L. F. Domino reactions in organic synthesis. *Chem. Rev.* **96**, 115–136 (1996).

28. Nicolaou, K. C., David, J. E. & Paul, G. B. Cascade reactions in total synthesis. *Angew. Chem. Int. Ed. Engl.* **45**, 7134–7186 (2006).

29. Liu, W., Khedkar, V., Baskar, B., Schurmann, M. & Kumar, K. Branching cascades: a concise synthetic strategy targeting diverse and complex molecular frameworks. *Angew. Chem. Int. Ed. Engl.* **50**, 6900–6905 (2011).

30. Padwa, A., Kline, D. N. & Norman, B. H. Synthesis of the benzazepin-4-one ring-system via dipolar cyclo-addition of n-phenylnitrones with activated allenes. *J. Org. Chem.* **54**, 810–817 (1989).

31. Padwa, A., Bullock, W. H., Kline, D. N. & Perumattam, J. Heterocyclic synthesis via the reaction of nitrones and hydroxylamines with substituted allenes. *J. Org. Chem.* **54**, 2862–2869 (1989).

32. Wilkens, J., Kuhling, A. & Blechert, S. Hetero-cope rearrangements.6. short and stereoselective syntheses of 2-vinylindoles by a tandem-process. *Tetrahedron* **43**, 3237–3246 (1987).

33. Mo, D. L., Wink, D. J. & Anderson, L. L. Solvent-controlled bifurcated cascade process for the selective preparation of dihydrocarbazoles or dihydropyridoindoles. *Chemistry* **20**, 13217–13225 (2014).

34. Pecak, W. H., Son, J., Burnstine, A. J. & Anderson, L. L. Synthesis of 1,4-enamino ketones by [3,3]-rearrangements of dialkenylhydroxylamines. *Org. Lett.* **16**, 3440–3443 (2014).

35. Wirth, T. & Blechert, S. Synthesis of 2,3-disubstituted indoles. *Synlett* 717–718 (1994).

36. Nguyen, T. B., Martel, A., Dhal, R. & Dujardin, G. N-benzyl aspartate nitrones: unprecedented single-step synthesis and [3 + 2] cycloaddition reactions with alkenes. *Org. Lett.* **10**, 4493–4496 (2008).

37. Zhang, X. *et al.* Asymmetric synthesis of α,α-disubstituted amino acids by cycloaddition of (e)-ketonitrones with vinyl ethers. *Org. Lett.* **16**, 1936–1939 (2014).

38. Devkota, K. P. *et al.* Compounds from Simarouba berteroana which inhibit proliferation of NF1-defective cancer cells. *Phytochem. Lett.* **7**, 42–45 (2014).

39. Cebrian-Torrejon, G. *et al.* Alkaloids from Rutaceae: activities of canthin-6-one alkaloids and synthetic analogues on glioblastoma stems cells. *Medchemcomm* **3**, 771–774 (2012).

40. Srihari, P., Yaragorla, S. R., Basu, D. & Chandrasekhar, S. Tris(pentafluorophenyl)borane-catalyzed synthesis of N-benzyl pyrrolidines. *Synthesis* 2646–2648 (2006).

41. Singh, A., Loomer, A. L. & Roth, G. P. Synthesis of oxindolyl pyrazolines and 3-amino oxindole building blocks via a nitrile imine [3 + 2] cycloaddition strategy. *Org. Lett.* **14**, 5266–5269 (2012).

42. Dakas, P. Y., Parga, J. A., Hoing, S., Scholer, H. R., Sterneckert, J. & Kumar, K. *et al.* Discovery of neuritogenic compound classes inspired by natural products. *Angew. Chem. Int. Ed. Engl.* **52**, 9576–9581 (2013).

43. Fan, Y. C. & Kwon, O. Advances in nucleophilic phosphine catalysis of alkenes, allenes, alkynes, and MBHADs. *Chem. Commun.* **49**, 11588–11619 (2013).

44. Lu, Z., Zheng, S. Q., Zhang, X. M. & Lu, X. Y. An unexpected phosphine-catalyzed [3 + 2] annulation. Synthesis of highly functionalized cyclopentenes. *Org. Lett.* **10**, 3267–3270 (2008).

45. Langdon, S. R., Brown, N. & Blagg, J. Scaffold diversity of exemplified medicinal chemistry space. *J. Chem. Inf. Model.* **51**, 2174–2185 (2011).

46. Rogers, D. J. & Tanimoto, T. T. A computer program for classifying plants. *Science* **132**, 1115–1118 (1960).

47. Rogers, D. & Hahn, M. Extended-connectivity fingerprints. *J. Chem. Inf. Model.* **50**, 742–754 (2010).

48. Lipinski, C. A., Lombardo, F., Dominy, B. W. & Feeney, P. J. Experimental and computational approaches to estimate solubility and permeability in drug discovery and development settings. *Adv. Drug Deliv. Rev.* **23**, 3–25 (1997).

49. Yan, A. X. & Gasteiger, J. Prediction of aqueous solubility of organic compounds by topological descriptors. *Qsar Comb. Sci.* **22**, 821–829 (2003).

50. Sauer, W. H. B. & Schwarz, M. K. Molecular shape diversity of combinatorial libraries: a prerequisite for broad bioactivity. *J. Chem. Inf. Comp. Sci.* **43**, 987–1003 (2003).

51. Duckert, H. *et al.* Natural product-inspired cascade synthesis yields modulators of centrosome integrity. *Nat. Chem. Biol.* **8**, 179–184 (2012).

52. Dumontet, C. & Jordan, M. A. Microtubule-binding agents: a dynamic field of cancer therapeutics. *Nat. Rev. Drug Dis.* **9**, 790–803 (2010).

53. Atwood, S. X., Chang, A. L. S. & Oro, A. E. Hedgehog pathway inhibition and the race against tumor evolution. *J. Cell. Biol.* **199**, 193–197 (2012).

54. Kawaratani, Y. *et al.* New microtubule polymerization inhibitors comprising a nitrooxymethylphenyl group. *Bioorg. Med. Chem.* **19**, 3995–4003 (2011).

55. Kavallaris, M. Microtubules and resistance to tubulin-binding agents. *Nat. Rev. Cancer* **10**, 194–204 (2010).

56. Bhattacharyya, B. & Wolff, J. Promotion of fluorescence upon binding of colchicine to tubulin. *Proc. Natl Acad. Sci. USA* **71**, 2627–2631 (1974).

57. Mullor, J. L., Sanchez, P. & Altaba, A. R. Pathways and consequences: Hedgehog signaling in human disease. *Trends Cell Biol.* **12**, 562–569 (2002).

58. McMillan, R. & Matsui, W. Molecular pathways: the hedgehog signaling pathway in cancer. *Clin. Cancer Res.* **18**, 4883–4888 (2012).

59. Nakamura, T. *et al.* Induction of osteogenic differentiation by hedgehog proteins. *Biochem. Bioph. Res. Commun.* **237**, 465–469 (1997).

60. Wu, X., Walker, J., Zhang, J., Ding, S. & Schultz, P. G. Purmorphamine induces osteogenesis by activation of the hedgehog signaling pathway. *Chem. Biol.* **11**, 1229–1238 (2004).

61. Sasaki, H., Hui, C. C., Nakafuku, M. & Kondoh, H. A binding site for Gli proteins is essential for HNF-3 beta floor plate enhancer activity in transgenics and can respond to Shh in vitro. *Development* **124**, 1313–1322 (1997).

62. Lipinski, R. J., Gipp, J. J., Zhang, J., Doles, J. D. & Bushman, W. Unique and complementary activities of the Gli transcription factors in Hedgehog signaling. *Exp. Cell Res.* **312**, 1925–1938 (2006).

63. Taipale, J. *et al.* Effects of oncogenic mutations in Smoothened and Patched can be reversed by cyclopamine. *Nature* **406**, 1005–1009 (2000).

64. Amakye, D., Jagani, Z. & Dorsch, M. Unraveling the therapeutic potential of the Hedgehog pathway in cancer. *Nat. Med.* **19**, 1410–1422 (2013).

65. Lipkus, A. H. *et al.* Structural diversity of organic chemistry. A scaffold analysis of the CAS Registry. *J. Org. Chem.* **73**, 4443–4451 (2008).

66. Chen, J. K., Taipale, J., Cooper, M. K. & Beachy, P. A. Inhibition of Hedgehog signaling by direct binding of cyclopamine to Smoothened. *Gene Dev.* **16**, 2743–2748 (2002).

## Acknowledgements

This work was supported by research funds from the Max Planck Society. We thank Prof. Dr H. Waldmann (MPI, Dortmund) for his support and encouragement.

## Author contributions

M.G.-C. designed and carried out the synthesis experiments and PMI analysis. L.K. and C.D.R. performed the biological and biochemical experiments. C.U. and C.S. performed the X-ray analysis of single crystals. C.O. performed the cheminformatics analysis to calculate the scaffold diversity. K.K. and S.Z. designed the experiments and supervised the project. All authors discussed the results and commented on the manuscript. K.K., S.Z. and L.K. wrote the manuscript.

# Cellular delivery and photochemical release of a caged inositol-pyrophosphate induces PH-domain translocation *in cellulo*

Igor Pavlovic[1], Divyeshsinh T. Thakor[1], Jessica R. Vargas[2], Colin J. McKinlay[2], Sebastian Hauke[3], Philipp Anstaett[1], Rafael C. Camuña[4], Laurent Bigler[1], Gilles Gasser[1], Carsten Schultz[3], Paul A. Wender[2] & Henning J. Jessen[5]

Inositol pyrophosphates, such as diphospho-myo-inositol pentakisphosphates (InsP$_7$), are an important family of signalling molecules, implicated in many cellular processes and therapeutic indications including insulin secretion, glucose homeostasis and weight gain. To understand their cellular functions, chemical tools such as photocaged analogues for their real-time modulation in cells are required. Here we describe a concise, modular synthesis of InsP$_7$ and caged InsP$_7$. The caged molecule is stable and releases InsP$_7$ only on irradiation. While photocaged InsP$_7$ does not enter cells, its cellular uptake is achieved using nanoparticles formed by association with a guanidinium-rich molecular transporter. This novel synthesis and unprecedented polyphosphate delivery strategy enable the first studies required to understand InsP$_7$ signalling in cells with controlled spatiotemporal resolution. It is shown herein that cytoplasmic photouncaging of InsP$_7$ leads to translocation of the PH-domain of Akt, an important signalling-node kinase involved in glucose homeostasis, from the membrane into the cytoplasm.

[1] Department of Chemistry, University of Zurich, Winterthurerstrasse 190, Zurich 8057, Switzerland. [2] Departments of Chemistry and Chemical and Systems Biology, Stanford University, Stanford, California 94305, USA. [3] European Molecular Biology Laboratory (EMBL), Cell Biology & Biophysics Unit, Meyerhofstrasse 1, 69117 Heidelberg, Germany. [4] Departamento de Química Orgánica, Facultad de Ciencias, Universidad de Málaga, Malaga 29071, Spain. [5] Department of Chemistry and Pharmacy, Albert-Ludwigs University Freiburg, Albertstrasse 21, 79104 Freiburg, Germany. Correspondence and requests for materials should be addressed to P.A.W. (email: Wenderp@stanford.edu) or to H.J.J. (email: henning.jessen@oc.uni-freiburg.de).

**D**iphospho-inositol polyphosphates (InsP$_7$) are second messengers involved in essential cell signalling pathways[1-4]. A distinct difference of InsP$_7$ compared with other inositol polyphosphates is the presence of a phosphoanhydride bond in, for example, the 5-position (5-InsP$_7$, Fig. 1), rendering them a structurally unique class of second messengers. This special feature is also the reason for their nickname 'inositol pyrophosphates'. InsP$_7$ are implicated in the regulation of diverse cellular and metabolic functions in different kingdoms of life[1-8]. It has been proposed that InsP$_7$ bind to the pleckstrin homology (PH) domain of protein kinase B (Akt), and competitively suppress its specific phosphatidylinositol 3,4,5-trisphosphate (PIP$_3$) association at the plasma membrane, thereby inhibiting phosphoinositide-dependent kinase 1 (PDK1)-mediated phosphorylation of Akt[9,10]. However, there remains uncertainty as to whether the reduced phosphorylation of Akt is a result of the inhibition of its membrane association via its PH-domain, since the *in vitro* assays that have been performed do not contain any membrane or membrane mimics. In addition, InsP$_7$ might act either as allosteric inhibitors or as non-enzymatic phosphorylating agents or both[3,11]. Notwithstanding, inhibition of the Akt pathway by InsP$_7$ has an impact on glucose uptake and insulin sensitivity, as exemplified by a mouse model that lacks inositol hexakisphosphate-kinase 1 (IP6K1). These knockout mice have reduced levels of InsP$_7$ and show a lean phenotype on high-fat diet concomitant with increased insulin sensitivity[9]. As a consequence, IP6K1 has recently been proposed as a novel target in the treatment of diabetes and obesity[12]. To address fundamental questions about the mechanism of action of these potent signalling molecules and their subcellular localization, the development of new chemical tools is required.

To understand cellular signalling mediated by second messengers, photocaged analogues that can be activated on demand inside living cells with spatiotemporal resolution have attracted great interest[13]. Unfortunately, preparation of such analogues often requires lengthy synthetic sequences. Phosphorylated second messengers derived from *myo*-inositol such as *myo*-inositol 1,4,5-trisphosphate (Fig. 1) present additional challenges as their polyanionic nature precludes efficient cellular uptake[14]. For example, while cell-permeable and photocaged analogues of different *myo*-inositol polyphosphates (InsP$_x$) and phosphatidyl *myo*-inositol polyphosphates (PtdIns-P$_x$) have been reported[15-18], the phosphate groups typically need to be reversibly masked. This complicates their use, as multiple intracellular hydrolysis events must occur before the free polyphosphate is formed[19]. Even so, no such photocaged derivatives are currently available for the more complex

diphospho-*myo*-inositol pentakisphosphates as, for example, 5-InsP$_7$ (Fig. 1)[20].

Here we report the design, step-economical synthesis, photophysical and metabolic evaluation of photocaged 5-InsP$_7$, and significantly, a general solution to the delivery of unmodified polyphosphate probes into cells using guanidinium-rich molecular transporters[21]. On cytoplasmic uncaging, 5-InsP$_7$-mediated PH-domain translocation from the membrane into the cytosol in living cells is demonstrated for the first time on a 15-min timescale.

## Results

**Synthesis.** All current synthetic approaches to access any InsP$_7$ isomer rely on a global hydrogenation in the last step, during which up to 13 protecting groups need to be removed[22-30]. Significantly, however, hydrogenation is incompatible with photocaging groups and many other functional moieties like, for example, fluorophores. To address this problem, a novel strategy based on the development of a levulinate benzyl ester adaptor (LevB) is described. The introduction of this new protecting group and its combination with fluorenylmethyl (Fm) protection[31-33] and photocage introduction enables the previously inaccessible synthesis of the first photoactivatable diphospho-inositol InsP$_7$ probe **9** equipped with a [7-(diethylamino)-coumarin-4-yl]methyl (DEACM) photocage (Fig. 2a). It is noteworthy that this strategy potentially facilitates the introduction of other tags, such as, for example, photoaffinity labels and fluorophores.

The synthesis commenced with benzylidene protected **2** prepared as previously described (Fig. 2a)[23]. The 5-OH position of **2** is available for phosphitylation, allowing virtually any protected phosphate to be introduced. However, none of the existing protecting groups are compatible with the subsequent introduction of the coumarin cage. Such protecting groups would need to be stable under acidic and basic conditions and must enable double deprotection under very mild conditions. To meet these stringent requirements, a new phosphate-protecting group is required. The approach described herein is based on an Umpolung strategy that had been exploited in prodrug design for nucleotides[34,35]. Conceptually, this strategy is useful to generally couple phenol or alcohol protecting groups to phosphates via a benzyl adaptor, greatly enhancing the available protecting group strategies for phosphates (Fig. 2b). A novel P-amidite **3** was developed (Fig. 2, Supplementary Fig. 1), which is connected via a benzyl ester to a levulinate group (LevB). After oxidation to the phosphate triester **6**, the LevB group can be cleaved by hydrazone formation, initiating cyclization and finally a Grob-type fragmentation to give the unprotected phosphates (Fig. 2b).

After coupling P-amidite **3** to alcohol **2** and oxidation, all inositol-protecting groups were cleaved with TFA. The resultant protected monophosphate **4** was phosphitylated with (Fm)$_2$-P-amidite **5** and oxidized. Notwithstanding the significant molecular crowding of hexakisphosphate **6**, both LevB groups were efficiently cleaved under mild conditions (hydrazine*AcOH/TFA), releasing phosphate **7**. Next, the P-anhydride was formed using Fm-protected photocaged P-amidite **8** (ref. 33). All 11 Fm-protecting groups were then cleaved with piperidine resulting in highly pure photocaged 5-InsP$_7$ **9** in 45% yield over 2 steps (11% overall yield from **1**) on precipitation of the compound as the dodeca-piperidinium salt. The piperidinium ions can also be exchanged with sodium ions by precipitation. In addition, natural 5-InsP$_7$ **10** can be prepared following the same strategy by using (Fm)$_2$-P-amidite **5** in the anhydride forming reaction (8% overall yield from **1**)[36,37]. This eight-step synthesis represents a general strategy to access 5-InsP$_7$ **10** and caged analogues in scalable amounts (30 mg of **9** have been prepared) and very high quality

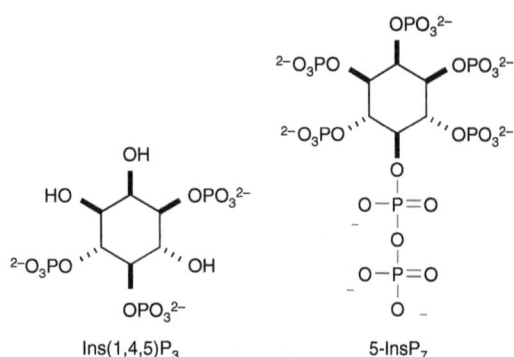

**Figure 1 | Phosphorylated second messengers derived from *myo*-inositol.** Chemical structures of *myo*-Inositol 1,4,5-trisphosphate and 5-diphospho-*myo*-inositol pentakisphosphate (5-InsP$_7$).

**a**

**b**

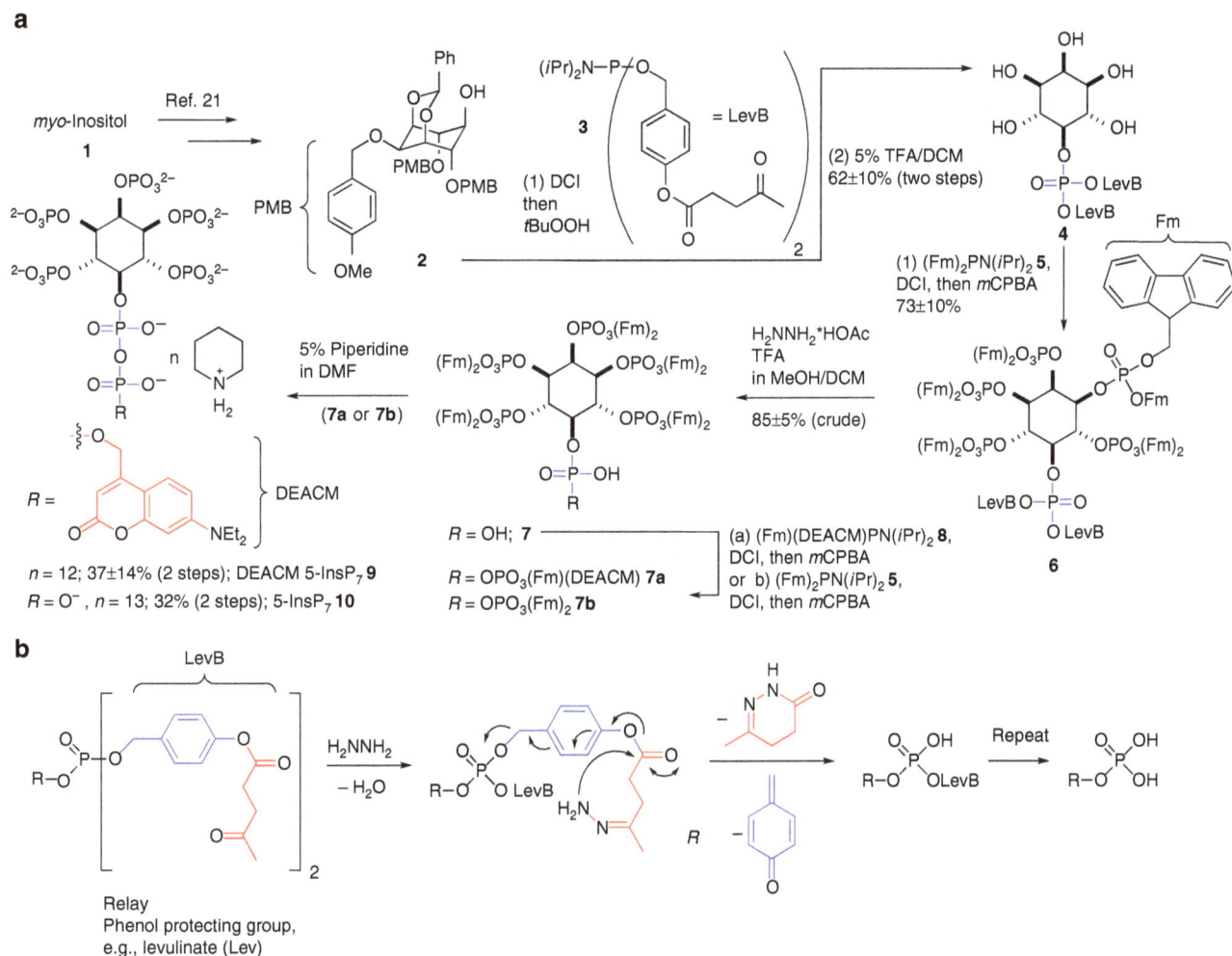

**Figure 2 | Synthesis of photocaged 5-InsP₇ and mechanism of LevB cleavage.** (a) Synthesis of DEACM 5-InsP$_7$ **9** and 5-InsP$_7$ **10** based on fluorenylmethyl (Fm) protection and a novel phosphate-protecting group (LevB). DCI, 4,5-dicyanoimidazole; DEACM, [7-(diethylamino)-coumarin-4-yl]methyl; Fm, fluorenylmethyl; mCPBA, metachloro perbenzoic acid; TFA, trifluoroacetic acid. (b) An adaptor strategy for phosphate release: hydrazine triggers levulinate (red) cleavage and 1,6-elimination (blue) to release free phosphate. Generally, levulinate could also be replaced with other phenol protecting groups.

without the need for a final hydrogenation under aqueous conditions.

***In vitro* stability and photophysical properties.** To serve as a useful tool, DEACM 5–InsP$_7$ **9** must be stable towards enzymatic digestion to enable cellular uptake and release only on photolysis. To test its stability, **9** was incubated in tissue homogenate (brain, Fig. 3a; liver, Supplementary Fig. 3 and Supplementary Methods) and cell extract (Supplementary Figs 4–5 and Supplementary Methods). Readout was achieved by resolution on polyacrylamide gels (35%, Fig. 3 and Supplementary Methods)[38]. DEACM 5–InsP$_7$ **9** did not decompose under these conditions over incubation times up to 5 h (Fig. 3a, Lanes III–V and Supplementary Figs 3–5). Thus, **9** is a probe that has the potential to be broadly applied in different cell and tissue types. Importantly, on exposure to ultraviolet light (366 nm, 4 W, distance 10 cm) in extracts, it was cleanly converted into 5-InsP$_7$ **10**, as verified by PAGE (Fig. 3a, Lanes VI and VII and Supplementary Figs 3–4) and HPLC analysis (Supplementary Fig. 6) with **10** as a standard.

Next, the photophysical properties of DEACM 5–InsP$_7$ **9** were characterized. The quantum yield for the disappearance of **9** $\Delta\varphi_{chem}$ is 0.71% at 355 nm as determined by actinometry following a novel protocol (Supplementary Methods)[39–41]. The fluorescence quantum yield $\varphi_f$ is 6.2% and the lifetime $\tau_f$ is 1.2 ns. Notably, **9** also exhibits typical coumarin fluorescence at 500 nm (excitation at 386 nm; Supplementary Figs 7–10 and Supplementary Methods)[33,42].

**Cellular delivery and uncaging.** Notwithstanding the efficiency of this synthesis, it was found as expected that DEACM 5–InsP$_7$ **9**, like other polyanions[11,14], does not readily cross the non-polar membrane of a cell (Fig. 4b,c). To address this problem, its non-covalent complexation, cell uptake and release using guanidinium-rich molecular transporters were studied[43]. **9** was mixed with amphipathic, guanidinium-rich transporter **11** (Fig. 4a), in an equimolar ratio to form nanoparticles. HeLa cells were treated with these complexes and analysed for coumarin fluorescence by flow cytometry. Significantly, while DEACM 5–InsP$_7$ **9** itself does not appreciably enter cells, the transporter-complexed **9** does, as demonstrated by a 10-fold increase in intracellular fluorescence (Fig. 4b). According to flow cytometry analysis, over 99% of cells display increased levels of **9** following treatment with the transporter/DEACM–InsP$_7$ complex (Fig. 4c). It is important to note that other transfection reagents like Lipofectamine 2000 do not efficiently deliver **9** into cells (Fig. 4b).

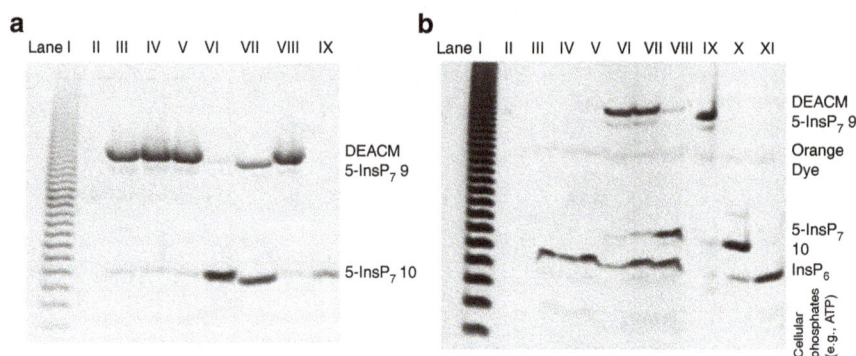

**Figure 3 | *In vitro* and *in cellulo* release of 5-InsP₇.** (**a**) Analysis of DEACM 5-InsP₇ **9** by gel electrophoresis (PAGE) and toluidine blue staining. **9** is stable for hours in rat brain extract (lanes III–V) and can be uncaged by ultraviolet irradiation (lane VI). Lane I: poly-P marker. Lane II: empty. Lane III: **9** in brain extract (3 h). Lane IV: **9** in brain extract (2 h). Lane V: **9** in brain extract (1 h). Lane VI: **9** in brain extract (1 h), then ultraviolet irradiation (15 min). Lane VII: **9** in distilled water, then ultraviolet irradiation (15 min). Lane VIII: **9**. Lane IX: 5-InsP₇ **10**. (**b**) Analysis of cellular uptake and *in cellulo* photouncaging with and without MoTr **11** after TiO₂ microsphere extraction followed by gel electrophoresis (PAGE). Bands containing **9** and **10** were additionally extracted and analysed by MALDI mass spectrometry. **9** only enters cells in the presence of MoTr **11** (lanes VI, VII) and can be uncaged in living cells (lane VIII). Lane I: Poly-P marker to assess quality of separation. Lane II: empty. Lane III: HeLa cells (control). Lane IV: HeLa cells + **11** (control). Lane V: HeLa cells + **9** (5 h). Lane VI: HeLa cells + **9** + **11** (5 h), Lane VII: HeLa cells + **9** + **11** (16 h). Lane VIII: HeLa cells + **9** + **11** (16 h, then 10 min irradiation 366 nm, 4W). Lane IX: DEACM 5-InsP₇ **9** (control). Lane X: 5-InsP₇ **10** (control). Lane XI: InsP₆ (control).

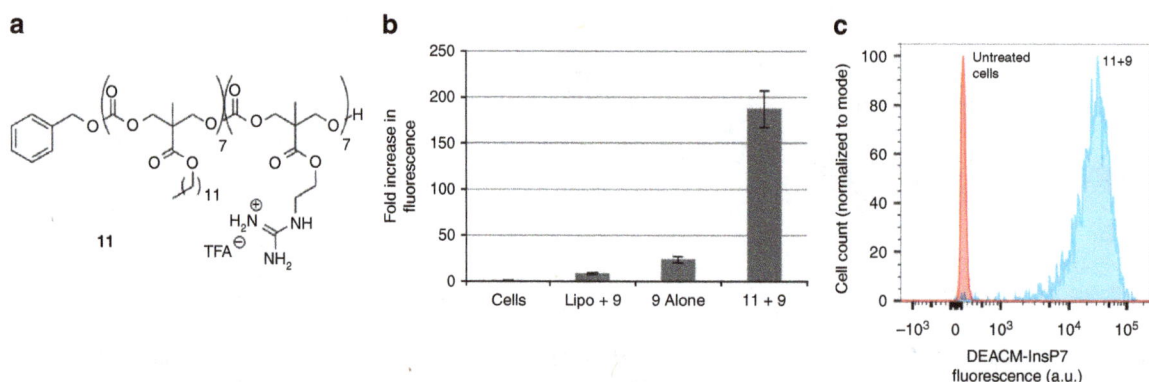

**Figure 4 | Intracellular delivery of photocaged 5-InsP₇ to HeLa cells with a guanidinium-rich transporter.** (**a**) Structure of amphipathic oligocarbonate transporter, **11**. (**b**) Cellular uptake of **9** as determined by flow cytometry. Complexes were formulated at a 1:1 mole ratio of **9** (5 μM) to **11**. Values reported are normalized to the autofluorescence of untreated cells. (**c**) Histogram plot of intracellular fluorescence demonstrates >99% delivery efficiency to cells.

Delivery and intracellular distribution of **9** were further analysed in HeLa cells by confocal microscopy after 4 and 16 h (Fig. 5). The z-stack analysis shows DEACM 5–InsP₇ **9** distributed throughout the cytoplasm at both time points (Fig. 5, single z-slice shown). Both diffuse fluorescence and fluorescent puncta are observed, consistent with mixed diffusion or endosomal uptake and release[44–46]. This is additionally supported by a 65% reduction in cellular uptake when cells were treated at 4 °C, a condition known to inhibit most endocytotic processes (Supplementary Fig. 11).

Cellular uptake, stability and efficient uncaging in living cells was additionally verified by extraction of diphospho-inositol polyphosphates and other cellular phosphates based on a recently published TiO₂ microsphere enrichment method[47]. Here it is shown that this method can also be used to extract analogues such as **9** from complex cell and tissue lysates (Fig. 3b and Supplementary Methods) enabling studies concerning its intracellular fate after delivery. After incubation of DEACM 5–InsP₇ **9** with HeLa cells in the presence or absence of MoTr **11** and repeated washings to remove external **9**, the extracts prepared from those cells (1 million cells) clearly showed a distinct novel band corresponding to **9** in the PAGE analysis after enrichment with TiO₂ and elution (Fig. 3b, lanes VI–VII), whereas no such

uptake could be detected in the control experiment without transporter (Fig. 3b, lane V). To verify its identity, the band corresponding to DEACM 5–InsP₇ **9** was extracted from the gel and analysed by MALDI mass spectrometry, demonstrating its intracellular stability (Supplementary Fig. 12 and Supplementary Methods) for multiple hours. Moreover, efficient intracellular uncaging by irradiation at 366 nm was proven using the same extraction and resolution method (TiO₂ enrichment, then PAGE) in combination with mass spectrometry after extraction of Lane VIII (Fig. 3b). These conditions were found to be of no immediate toxicity (Supplementary Fig. 13 and Supplementary Methods). In summary, the photocaged molecule **9** is efficiently taken up by cells in the presence of MoTr **11**, evenly distributed throughout the cytoplasm, stable for multiple hours in its caged form and can be selectively uncaged to 5-InsP₇ **10**, thus fulfilling the stringent requirements imposed on an intracellular signalling probe.

**PH-domain translocation.** To determine the suitability of the combined delivery and uncaging strategy for a deeper understanding of the effect of InsP₇ fluctuations, PH-domain translocation on cytoplasmic InsP₇ release was studied. The rationale for this experiment is provided by the lean phenotype displayed by IP6K1 knockout mice on high-fat diet and the observation that

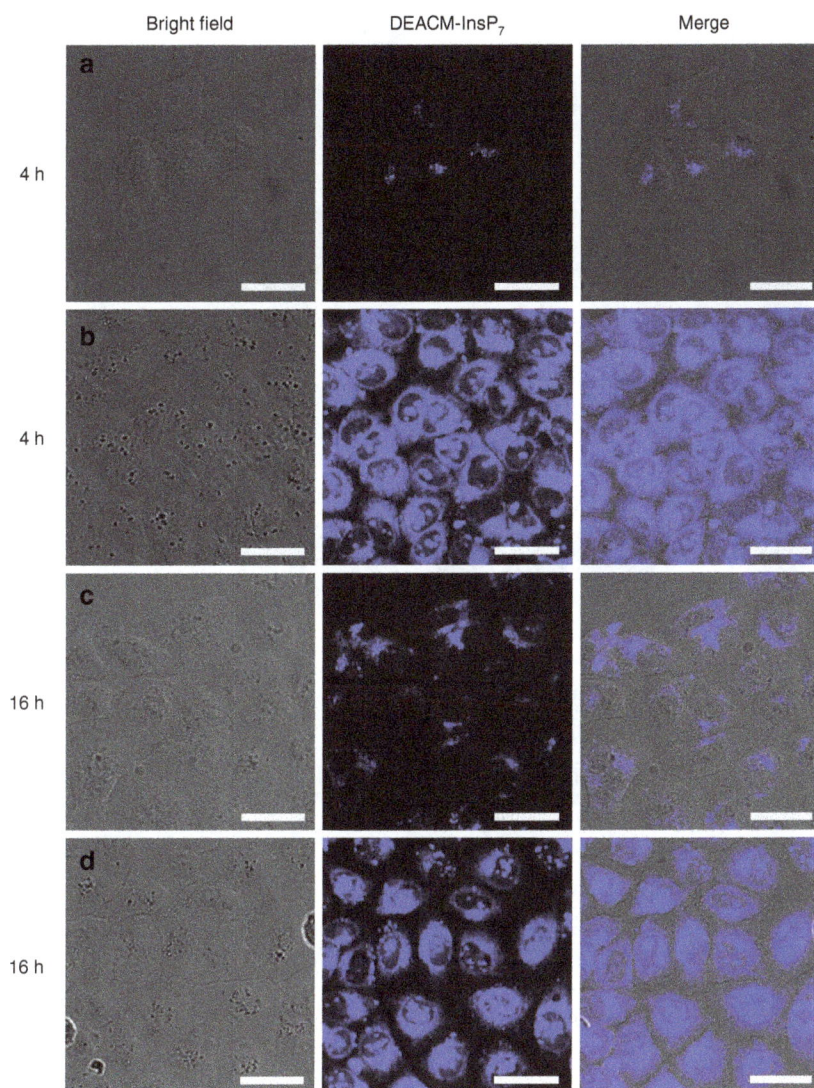

**Figure 5 | The molecular transporter 11 delivers photocaged 5-InsP₇9 into the cytoplasm.** Confocal microscopy analysis of HeLa cells incubated with 5 µM DEACM-InsP₇ **9** in the presence or absence of transporter **11** after 4 and 16 h of incubation. Efficient uptake of **9** is demonstrated by blue coumarin fluorescence emitted from the compound. Cells treated with (**a**) **9** alone for 4 h, (**b**) **9** + **11** for 4 h, (**c**) **9** alone for 16 h and (**d**) **9** + **11** for 16 h. Scale bars, 35 µm.

InsP₇ inhibit Akt phosphorylation *in vitro* and *in vivo* by binding to the PH-domain[9]. Collectively, these findings suggest an effect of 5-InsP₇ on membrane localization of Akt. However, no tool to augment any InsP₇ within seconds in living cells was previously available. With the new tools in hand, HeLa cells were transiently transfected with a plasmid expressing the PH-domain of Akt fused to an enhanced green fluorescent reporter protein (eGFP)[48,49]. Cells were serum-starved to induce cytoplasmic localization of the PH-domain due to absence of growth factors and therefore inactivation of the PI3K/Akt/mTOR pathway[50]. PH–eGFP plasma membrane association was then efficiently induced within 10 min on external addition of a combination of growth factors (insulin-like growth factor (IGF); endothelial growth factor (EGF)) into the medium. During the starvation period, cells were loaded with caged InsP₇ **9**/MoTr **11** nanoparticles for 4 h. This treatment alone had no effect on PH-domain localization. Next, cells were irradiated under a confocal laser-scanning microscope with short laser pulses (375 nm, 10 MHz, 30 s) in different areas (Fig. 6 dotted circle, and Supplementary Figs 14–23), and PH-domain localization was traced using the green channel. After photouncaging, a delayed

but complete PH-domain translocation from the plasma membrane into the cytoplasm was observed, and these results were repeated several times ($n = 4$). Significantly, translocation did not occur when cells were incubated with photocaged InsP₇ **9** or MoTr **11** only (Supplementary Figs 18–23). In these cases, the PH-domains remained localized on the membrane for several hours, demonstrating the need for the presence of all components and ruling out photobleaching of eGFP in the irradiated areas. A detailed analysis of additional micrographs in pseudo-colour with ratiometric changes is shown in the Supplementary Information (Supplementary Figs 14–23). This is the first example of controlled 5-InsP₇ **10** augmentation inside of a living cell within a few seconds timeframe coupled to a microscopic readout on the single cell level. We posit that this strategy will be useful to understand InsP₇ signalling in more detail as previously possible, as evidenced by the delayed PH-domain translocation observed for the first time in our experiments.

## Discussion

This study provides a new strategy to synthesize InsP₇ that enables introduction of caging subunits. The potential utility of

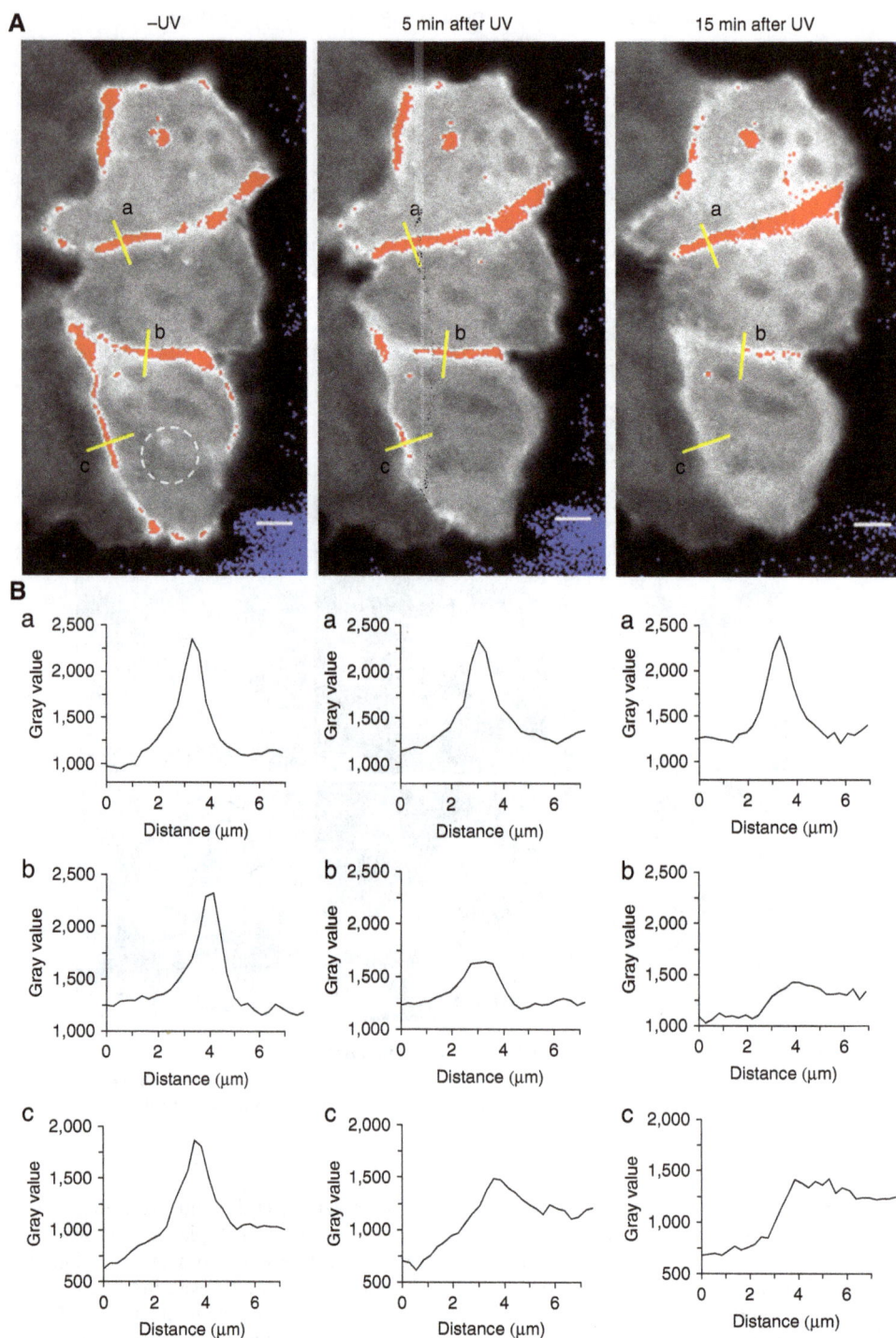

**Figure 6 | PH-domain translocation in irradiated and control cells.** Confocal fluorescence microscopy analysis of PH–eGFP translocation in HeLa Kyoto cells after photouncaging in defined areas (dotted circle). (**A**) Serum-starved cells were loaded with 5 μM **9** + **11** for 4 h and then stimulated with IGF and EGF (100 ng ml$^{-1}$). Robust recruitment of the PH-domain to the membrane is observed. Photouncaging in the dotted area (white circle) is achieved by short ultraviolet laser pulses and the change of fluorescence intensity followed over time (0, 5 and 15 min). (**B**) Development of the fluorescence intensity (indicated as gray value) over time (0, 5, 15 min) in three different membrane sections (a, b, c; distance in μm). Photouncaging leads to translocation of the PH–eGFP construct into the cytoplasm after 5 min from the membrane of the irradiated cell. After 15 min, complete translocation of the PH-domain into the cytoplasm is observed (**B**, b and c), whereas in the non-irradiated control cell the PH–eGFP construct remains localized on the membrane (**B**, a). Images are presented in pseudo-colour, normalized over time. Intensities were acquired pre-saturated, with the entire dynamic range of intensity available. Scale bars, 5 μM.

photocaged 5-InsP$_7$ **9** was demonstrated by photon-triggered uncaging in rat brain homogenate and other cell extracts. A complex of **9** with molecular transporter **11** was then shown to efficiently enter cells after non-covalent nanoparticle assembly.

Collectively, these results provide the first example of the synthesis of a photocaged analogue of InsP$_7$ and of its subsequent delivery into cells using non-covalent complexation with a guanidinium-rich molecular transporter. A recently developed

TiO$_2$ microsphere enrichment method was applied to study the *in cellulo* stability of 5-InsP$_7$ analogues and their efficient photochemical release in combination with MALDI mass spectrometry. We expect that this combined synthesis, delivery and analytical strategy will find widespread and general application in cell signalling studies as a convenient way to rapidly augment 5-InsP$_7$ **10** with spatiotemporal resolution. Along these lines, it was shown that cytoplasmic release of 5-InsP$_7$ triggers delayed but complete membrane desorption of the PH-domain of Akt within 15 min. In the human proteome, PH-domains are the 11th most common domain[51], and the new approach described in this publication will enable a systematic understanding of the effect of inositol-pyrophosphate augmentation on protein localization.

## Methods

**Experimental data of synthetic compounds.** For $^1$H, $^{13}$C and $^{31}$P NMR spectra of compounds and MALDI and HR-ESI MS spectra see Supplementary Figs 24–70. $^1$H NMR spectra were recorded on Bruker 400 MHz spectrometers or Bruker 500 MHz spectrometers (equipped with a cryo platform) at 298 K in the indicated deuterated solvent. $^{31}$P[$^1$H]-NMR spectra and $^{31}$P NMR spectra were recorded with $^1$H-decoupling or $^1$H coupling on Bruker 162 MHz or Bruker 202 MHz spectrometers (equipped with a cryo platform) at 298 K in the indicated deuterated solvent. All signals were referenced to an internal standard (PPP). $^{13}$C[$^1$H]-NMR spectra were recorded with $^1$H-decoupling on Bruker 101 MHz or Bruker 125 MHz spectrometers (equipped with a cryo platform) at 298 K in the indicated deuterated solvent. All signals were referenced to the internal solvent signal as standard (CDCl$_3$, $\delta$ 77.0; CD$_3$OD, $\delta$ 49.0; DMSO-d$_6$, $\delta$ 39.5).

Detailed synthetic procedures for all new compounds are provided in the Supplementary Information (see Supplementary Methods, chemical synthesis).

**Cellular uptake by flow cytometry.** Caged-IP7 **9** and oligomer **11** were brought up in pH 7.4 PBS buffer at 1 mM concentrations. HeLa cells were seeded at 40,000 cells per well in a 24-well plate and allowed to adhere overnight. The IP7:co-oligomer complexes were formed at a 1:1 molar ratio by mixing 8 µl of 1 mM oligomer stock with 8 µl of 1 mM caged-IP7 **9** stock in 184 µl PBS pH 7.4. For conditions with caged-IP7 **9** alone, 8 µl of 1 mM caged-IP7 **9** stock was added to 192 µl PBS pH 7.4. The complexes were allowed to incubate for 30 min at room temperature. The Lipofectamine 2000 control was prepared in OptiMEM according to the manufacturer's instructions (0.75 µl Lipofectamine into 62.5 µl OptiMEM, 8 µl caged-IP7 **9** stock into 54.5 µl OptiMEM). The cells were washed with ∼0.5 ml serum-free DMEM medium, then 400 µl serum-free DMEM was added to wells with untreated cells; 368.75 µl to wells treated with Lipofectamine 2000; 350 µl to treated wells. Then 31.25 µl Lipofectamine:caged-IP7 **9** and 50 µl of the caged-IP7:co-oligomer complexes were added to each respective well for a final caged-IP7 **9** concentration of 5 µM; all conditions were performed in triplicate. The cells were incubated at 37 °C for 3 h. The medium was then removed and the cells were washed with 1.0 ml PBS. About 0.4 ml EDTA trypsin was added, and the cells incubated for 8 min at 37 °C. Next, 0.6 ml of serum-containing DMEM medium was added, and the contents of each well were transferred to a 15 ml centrifuge tube and centrifuged (1,200 r.p.m. for 5 min). The cells were collected and re-dispersed in 200 µl PBS, transferred to FACS tubes, and read on a flow cytometry analyser. Results were analysed using FlowJo software. The data presented are the mean fluorescent signals from 10,000 cells analysed.

**Fluorescence microscopy.** Confocal microscopy was conducted on a CLSM SP5 Mid ultraviolet–visible Leica inverted confocal laser-scanning microscope equipped with a 15 W MaiTai DeepSee twophoton laser (Stanford Cell Sciences Imaging Facility, Award #S10RR02557401 from the National Center for Research Resources). HeLa cells were incubated in serum-free DMEM with 5 µM **9** or **9 + 11** (equimolar), for 4 or 16 h. After incubation, media were removed and cells washed 3 × with 1.0 mg ml$^{-1}$ heparin (180 U mg$^{-1}$, Aldrich) solution in PBS, then imaged in clear serum-free DMEM. Pictures were recorded with a HCX APO L × 20/1.00 water immersion objective and photomultiplier tube detector.

**eGFP–Akt-PH-domain translocation.** Mammalian cells (HeLa Kyoto) were grown up to 50–60% confluence (eight wells Lab-Tek Chamber Slide) in 250 µl DMEM (10% FBS, 10% Pen/Strep, high glucose (4.5 g l$^{-1}$)) at 37 °C, 5% CO$_2$ for 16 h. Cells were then washed with 250 µl PBS followed by a wash with serum-free DMEM (-FBS, -Pen/Step). About 230 µl serum-free DMEM medium were then added to each well.

Transfection of HELA cells with the eGFP Akt-PH-domain for imaging: The transfection mixture was prepared for eight wells. About 200 ng DNA (eGFP–Akt-PH) in 150 µl DMEM (-FBS) were incubated for 10 min at room temperature. Then, 10 µl FuGENE transfecting reagent were added to DNA

containing DMEM (-FBS) medium and incubated for 20 min at room temperature. About 20 µl of the transfection mixture were then added to each well. Cells were incubated for 6–8 h at 37 °C (5% CO$_2$.). After this period, the transfection mixture was removed by washing the cells with 250 µl PBS and DMEM (-FBS). Cells were starved by adding 250 µl DMEM (-FBS) and incubated for 14–16 h at 37 °C (5% CO$_2$).

After overnight starvation, the cells were washed with PBS (250 µl) and 230 µl DMEM (-FBS) were added. Afterwards, 20 µl sample mixture (consisting of caged InsP$_7$:Co-oligomer (D:G7:7) complexes) were added to each well and distributed equally. The final concentration of the complex was 5 µM for each well. Cells were then incubated at 37 °C (5% CO$_2$) for 4 h.

**Imaging of Akt-PH translocation.** After 4-h incubation with 9 and 11, HeLa cells were washed with 250 µl (2 × ) imaging buffer (115 mM NaCl, 1.2 mM CaCl$_2$, 1.2 mM MgCl$_2$, 1.2 mM K$_2$HPO$_4$, 20 mM HEPES). About 200 µl imaging buffer were added to the wells. Translocation of eGFP–Akt-PH to the membrane was stimulated by adding a mixture of growth factors EGF and IGF (100 ng ml$^{-1}$ each). Translocation was monitored for 10 min at 37 °C (5% CO$_2$).

Cells that had responded to treatment with the growth factors and that had the PH-domain localized at the membrane were used in the uncaging experiments by confocal laser-scanning microscopy. Non-illuminated neighboring cells were used as controls.

**Confocal laser-scanning microscopy.** Imaging was performed on an Olympus IX83 confocal laser-scanning microscope at 37 °C in a 5% CO$_2$ high humidity atmosphere (EMBL incubation box). Imaging was performed using an Olympus Plan-APON × 60 (numerical aperture 1.4, oil) objective. The images were acquired utilizing a Hamamatsu C9100-50 EM CCD camera. Image acquisition was performed via FluoView imaging software, version 4.2. The green channel was imaged using the 488 nm laser line (120 mW cm$^{-2}$) at 3% laser power and a 525/50 emission mirror. The red channel was imaged using the 559 nm laser (120 mW cm$^{-2}$) at 2.0% laser power and a 643/50 emission filter. A pulsed 375 nm laser line (10 MHz) was applied for uncaging experiments. For uncaging experiments, circular regions of interest of 4–10 µm diameter were pre-defined. Pre-activation images were captured for five frames (5 s per frame), followed by 30 s of activation within the regions of interest. Recovery images were captured for 35 min at a frame rate of 5 s per frame.

**Image analysis.** Image analysis was conducted utilizing Fiji open source image analysis software tool[52]. Lookup tables were applied to match the colour within the recorded image with the wavelengths of detected light. For comparability, the lookup tables of pre- and postactivated images were set the same weighting.

## References

1. Bennett, M., Onnebo, S. M. N., Azevedo, C. & Saiardi, A. Inositol pyrophosphates: metabolism and signalling. *Cell. Mol. Life Sci.* **63**, 552–564 (2006).
2. Monserrate, J. P. & York, J. D. Inositol phosphate synthesis and the nuclear processes they affect. *Curr. Opin. Cell Biol.* **22**, 365–373 (2010).
3. Shears, S. B., Weaver, J. D. & Wang, H. Structural insight into inositol pyrophosphate turnover. *Adv. Biol. Regul.* **53**, 19–27 (2013).
4. Wilson, M. S. C., Livermore, T. M. & Saiardi, A. Inositol pyrophosphates: between signalling and metabolism. *Biochem. J.* **452**, 369–379 (2013).
5. Rao, F. *et al.* Inositol pyrophosphates promote tumor growth and metastasis by antagonizing liver kinase B1. *Proc. Natl Acad. Sci. USA* **112**, 1773–1778 (2015).
6. Rao, F. *et al.* Inositol hexakisphosphate kinase-1 mediates assembly/disassembly of the CRL4-signalosome complex to regulate DNA repair and cell death. *Proc. Natl Acad. Sci. USA* **111**, 16005–16010 (2014).
7. Laha, D. *et al.* VIH2 regulates the synthesis of inositol pyrophosphate InsP8 and jasmonate-dependent defenses in Arabidopsis. *Plant Cell* **27**, 1082–1097 (2015).
8. Illies, C. *et al.* Requirement of inositol pyrophosphates for full exocytotic capacity in pancreatic beta cells. *Science* **318**, 1299–1302 (2007).
9. Chakraborty, A. *et al.* Inositol pyrophosphates inhibit akt signalling, thereby regulating insulin sensitivity and weight gain. *Cell* **143**, 897–910 (2010).
10. Luo, H. R. *et al.* Inositol pyrophosphates mediate chemotaxis in dictyostelium via pleckstrin homology domain-PtdIns(3,4,5)P3 interactions. *Cell* **114**, 559–572 (2003).
11. Wu, M. X. *et al.* Elucidating diphosphoinositol polyphosphate function with nonhydrolyzable analogues. *Angew. Chem. Int. Ed.* **53**, 7192–7197 (2014).
12. Mackenzie, R. W. & Elliott, B. T. Akt/PKB activation and insulin signalling: a novel insulin signalling pathway in the treatment of type 2 diabetes. *Diabetes Metab. Syndr. Obes.* **7**, 55–64 (2014).
13. Yu, H. T., Li, J. B., Wu, D. D., Qiu, Z. J. & Zhang, Y. Chemistry and biological applications of photo-labile organic molecules. *Chem. Soc. Rev.* **39**, 464–473 (2010).

14. Ozaki, S., DeWald, D. B., Shope, J. C., Chen, J. & Prestwich, G. D. Intracellular delivery of phosphoinositides and inositol phosphates using polyamine carriers. *Proc. Natl Acad. Sci. USA* **97,** 11286–11291 (2000).

15. Best, M. D., Zhang, H. L. & Prestwich, G. D. Inositol polyphosphates, diphosphoinositol polyphosphates and phosphatidylinositol polyphosphate lipids: Structure, synthesis, and development of probes for studying biological activity. *Nat. Prod. Rep.* **27,** 1403–1430 (2010).

16. Hoglinger, D., Nadler, A. & Schultz, C. Caged lipids as tools for investigating cellular signalling. *Biochim. Biophys. Acta* **1841,** 1085–1096 (2014).

17. Kantevari, S., Gordon, G. R. J., MacVicar, B. A. & Ellis-Davies, G. C. R. A practical guide to the synthesis and use of membrane-permeant acetoxymethyl esters of caged inositol polyphosphates. *Nat. Protoc.* **6,** 327–337 (2011).

18. Li, W. H., Llopis, J., Whitney, M., Zlokarnik, G. & Tsien, R. Y. Cell-permeant caged InsP(3) ester shows that $Ca^{2+}$ spike frequency can optimize gene expression. *Nature* **392,** 936–941 (1998).

19. Pavlovic, I. et al. Prometabolites of 5-Diphospho-myo-inositol Pentakisphosphate. *Angew. Chem. Int. Ed.* **54,** 9622–9626 (2015).

20. Glennon, M. C. & Shears, S. B. Turnover of inositol pentakisphosphates, inositol hexakisphosphate and diphosphoinositol polyphosphates in primary cultured-hepatocytes. *Biochem. J.* **293,** 583–590 (1993).

21. Geihe, E. I. et al. Designed guanidinium-rich amphipathic oligocarbonate molecular transporters complex, deliver and release siRNA in cells. *Proc. Natl Acad. Sci. USA* **109,** 13171–13176 (2012).

22. Albert, C. et al. Biological variability in the structures of diphosphoinositol polyphosphates in Dictyostelium discoideum and mammalian cells. *Biochem. J.* **327,** 553–560 (1997).

23. Capolicchio, S., Thakor, D. T., Linden, A. & Jessen, H. J. Synthesis of unsymmetric diphospho-inositol polyphosphates. *Angew. Chem. Int. Ed.* **52,** 6912–6916 (2013).

24. Capolicchio, S., Wang, H. C., Thakor, D. T., Shears, S. B. & Jessen, H. J. Synthesis of densely phosphorylated Bis-1,5-diphospho-myo-inositol tetrakisphosphate and its enantiomer by bidirectional P-anhydride formation. *Angew. Chem. Int. Ed.* **53,** 9508–9511 (2014).

25. Falck, J. R. et al. Synthesis and structure of cellular mediators—inositol polyphosphate diphosphates. *J. Am. Chem. Soc.* **117,** 12172–12175 (1995).

26. Laussmann, T., Reddy, K. M., Reddy, K. K., Falck, J. R. & Vogel, G. Diphospho-myo-inositol phosphates from Dictyostelium identified as D-6-diphospho-myo-inositol pentakisphosphate and D-5,6-bisdiphospho-myo-inositol tetrakisphosphate. *Biochem. J.* **322,** 31–33 (1997).

27. Reddy, K. M., Reddy, K. K. & Falck, J. R. Synthesis of 2- and 5-diphospho-myo-inositol pentakisphosphate (2- and 5-PP-InsP(5)), intracellular mediators. *Tetrahedron Lett.* **38,** 4951–4952 (1997).

28. Wang, H. C. et al. Synthetic inositol phosphate analogues reveal that PPIP5K2 has a surface-mounted substrate capture site that is a target for drug discovery. *Chem. Biol.* **21,** 689–699 (2014).

29. Wu, M., Dul, B. E., Trevisan, A. J. & Fiedler, D. Synthesis and characterization of non-hydrolysable diphosphoinositol polyphosphate messengers. *Chem. Sci.* **4,** 405–410 (2013).

30. Zhang, H., Thompson, J. & Prestwich, G. D. A scalable synthesis of the IP7Isomer, 5-PP-Ins(1,2,3,4,6)P5. *Org. Lett.* **11,** 1551–1554 (2009).

31. Watanabe, Y., Nakamura, T. & Mitsumoto, H. Protection of phosphate with the 9-fluorenylmethyl group. Synthesis of unsaturated-acyl phosphatidylinositol 4,5-bisphosphate. *Tetrahedron Lett.* **38,** 7407–7410 (1997).

32. Mentel, M., Laketa, V., Subramanian, D., Gillandt, H. & Schultz, C. Photoactivatable and cell-membrane-permeable phosphatidylinositol 3,4,5-trisphosphate. *Angew. Chem. Int. Ed.* **50,** 3811–3814 (2011).

33. Subramanian, D. et al. Activation of membrane-permeant caged PtdIns(3)P induces endosomal fusion in cells. *Nat. Chem. Biol.* **6,** 324–326 (2010).

34. Jessen, H. J., Schulz, T., Balzarini, J. & Meier, C. Bioreversible protection of nucleoside diphosphates. *Angew. Chem. Int. Ed.* **47,** 8719–8722 (2008).

35. Thomson, W. et al. Synthesis, bioactivation and anti-hiv activity of the Bis (4-Acyloxybenzyl) and Mono(4-Acyloxybenzyl) esters of the 5'-monophosphate of Azt. *J. Chem. Soc. Perkin Trans.* **1,** 1239–1245 (1993).

36. Cremosnik, G. S., Hofer, A. & Jessen, H. J. Iterative synthesis of nucleoside oligophosphates with phosphoramidites. *Angew. Chem. Int. Ed.* **53,** 286–289 (2014).

37. Hofer, A. et al. A Modular synthesis of modified phosphoanhydrides. *Chem. Eur. J.* **21,** 10116–10122 (2015).

38. Losito, O., Szijgyarto, Z., Resnick, A. C. & Saiardi, A. Inositol pyrophosphates and their unique metabolic complexity: analysis by gel electrophoresis. *PLoS ONE* **4,** e5580 (2009).

39. Anstaett, P., Leonidova, A. & Gasser, G. Caged phosphate and the slips and misses in determination of quantum yields for ultraviolet-A-induced photouncaging. *ChemPhysChem* **16,** 1857–1860 (2015).

40. Anstaett, P., Leonidova, A., Janett, E., Bochet, C. G. & Gasser, G. Reply to commentary by trentham et al. on "caged phosphate and the slips and misses in

41. Corrie, J. E., Kaplan, J. H., Forbush, B., Ogden, D. C. & Trentham, D. R. Commentary on "caged phosphate and the slips and misses in determination of quantum yields for ultraviolet-a-induced photouncaging" by G. Gasser and Co-Workers. *ChemPhysChem* **16,** 1861–1862 (2015).

42. Schonleber, R. O., Bendig, J., Hagen, V. & Giese, B. Rapid photolytic release of cytidine 5-diphosphate from a coumarin derivative: A new tool for the investigation of ribonucleotide reductases. *Bioorg. Med. Chem.* **10,** 97–101 (2002).

43. Stanzl, E. G., Trantow, B. M., Vargas, J. R. & Wender, P. A. Fifteen years of cell-penetrating guanidinium-rich molecular transporters: basic science, research tools, and clinical applications. *Acc. Chem. Res.* **46,** 2944–2954 (2013).

44. Lee, H. L. et al. Single-molecule motions of oligoarginine transporter conjugates on the plasma membrane of Chinese hamster ovary cells. *J. Am. Chem. Soc.* **130,** 9364–9370 (2008).

45. Rothbard, J. B., Jessop, T. C., Lewis, R. S., Murray, B. A. & Wender, P. A. Role of membrane potential and hydrogen bonding in the mechanism of translocation of guanidinium-rich peptides into cells. *J. Am. Chem. Soc.* **126,** 9506–9507 (2004).

46. Wender, P. A., Galliher, W. C., Goun, E. A., Jones, L. R. & Pillow, T. H. The design of guanidinium-rich transporters and their internalization mechanisms. *Adv. Drug Deliv. Rev.* **60,** 452–472 (2008).

47. Wilson, M. S. C., Bulley, S. J., Pisani, F., Irvine, R. F. & Saiardi, A. A novel method for the purification of inositol phosphates from biological samples reveals that no phytate is present in human plasma or urine. *Open Biol.* **5,** 150014 (2015).

48. Cormack, B. P., Valdivia, R. H. & Falkow, S. FACS-optimized mutants of the green fluorescent protein (GFP). *Gene* **173,** 33–38 (1996).

49. Laketa, V. et al. PIP(3) induces the recycling of receptor tyrosine kinases. *Sci. Signal.* **7,** ra5 (2014).

50. Jo, H. et al. Small molecule-induced cytosolic activation of protein kinase Akt rescues ischemia-elicited neuronal death. *Proc. Natl Acad. Sci. USA* **109,** 10581–10586 (2012).

51. Lemmon, M. A. Pleckstrin homology (PH) domains and phosphoinositides. *Biochem. Soc. Symp.* **74,** 81–93 (2007).

52. Schindelin, J. et al. Fiji: an open-source platform for biological-image analysis. *Nat. Methods* **9,** 676–682 (2012).

## Acknowledgements

Imaging was performed with support of the Center for Microscopy and Image Analysis at UZH and the Stanford Shared FACS Facility. We thank Professors Jay Siegel, John Robinson, Adolfo Saiardi, Robert Waymouth, Chris Contag, Lynette Cegelski, Dr. Vanessa Pierroz, Dr. Riccardo Rubbiani and Dr. Vibor Laketa for discussions, materials and procedures. This work was supported by The Swiss National Science Foundation (PP00P2_157607 to H.J.J. and PP00P2_133568 & PP00P2_157545 to G.G.), the National Institutes of Health (NIH-CA031841, NIH-S10RR027431-01 and NIH-CA031845 to P.A.W.), and the Deutsche Forschungsgemeinschaft (DFG, TRR83 to C.S.). Fellowships: National Science Foundation and the Stanford Center for Molecular Analysis and Design.

## Author contributions

I.P., D.T.T., R.C.C., J.R.V., C.J.M., P.A. and S.H. conducted the experiments. H.J.J., P.A.W., C.S., L.B. and G.G. planned the experiments. H.J.J., P.A.W., J.R.V. and C.J.M. wrote the manuscript. All authors discussed the results and revised the manuscript.

# 13

# Self-assembly of dynamic orthoester cryptates

René-Chris Brachvogel[1], Frank Hampel[1] & Max von Delius[1]

The discovery of coronands and cryptands, organic compounds that can accommodate metal ions in a preorganized two- or three-dimensional environment, was a milestone in supramolecular chemistry, leading to countless applications from organic synthesis to metallurgy and medicine. These compounds are typically prepared via multistep organic synthesis and one of their characteristic features is the high stability of their covalent framework. Here we report the use of a dynamic covalent exchange reaction for the one-pot template synthesis of a new class of coronates and cryptates, in which acid-labile $O,O,O$-orthoesters serve as bridgeheads. In contrast to their classic analogues, the compounds described herein are constitutionally dynamic in the presence of acid and can be induced to release their guest via irreversible deconstruction of the cage. These properties open up a wide range of application opportunities, from systems chemistry to molecular sensing and drug delivery.

[1] Department of Chemistry and Pharmacy, Friedrich-Alexander-University Erlangen-Nürnberg (FAU), Henkestrasse 42, 91054 Erlangen, Germany. Correspondence and requests for materials should be addressed to M.v.D. (email: max.vondelius@fau.de).

Past progress in supramolecular chemistry has been driven chiefly by the development of new macrocyclic molecules[1]. Pedersen's crown ethers (also called coronands)[2] and Lehn's cryptands[3,4] (Fig. 1) are excellent examples of compounds that initially served as platforms for studying non-covalent interactions, but have ultimately found widespread application in industry and medicine[5].

In the last decade, rationally designed three-dimensional cage compounds[6–8] have become larger and larger[9–13], enabling in one extreme case even the accommodation of a small protein[14]. To achieve the self-assembly of such large structures, the method of choice is dynamic constitutional/covalent chemistry (DCC)[15–17], which offers the essential feature of error correction that is needed to avoid significant side-product formation. Besides providing high-yielding syntheses, DCC generally gives rise to target structures that are dynamic and responsive to external stimuli under the conditions of their preparation. In the context of the emerging field of systems chemistry[18,19], dynamic macrocycles and cages have served as a valuable testing ground for the investigation and manipulation of complex molecular networks[20–25]. Constitutionally dynamic cryptates[26,27] would represent the smallest and simplest conceivable three-dimensional platform for studying molecular complexity; however, to the best of our knowledge, there are no reports of monometallic cryptates yet, which can be prepared and manipulated based on a dynamic covalent exchange reaction[28–31].

Here we describe how orthoester exchange, a previously ignored dynamic covalent reaction[32], can be used for the one-pot synthesis of monometallic cryptates (for example, see Fig. 1) from strikingly simple starting materials. We provide comprehensive characterization data (including an X-ray structure) for this new class of compounds and report on their dynamic properties, as well as on the formation of orthoester crown ethers as reaction intermediates and the unexpected finding that 4 Å molecular sieves (MS) can act as a source of sodium guest.

## Results

**Template synthesis of a dynamic orthoester cryptate.** Inspired by reports on dynamic 'scaffolding ligands' (O,N,P-ortho-esters)[33,34], we have recently investigated the exchange reaction of carboxylic O,O,O-orthoesters with simple alcohols from a DCC perspective[32]. We realized during the course of these studies that the tripodal architecture and dynamic chemistry of orthoesters[35] could be well suited for establishing two bridgeheads in macrobicyclic compounds (Fig. 1). The synthesis of such orthoester cryptands could be carried out in one step and under thermodynamic control, while a suitable metal ion could serve as a template.

To test these hypotheses, we treated a chloroform solution of two bulk chemicals, trimethyl orthoacetate (**1**) and diethylene glycol (**2**), with catalyst trifluoroacetic acid (TFA) and a stoichiometric metal template (Fig. 2). Analysis of these initial experiments by [1]H nuclear magnetic resonance (NMR) spectroscopy and electrospray ionization mass spectrometry revealed that complex mixtures, containing the desired cryptate among other exchange products, had formed. Careful optimization of the reaction conditions (use of MS as a thermodynamic sink for water and methanol; use of the 'non-coordinating' counteranion tetrakis[3,5-bis(trifluoromethyl)phenyl]borate (BArF$^-$))[36–40] eventually led to the formation of cryptate [Na$^+ \subset o$-Me$_2$-1.1.1] BArF$^-$ (named in loose analogy to Lehn's classic cryptates; '$o$' stands for orthoester) as the predominant reaction product (isolated yields typically 60%–70%).

As shown in the [1]H NMR spectra presented in Fig. 2, the reversible reaction between **1** and **2** initially generates a remarkable diversity of different exchange products (various degrees of replacement of MeOH by **2**, as well as formation of macrocyclic and oligomeric products (Fig. 2b)) and it is only on slow removal of MeOH by MS (4 Å) that the system converges to the final reaction product (Fig. 2c).

It should be noted that during the exchange process, water has to be excluded from the reaction mixture, which is not trivial to achieve, because even rigorously dried MS tend to slowly release residual water. As a consequence, hydrolysis of **1** can lead to the slow formation of methyl acetate (MeOAc) as a side product (Fig. 2c). From a preparative standpoint, the formation of MeOAc is not a problem, because it can easily be removed under reduced pressure (Fig. 2d). In addition, its formation as the sole side product provides two valuable pieces of information regarding the dynamic system under study. First, it is remarkable that we find the cryptate as the exclusive reaction product, even though the partial decomposition of orthoester **1** leads to a non-ideal ratio of orthoester to diol (ideal value 2:3). This result indicates that there is a thermodynamic bias for the formation of the final cryptate in the presence of sodium template. Second, the fact that we observe only one type of ester (MeOAc) suggests that exchange products incorporating one or more diethylene glycol chains (generated much more rapidly than MeOAc) are kinetically stabilized against hydrolysis, presumably due to (chelate) binding of sodium.

This kinetic stabilization due to metal binding[41] is most pronounced in pristine cryptate [Na$^+ \subset o$-Me$_2$-1.1.1]BArF$^-$. For example, when we mixed [Na$^+ \subset o$-Me$_2$-1.1.1]BArF$^-$ with trimethyl orthoacetate (**1**) in water-saturated chloroform, only the simple orthoester **1** was found to hydrolyse, whereas cryptate [Na$^+ \subset o$-Me$_2$-1.1.1]BArF$^-$ remained stable for 7 days (Supplementary Fig. 2). In the absence of acid, the cage is in fact stable in dimethyl sulfoxide/water mixtures and can be purified by silica gel chromatography (Fig. 2e). These observations are highly unusual for O,O,O-orthoesters that are not stabilized by the presence of five- or six-membered rings (as in Corey's OBO protecting group)[42]. These properties imply that orthoester-based hosts could have a unique advantage over existing coronands and cryptands: charged guests could be transported across lipophilic membranes[43] and subsequent hydrolysis would trigger the release of the guest ([Na$^+ \subset o$-Me$_2$-1.1.1]BArF$^-$ hydrolyses readily in the presence of excess water and acid; Supplementary Fig. 3). The high potential for such a supramolecular approach for drug formulation and delivery is underscored by a recent patent publication, which describes related hydrolysis-prone crown ether compounds (scheduled for phase 1 clinical trials in 2015)[44].

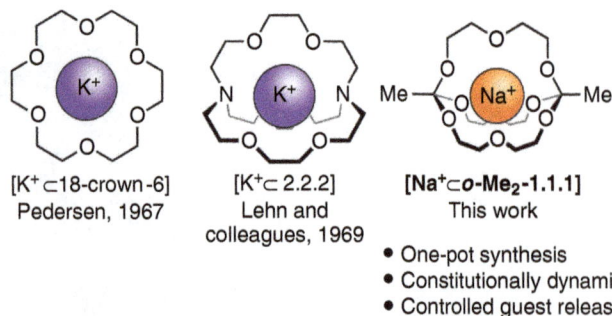

[K$^+ \subset$18-crown-6]
Pedersen, 1967

[K$^+ \subset$ 2.2.2]
Lehn and
colleagues, 1969

[Na$^+ \subset o$-Me$_2$-1.1.1]
This work

- One-pot synthesis
- Constitutionally dynamic
- Controlled guest release

**Figure 1 | Coronates and cryptates.** Comparison of a classic coronate and cryptate with one of the orthoester-based, constitutionally dynamic cryptates described in this work.

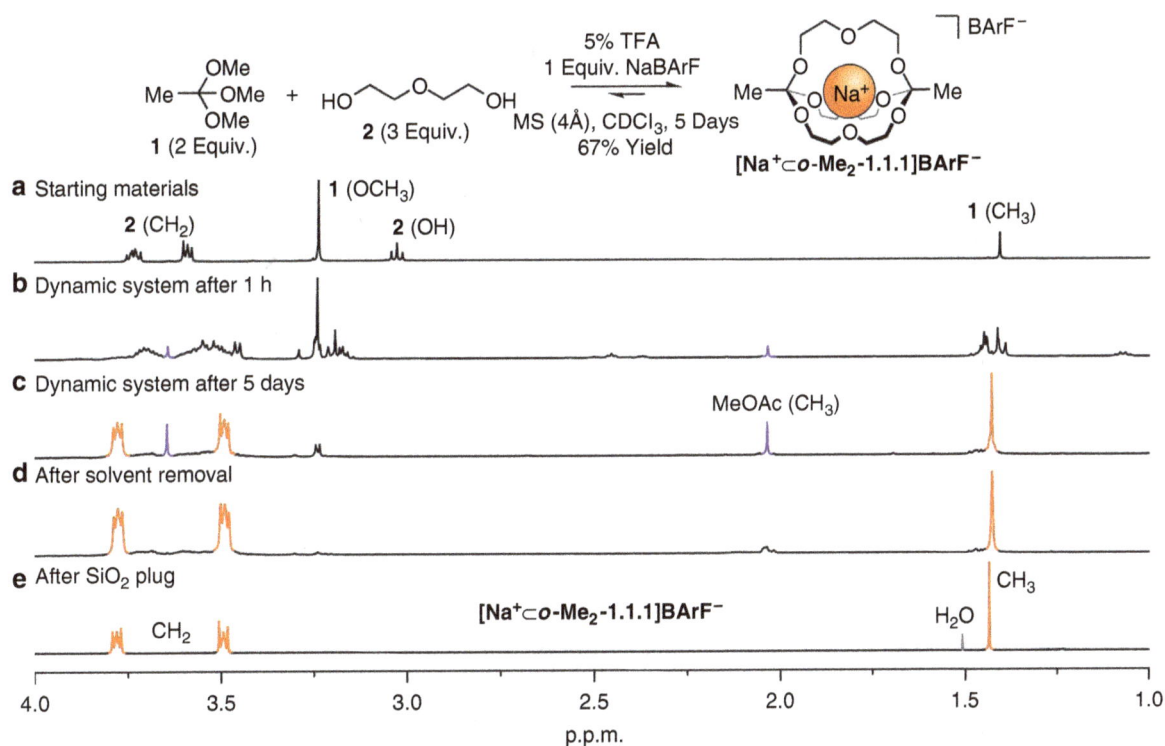

**Figure 2 | One-pot self-assembly of dynamic cryptate [Na$^+$ $\subset$ o-Me$_2$-1.1.1]BArF$^-$.** Partial $^1$H NMR spectra (400 MHz, CDCl$_3$, 298 K) showing the evolution of the dynamic system over time (see Supplementary Fig. 1 for full spectra). (**a**) Starting materials. (**b**) Complex mixture that forms rapidly after addition of acid catalyst (1 h). (**c**) After 5 days [Na$^+$ $\subset$ o-Me$_2$-1.1.1]BArF$^-$ is formed as major product, alongside hydrolysis product MeOAc (singlets at 3.6 and 2.0 p.p.m.). (**d**) The crude product is obtained by removal of MeOAc under reduced pressure. (**e**) Further purification is conveniently achieved by passing the crude product through a short plug of silica gel. Reaction conditions: trimethyl orthoacetate (120 μmol), diethylene glycol (180 μmol), NaBArF (60 μmol), TFA (3.0 μmol; added over 5 days), MS (4 Å, 1 g), CDCl$_3$ (6.0 ml), 5 days, room temperature.

**Solid-state structure.** Following the comprehensive characterization of [Na$^+$ $\subset$ o-Me$_2$-1.1.1]BArF$^-$ by NMR spectroscopy and mass spectrometry (Supplementary Figs 4–13), we turned our attention towards obtaining further structural and dynamic insights on this compound. To our delight, single crystals of [Na$^+$ $\subset$ o-Me$_2$-1.1.1]BArF$^-$ suitable for X-ray crystallography could be obtained by slow diffusion of cyclopentane into dilute solutions in chloroform or dichloromethane. The solid-state structure (Fig. 3a, Supplementary Fig. 14, Supplementary Tables 1 and 2, and Supplementary Data 1) shows that the sodium ion is bound to all nine surrounding oxygen atoms with a mean Na–O bond length of 2.56 Å, which is very close to the mean Na–O distance of 2.57 Å found in the solid-state structure of Lehn's classic cryptate [Na$^+$ $\subset$ 2.2.2]I$^-$ (ref. 45). An interesting question arises from the relatively small distances between the three diethylene glycol chains (O–O distance between two chains: 4.5 Å; see space-filling model in Fig. 3a) and the relatively rigid architecture of the cage (in contrast to classic cryptates, no inversion is possible at the terminus of the cage): can the metal ion exit from the o-Me$_2$-1.1.1 cage under ambient conditions? Or in other words, is [Na$^+$ $\subset$ o-Me$_2$-1.1.1]BArF$^-$ in fact a carceplex, not a cryptate?

**Thermodynamics and kinetics of guest exchange.** To answer this question, we carried out competition experiments in which complexation agents such as Lehn's cryptate 2.2.1 (Na$^+$ binding constant $K_A = 10^{13}$ M$^{-1}$ in D$_2$O-saturated CDCl$_3$)$^{46}$ were titrated to a freshly deacidified solution of [Na$^+$ $\subset$ o-Me$_2$-1.1.1]BArF$^-$ in chloroform. As shown in Fig. 3b (top), addition of classic cryptand 2.2.1 to our orthoester cryptate led to quantitative formation of cryptate [Na$^+$ $\subset$ 2.2.1]BArF$^-$ and orthoester cryptand o-Me$_2$-1.1.1, indicating that 2.2.1 has a significantly higher binding constant under these conditions. The reaction outcome also confirms that sodium can exit from the orthoester cage and the observed $^1$H NMR spectra demonstrate that in this experiment sodium ion exchange is slow between the two competing cryptands. Following such a titration, we treated a 1:1 mixture of [Na$^+$ $\subset$ 2.2.1]BArF$^-$ and o-Me$_2$-1.1.1 with catalytic TFA, resulting in the complete conversion of o-Me$_2$-1.1.1 into orthoester products featuring eight-membered rings (Supplementary Fig. 19). This experiment demonstrates that o-Me$_2$-1.1.1, unlike [Na$^+$ $\subset$ o-Me$_2$-1.1.1]BArF$^-$, does not represent a thermodynamic minimum and thus cannot be prepared without template from compounds **1** and **2** via reversible orthoester exchange.

In a second competition experiment, we titrated weaker complexation agent 15-crown-5 (Na$^+$ binding constant $K_A = 10^5$ M$^{-1}$ in acetonitrile)$^{46}$ to cryptate [Na$^+$ $\subset$ o-Me$_2$-1.1.1]BArF$^-$. As evident from the series of $^1$H NMR spectra (Fig. 3b, bottom), at equimolar addition of 15-crown-5 the equilibrium lies on the side of the orthoester cryptate, although broadening of the peaks indicates that exchange of sodium is fast in this case. Titration with up to 20 equivalents of 15-crown-5 gave rise to binding isotherms, from which we could deduce that the binding constant of o-Me$_2$-1.1.1 is about one order of magnitude higher than that of 15-crown-5 (see Supplementary Figs 20–22 for further thermodynamic data). In a pristine mixture of cryptate [Na$^+$ $\subset$ o-Me$_2$-1.1.1]BArF$^-$ and cryptand o-Me$_2$-1.1.1 (vide infra for preparation method), we were able to determine an exchange rate of 0.6 s$^{-1}$ for sodium exchange between degenerate orthoester cryptands (NMR exchange spectroscopy (EXSY); Supplementary Fig. 23 and Supplementary Note 1).

**a**

X-ray structure:

Front view

Side view

Space filling

**b**

Competition for binding of Na$^+$:

[Na$^+$ ⊂ o-Me$_2$-1.1.1]BArF$^-$        2.2.1

Slow exchange

100%

0%

3.85  3.75  3.65  3.55  3.45  3.35   2.10  2.00   1.50  1.40  1.30
                        p.p.m.

[Na$^+$⊂o-Me$_2$-1.1.1]BArF$^-$        15-crown-5

Fast exchange

100%

0%

3.85  3.75  3.65  3.55  3.45  3.35   2.10  2.00   1.50  1.40  1.30
                        p.p.m.

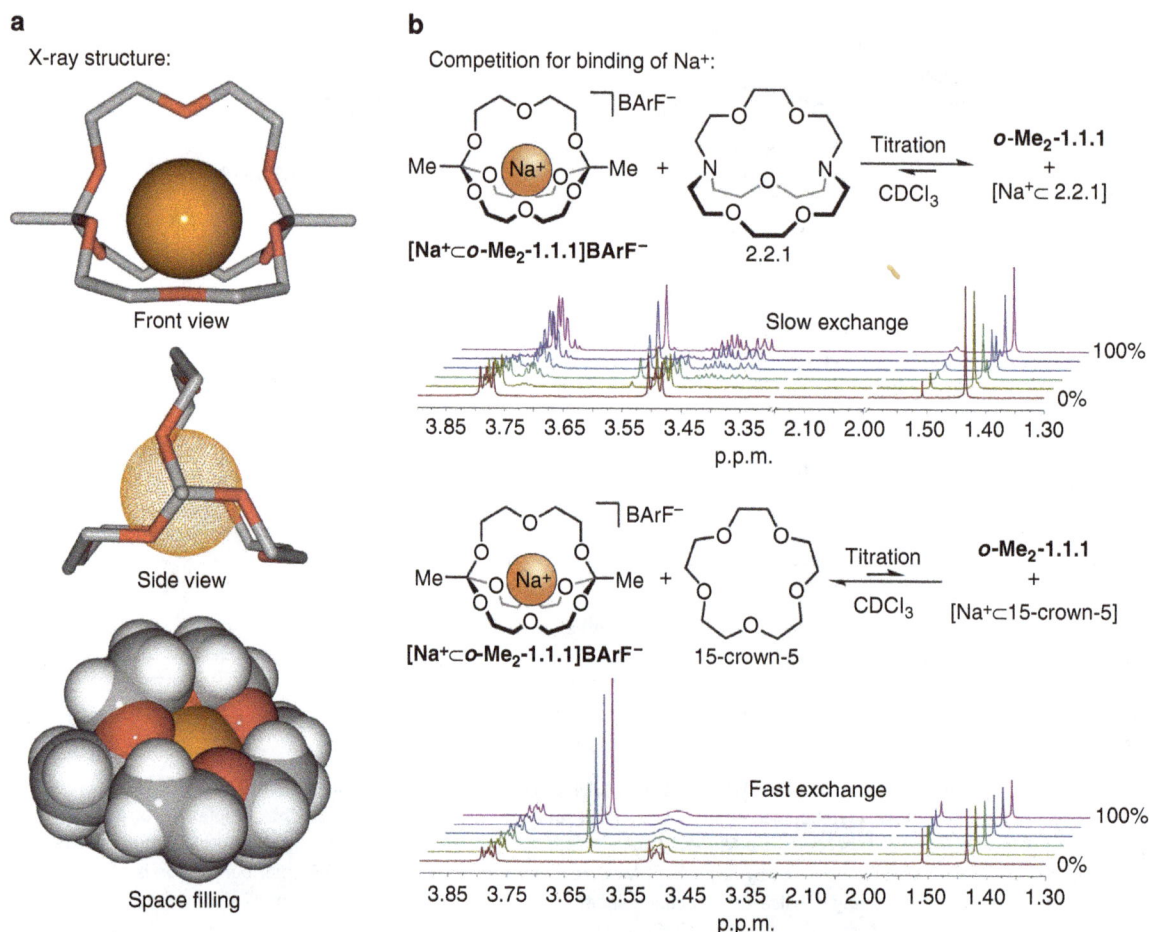

**Figure 3 | Solid-state structure of [Na$^+$ ⊂ o-Me$_2$-1.1.1]BArF$^-$ and competition experiments with classic hosts for Na$^+$. (a)** Solid-state structure determined by single-crystal X-ray diffraction. Crystals were obtained by slow diffusion of cyclopentane into dichloromethane. Counteranion, hydrogen atoms and disorder (along the CH$_2$-CH$_2$-O-CH$_2$-CH$_2$ chains, disorder with a population of 1:1 was observed; see Supplementary Fig. 14) are omitted for clarity. Oxygen atoms are shown in red, carbon atoms in grey. Sodium (orange) is shown at 65% of the van der Waals radius in stick model representations. Na–O bond lengths: 2.51–2.58 Å (six orthoester oxygens), 2.55–2.62 Å (three chain oxygens). **(b)** $^1$H NMR titration experiments (400 MHz, CDCl$_3$, 298 K) using classic complexation agents 2.2.1 (top) and 15-crown-5 (bottom). Pristine orthoester cryptate is shown at the front, 100% addition of competing host is shown at the back. The region around 2.1 p.p.m. is included to demonstrate that orthoester hydrolysis is not occurring during these experiments (despite the presence of water; singlet at 1.6 p.p.m.). For details, see Supplementary Figs 15–18.

**Other metal templates and orthoester crown ethers.** To confirm that a metal template effect is responsible for the remarkably clean formation of [Na$^+$ ⊂ o-Me$_2$-1.1.1]BArF$^-$ (Fig. 1), we studied the exchange reaction between **1** and **2** in the absence of template and in the presence of different metal templates under otherwise identical reaction conditions. As shown in Fig. 4 (top), the exchange reaction without template mainly gave rise to exchange product **3**, a monomeric orthoester featuring an eight-membered ring that results from one molecule of diethyleneglycol (**2**) having displaced two molecules of methanol (Supplementary Fig. 24). Theoretically, it should be possible to remove the last equivalent of methanol by increasing the time during which the mixture is exposed to MS, but we found that the system has a strong tendency to remain at this particular state (that is, product **3**). However, when such a dynamic mixture was treated with one equivalent of Sodium tetrakis[3,5-bis(trifluoromethyl)phenyl]borate (NaBArF), the dynamic system responded by forming cryptate [Na$^+$ ⊂ o-Me$_2$-1.1.1]BArF$^-$ quantitatively and in a relatively short time (Fig. 4, right-hand side, and Supplementary Fig. 25).

When we used metal salts LiTPFPB, NaBArF or KBArF for the self-assembly reaction, we observed that after 2–3 days reaction time two distinct reaction products had formed in a 1:1 ratio.

Careful analysis of one- and two-dimensional NMR spectra, as well as high-resolution mass spectrometry, indicated that unprecedented orthoester crown ethers[1] o-Me$_2$-(OMe)$_2$-16-crown-6 had formed as a mixture of *syn* and *anti* diastereomers (Fig. 4, centre). These crown ethers are the main products at a stage of the reaction where two equivalents of methanol have been removed through the effect of MS (2–3 days reaction time). The dynamic chemical system is thus not only responsive to the described metal template effect, but also to the precise quantity of available methanol.

**Molecular sieves (4 Å) as an unexpected sodium source.** To our initial surprise, the crown ethers originating from the lithium and potassium salts eventually transformed into the corresponding sodium cryptates [Na$^+$ ⊂ o-Me$_2$-1.1.1]X$^-$ (Fig. 4, X$^-$ = tetraarylborate anion). Using mass spectrometry, $^{23}$Na and $^7$Li NMR spectroscopy and, most notably, atom absorption and emission spectroscopy, we were able to confirm that these cage compounds were indeed the sodium cryptates [Na$^+$ ⊂ o-Me$_2$-1.1.1]X$^-$ (Supplementary Figs 26 and 27), and that the sodium source is type A zeolite (4 Å MS)[47], a porous framework material that contains accessible sodium ions. A search

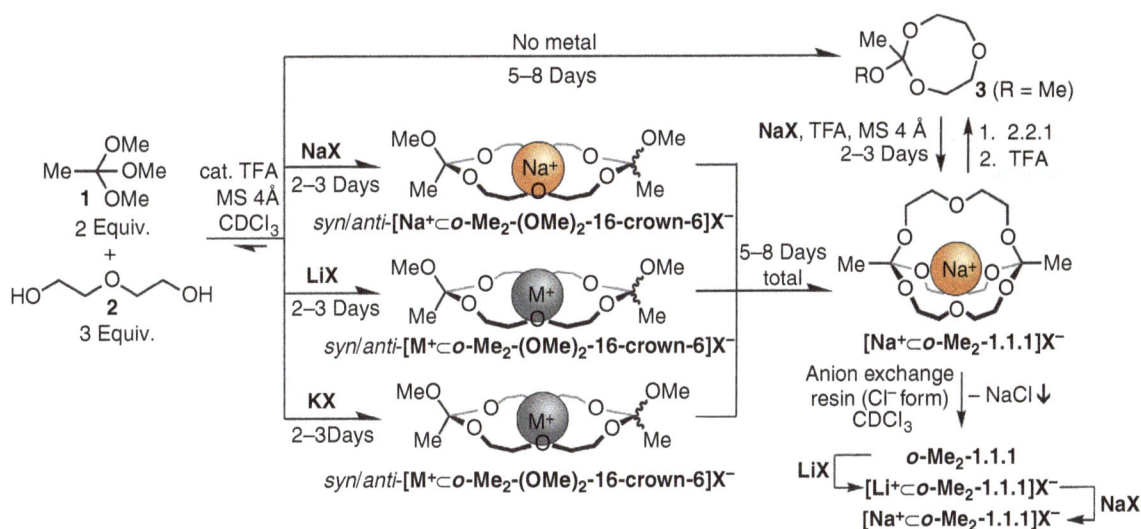

**Figure 4 | Complex behaviour of a dynamic orthoester system.** Structures of predominant products as a function of time, template and other chemical stimuli. Reaction conditions: trimethyl orthoacetate (120 μmol), diethylene glycol (180 μmol), metal salt (60 μmol), TFA (0.6 μmol added per day), MS (4 Å, 1 g), CDCl$_3$ (6.0 ml), room temperature. Eight-membered ring products (**3** for R = Me) could be quantitatively converted into cryptate [**Na$^+$ ⊂ o-Me$_2$-1.1.1**]X$^-$ and vice versa. Pristine cryptand **o-Me$_2$-1.1.1** could be prepared using anion exchange resin Lewatit MP64 (chloride form). Cryptand **o-Me$_2$-1.1.1** could be transformed into lithium cryptate [**Li$^+$ ⊂ o-Me$_2$-1.1.1**]X$^-$ and from there back into [**Na$^+$ ⊂ o-Me$_2$-1.1.1**]X$^-$. NaX: NaBArF; LiX: LiTPFPB (lithium tetrakis(pentafluorophenyl)borate); KX: KBArF; M$^+$: mixture of Na$^+$ with residual Li$^+$ or K$^+$ due to cation exchange with MS.

of the literature revealed several reports on this type of ion exchange in aqueous[47,48] and one report in organic medium[49]. When we used metal salt KBArF in conjunction with MS 3 Å, which contain potassium instead of sodium ions, we did not observe self-assembly of a potassium cryptate from starting materials **1** and **2**. Collectively, our experiments with different metal ions point towards a pronounced preference for sodium cryptate [**Na$^+$ ⊂ o-Me$_2$-1.1.1**] over the potential lithium or potassium cryptates, which could be explained by the differences in effective ionic radii between lithium (0.9 Å), sodium (1.2 Å) and potassium (1.6 Å)[50]. Only sodium appears to have the right size for forming nine efficient metal–oxygen bonds, while not inducing energetically costly conformations within the organic host.

**Preliminary experiments on scope and limitations.** We conducted preliminary studies on the scope of the described orthoester cryptates and coronates. A cryptate derived from orthoester trimethyl orthopropanoate, [**Na$^+$ ⊂ o-Et$_2$-1.1.1**]BArF$^-$ (terminal substituent: Et), could be prepared without difficulty, following a similar procedure to that used for cryptate [**Na$^+$ ⊂ o-Me$_2$-1.1.1**]BArF$^-$. Using trimethyl orthoformate as the starting material (terminal substituent: H) led to the formation of remarkably stable and pure crown ether complexes [**Na$^+$ ⊂ o-H$_2$-(OMe)$_2$-16-crown-6**], which did not further react to the corresponding cryptates under our standard conditions. A notable limitation of the self-assembly reaction concerns the counteranion of the sodium template. Thus far, we were able to prepare cryptate [**Na$^+$ ⊂ o-Me$_2$-1.1.1**]X$^-$ with three different tetra-arylborate anions (X$^-$ = BArF$^-$, TPFPB$^-$ and tetrakis(4-chlorophenyl)borate), while simpler anions such as PF$_6^-$ or BF$_4^-$ have failed, suggesting that a truly 'non-coordinating'[37] anion needs to be present during self-assembly.

In an attempt to exchange the counteranion in pristine [**Na$^+$ ⊂ o-Me$_2$-1.1.1**]BArF$^-$ to chloride, we discovered a convenient method for preparing cryptand **o-Me$_2$-1.1.1** (Fig. 4, bottom right). When a solution of [**Na$^+$ ⊂ o-Me$_2$-1.1.1**]BArF$^-$ in CDCl$_3$ was treated with anion exchange resin Lewatit MP-64 (Cl form), we observed the clean formation of **o-Me$_2$-1.1.1**, for

which the precipitation of NaCl is presumably the driving force. With pristine **o-Me$_2$-1.1.1** in our hands, we were able to prepare cryptate [**Li$^+$ ⊂ o-Me$_2$-1.1.1**]TPFPB$^-$ by exposing the cryptand to a solution of the lithium salt (Fig. 4, bottom right; structure confirmed by $^1$H/$^7$Li hetero nuclear overhauser effect NMR spectroscopy; Supplementary Fig. 28). The lithium cryptate could be transformed back into [**Na$^+$ ⊂ o-Me$_2$-1.1.1**]BArF$^-$ by addition of one equivalent of NaBArF, confirming the preference of this cryptand for Na$^+$. Based on preliminary NMR data, K$^+$ (salt: KBArF) appears to not enter the cage, but presumably 'nests' on the crown-ether-type faces of the cryptand. Further experiments to such ends are ongoing in our laboratory.

## Discussion

We have shown that a dynamic system based on two strikingly simple organic starting materials converges to three distinct types of exchange products under the influence of dry MS: (i) in the absence of a template, a simple exchange product featuring an eight-membered ring is formed; (ii) in the presence of a sodium template, an unprecedented dynamic orthoester cryptate is formed, in which nine oxygen donors are bound to the metal ion; (iii) *en route* to the sodium cryptates, novel orthoester coronates can be observed and, in some cases, isolated as a mixture of *syn* and *anti* isomers. Sodium cryptate [**Na$^+$ ⊂ o-Me$_2$-1.1.1**]BArF$^-$ was found to be surprisingly stable against water in neutral solution, but susceptible to hydrolysis in the presence of water and acid. We believe that this property will make orthoester cages useful for the traceless delivery of cations into biological systems[44]. Competition experiments in solution suggest that the encapsulated metal ion is in slow exchange with the bulk ($k_{obs} = 0.6$ s$^{-1}$) and the binding constant for Na$^+$ lies between the classic complexation agents 15-crown-5 and 2.2.1. Besides their interesting structural and dynamic properties, orthoester cryptates offer preparative advantages over their classic analogues: their synthesis relies on a one-pot, dynamic covalent ring-closing reaction, and substituted cages **o-R$_2$-1.1.1** are accessible simply by using different orthoesters as starting materials. We are currently working towards further diversifying the target structures, as well as increasing the dynamic system's complexity (for example, self-sorting and response to external stimuli).

## Methods

**Preparation of stock solutions.** NaBArF (0.14 mmol, 127.8 mg) and diethylene glycol (0.42 mmol, 39.9 µl) were dissolved in 14 ml CDCl$_3$ and the mixture was dried over MS (4 Å, 1 g) for 3 days. TFA (0.24 mmol, 18.4 µl) was dissolved in CDCl$_3$ (total volume 2.00 ml).

**Self-assembly of cryptate [Na$^+$ ⊂ o-Me$_2$-1.1.1]BArF$^-$.** To 6 ml of stock solution were added fresh MS (4 Å, 1 g) and the reaction mixture was left to stand at room temperature. After 16 h, 1 mol% TFA (10 µl) was added from stock solution, the mixture was shaken and trimethyl orthoacetate (0.12 mmol, 15.4 µl) was added. Every 24 h, 1 mol% TFA was added to keep the exchange reaction active (MS slowly transform the acid catalyst into inactive anhydride and/or esters). The reaction progress was monitored regularly by $^1$H NMR spectroscopy. After 5 days, the solvent was removed under reduced pressure and [Na$^+$ ⊂ o-Me$_2$-1.1.1]BArF$^-$ was obtained as a colourless solid (67% yield). Characterization data: M.p. 124 °C - 128 °C. $^1$H NMR (400 MHz, CDCl$_3$, 298 K): $\delta$ = 7.68 (t, J = 2.8 Hz, 8H), 7.51 (s, 4H), 3.79–3.77 (m, 12H), 3.50–3.48 (m, 12H), 1.43 p.p.m. (s, 6H). $^{13}$C NMR (100 MHz, CDCl$_3$, 298 K): $\delta$ = 162.8, 162.3, 161.8, 161.3, 135.1, 129.7, 129.4, 129.1, 128.9, 128.8, 126.2, 123.5, 120.8, 117.7, 113.0, 69.2, 62.0, 17.7 p.p.m. $^{11}$B NMR (128 MHz, CDCl$_3$, 298 K): $\delta$ = − 6.7 p.p.m. $^{19}$F NMR (282 MHz, CDCl$_3$, 298 K): $\delta$ = − 62.1 p.p.m. $^{23}$Na NMR (132 MHz, CDCl$_3$, 298 K): $\delta$ = − 5.9 p.p.m. HRMS (ESI$^+$): $m/z$ = 389.1794 [M + Na]$^+$ (calcd. 389.1782 for C$_{16}$H$_{30}$O$_9$Na). For further experimental details and characterization data, see Supplementary Methods and Supplementary Figs 29–47.

**Exclusion of moisture.** Molecular sieves were dried by heating for 3 days at 150 °C under reduced pressure ($10^{-2}$ mbar). All solvents were dried over MS for at least 24 h. All orthoester exchange reactions (catalysed by TFA) were carried out under nitrogen. After the acid was quenched (for example, by addition of triethylamine or basic aluminum oxide), most orthoester complexes described herein were found to be unusually stable against water and could be handled on the benchtop without further precautions (Supplementary Fig. 2).

## References

1. Steed, J. W. & Gale, P. A. (eds.) *Supramolecular Chemistry* (Wiley-VCH, 2012).
2. Pedersen, C. J. Cyclic polyethers and their complexes with metal salts. *J. Am. Chem. Soc.* **89**, 7017–7036 (1967).
3. Dietrich, B., Lehn, J. M. & Sauvage, J. P. Diaza-polyoxa-macrocycles et macrobicycles. *Tetrahedron Lett.* **10**, 2885–2888 (1969).
4. Lehn, J. M. Cryptates: the chemistry of macropolycyclic inclusion complexes. *Acc. Chem. Res.* **11**, 49–57 (1978).
5. Schneider, H.-J. (eds.) *Applications of Supramolecular Chemistry* (CRC Press, 2012).
6. Harris, K., Fujita, D. & Fujita, M. Giant hollow M$_n$L$_{2n}$ spherical complexes: structure, functionalisation and applications. *Chem. Commun.* **49**, 6703–6712 (2013).
7. Han, M., Engelhard, D. M. & Clever, G. H. Self-assembled coordination cages based on banana-shaped ligands. *Chem. Soc. Rev.* **43**, 1848–1860 (2014).
8. Zhang, G. & Mastalerz, M. Organic cage compounds - from shape-persistency to function. *Chem. Soc. Rev.* **43**, 1934–1947 (2014).
9. Tozawa, T. *et al.* Porous organic cages. *Nat. Mater.* **8**, 973–978 (2009).
10. Sun, Q. -F. *et al.* Self-assembled M$_{24}$L$_{48}$ polyhedra and their sharp structural switch upon subtle ligand variation. *Science* **328**, 1144–1147 (2010).
11. Granzhan, A., Schouwey, C., Riis-Johannessen, T., Scopelliti, R. & Severin, K. Connection of metallamacrocycles via dynamic covalent chemistry: a versatile method for the synthesis of molecular cages. *J. Am. Chem. Soc.* **133**, 7106–7115 (2011).
12. Zhang, G., Presly, O., White, F., Oppel, I. M. & Mastalerz, M. A permanent mesoporous organic cage with an exceptionally high surface area. *Angew. Chem. Int. Ed.* **53**, 1516–1520 (2014).
13. Kim, J. *et al.* Reversible morphological transformation between polymer nanocapsules and thin films through dynamic covalent self-assembly. *Angew. Chem. Int. Ed.* **54**, 2693–2697 (2015).
14. Fujita, D. *et al.* Protein encapsulation within synthetic molecular hosts. *Nat. Commun.* **3**, 1093 (2012).
15. Corbett, P. T. *et al.* Dynamic combinatorial chemistry. *Chem. Rev.* **106**, 3652–3711 (2006).
16. Jin, Y., Yu, C., Denman, R. J. & Zhang, W. Recent advances in dynamic covalent chemistry. *Chem. Soc. Rev.* **42**, 6634–6654 (2013).
17. Herrmann, A. Dynamic combinatorial/covalent chemistry: a tool to read, generate and modulate the bioactivity of compounds and compound mixtures. *Chem. Soc. Rev.* **43**, 1899–1933 (2014).
18. Ludlow, R. F. & Otto, S. Systems chemistry. *Chem. Soc. Rev.* **37**, 101–108 (2008).
19. Li, J., Nowak, P. & Otto, S. Dynamic combinatorial libraries: from exploring molecular recognition to systems chemistry. *J. Am. Chem. Soc.* **135**, 9222–9239 (2013).
20. Hiraoka, S., Harano, K., Shiro, M. & Shionoya, M. Quantitative dynamic interconversion between AgI-mediated capsule and cage complexes accompanying guest encapsulation/release. *Angew. Chem. Int. Ed.* **44**, 2727–2731 (2005).
21. Carnall, J. M. A. *et al.* Mechanosensitive self-replication driven by self-organization. *Science* **327**, 1502–1506 (2010).
22. Riddell, I. A. *et al.* Anion-induced reconstitution of a self-assembling system to express a chloride-binding Co$_{10}$L$_{15}$ pentagonal prism. *Nat. Chem.* **4**, 751–756 (2012).
23. Zarra, S., Wood, D. M., Roberts, D. A. & Nitschke, J. R. Molecular containers in complex chemical systems. *Chem. Soc. Rev.* **44**, 419–432 (2015).
24. Stefankiewicz, A. R., Sambrook, M. R. & Sanders, J. K. M. Template-directed synthesis of multi-component organic cages in water. *Chem. Sci.* **3**, 2326–2329 (2012).
25. Ayme, J. -F., Beves, J. E., Campbell, C. J. & Leigh, D. A. The self-sorting behavior of circular helicates and molecular knots and links. *Angew. Chem. Int. Ed.* **53**, 7823–7827 (2014).
26. Saalfrank, R. W. *et al.* Topologic equivalents of coronands, cryptands and their inclusion complexes: synthesis, structure and properties of {2}-metallacryptands and {2}-metallacryptates. *Chem. Eur. J.* **3**, 2058–2062 (1997).
27. Saalfrank, R. W., Maid, H. & Scheurer, A. Supramolecular coordination chemistry: the synergistic effect of serendipity and rational design. *Angew. Chem. Int. Ed.* **47**, 8794–8824 (2008).
28. Jazwinski, J. *et al.* Polyaza macrobicyclic cryptands: synthesis, crystal structures of a cyclophane type macrobicyclic cryptand and of its dinuclear copper(I) cryptate, and anion binding features. *J. Chem. Soc., Chem. Commun.* 1691–1694 (1987).
29. MacDowell, D. & Nelson, J. Facile synthesis of a new family of cage molecules. *Tetrahedron Lett.* **29**, 385–386 (1988).
30. Voloshin, Y. Z. *et al.* Template synthesis, structure and unusual series of phase transitions in clathrochelate iron(II) α-dioximates and oximehydrazonates formed by capping with functionalized boron-containing agents. *Polyhedron* **20**, 2721–2733 (2001).
31. Wise, M. D. *et al.* Large, heterometallic coordination cages based on ditopic metallo-ligands with 3-pyridyl donor groups. *Chem. Sci.* **6**, 1004–1010 (2015).
32. Brachvogel, R. -C. & von Delius, M. Orthoester exchange: a tripodal tool for dynamic covalent and systems chemistry. *Chem. Sci.* **6**, 1399–1403 (2015).
33. Lightburn, T. E., Dombrowski, M. T. & Tan, K. L. Catalytic scaffolding ligands: an efficient strategy for directing reactions. *J. Am. Chem. Soc.* **130**, 9210–9211 (2008).
34. Tan, K. L. Induced intramolecularity: an effective strategy in catalysis. *ACS Catal.* **1**, 877–886 (2011).
35. DeWolfe, R. H. Synthesis of carboxylic and carbonic ortho esters. *Synthesis* 153–172 (1974).
36. Rosenthal, M. R. The myth of the non-coordinating anion. *J. Chem. Educ.* **50**, 331 (1973).
37. Krossing, I. & Raabe, I. Noncoordinating anions—fact or fiction? a survey of likely candidates. *Angew. Chem. Int. Ed.* **43**, 2066–2090 (2004).
38. Lee, S., Chen, C. -H. & Flood, A. H. A pentagonal cyanostar macrocycle with cyanostilbene CH donors binds anions and forms dialkylphosphate [3]rotaxanes. *Nat. Chem.* **5**, 704–710 (2013).
39. Lin, Y. -H., Lai, C. -C., Liu, Y. -H., Peng, S. -M. & Chiu, S. -H. Sodium Ions template the formation of rotaxanes from BPX26C6 and nonconjugated amide and urea functionalities. *Angew. Chem. Int. Ed.* **52**, 10231–10236 (2013).
40. Wu, K. -D., Lin, Y. -H., Lai, C. -C. & Chiu, S. -H. Na$^+$ ion templated threading of oligo(ethylene glycol) chains through BPX26C6 allows synthesis of [2]Rotaxanes under solvent-free conditions. *Org. Lett.* **16**, 1068–1071 (2014).
41. Gold, V. & Sghibartz, C. M. Crown ether acetals: detection of cation binding by kinetic measurements. *J. Chem. Soc., Chem. Commun.* 507–508 (1978).
42. Corey, E. J. & Raju, N. A new general synthetic route to bridged carboxylic ortho esters. *Tetrahedron Lett.* **24**, 5571–5574 (1983).
43. Kirch, M. & Lehn, J.-M. Selective transport of alkali metal cations through a liquid membrane by macrobicyclic carriers. *Angew. Chem. Int. Ed.* **14**, 555–556 (1975).
44. Botti, P., Tchertchian, S. & Theurillat, D. Orthoester derivatives of crown ethers as carriers for pharmaceutical and diagnostic compositions. *Eur. Pat. Appl.* EP 2 332 929 A1 (2009).
45. Moras, P. D. & Weiss, R. Etude structurale des cryptates. III. Structure cristalline et moléculaire du cryptate de sodium C$_{18}$H$_{36}$N$_2$O$_6$.NaI. *Acta Cryst.* **B29**, 396–399 (1973).
46. Izatt, R. M., Pawlak, K., Bradshaw, J. S. & Bruening, R. L. Thermodynamic and kinetic data for macrocycle interactions with cations and anions. *Chem. Rev.* **91**, 1721–2085 (1991).
47. Breck, D. W., Eversole, W. G., Milton, R. M., Reed, T. B. & Thomas, T. L. Crystalline zeolites. I. The properties of a new synthetic zeolite, type A. *J. Am. Chem. Soc.* **78**, 5963–5972 (1956).

48. Breck, D. W. Crystalline molecular sieves. *J. Chem. Educ.* **41,** 678 (1964).

49. Golden, J. H., Mutolo, P. F., Lobkovsky, E. B. & DiSalvo, F. J. Lithium-mediated organofluorine hydrogen bonding: structure of lithium tetrakis(3,5-bis(trifluoromethyl)phenyl)borate tetrahydrate. *Inorg. Chem.* **33,** 5374–5375 (1994).

50. Shannon, R. D. Revised effective ionic radii and systematic studies of interatomic distances in halides and chalcogenides. *Acta Cryst.* **A32,** 751–767 (1976).

## Acknowledgements

This paper is dedicated to Professor Jean-Marie Lehn on the occasion of his 75th birthday. We gratefully acknowledged funding from the Fonds der Chemischen Industrie (FCI, Liebig Fellowship Li 191/01 for M.v.D., doctoral fellowship for R.-C.B.) and the Deutsche Forschungsgemeinschaft (DFG, Emmy-Noether Programme DE 1830/2-1). We thank Dr Harald Maid for help with visualizing the solid-state structures, Professor Walter Bauer for measuring $^{23}$Na and $^{7}$Li NMR spectra, and Dr Ralph Puchta for preliminary DFT calculations. Dr Jörg Sutter, Stefanie Zürl and Franziska Popp are acknowledged for recording atom absorption and atom emission spectra.

## Author contributions

R.-C.B. and M.v.D. performed the experiments, contributed to the design of the experiments and the analysis of the data. F.H. solved the crystal structure. M.v.D. conceived the project and wrote the manuscript.

# 14

# Unique distal size selectivity with a digold catalyst during alkyne homocoupling

Antonio Leyva-Pérez[1], Antonio Doménech-Carbó[2] & Avelino Corma[1]

Metal-catalysed chemical reactions are often controlled by steric hindrance around the metal atom and it is rare that substituents far away of the reaction site could be differentiated during reaction, particularly if they are simple alkyl groups. Here we show that a gold catalyst is able to discriminate between linear carbon alkynes with 10 or 12 atoms in the chain during the oxidative homocoupling of alkynes: the former is fully reactive and the latter is practically unreactive. We present experimental evidences, which support that the distal size selectivity occurs by the impossibility of transmetallating two long alkyl chains in an A-framed, mixed-valence digold (I, III) acetylide complex. We also show that the reductive elimination of two alkyne molecules from a single Au(III) atom occurs extremely fast, in $<1\,min$ at $-78\,°C$ (turnover frequency $> 0.016\,s^{-1}$).

[1]Instituto de Tecnología Química, Universidad Politécnica de Valencia-Consejo Superior de Investigaciones Científicas, Avda. de los Naranjos s/n, 46022 Valencia, Spain. [2]Departament de Química Analítica, Universitat de Valencia, Dr Moliner, 50, 46100 Burjassot, Valencia, Spain. Correspondence and requests for materials should be addressed to A.L.-P. (email: anleyva@itq.upv.es) or to A.C. (email: acorma@itq.upv.es).

Organic synthesis takes advantage of the predictable outcome of reactions where bulky groups are involved. Metal-catalysed carbon–carbon bond-forming reactions are not an exception since they are generally controlled by the steric hindrance around the catalytic site, which is given by the bulkiness of the ligands or the reactants, rather than for substituents far away from the reactive site. It is rare that substituents at more than six atoms distance from the reactive site can be sterically differentiated during catalysis[1,2]. Here we show a gold catalyst that is able to discriminate between linear chain alkynes only differing in one ethylene group at an eight-carbon distance of the catalytic site. This level of discrimination is extraordinary and difficult to achieve even for rigid microporous solid frameworks[3–5].

Gold catalysis has been intensively studied during the last decade and, in particular, the use of the redox pair Au(I)/Au(III) has arisen much interest among chemists as a catalyst for new carbon–carbon and carbon–heteroatom bond-forming reactions[6–10]. Despite the increasing number of new transformations reported, particularly with alkynes[11,12], fundamental studies covering the elemental steps of the gold-catalysed redox cycle remain scarce[13,14].

Here we dissect the fundamental steps during the oxidative coupling of alkynes catalysed by Au(I)/Au(III) to find a reasonable explanation to the distal size selectivity observed.

## Results

**Distal size selectivity of alkynes with gold catalysts.** Fig. 1 shows the results for the homocoupling of 1-decyne **1a** and 1-dodecyne **1b** under gold-catalysed conditions. We found that 1-decyne **1a** ($C_{10}$) reacts smoothly under the gold-catalysed conditions, while 1-dodecyne **1b** is nearly unreactive for homocoupling (<5% yield after 24 h). Kinetic measurements by gas chromatography–mass spectrometry (GC–MS, see Supplementary Fig. 1) show that the initial rate of 1-decyne **1a** is 3.5 times higher than 1-dodecyne **1b,** and that while **1a** smoothly converts over the time to the homocoupling product **2a** 1-dodecyne **1b** rapidly stops converting to the homocoupled product **2b**. Notice that a minor non-catalysed polymerization of the alkyne occurs together with the homocoupling reaction.

Other linear terminal alkynes shorter than 1-decyne **1a** ($C_6$–$C_9$) react at a similar rate than **1a** under gold-catalysed conditions, with final product yields between 60–70%, while 1-undecyne ($C_{11}$) gives a low yield of homocoupling product (19%) and a longer linear terminal alkyne ($C_{14}$) is unreactive towards homocoupling. The use of freshly-prepared or just-open commercial bottles of Au catalyst and selectfluor assures reproducible results, otherwise lower yields can be obtained

although with still a similar distal size selectivity. This change based on the chain size is remarkable, when moving from 1-decyne **1a** (65–80%), then to 1-undecyne (19%) and finally to 1-dodecyne **1b** and longer linear alkynes (<5%). It must be highlighted that this distal size selectivity seems to be quite unique with gold, since when the homocoupling of alkynes was carried out under typical copper-mediated (Glaser coupling, that is, the most used protocol for the homocoupling of alkynes in organic synthesis)[15,16] or copper-catalysed (Glaser–Hay coupling, with $O_2$ as the final oxidant) conditions the reaction rate and final yields of 1-decyne **2a** and 1-dodecyne **2b** are very similar for both alkynes (see Fig. 1). These results suggest a profound difference in the mechanism of homocoupling of linear alkynes with gold or copper catalysts.

The nearly atomically precise selection of long linear alkynes by gold during the oxidative homocoupling of alkynes is perhaps more striking when we consider alkynes containing bulky substituents at the contiguous positions of the triple bond. For instance, *tert*-butylacetylene **1c** reacts smoothly (91% yield of **2c**) while cyclohexylacetylene **1d** gives only a moderate yield of homocoupling product **2d** (37% yield). These results clearly contrast with those obtained for copper: while *tert*-butylacetylene **1c** is poorly reactive in the Glaser and Hay couplings (11 and 34% yield after 24 h), cyclohexylacetylene **1d** shows a higher reactivity (46% yield after 4 days for the Glaser coupling and 53% yield after 24 h for the Hay coupling) closer to that of long linear alkynes **1a–b**. Overall, the results in Fig. 1 clearly show that the gold-catalysed oxidative homocoupling of alkynes occurs with size selectivity along the whole substrate rather than with the classical size selectivity around the reactive site of the substrate, the latter operating in the copper-catalysed reactions.

**Study of the origin of the distal size selectivity with gold catalysts.** The elemental steps of the homocoupling of alkynes on gold catalysts (supposing a mixed-valence gold complex) are shown in Fig. 2 and include oxidation of gold(I) to gold (III), formation of the digold complex (approach), transmetallation of the alkyne and reductive elimination[17,18].

To determine if the first step, that is, the oxidation by selectfluor of the *in situ* formed gold(I)-acetylide[17], is responsible for the distal size selectivity observed, the gold(I) acetylides of 1-decyne **1a** (**3a**) and 1-dodecyne **1b** (**3b**) were prepared and a stoichiometric amount of oxidant was added to each of them under the reaction conditions given in Fig. 3.

The results show that selectfluor is consumed within 45 min with an initial rate of $250\,h^{-1}$ for acetylide **3a** and $220\,h^{-1}$ for **3b** (measured from the initial slope of the curve and within the timing limitations of the $^1H$ NMR technique). The reaction rate is

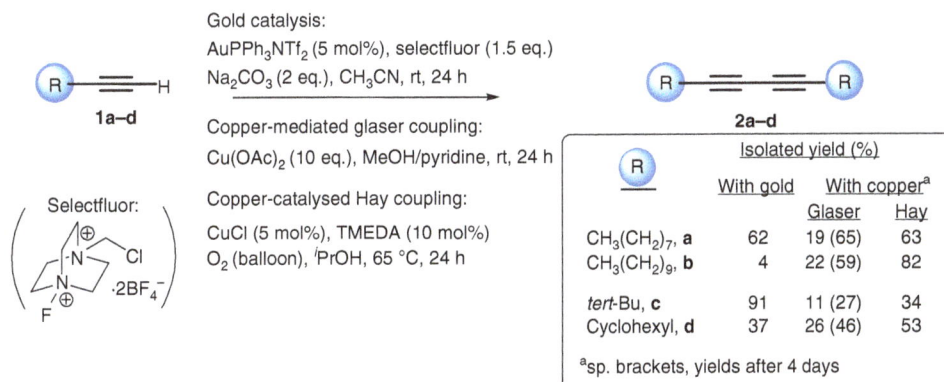

Gold catalysis:
AuPPh₃NTf₂ (5 mol%), selectfluor (1.5 eq.)
Na₂CO₃ (2 eq.), CH₃CN, rt, 24 h

Copper-mediated glaser coupling:
Cu(OAc)₂ (10 eq.), MeOH/pyridine, rt, 24 h

Copper-catalysed Hay coupling:
CuCl (5 mol%), TMEDA (10 mol%)
O₂ (balloon), ⁱPrOH, 65 °C, 24 h

| R | Isolated yield (%) | | |
|---|---|---|---|
| | With gold | With copper[a] | |
| | | Glaser | Hay |
| CH₃(CH₂)₇, **a** | 62 | 19 (65) | 63 |
| CH₃(CH₂)₉, **b** | 4 | 22 (59) | 82 |
| *tert*-Bu, **c** | 91 | 11 (27) | 34 |
| Cyclohexyl, **d** | 37 | 26 (46) | 53 |

[a]sp. brackets, yields after 4 days

**Figure 1 | Gold or copper-catalysed oxidative homocoupling of terminal alkynes.** Notice the different reactivity of alkynes **1a–d** under the gold-catalysed conditions reported here and typical copper-catalysed conditions (Glaser or Hay conditions). Isolated yields are the average of two runs.

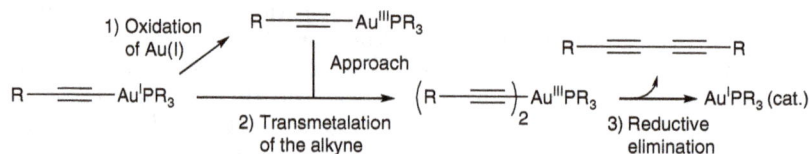

**Figure 2 | Proposed reaction steps for the homocoupling of alkynes on gold catalysts.** The reaction pathway includes oxidation of gold(I) to gold (III), formation of the digold complex (approach), transmetallation of the alkyne and reductive elimination[17].

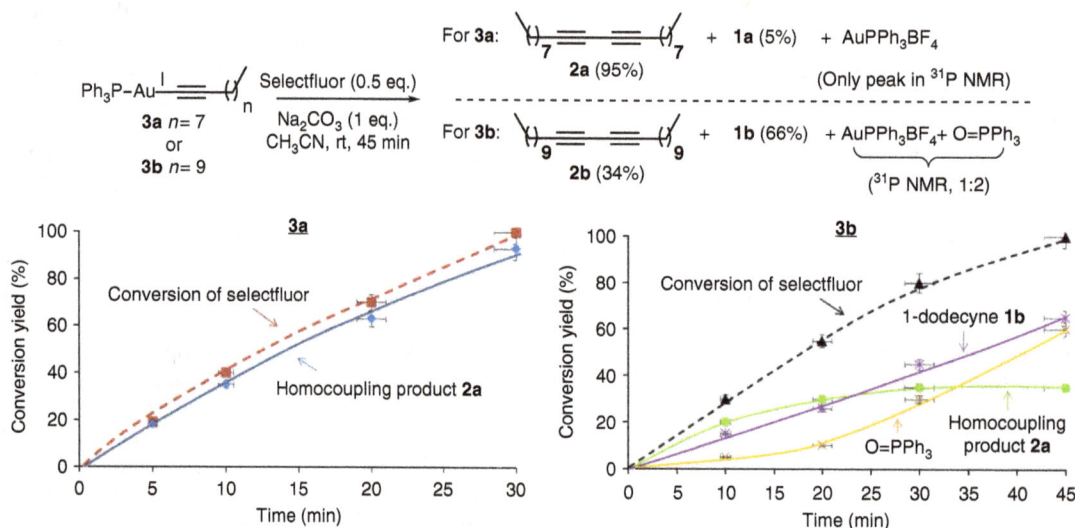

**Figure 3 | Oxidation of gold(I) acetylides by selectfluor.** Conversion and yields are calculated on the basis of $^1H$ and $^{19}F$ NMR spectroscopy for selectfluor, $^1H$ NMR spectroscopy and GC-MS (after extraction of the reaction mixture with n-hexane and addition of dodecane as an external standard) for alkynes and $^{31}P$ NMR spectroscopy for phosphines. Error bars represent an uncertainty of 5% in the value.

expressed in $h^{-1}$ and increases with the concentration of selectfluor (Supplementary Fig. 2)[17]. While the consumption of selectfluor is similar for the two acetylides 3a–b, the appearance of the homocoupling product is not. For 3a ($C_{10}$), a new triplet at ~2.1 p.p.m. corresponding to the homocoupling product 2a appears in the $^1H$ NMR spectra, and this peak increases as selectfluor converts. In contrast, 3b ($C_{12}$) gives two new signals in the $^1H$ NMR spectra at ~2.3–2.1 p.p.m., and those signals correspond to the homocoupling product 2b and 1-dodecyne 1b. Extraction of the reaction mixture with n-hexane and analysis by GC–MS and also by $^1H$ and $^{13}C$ NMR spectroscopy confirms that the homocoupling product 2a is the only alkyne product after oxidation of the acetylide of 3a, and that the homocoupling product 2b and free 1-dodecyne 1b in a 1:2 molar ratio are the alkyne products from the acetylide of 3b. The appearance of free alkyne 1b from 3b prompted us to check carefully the spectra of 3a during oxidation, and small amounts (<5%) of free 1-decyne 1a were detected in solution by GC–MS. These results show that the free alkynes 1a or 1b are somehow formed during the oxidation of the Au(I) acetylide complexes 3a–b, respectively.

To check if the free alkyne is able or not to enter back to the catalytic cycle, an equimolecular amount of 1-dodecyne 1b was added to the $C_{10}$ acetylide 3a and then we proceed to oxidize with selectfluor. The results (Supplementary Fig. 3) show that the heterocoupling product $C_{10}$–$C_{12}$ 4a was formed together with the homocoupling product $C_{10}$–$C_{10}$ 2a. Complementary 1-decyne 1a was added during the oxidation of the $C_{12}$ acetylide 3b, and we found again that the heterocoupling product $C_{10}$–$C_{12}$ 4a was formed together with the corresponding homocoupling product $C_{12}$–$C_{12}$ 2b. These results confirm the scrambling of alkynes during reaction. Notice that the homocoupling product of the free alkyne added externally is not obtained in any case, which is in

accordance with blank experiments that show that alkyne scrambling between the gold(I) acetylide and a second free alkyne does not occur without any oxidant in the reaction medium. When the two gold(I) acetylides 3a and 3b are combined in a single flask and selectfluor is added, the heterocoupling product 4a is formed in 37% yield, the $C_{10}$–$C_{10}$ homocoupling product 2a is formed in 58% yield, the $C_{12}$–$C_{12}$ 2b is formed in only 5% yield and the free alkyne 1b (50%) persists in solution. These results support that free alkyne is released after oxidation by selectfluor of the acetylide Au(I) complex, that the amount of free alkyne remaining in solution during the coupling is much higher for the $C_{12}$ than for the $C_{10}$ gold(I) acetylide, and that the released alkyne re-enters into the catalytic cycle. If anhydrous acetonitrile with a 0.5% (v:v) of $D_2O$ is used as a reaction solvent (instead of non-dried acetonitrile) for the oxidation of 3b, much of the alkyne released (~80%) is $d^1$-1-dodecyne $1b$-$d^1$, which suggests that the released alkyne takes the proton from the solvent.

$^{31}P$ NMR spectroscopy also shows significant differences during the oxidation by selectfluor of 3a and 3b. For the former ($C_{10}$), the original peak of phosphine 3a at ~42 p.p.m. disappears as the oxidation converts and the generation of a single new peak at ~32 p.p.m. corresponding to $PPh_3AuBF_4$ is observed. In contrast, 3b ($C_{10}$) shows that the original peak (~42 p.p.m.) is transformed into $PPh_3AuBF_4$ (32 p.p.m.) and also into free $O=PPh_3$ (28 p.p.m.) in a 1:2 molar ratio. Despite homocoupling of the alkyne being incomplete, $^1H$ and $^{19}F$ spectroscopy clearly show the total consumption of selectfluor. Kinetic measurements (see Fig. 3 above) show that as soon as the formation of the homocoupling $C_{12}$ product 2b stops during the oxidation of 3b with selectfluor, the formation of free $O=PPh_3$ increases. A new downshifted peak attributable to a phosphine oxide-gold complex was found during reaction by $^{31}P$ NMR spectroscopy

(Supplementary Fig. 3). No additional $^{31}$P peaks were found during oxidation, which suggests that phosphine ligands are not detached during reaction, at least at a measurable rate, and that transmetallation (if occurs) and reductive elimination are much faster processes than gold(I) oxidation since no phosphine intermediates are detected. These results indicate that the oxidation of the phosphine follows the oxidation of gold(I) to gold(III), and when the former occurs the active acetylide gold species decompose to $O = PPh_3$, free terminal alkyne and gold[19,20].

In any case, the results showed above are consistent with the oxidation by selectfluor of the gold(I) acetylide complex independently of the size of the alkyne, with release of homocoupling product and also of free alkyne into the solution. This free alkyne is able to re-enter into the catalytic cycle provided the gold active species is still active. In the case of the C$_{10}$ acetylide **3a**, the amount of free alkyne **1a** detected in the solution is very low and the main alkyne product is the homocoupling product **2a**. In contrast, the C$_{12}$ acetylide **3b** leaves significant amounts of 1-docecyne **1b** in solution which, at the end, does not form back productive acetylide gold complex for further homocoupling, at least at a rate that could compete with the destruction of the gold catalyst by oxidation of the metal ligand PPh$_3$ to $O = PPh_3$.

The results in Figs 1 and 3 suggest that the distal size selectivity observed for gold occurs after oxidation, when both species Au(I) and Au(III) are already present in the reaction medium. Previous mechanistic studies unveiled that the gold-catalysed oxidative homocoupling of alkynes in solution proceeds through a bimetallic Au(I)/Au(III) acetylide transition state[17], and density functional theory calculations confirm the feasibility of a digold intermediate species (Supplementary Fig. 4). The digold

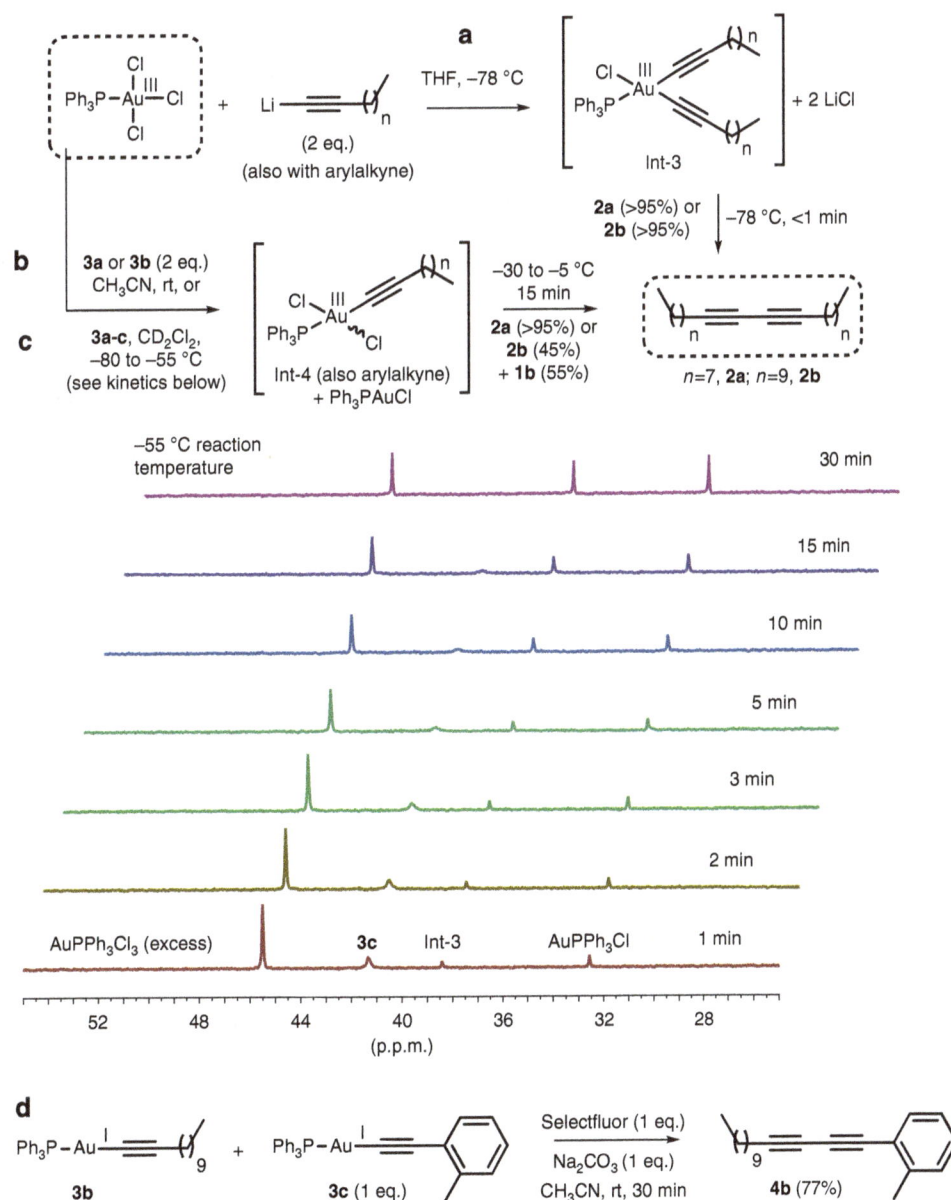

**Figure 4 | Studies on the reductive elimination step.** The top schemes (**a**-**c**) show the reductive elimination or transmetallation between gold and/or lithium acetylides in stoichiometric amounts. GC yields refer after extraction of the reaction mixture with *n*-hexane and addition of dodecane as an external standard. The kinetics at low-temperature of the transmetallation of tolylacetylene from **3c** to an excess (3 equiv.) of AuPPh$_3$Cl$_3$, followed by $^{31}$P NMR, is also included. Spectra were recorded every 30 s after 1 scan (2.5 s), and a total of 60 measurements were taken. The *in situ* oxidation and transmetallation of gold(I) alkyl and aryl acetylides is shown at the bottom (**d**).

intermediate sets the Au(I)/Au(III) cations in position and, as a consequence, the alkynes are oriented in a *cisoid* conformation. If one compares this reaction intermediate for gold with that widely accepted for copper[21] consisting in a tetrahedrically coordinated dimeric Cu(II) acetylide species with the alkynes oriented in opposite directions, one can attempt to explain the different behaviour of gold and copper on the basis of the different geometries.

To assess if the spatial approaching of the two gold atoms has any influence on the reaction outcome, three different diphenylphosphine-bridged digold(I) complexes 5a–c with one (methylene), two (ethylene) and three (propylene) methylene units in the tether were prepared (Supplementary Fig. 5 and Supplementary Discussion)[22,23]. X-ray Diffraction together with cyclic voltammetry studies showed that while 5a presents aurophilic bonding between two gold atoms[14], 5c has no aurophilic bonds at all[14] and 5b has an intermediate behaviour, depending on the counteranion[24,25]. The results indicate that aurophilic bonding has a positive influence on the reaction rate but none on the distal size selectivity. Since previous kinetic studies have demonstrated that the rate-determining step of the gold-catalysed homocoupling of alkynes in solution is the oxidation of gold(I)[17], it is not surprising that aurophilic bonding improves the reaction rate by neighbouring gold atom-assisted oxidation of gold(I)[26,27]. To test if aurophilic bonding has also any influence in the distal size selectivity, we prepared the digold acetylide complexes 6a [$Au_2dppe(C_{10})_2$] and 6b [$Au_2dppe(C_{12})_2$], and the results after oxidation with selectfluor were compared with those obtained before for the corresponding monogold acetylides 3a–b. The results were nearly identical, with the same amounts of homocoupling products 2a (>95%) or 2b (45%) obtained, and free alkyne 1-dodecyne 1b (~50%) released into solution for the case of the dodecynyl acetylide 6b. Cyclic voltammetry (Supplementary Figs 6 and 7 and Supplementary Discussion) confirmed that, in all the complexes tested, oxidation of Au(I) to Au(III) occurs without detecting any Au(II) species[9,16,23,28]. These results validate the mixed-valence digold intermediate Int–1 and clearly show that aurophilic bonding improves the reaction rate but that it is not involved in the distal size selectivity.

The reductive elimination from Au(III) complexes to form new $C_{sp3}$–X and $C_{sp2}$–X bonds (X = heteroatom, $C_{sp3}$, $C_{sp2}$, . . .) is a well-known process[13,29–32] that has been recently reported to occur very rapidly (even at $-52\,°C$)[33], faster than in palladium (II). Au(III) intermediates have been isolated and characterized[34–36]. Despite the great body of work reported in the literature for alkynes and gold, the reductive elimination of $C_{sp}$ bonds from Au(III) complexes has not been reported yet as far as we know[8], and very few examples of stable Au(III)-diacetylide complexes appear in the literature[37,38]. This lack of precedents for Au(III)-diacetylides might indicate that the reactivity of the alkynes once σ-bound to Au(III) is very high, including a potential reductive elimination. It has been proposed that the reductive elimination from a tetracoordinated planar Au(III) complex is favoured respect to other metals[33] since it leaves behind a two-coordinated Au(I) complex stabilized by relativistic effects[39–41]. To study the reductive elimination of alkynes from Au(III) and also if the distal size selectivity occurs at this stage, the Au(III) complex PPh₃AuCl₃ was prepared[42] and the corresponding Au(III) acetylide complex of 1-dodecyne 1b was forced to be formed[43]. To do that, n-BuLi was added to 1-dodecyne 1b in tetrahydrofurane (THF)-$d^8$ at $-78\,°C$ and the resulting solution was added to PPh₃AuCl₃ suspended (not soluble) in THF-$d^8$ at room temperature (2 equiv. of lithium acetylide with respect to gold). The result is shown in Fig. 4a. The bright-yellow complex PPh₃AuCl₃ became immediately

(<1 min) colourless and soluble. ³¹P NMR spectroscopy of the solution just after addition of the acetylide solution shows the complete disappearance of the original signal of PPh₃AuCl₃ (42 p.p.m.) and the appearance of a new single peak at 33 p.p.m. that could correspond to AuPPh₃Cl. Addition of n-hexane to the THF solution precipitates a white solid that by isolation and characterization with ¹H and ³¹P NMR spectroscopy was confirmed to be AuPPh₃Cl, the corresponding phosphine peak being observed during reaction. Analysis by GC–MS of the n-hexane supernatant shows that the homocoupling $C_{12}$ product 2b is the only alkyne product of the reaction, no traces of free 1-dodecyne 1b being present. A similar result is obtained when PPh₃AuCl₃ is treated with the lithium acetylide of 1-decyne 1a since AuPPh₃Cl and the $C_{10}$ homocoupling product 2a are the only products of the reaction. These results indicate that the reductive elimination of two acetylide fragments from a single Au(III) atom is very fast, in accordance with recent precedents in the literature that reports the efficient formation of $C_{sp2}$–$C_{sp2}$ bond with Au(III) via reductive elimination at $-52\,°C$ reaction temperature[30]. Thus, we must conclude that the distal size selectivity does not occur during the final reductive elimination but occurs in an intermediate step between oxidation and reductive elimination, since reductive elimination of the two alkyne molecules from Au(III) is extremely fast and does not give any distal size selectivity.

Gold is the most electronegative metal of the periodic table and, consequently, transmetallation of carbon ligands from Au(I)–C and Au(III)–C bonds to other metals is favoured and well-reported in the literature[44–46]. However, transmetallation from Au(I) to Au(III) is scarcely reported[29] although the migration of a carbon ligand from the more electronegative Au(I) to the more electropositive Au(III) cation should be favoured. Figure 4b shows the results obtained when the $C_{10}$ Au(I) acetylide 3a was mixed with the Au(III) complex PPh₃AuCl₃ in acetonitrile. In this case, the homocoupling product 2a is formed as only alkyne product in <5 min at room temperature. This result suggests that the transmetallation of the $C_{10}$ acetylide fragment of Au(I) complex 3a to the Au(III) cation readily occurs. When the $C_{12}$ Au(I) acetylide 3b was used, the homocoupling product 2b was obtained together with free 1-dodecyne 1b, as described before. To better study this step, we followed the reaction of PPh₃AuCl₃ with 3a ($C_{10}$) or 3b ($C_{12}$), and also with Au(I) arylacetylide 3c (PPh₃Au-*ortho*-tolylacetylene), by means of low-temperature ³¹P NMR spectroscopy (Fig. 4c and kinetics). For doing that, the NMR tube containing a solution of PPh₃AuCl₃ in CD₂Cl₂ was placed in liquid nitrogen ($-196\,°C$) and then a solution of the corresponding Au(I) acetylide 3a–c in CD₂Cl₂, cooled at $-78\,°C$ (acetone-dry ice), was slowly added into the NMR tube. After that, the tube was rapidly transferred to the NMR equipment at $-80\,°C$, and measurements were started after shimming. It was found that 3a or 3b react with PPh₃AuCl₃ in <5 min at $-80\,°C$ to give 0.5 equiv. of PPh₃AuCl and 0.5 equiv. of an unknown phosphine product at 38 p.p.m. and that this intermediate (Int–4) is stable at least up to $-30\,°C$. No differences between 3a ($C_{10}$) and 3b ($C_{12}$) were found here. When the NMR tube was left to warm, the only peak found by ³¹P NMR was that corresponding to PPh₃AuCl. ¹H and¹³C NMR confirmed the formation of the homocoupling product 2a from 3a and the mixture of 1b and 2b from 3b.

The reaction of PPh₃AuCl₃ with 3c proceeds slower and a more accurate kinetics could be carried out at $-55\,°C$. Again, PPh₃AuCl and the unknown peak at 38 p.p.m. were the only products formed in equimolecular amounts. *In situ* ¹H measurements of the reaction mixture at $-55\,°C$ showed a new peak at 2.40 p.p.m. downshifted $-0.05$ p.p.m. from the original methyl group of the Au(I) tolylacetylide 3c (2.45) and also downshifted

(−0.11 p.p.m.) from the homocoupled product (2.51). It is perhaps more informative the result obtained by *in situ* $^{13}$C NMR measurement at −55 °C. It was found that new peaks corresponding to the arene ring of the alkyne appear and that they present the typical $J_{C-P}$ of the phosphine Au-acetylide complexes and, consistently, that the methyl peak is also shifted with respect to the starting material. The $^{13}$C NMR spectrum strongly suggests that the unknown species at 38 p.p.m. is a new phosphine Au-acetylide that, according to the formation of 1 equiv. of AuPPh$_3$Cl per molecule of **3c** consumed, it might correspond to PPh$_3$Au(III)-tolylacetylene. A similar result is found when lithium tolylacetylide is used. The downshift of the methyl group in $^1$H NMR points also in that direction. To further assess if **Int-4** is the result of one alkyne transmetallation from **3c** to PPh$_3$AuCl$_3$, the reaction was carried out with an excess (3 equiv.) of the latter to maximize monotransmetallation. Kinetics of the reaction in Fig. 4 gives an initial rate of 0.28 s$^{-1}$. *In situ* $^1$H and $^{13}$C measurements of the reaction mixture at −55 °C after consumption of the Au(I) acetylide **3c** gave neat spectra with the same phosphine intermediate found above. As for **3a** and **3b**, this intermediate is stable up to −5 °C and only at this temperature it starts to further convert into PPh$_3$AuCl. The kinetic results together with the stoichiometry of the reaction suggest that the gold(I)-acetylides **3a–c** transfer the first alkyne group to the Au(III) atom of PPh$_3$AuCl$_3$ at low temperature (−80 °C for alkyl, −55 °C for aryl) to give the intermediate **Int-4** PPh$_3$Au(III)-alkyne and that the second alkyne transfers at much higher temperature (−30 to −5 °C) to give **Int-3**. As soon as **Int-3** is formed, reductive elimination occurs and the homocoupling product is released. This means that the transmetallation of the second alkyne is much slower than the transmetallation of the first alkyne, since the former occurs at much higher temperature, and that the reductive elimination is in turn much faster than the second alkyne transmetallation, which correlates well with the results observed with lithium acetylides. In short, these results show that **Int-4** is an intermediate of the reaction and that it could very well occur that the mixture of Au(I) acetylide and **Int-4** is responsible for the distal size selectivity observed.

To mimic better the gold-catalysed system when having separated Au(III) and Au(I) acetylides, and to avoid the use of PPh$_3$AuCl$_3$, we made use of the different rates of oxidation with selectfluor of alkyl and aryl acetylides. It was shown above that the oxidation of the alkyl acetylide complexes **3a–b** (monogold) and **6a–b** (digold) by selectfluor proceeds in ∼30 min, and we

have previously shown that the same oxidation for PPh$_3$Au-*ortho*-tolylacetylene **3c** takes place in ∼4 h$^{17}$. Thus, in principle, the addition of selectfluor to a mixture of an Au(I) alkyl acetylide and an Au(I) aryl acetylide should oxidize first the alkyl Au(I) complex and form the corresponding Au(III) alkyl acetylide. Once the Au(III) alkyl acetylide is formed, the Au(I) aryl acetylide can transmetalate. It is true that while the oxidation with selectfluor occurs, the remaining Au(I) alkyl acetylide (not oxidized yet) could compete with the Au(I) aryl acetylide to transmetalate. To avoid the self-coupling of the alkyl acetylide, we employed **3b** (C$_{12}$), since the rate of homocoupling is the lowest in the alkyl series and it should not compete with the aryl acetylide. Thus, the Au(I) alkyl acetylide **3b** and PPh$_3$Au-*ortho*-tolylacetylene **3c** were mixed in the same flask and then selectfluor was added. The result in Fig. 4d shows that the heterocoupled product **4b** is the major product of the reaction with a 77% yield, indicating that the transmetallation from Au(I) aryl acetylides to Au(III) alkyl acetylides readily occurs.

With the hope that the digold intermediate of unreactive 1-dodecyne could be trapped and characterized, we carried out the oxidation of complex **6b** with selectfluor at −78 °C in a CD$_2$Cl$_2$/CD$_3$CN solvent mixture. The results (Supplementary Fig. 8) show the desymmetrization of the original H and P peaks of **6b** into two new signals, which points to the formation of the divalent-mixed Au(I,III) complex. Complex **6b** was independently prepared without one of the dodecylide fragments and its $^{31}$P NMR signal fits well with the one detected after oxidation of **6b** by selectfluor, which supports that one of the dodecylide fragments of **6b** leaves as soon as one Au(I) atom is oxidized to Au(III).

With all the above results, we can suggest that the distal size selectivity occurs just before the transmetallation step of one alkyl acetylide fragment from Au(I) to Au(III), during the formation of the mixed-valence digold (I,III) complex. The tetracoordinated planar Au(III) site and the linear Au(I) site impose a parallel disposition of the aliphatic chains of each acetylene, which resembles the formation of a highly stabilized self-assembled monolayer, in this case just via two contiguous aliphatic chains. For longer chains, their interaction is enhanced thus blocking the coupling of the alkynes, while for shorter chains the lateral interaction is weak and the reaction readily occurs.

**Mechanism of gold-catalysed distal-selective alkyne coupling.** Figure 5 shows the proposed mechanism to explain the origin

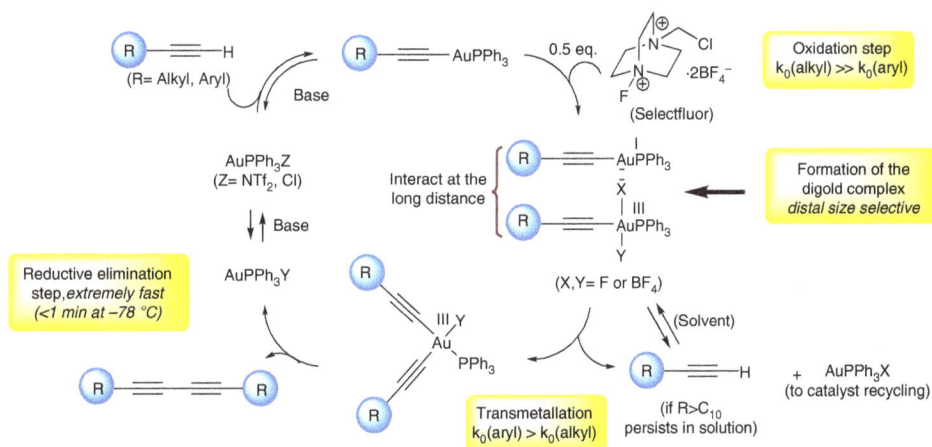

**Figure 5 | Proposed mechanism for the gold-catalysed homocoupling of alkynes with distal size selectivity.** The formation of a Au(I)/Au(III) digold intermediate is responsible for the distal steric differentiation.

of size selectivity during the gold-catalysed homocoupling of alkynes.

First, the linear Au(I) acetylide complex suffers the oxidation by selectfluor to give the corresponding Au(III) complex. This oxidation is favoured by a second neighbouring Au(I) acetylide molecule in solution but no size selectivity is found for this process, since selectfluor oxidation occurs at a similar rate for both $C_{10}$ and $C_{12}$ gold acetylides. At this point, we propose an Au(I)/Au(III) intermediate with a congested structure as responsible for the distal size selectivity observed, where the tetracoordinated planar Au(III) acetylide and the dicoordinated linear Au(I) acetylide present a parallel disposition of the aliphatic chains of each acetylene, which resembles the formation of a highly stabilized self-assembled monolayer. This intermediate is supported here experimentally by the similarity of the results obtained with the digold complexes **5a–b** and the corresponding acetylides of **5b** (complexes **6a–b**), since these complexes have a locked cisoid conformation of acetylides and phosphines. Notice that this particular steric scenario, created by the interaction of a tetracoordinated planar metal complex and a linear metal complex, cannot be found in any metal redox pair of the periodic table but only in Au(I)/Au(III)[40]. In addition, this Au(I)/Au(III) intermediate suitably accommodates acetylides fully substituted in contiguous positions to the triple bond such as *tert*-butylacetylene **1c**, which explains the feasibility of couplings with hindered alkynes around rather than along the triple bond. The experimental results also suggest that protodeauration competes with the transmetallation of the linear alkyl acetylide from Au(I) to Au(III), and that free acetylene is also released after decomposition of the fully oxidized gold acetylide. The alkyne can rebind to the Au(I) atom and enter back into the catalytic cycle provided the acetylide is still active.

**Gold-catalysed heterocoupling of alkynes.** The heterocoupling of terminal alkynes has been reported in the literature with different copper catalysts but, in all cases, a five-times amount of one of the alkynes must be added to achieve good yields of heterocoupling product[47–50]. Otherwise, with a one-to-one molar ratio of alkynes, a statistical 1:2:1 molar ratio of homo:hetero:homocoupled products, or a 1:1:1 molar ratio in some cases, is obtained[47–50]. In addition, the two alkynes must be differentiated either electronically (aryl or heteropropargyl versus alkyl alkynes, activated versus deactivated aryl alkynes) or sterically near the triple bond (linear versus bulky alkynes). The high excess of one of the alkyne reactants together with the need of electronic and/or steric differentiation of the terminal triple bond limits severely the application of these methodologies in organic synthesis. A much better methodology would consist in a catalytic system able to couple two different terminal alkynes regardless of the nature of the terminal triple bond and without the need of adding an excess of one of them. When we carried out the heterocoupling of different alkynes in equimolecular amounts under the present gold-catalysed conditions, we found that the heterocoupling products were formed in moderate to good yields and with high selectivity (Supplementary Fig. 9)[51,52].

One must not be surprised with these results when considering the high yield of heterocoupling obtained when we mix the $C_{12}$ Au(I) alkyl acetylide **3b** and PPh₃Au-*ortho*-tolylacetylene **3c** (see Fig. 4d above). If we compare the initial rate of oxidation for the Au(I) alkyl acetylides ($220\,h^{-1}$, see above) with that of the aryl acetylide ($45\,h^{-1}$)[17], a five-times relationship $k_0(\text{alkyl})/k_0(\text{aryl}) = k_{rel}(\text{oxidation}) \approx 5$ is found. However, the initial reaction rate for the homocoupling of 1-decyne **1a** under the gold-catalysed reaction conditions indicated above (Fig. 1) is approximately three times lower than that for *ortho*-tolylacetylene

**1e** (ref. 17), $k_0(\text{alkyl})/k_0(\text{aryl}) = k_{rel}(\text{homocoupling}) = 0.36$. Therefore, the rate of oxidation of Au(I)-acetylide and the rate of formation of homocoupling product are decoupled for alkyl and aryl alkynes, with an estimated value of $k_{rel}(\text{oxidation})/k_{rel}(\text{homocoupling}) \approx 14$. Since the last step, the reductive elimination from Au(III) is extremely fast (less than a minute at $-78\,°C$) for both alkyl and aryl alkynes, the cross-coupling of aryl and alkyl terminal alkynes is favoured under the present gold-catalysed conditions. Not only that, but the cross-coupling between two different aryl alkynes and also between two different alkyl alkynes is also performed. The coupling of two different alkynes based on gold-catalysed conditions opens a reactivity window for carrying out oxidative heterocouplings of terminal alkynes in 1:1 molar ratio with yield and selectivity clearly beyond the statistical range[53,54].

## Discussion

A near atomically precise distal size selectivity occurs during the oxidative coupling of terminal alkynes under gold-catalysed conditions. The formation of a crowded Au(I)/Au(III) digold intermediate is responsible for the distal steric differentiation. Oxidation by selectfluor, transmetallation and reductive elimination do not produce distal size selectivity. The reductive elimination of two alkynes from a single Au(III) atom is extremely fast and occurs in less than a minute at $-78\,°C$. The subtle steric and electronic discrimination of alkynes by this gold-catalysed system allows the heterocoupling of two different alkynes in equimolecular amounts regardless of the nature of the terminal triple bond.

## References

1. Leow, D., Li, G., Mei, T.-S. & Yu, J.-Q. Activation of remote meta-C–H bonds assisted by an end-on template. *Nature* **486**, 518–522 (2012).
2. Tang, R.-Y., Li, G. & Yu, J.-Q. Conformation–induced remote meta-C–H activation of amines. *Nature* **507**, 215–220 (2014).
3. Denayer, J. F. M. *et al.* Rotational entropy driven separation of alkane/isoalkane mixtures in zeolite cages. *Angew. Chem. Int. Ed.* **44**, 400–403 (2005).
4. Corma, A., Rey, F., Rius, J., Sabater, M. J. & Valencia, S. Supramolecular self-assembled molecules as organic directing agent for synthesis of zeolites. *Nature* **431**, 287–290 (2004).
5. Cantín, A. *et al.* Synthesis and structure of the bidimensional zeolite ITQ-32 with small and large pores. *J. Am. Chem. Soc.* **127**, 11560–11561 (2005).
6. Hashmi, A. S. K. Gold-catalysed organic reactions. *Chem. Rev.* **107**, 3180–3211 (2007).
7. de Haro, T. & Nevado, C. On gold-mediated C–H activation processes. *Synthesis* **16**, 2530–2539 (2011).
8. Engle, K. M., Mei, T.-S., Wang, X. & Yu, J.-Q. Bystanding F + oxidants enable selective reductive elimination from high-valent metal centers in catalysis. *Angew. Chem. Int. Ed.* **50**, 1478–1491 (2011).
9. Zhang, L. A Non-diazo approach to α-oxo gold carbenes via gold-catalysed alkyne oxidation. *Acc. Chem. Res.* **47**, 877–888 (2014).
10. Boronat, M., Leyva-Pérez, A. & Corma, A. Theoretical and experimental insights into the origin of the catalytic activity of subnanometric gold clusters: attempts to predict reactivity with clusters and nanoparticles of gold. *Acc. Chem. Res.* **47**, 834–844 (2014).
11. Corma, A., Leyva-Pérez, A. & Sabater, M. J. Gold-catalysed carbon–heteroatom bond–forming reactions. *Chem. Rev.* **111**, 1657–1712 (2011).
12. Hashmi, A. S. K. Dual gold catalysis. *Acc. Chem. Res.* **47**, 864–876 (2014).
13. Brenzovich, Jr W. E. *et al.* Gold-catalysed intramolecular aminoarylation of alkenes: C-C bond formation through bimolecular reductive elimination. *Angew. Chem. Int. Ed.* **49**, 5519–5522 (2010).
14. Tkatchouk, E. *et al.* Two metals are better than one in the gold catalysed oxidative heteroarylation of alkenes. *J. Am. Chem. Soc.* **133**, 14293–14300 (2011).
15. Siemsen, P., Livingston, R. C. & Diederich, F. Acetylenic coupling: A powerful tool in molecular construction. *Angew. Chem. Int. Ed.* **39**, 2632–2657 (2000).
16. Stefani, H. A., Guarezemini, A. S. & Cella, R. Homocoupling reactions of alkynes, alkenes and alkyl compounds. *Tetrahedron* **66**, 7871–7918 (2010).

17. Leyva-Pérez, A., Doménech, A., Al-Resayes, S. I. & Corma, A. Gold redox catalytic cycles for the oxidative coupling of alkynes. *ACS Catal.* **2**, 121–126 (2012).
18. Hopkinson, M. N., Ross, J. E., Giuffredi, G. T., Gee, A. D. & Gouverneur, V. Gold-catalysed cascade cyclization-oxidative alkynylation of allenoates. *Org. Lett.* **12**, 4904–4907 (2010).
19. Liu, L.-P., Xu, B., Mashuta, M. S. & Hammond, G. B. Synthesis and structural characterization of stable organogold(I) compounds. Evidence for the mechanism of gold-catalysed cyclizations. *J. Am. Chem. Soc.* **130**, 17642–17643 (2008).
20. Ball, L. T., Lloyd-Jones, G. C. & Russell, C. A. Gold-catalysed oxidative coupling of arylsilanes and arenes: origin of selectivity and improved precatalyst. *J. Am. Chem. Soc.* **136**, 254–264 (2014).
21. Kürti, L. & Czakó, B. *Strategic Applications of Named Reactions in Organic Synthesis* 186 (Elsevier Academic Press, 2005).
22. Berners-Price, S. J. & Sadler, P. J. Gold(I) complexes with bidentate tertiary phosphine ligands: formation of annular vs. tetrahedral chelated complexes. *Inorg. Chem.* **25**, 3822–3827 (1986).
23. Mirabelli, C. K. *et al.* Antitumor activity of bis(diphenylphosphino)alkanes, their gold(I) coordination complexes, and related compounds. *J. Med. Chem.* **30**, 2181–2190 (1987).
24. Li, D., Hang, X., Che, C.-M., Lo, W.-C. & Peng, S.-M. Luminescent gold(I) acetylide complexes. Photophysical and photoredox properties and crystal structure of [{Au(C≡CPh)}$_2$($\mu$-PPh$_2$CH$_2$CH$_2$PPh$_2$)]. *J. Chem. Soc. Dalton Trans.* **19**, 2929–2932 (1993).
25. Brandys, M.-C., Jennings, M. C. & Puddephatt, R. J. Luminescent gold(I) macrocycles with diphosphine and 4,4-bipyridyl ligands. *J. Chem. Soc. Dalton Trans.* **24**, 4601–4606 (2000).
26. Fackler, J. & John, P. Metal–metal bond formation in the oxidative addition to dinuclear gold(I) species. Implications from dinuclear and trinuclear gold chemistry for the oxidative addition process generally. *Polyhedron* **16**, 1–17 (1997).
27. Fackler, J. & John, P. Forty-five years of chemical discovery including a golden quarter-century. *Inorg. Chem.* **41**, 6959–6972 (2002).
28. Doménech, A., Leyva-Pérez, A., Al-Resayes, S. I. & Corma, A. Electrochemical monitoring of the oxidative coupling of alkynes catalysed by triphenylphosphine gold complexes. *Electrochem. Commun.* **19**, 145–148 (2012).
29. Cui, L., Zhang, G. & Zhang, L. Homogeneous gold-catalysed efficient oxidative dimerization of propargylic acetates. *Bioorg. Med. Chem. Lett.* **19**, 3884–3887 (2009).
30. Zhang, G., Peng, Y., Cui, L. & Zhang, L. Gold-catalysed homogeneous oxidative cross-coupling reactions. *Angew. Chem. Int. Ed.* **48**, 3112–3115 (2009).
31. Hopkinson, M. N. *et al.* Gold-catalysed intramolecular oxidative cross-coupling of nonactivated arenes. *Chem. Eur. J.* **16**, 4739–4743 (2010).
32. Zhang, G., Cui, L., Wang, Y. & Zhang, L. Homogeneous gold-catalysed oxidative carboheterofunctionalization of alkenes. *J. Am. Chem. Soc.* **132**, 1474–1475 (2010).
33. Wolf, W. J., Winston, M. S. & Toste, F. D. Exceptionally fast carbon–carbon bond reductive elimination from gold(III). *Nat. Chem.* **6**, 159–164 (2014).
34. Hashmi, A. S. K. Homogeneous gold catalysis beyond assumptions and proposals-characterized intermediates. *Angew. Chem. Int. Ed.* **49**, 5232–5241 (2010).
35. Hofer, M. & Nevado, C. Unexpected outcomes of the oxidation of (pentafluorophenyl)triphenylphosphanegold(I). *Eur. J. Inorg. Chem.* **9**, 1338–1341 (2012).
36. Hashmi, A. S. K. *et al.* Dual gold catalysis: σ,π-propyne acetylide and hydroxyl-bridged digold complexes as easy-to-prepare and easy-to-handle precatalysts. *Chemistry* **19**, 1058–1065 (2013).
37. Méndez, L. A., Jiménez, J., Cerrada, E., Mohr, F. & Laguna, M. A Family of alkynylgold(III) complexes [Au$^I$($\mu$-{CH$_2$}$_2$PPh$_2$)$_2$Au$^{III}$(C≡CR)$_2$] (R = Ph, tBu, Me$_3$Si): facile and reversible comproportionation of gold(I)/gold(III) to digold(II). *J. Am. Chem. Soc.* **127**, 852–853 (2005).
38. Au, V. K.-M., Wong, K. M.-C., Zhu, N. & Yam, V. W.-W. Luminescent cyclometalated dialkynylgold(III) complexes of 2-phenylpyridine-type derivatives with readily tunable emission properties. *Chem. Eur. J.* **17**, 130–142 (2011).
39. Schwerdtfeger, P. Relativistic effects in gold chemistry. 2. The stability of complex halides of gold(III). *J. Am. Chem. Soc.* **111**, 7261–7262 (1989).
40. Gorin, D. J. & Toste, F. D. Relativistic effects in homogeneous gold catalysis. *Nature* **446**, 395–403 (2007).
41. Leyva-Pérez, A. & Corma, A. Similarities and differences between Gold, Platinum and Mercury "relativistic" triad in catalysis. *Angew. Chem. Int. Ed.* **51**, 614–635 (2011).
42. Leyva, A., Zhang, X. & Corma, A. Chemoselective hydroboration of alkynes vs. alkenes over gold catalysts. *Chem. Commun.* **33**, 4897–5044 (2009).
43. Usón, R., Laguna, A. & Vicente, J. Novel anionic gold(I) and gold(III) organocomplexes. *J. Organomet. Chem.* **131**, 471–475 (1977).
44. Khairul, W. M. *et al.* Transition metal alkynyl complexes by transmetallation from Au(C≡CAr)(PPh$_3$) (Ar = C$_6$H$_5$ or C$_6$H$_4$Me-$_4$). *Dalton Trans.* **4**, 610–620 (2009).
45. Chen, Y., Chen, M. & Liu, Y. Gold-catalysed cyclization of 1,6-diyne-4-en-3-ols: stannyl transfer from 2-tributylstannylfuran through Au/Sn transmetallation. *Angew. Chem. Int. Ed.* **51**, 6181–6186 (2012).
46. Hofer, M., Gomez-Bengoa, E. & Nevado, C. A Neutral gold(III) – boron transmetallation. *Organometallics* **33**, 1328–1332 (2014).
47. Yin, W., He, C., Chen, M., Zhang, H. & Lei, A. Nickel-catalysed oxidative coupling reactions of two different terminal alkynes using O$_2$ as the oxidant at room temperature: facile syntheses of unsymmetric 1,3-diynes. *Org. Lett.* **11**, 709–712 (2009).
48. Balaraman, K. & Kesavan, V. Efficient copper(II) acetate catalysed homo and heterocoupling of terminal alkynes at ambient conditions. *Synthesis* **20**, 3461–3466 (2010).
49. Xiao, R., Yao, R. & Cai, M. Practical oxidative homo- and heterocoupling of terminal alkynes catalysed by immobilized copper in MCM-41. *Eur. J. Org. Chem.* **22**, 4178–4184 (2012).
50. Navale, B. S. & Bhat, R. G. Copper(I) iodide-DMAP catalysed homo- and heterocoupling of terminal alkynes. *RSC Adv.* **3**, 5220–5226 (2013).
51. Ohashi, K. *et al.* Indonesian medicinal plants. XXV.1) Cancer cell invasion inhibitory effects of chemical constituents in the parasitic plant scurrula atropurpurea (Loranthaceae). *Chem. Pharm. Bull.* **51**, 343–345 (2003).
52. Xu, Z., Byun, H.-S. & Bittman, R. Synthesis of photopolymerizable long-chain conjugated diacetylenic acids and alcohols from butadiyne synthons. *J. Org. Chem.* **56**, 7183–7186 (1991).
53. Lee, S., Lee, T., Lee, Y. M., Kim, D. & Kim, S. Solid-phase library synthesis of polyynes similar to natural products. *Angew. Chem. Int. Ed.* **46**, 8422–8425 (2007).
54. Liu, J., Lam, J. W. Y. & Tang, B. Z. Acetylenic polymers: syntheses, structures, and functions. *Chem. Rev.* **109**, 5799–5867 (2009).

## Acknowledgements

Financial support by Consolider-Ingenio 2010 (proyecto MULTICAT) and Severo Ochoa programs from MCIINN and Prometeo program from Generalitat Valenciana is acknowledged. A. L.-P. thanks ITQ for the concession of a contract. We thank Dr J.A. Vidal for assistance with the low-temperature NMR experiments, and Dr M. Boronat for the DFT calculations.

## Author contributions

A.L.-P. carried out the experiments and wrote the manuscript. A.D.-C. carried out the electrochemistry and wrote the conclusions thereof. A.C. wrote the manuscript.

# 15

# Transient signal generation in a self-assembled nanosystem fueled by ATP

Cristian Pezzato[1] & Leonard J. Prins[1]

A fundamental difference exists in the way signal generation is dealt with in natural and synthetic systems. While nature uses the transient activation of signalling pathways to regulate all cellular functions, chemists rely on sensory devices that convert the presence of an analyte into a steady output signal. The development of chemical systems that bear a closer analogy to living ones (that is, require energy for functioning, are transient in nature and operate out-of-equilibrium) requires a paradigm shift in the design of such systems. Here we report a straightforward strategy that enables transient signal generation in a self-assembled system and show that it can be used to mimic key features of natural signalling pathways, which are control over the output signal intensity and decay rate, the concentration-dependent activation of different signalling pathways and the transient downregulation of catalytic activity. Overall, the reported methodology provides temporal control over supramolecular processes.

[1] Department of Chemical Sciences, University of Padova, Via Marzolo 1, 35131 Padova, Italy. Correspondence and requests for materials should be addressed to L.J.P. (email: leonard.prins@unipd.it).

Communication is an essential feature of all living systems as it permits coordination, organization and adaptation[1]. To achieve this, nature has evolved elaborate signalling pathways relying on circular enzymatic networks to regulate all intra and extracellular functions[2,3]. In such networks, a trigger can up or downregulate an enzymatic cascade reaction leading to the transient generation of an output signal, after which the system returns to the original state. The circular nature of enzymatic networks implies that signal generation is a process that requires energy consumption. Over the past decades, chemists have developed a wealth of supramolecular sensory systems able to generate an output signal in response to an external trigger[4–8]. A common feature of these systems is that signal generation is thermodynamically driven, that is, the system adapts to a trigger-induced change in the energetic landscape developing into a new, energetically more favourable, resting state. This change is accompanied with a change in a property (fluorescence, current, catalysis, solubility and so on) that can be measured and correlated to the intensity of the trigger. This approach is exemplified in Fig. 1a (step i) for an indicator-displacement assay in which an analyte displaces a receptor-bound indicator[9]. The energy cost required for dissociation of the indicator ($\Delta G^{\circ}_{diss}$) is more than compensated for by the energy gain resulting from formation of the receptor–analyte complex ($\Delta G^{\circ}_{ass}$) (Fig. 1b). Overall this process is energetically downhill and generates a signal that is stable in time. For applications in sensing, this is obviously a highly attractive property and, consequently, a wide variety of supramolecular sensing systems have been developed based on this principle[4–8,10]. However, the absence of a return to the original state, which would require an amount of energy equal to $\Delta G^{\circ}_{ass} - \Delta G^{\circ}_{diss}$, makes it fundamentally different from the way nature deals with signal generation. Indeed, there is currently an enormous interest in the development of synthetic systems that

require energy consumption to remain in a functional state[11]. Such systems have properties that are more similar to living systems and are expected to offer new applications in the field of materials and nanotechnology[12]. Examples of dissipative self-assembled systems[13–15], transient catalysts[16], molecular transport systems[17] and molecular motors[18–20] are emerging.

Here we demonstrate a general strategy that permits transient signal generation in a self-assembled system. The approach relies on the addition of an additional step to the scheme depicted in Fig. 1a (step ii) that provides the necessary energy ($\Delta G^{\circ}_{ass} - \Delta G^{\circ}_{diss}$) to convert the system back to the original state. It is shown that this gives the possibility to perform multiple signalling cycles with the same system and, importantly, allows for temporal control over signal intensity. It is then demonstrated that this approach can be used to mimic features of natural signalling pathways, such as the activation of one or two signalling pathways depending on the initial trigger concentration and the transient regulation of catalytic activity. A related strategy has been previously applied in the context of enzyme assay development[21] and for studying the conformational dynamics of a protein[22]. Also an entirely different approach towards (irreversible) transient signal generation has been reported[23].

## Results

**Transient signal generation fuelled by ATP.** Our system relies on the strong interaction between oligoanions and the cationic surface of Au NP **1**, which are gold nanoparticles ($d = 1.8 \pm 0.4\,nm$) covered with hydrophobic C9-thiols terminating with a 1,4,7-triazacyclononane (TACN)•$Zn^{2+}$ head group (Fig. 2a)[24]. Previous studies have shown that binding occurs under saturation conditions even at low micromolar concentrations in aqueous buffer (see the Methods section and Supplementary Fig. 1 for more details)[25,26]. These studies also showed that the strength of interaction strongly depends on the number of negative charges present in the oligoanions. This emerges clearly from a series of displacement experiments in which increasing amounts of A$X$P ($X$ = T, D, M) are added to Au NP **1** ([TACN•$Zn^{2+}$] = $10 \pm 1\,\mu M$) saturated with fluorescent probe **A** (3.7 $\mu M$, $\lambda_{ex} = 450\,nm$, $\lambda_{em} = 493\,nm$) (Fig. 2b). When bound to Au NP **1**, the fluorescence of **A** is quenched by the gold nanoparticle, but is turned on upon its displacement by nucleotides A$X$P ($X$ = T, D, M). This attractive property of gold nanoparticles has been at the basis of numerous indicator-displacement assays for a variety of (bio)analytes[27–30]. Measurement of the fluorescence intensity (FI) as a function of the concentration of A$X$P provides information on the relative affinity, $K_{rel}$, of A$X$P for Au NP **1** compared with **A** (Fig. 2c). Fitting of the measured curves yielded $K_{rel}$ values of 1.8, $9.2 \times 10^{-2}$ and $2.8 \times 10^{-4}$ for ATP, ADP and AMP, respectively, demonstrating the importance of multivalent interactions. In the past, we have exploited this for the development of assays based on an analyte-induced change in the thermodynamic equilibrium (analogous to step i in Fig. 1)[31,32]. However, we were intrigued by the question of whether this system would be able to generate a fluorescent signal that is transient in nature. This would rely on the initial displacement of **A** by the strong competitor ATP after which the irreversible destruction of ATP into weaker competitors would result in re-formation of the complex between **A** and Au NP **1**, and thus a disappearance of the signal (Fig. 3a). The enzyme potato apyrase very efficiently hydrolyses ATP into AMP and 2 equiv. of inorganic phosphate $P_i$, which makes it very suitable for the purpose given above[33]. We first verified that the presence of the enzyme had no effect on the stability of the complex between **A** and Au NP **1** (see Supplementary Fig. 2)

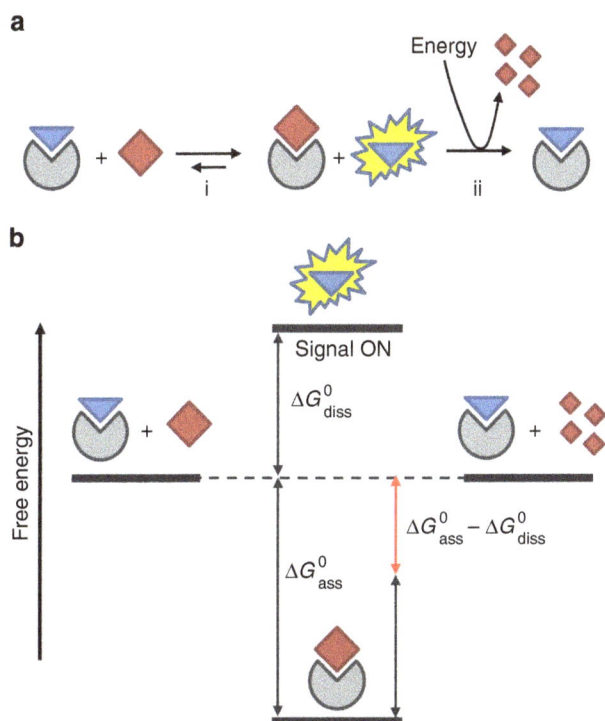

**Figure 1 | Transient signal generation.** (**a**) Schematic representation of an indicator-displacement assay (step i). An additional energy-consuming step (ii) is required to revert the system back to the original state. (**b**) Schematic energy diagram indicating the changes in free energy related to steps i and ii of the indicator-displacement assay depicted in **a**.

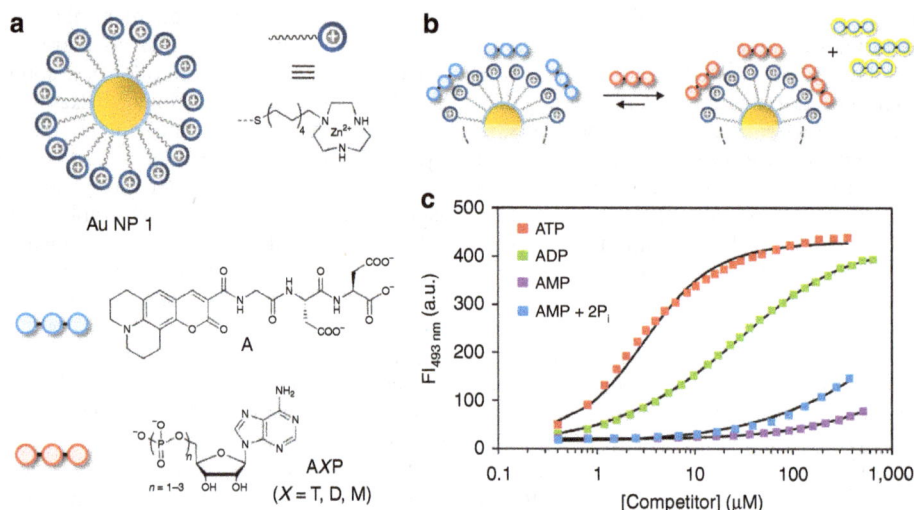

**Figure 2 | Displacement studies.** (**a**) Schematic representations of the molecules. (**b**) Displacement of probe **A** from Au NP **1** on the addition of a competitor. (**c**) Fluorescent intensity at 493 nm as a function of the concentration of competitor. The solid lines represent the best fit to model. Experimental conditions: $[\text{TACN}\bullet\text{Zn}^{2+}] = 10 \pm 1\,\mu\text{M}$, $[\textbf{A}] = 3.7\,\mu\text{M}$, $[\text{HEPES}] = 10\,\text{mM}$, pH 7.0, $[\text{CaCl}_2] = 1.0\,\text{mM}$, 37 °C, $\lambda_{\text{ex, }\textbf{A}} = 450\,\text{nm}$, $\lambda_{\text{em, }\textbf{A}} = 493\,\text{nm}$, slits = 2.5/5.0 nm.

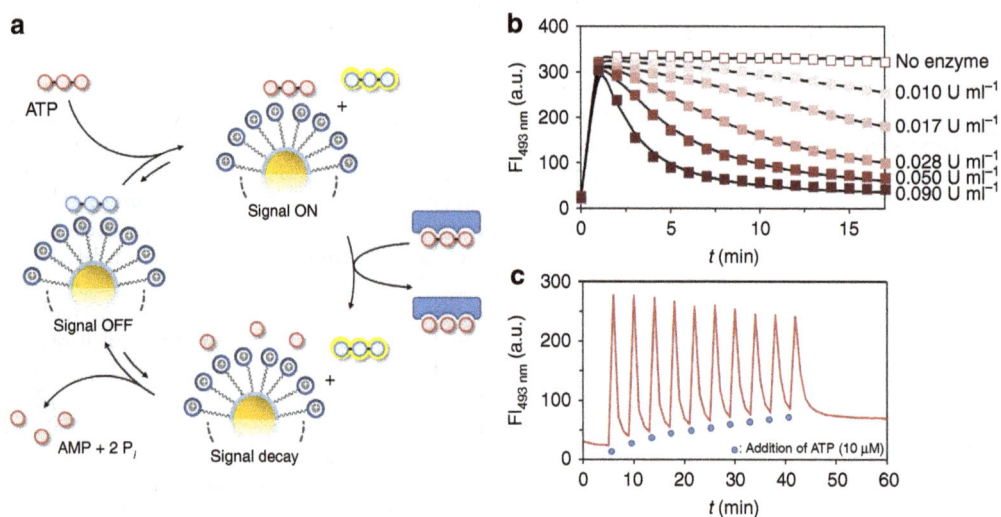

**Figure 3 | Transient signal generation fuelled by ATP.** (**a**) Schematic representation of the system for transient signal generation. (**b**) Fluorescent intensity at 493 nm as a function of time on the addition of ATP (10 μM) to Au NP **1** ($[\text{TACN}\bullet\text{Zn}^{2+}] = 10 \pm 1\,\mu\text{M}$) and **A** (3.7 μM) in the presence of different concentrations of potato apyrase. The solid lines represent fits to the kinetic model shown in Fig. 4. (**c**) Fluorescent intensity at 493 nm on 10 repetitive additions of ATP (10 μM) to a solution of Au NP **1** ($[\text{TACN}\bullet\text{Zn}^{2+}] = 10 \pm 1\,\mu\text{M}$) and **A** (3.7 μM) in the presence of potato apyrase (0.3 U ml$^{-1}$).

and determined the $K_{\text{rel}}$-value of the AMP + 2P$_i$ mixture rather than just AMP (Fig. 2c). As expected, a slightly higher value $(1.6 \times 10^{-3})$ was obtained, but still around 3 orders of magnitude lower than that of ATP (see the Methods section and Supplementary Figs 3 and 4 for full details about the displacement studies).

Kinetic measurements showed that in the absence of enzyme, the addition of ATP (10 μM) to a solution of **A** and Au NP **1** led to a rapid increase in FI, which then remained constant in time (Fig. 3b). We were very pleased to observe that repetition of the same experiment in the presence of increasing amounts of enzyme (from 0.01 up to 0.09 U ml$^{-1}$) resulted in the same initial increase in FI, but this time followed by a signal decay with rates depending on the concentration of enzyme (Fig. 3b). At the highest concentration of enzyme, the FI returned to the starting value after just 15 min. The reproducibility of transient signal generation by the system was demonstrated by performing 10

cycles with the same sample adding new batches of ATP (10 μM) each time the signal had returned to the starting value (Fig. 3c). It is noted that the higher fluorescent intensity after 10 cycles (69 a.u.) originates from the accumulation of waste (AMP + 2P$_i$) in the system (72 a.u. expected for 100 μM of AMP and 200 μM of P$_i$ based on the displacement curves).

**Signal–response curve and kinetic model.** The most important feature of any signalling system is the signal–response curve, as it correlates the output signal to the intensity of the input signal. We determined the response curves of our system at four different concentrations of ATP (10, 50, 100 and 150 μM) at a constant concentration of the enzyme (0.017 U ml$^{-1}$). Measurement of the FI as a function of time clearly showed strong differences between the four samples (Fig. 4b). In all cases, the addition of ATP caused a rapid increase of the FI. The

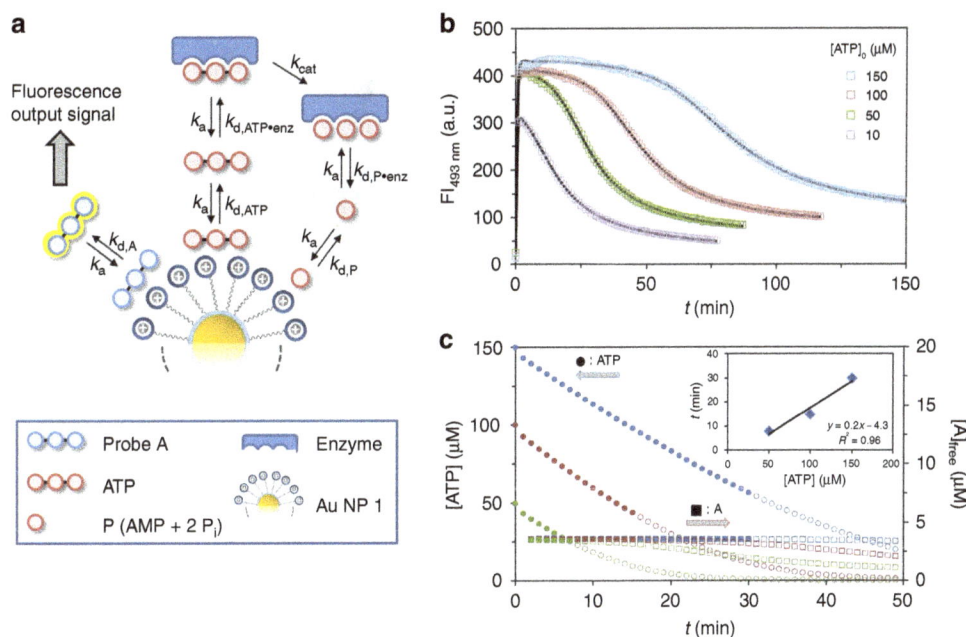

**Figure 4 | Signal–response curve and kinetic model.** (**a**) Kinetic model used for fitting the experimental data. (**b**) Fluorescent intensity at 493 nm as a function of time on the addition of different amounts of ATP to a solution containing Au NP **1** ([TACN•Zn$^{2+}$] = 10 ± 1 μM), probe **A** (3.7 μM) and a constant concentration of potato apyrase (0.017 U ml$^{-1}$). The solid lines represent the fit to a kinetic model. (**c**) Plots of the concentration of ATP (left axis) and free **A** (right axis) as a function of time for three different initial concentrations of ATP. The solid points indicate the period at which the concentration of free **A** is constant (<2% decrease with respect to the maximum value). The inset gives the duration of the time interval with a steady concentration of free **A** as a function of the initial amount of ATP.

maximum intensity was lower for [ATP]$_0$ = 10 μM (FI$_{493}$ = 306 a.u.), compared with the other three samples which reached a nearly identical intensity of ~410 a.u. These values correspond to the fluorescence intensities measured in the competition experiments between probe **A** and ATP (Fig. 2c). The nature of the displacement curve indeed confirms that a FI of ~410 a.u. is the maximum intensity of the output signal that can be generated by the system as it represents the situation at which (nearly) all of probe **A** is displaced from Au NP **1**. However, although the addition of 50, 100 or 150 μM of ATP gives the same initial signal, the signal decay is much different, being much slower for high concentrations of ATP. Although this can be intuitively explained based on the displacement curve, we were interested in a quantitative description of the signal generation process to have detailed information on the concentration of all species in the system. For that purpose, we developed a model taking into account all relevant kinetic processes that occur within the system (Fig. 4a, see Methods section and Supplementary Fig. 5 for more details). The relative affinities, $K_{rel}$, were used to define the dissociation rate constants of all complexes involving Au NP **1**, assuming that all association rate constants are very high. Dissociation rate constants for the complexes involving the enzyme were varied to fit the experimental data to the model. As illustrated by the solid lines the model neatly describes the experimental data (Fig. 4b). Analysis of the concentrations of ATP and **A** as a function of time illustrates the most interesting feature of this system, which is the possibility to develop a steady fluorescent signal with a duration that depends on the initial concentration of ATP. Interestingly, during the period in which the concentration of free **A**, and thus the output signal, remains constant (<2% decrease), a significant consumption of ATP takes place (see filled symbols in Fig. 4c). However, not until the concentration of ATP falls below a certain threshold value at which **A** can effectively compete with the remaining ATP for binding to Au NP **1**, the signal starts to decay. The correlation

between the duration of this lag time and the initial concentration of ATP demonstrates that it is possible to control this time interval, which may last up to 30 min (inset Fig. 4c). As will be discussed in the next section, signalling pathways in nature rely heavily on this kind of control mechanism.

**ATP-dependent activation of two different signals.** Having developed a robust system for transient signal generation, we were curious to find out whether it would be possible to mimic other features of natural signalling pathways. In particular, we were inspired by the purinergic signalling pathway as it bears a strong resemblance to our system. The purinergic signalling pathway is involved in the regulation of various physiological functions in the circulatory[34], immune[35] and nervous system[36], among others. It operates through the activation of purinergic receptors in the cell membrane by extracellular nucleotides and nucleosides. It is known that all receptors belonging to the P2X family, which is one out of the three known distinct classes of purinergic receptors, respond only to ATP[37]. However, they require different concentrations of ATP for activation and exhibit different desensitization kinetics. For example, the P2X1 receptor has a high affinity for ATP ($K_D \approx 0.01$–1 μM) and a desensitization rate <1 s, whereas activation of the P2X7 receptor requires ATP concentrations >100 μM but has a much slower desensitization rate (>20 s)[37]. Thus, the overall response of the signalling pathway (both in terms of output signal diversity and decay rate) is dependent on the intensity of the trigger (ATP). We argued that this property could be mimicked by adding a second fluorescent probe **B** (dipeptide Ac-WD-OH) to the system (Fig. 5a). Dipeptide **B** has a much lower affinity and surface saturation concentration (SSC) compared with **A** (0.6 ± 0.1 μM for [TACN•Zn$^{2+}$] = 10 ± 1 μM, see Supplementary Fig. 6) and, in addition, has a fluorescent tryptophan moiety ($\lambda_{ex}$ = 280 nm, $\lambda_{em}$ = 360 nm), which can be monitored independently from

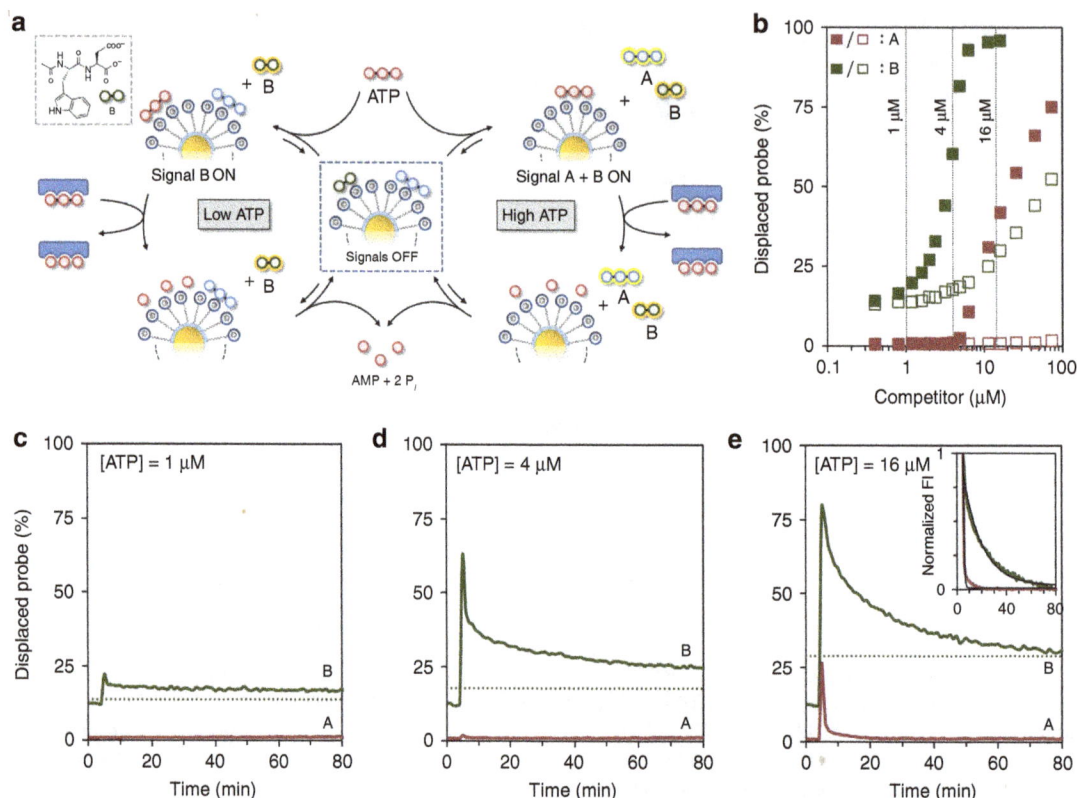

**Figure 5 | ATP-dependent activation of two different signals.** (**a**) Schematic representation of the ATP-dependent transient generation of one or two signals. (**b**) Amount of displaced probe **A** (red) or **B** (green) as a function of the concentration of either ATP (filled squares) or the waste mixture AMP + 2P$_i$ (empty squares). (**c–e**) Amount of displaced probe **A** (red) and **B** (green) as a function of time on the addition of ATP at 1 (**c**), 4 (**d**) or 16 μM (**e**) concentrations. The dotted lines mark the expected end value expected for probe **B** in the presence of the respective amount of waste formed. The inset in **e** gives the normalized change in FI as a function of time for the two probes and demonstrates the much faster decay of the signal originating from the high affinity prove **A**. In all these experiments probe **A** was excited at 370 nm (rather than at 450 nm) to obtain similar fluorescence intensities for probes **A** and **B**, thus permitting a monitoring of both probes with the same slit widths. Experimental conditions: [TACN•Zn$^{2+}$] = 20 ± 1 μM, [**A**] = [**B**] = 0.5 μM, [HEPES] = 10 mM, pH 7.0, [CaCl$_2$] = 1.0 mM, [potato apyrase] = 0.3 U ml$^{-1}$, 37 °C. $\lambda_{ex,\ \mathbf{A}}$ = 370 nm, $\lambda_{em,\ \mathbf{A}}$ = 493 nm, $\lambda_{ex,}$ $_{\mathbf{B}}$ = 280 nm, $\lambda_{em,\ \mathbf{B}}$ = 360 nm, slits = 10/10 nm.

probe **A**. The displacement of probes **A** (0.5 μM) and **B** (0.5 μM) from Au NP **1** by either ATP or the waste mixture AMP + 2P$_i$ was studied at a slightly higher concentration of Au NP **1** ([TACN•Zn$^{2+}$] = 20 ± 1 μM) to ensure that both probes were initially (nearly) quantitatively bound. As a result of the lower valency of probe **B** the margin between ATP and the waste mixture is much reduced. Nonetheless, the difference between probes **A** and **B** is sufficiently large to test the abovementioned hypothesis (Fig. 5b). In particular, three concentrations of ATP were identified at which the system was expected to generate a different response: 1 μM of ATP should not result in any signal, 4 μM of ATP should generate a signal from **B** but not **A**, whereas at 16 μM of ATP signals from both **A** and **B** should emerge. It is noted that higher concentrations of ATP could not be used because the accumulation of waste would block the re-formation of the complex between **B** and Au NP **1** after ATP hydrolysis. Also, it was found that high concentrations of ATP (>20 μM) caused a decrease in the fluorescence intensity of free **B**. Transient signal generation by the system was studied by measuring the fluorescence intensities of probe **A** and **B** in time on the addition of ATP at 1, 4 or 16 μM concentrations to the system in the presence of potato apyrase (Fig. 5c–e). The obtained curves clearly demonstrate that, just as in the purinergic signalling pathway, the initial concentration of ATP determines whether none, one or two signals are generated. It is noted that the amount of displaced **A** on the addition of 16 μM (27%) is lower than that expected based on the displacement assays (42%,

Fig. 5b). This is caused by the hydrolysis of ATP by the enzyme in the time between the addition of ATP and the continuation of the measurements. Importantly, the system also displays a correlation between probe affinity and signal duration (see inset of Fig. 5e). Thus, activation of probe **A** requires high concentrations of ATP, but the resulting signal decays rapidly ($k_{obs}$ = 1.5 min$^{-1}$). On the other hand, low concentrations of ATP are sufficient to activate probe **B**, but that signal persists for a much longer time (23×, $k_{obs}$ = 0.07 min$^{-1}$). The inverted relation between binding affinity and signal duration originates directly from the competition between the probes and ATP for binding to Au NP **1**. For example, low-affinity probe **B** requires a near-complete hydrolysis of ATP to be able to return to Au NP **1**. Here this takes a particular long time because the ATP hydrolysis by the enzyme slows down significantly on depletion of ATP.

**Transient downregulation of the catalytic activity of Au NP 1.** Another feature of natural signalling pathways is that they consist of cascades of enzymatic reactions that are either up or downregulated by an initial trigger. This permits the system to transform a weak input signal into a strong output signal by means of catalytic signal amplification. We were interested to find out whether we could exploit the potato apyrase-driven consumption of ATP for the transient regulation of the catalytic activity of Au NP **1**. Previous studies have shown that Au NP **1** and analogues are highly efficient catalysts for the

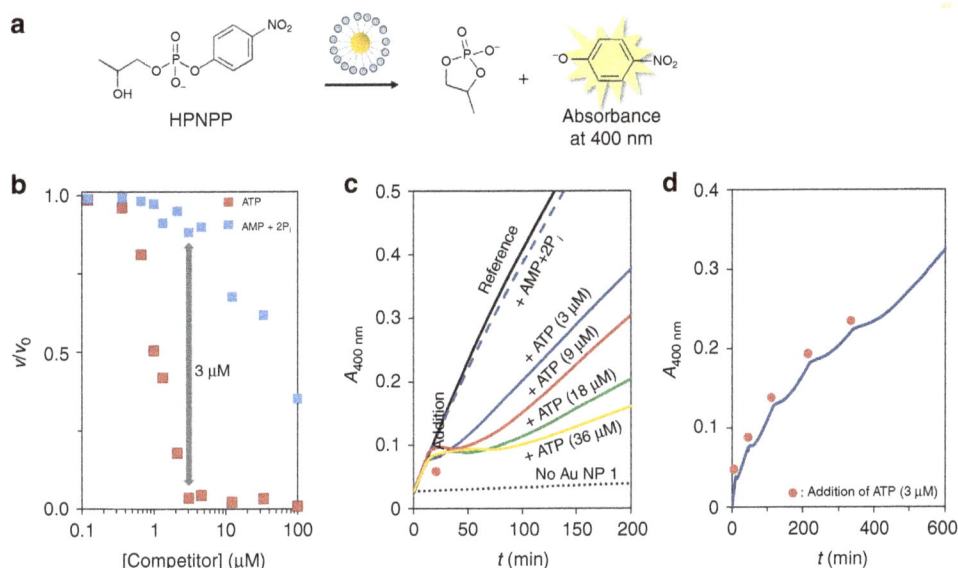

**Figure 6 | Transient downregulation of the catalytic activity of Au NP 1. (a)** Catalysis of the transphosphorylation of HPNPP by Au NP **1**. **(b)** Relative reaction rates ($v/v_0$) as a function of the concentration of either ATP or the mixture AMP + 2P$_i$. Experimental conditions: [TACN•Zn$^{2+}$] = 10 ± 1 µM, [HPNPP] = 1 mM, [HEPES] = 10 mM, pH 7.0, [CaCl$_2$] = 1.0 mM, 37 °C. **(c)** Absorbance at 400 nm (originating from the reaction product p-nitrophenolate) as a function of time for different mixtures. The red dot indicates the time at which inhibitors were added. The lower slope after reactivation of Au NP **1** at higher concentrations of ATP is caused by the higher concentrations of produced waste (see the inhibition by different concentrations of the waste mixture given in Supplementary Fig. 8). Experimental conditions: [TACN•Zn$^{2+}$] = 10 ± 1 µM, [HPNPP] = 1 mM, [HEPES] = 10 mM, pH 7.0, [CaCl$_2$] = 1.0 mM, [potato apyrase] = 0.06 U ml$^{-1}$, 37 °C. **(d)** Absorbance at 400 nm as a function of time on the sequential additions of ATP (5 × 3 µM). The red dots indicate the time of addition. Experimental conditions: same as for **c**.

transphosphorylation of 2-hydroxypropyl-4-nitrophenylphosphate (HPNPP), which is an RNA-model compound (Fig. 6a)[38,39]. These kinds of catalytic nanosystems have been referred to as nanozymes, for the fact that the reaction rate follows Michaelis–Menten reaction kinetics[38]. This implies that catalysis is described by a binding event (defined by the dissociation constant $K_M$) followed by a chemical reaction (defined by the first-order rate constant $k_{cat}$). For the transphosphorylation of HPNPP by Au NP **1**, values for $K_M$ and $k_{cat}$ of 0.25 mM and $1.4 \times 10^{-3}$ s$^{-1}$, respectively, were determined (see the Methods section and Supplementary Fig. 7). As for enzymes, the catalytic activity of Au NP **1** is inhibited in the presence of species able to compete with HPNPP for binding[25]. Indeed, kinetic studies at varying concentrations showed that ATP is a very effective inhibitor already at low micromolar concentrations (Fig. 6b). However, a significantly lower inhibitory effect of the AMP + 2P$_i$ waste mixture was observed (Fig. 6b). This difference raised the possibility to temporarily downregulate the catalytic activity of Au NP **1** by ATP in the presence of potato apyrase. This was verified by adding 3 µM of ATP to a solution containing Au NP **1** (10 µM), HPNPP (1 mM) and potato apyrase (0.06 U ml$^{-1}$). This caused an immediate inhibition of the transphosphorylation reaction evidenced by the constant absorbance at 400 nm (Fig. 6c). However, after a period of ∼5 min, the absorbance started to increase again indicating the reactivation of Au NP **1**. It is noted that even after prolonged times the catalytic activity of Au NP **1** was not restored to the level expected in the presence of AMP + 2P$_i$ (see dotted line in Fig. 6c). This could indicate that potato apyrase is inhibited by HPNPP causing a further reduction of the already slow hydrolysis rate of ATP at these concentrations. Nonetheless, the observation that the duration of the inhibition lag time depends on the amount of ATP added demonstrated that ATP indeed functions as transient regulatory element for the downregulation of Au NP **1** (Fig. 6c). The transient nature emerged also in a clear manner from a kinetic

experiment in which five cycles were performed by sequentially adding ATP (3 µM each) to the mixture of Au NP **1**, HPNPP and the enzyme (Fig. 6d).

## Discussion

In conclusion, we have developed a strategy that permits transient signal generation in a self-assembled system. Signal formation is triggered by an input in the form of ATP, the consumption of which returns the system back to the original state. The initial concentration of ATP regulates the intensity of the output signal and its duration and determines also whether one or two signals are generated in the case where two reporter molecules are used. The same mechanism is evoked when applied to the transient downregulation of the catalytic activity of the nanoparticles.

In more general terms, we have developed straightforward methodology to gain temporal control over supramolecular processes, that is, all these processes rely on noncovalent interactions between molecules. In our particular case, ATP shifts the equilibrium between Au NP **1** and **A** away from the free energy minimum and continues to do so as long as its concentration remains above a certain threshold value. Although we have focused on transient signal generation, one can envision that the same strategy may also be applied to control the temporal stability of self-assembled systems, such as nanoarchitectures or materials, in which the degradable compound plays an essential role as a building block or as a stabilizing unit. As such, it may provide an important new tool for the design of synthetic systems with 'life-like' properties, such as adaptation, self-healing and evolution.

## Methods

**General.** Au NPs **1** were synthesized and characterized as described in the literature[24] and stored at 4 °C in mQ water. The concentration of TACN-head groups was determined from kinetic titrations using either Zn(NO$_3$)$_2$ or Cu(NO$_3$)$_2$ as reported previously[25]. Zn(NO$_3$)$_2$ and Cu(NO$_3$)$_2$ were analytical grade products. The concentrations of metal ion stock solutions were determined by atomic absorption spectroscopy.

The synthesis and characterization of probes **A** (C343-GDD-OH)[31] and **B** (Ac-WD-OH)[40] have been reported previously. HPNPP was prepared as the sodium salt according to literature procedures[41,42]. The buffer, 4-(2-hydroxyethyl)-1-piperazineethanesulfonic acid (HEPES), ATP, ADP, AMP and potato apyrase were purchased from Sigma-Aldrich (product code A6237) and used without further purification. $Na_3PO_4$ (that is, $P_i$) and $CaCl_2$ were obtained from Carlo Erba Reagenti and BDH Prolabo, respectively, and used as received.

The enzyme potato apyrase was dissolved in 1.0 ml of mQ water and divided in 10 working aliquots of 100 µl, which were stored at $-20\,°C$. To avoid harmful freeze–thaw cycles, all experiments were carried out using fresh solutions, which were prepared by diluting one of the abovementioned aliquots to the desired concentration and used within 1 day.

The enzyme concentration is expressed in $U\,ml^{-1}$ and is based on the product information declared for the batch purchased (100 U). One unit is defined as the quantity of enzyme that will liberate 1.0 µmol of $P_i$ from ATP or ADP per minute at pH 6.5 and 30 °C.

Stock solution concentrations of probes and competitors were determined both by weight and ultraviolet–visible spectroscopy using the following molar extinction coefficients: $\varepsilon_{259\,(ATP,\,ADP,\,AMP)} = 15,400\,M^{-1}\,cm^{-1}$ and $\varepsilon_{450\,(C343)} = 45,000\,M^{-1}\,cm^{-1}$ at pH 7.0)[43,44].

Ultraviolet–visible measurements were performed on a Varian Cary 50 spectrophotometer, while fluorescence measurements were performed on a Varian Cary Eclipse fluorescence spectrophotometer. Both spectrophotometers were equipped with thermostatted cell holders. All experiments were performed starting from aqueous buffered solution ([HEPES] = 10 mM, pH 7.0) containing Au NP **1** ([TACN•$Zn^{2+}$] = 10 or $20 \pm 1$ µM). In a typical titration, consecutive amounts of stock solution of probes or competitors are added and the FI monitored as a function of time. Measurements were taken every minute. Optical parameters are reported in the figure captions.

**Complex formation between Au NP 1 and A or B.** The SSC of **A** on Au NP **1** was determined from a fluorescence titration of **A** to Au NP **1** (10 µM). The FI at 493 nm was measured as a function of the amount of **A** added (see Supplementary Fig. 1a). The resulting curve was fitted using a model in which the Au NP **1** was represented by a single binding site with an imposed very high affinity ($K_a = 1 \times 10^8\,M^{-1}$) for **A** to mimic binding under saturation conditions (see Supplementary Data 1). A factor X was used to correlate the FI to the concentration of **A**. Fitting of the experimental data yielded a SSC of 3.7 ($\pm$ 0.1) µM. The same value can also be obtained by extrapolation of the linear part to FI = 0. The same procedure has been used to determine the SSC of **B** (see Supplementary Fig. 6) yielding a value of 0.6 ($\pm$ 0.1) µM.

**Compatibility with potato apyrase and $Ca^{2+}$.** Potato apyrase requires metal ions for its hydrolytic activity (ATP → AMP + 2$P_i$), typically $Ca^{2+}$ ions. To test the stability of the complex between Au NP **1** and **A** in the presence of $Ca^{2+}$ and potato apyrase, the change in FI at 493 nm on the addition of increasing amounts of either $CaCl_2$ or potato apyrase to a solution of Au NP **1** (10 µM) and probe **A** (3.7 µM) was monitored. Au NP **1**•**A** complex is hardly affected by the presence of both $CaCl_2$ (up to 1.0 mM, see Supplementary Fig. 2a) and potato apyrase (at a fixed [$CaCl_2$] = 1.0 mM, see Supplementary Fig. 2b). Based on these results, it was decided to carry out all experiments at a constant concentration of $CaCl_2$ (1.0 mM), which is sufficient for potato apyrase.

**Displacement studies.** The relative affinities of probe **A** and a series of phosphate probes for Au NP **1** were determined from a series of competition experiments. Displacement studies were performed by measuring the FI at 493 nm on the addition of increasing amounts of either one of the competitors ATP, ADP, AMP, $P_i$ or the combined mixture AMP + 2$P_i$ to a solution of Au NP **1** (10 µM) and probe **A** (3.7 µM) (see Supplementary Fig. 3). A theoretical model was developed to describe the competition experiments between probe **A** and the competitors ATP, ADP, AMP, $P_i$ or the combined mixture AMP + 2$P_i$ (see Supplementary Fig. 4 and Supplementary Data 2). In the model, Au NP **1** is treated as having a single binding site for interaction with either one of the molecules. The total concentration of this binding site is equal to the SSC of **A** on Au NP **1** ([Au NP **1**]$_{tot}$ = 3.7 µM). The binding of probe **A** and the competitor to Au NP **1** is defined by the two thermodynamic equilibrium constants $K_a$ and $K_b$, respectively. A parameter X is used to correlate the concentration of unbound **A** to the FI (FI = $X \times$ **A**). As for the direct titrations, binding occurs under saturation conditions implying that the concentration of free Au NP **1** is very low. Consequently, the absolute values for $K_A$ and $K_B$ obtained from fitting the experimental data are not very informative. On the contrary, the ratio between $K_A$ and $K_B$ ($K_{rel}$) is obtained very precisely as it depends on the ratio between free **A** and the amount of competitor added.

**Kinetic model.** A theoretical model was developed to describe the kinetics of the system (see Supplementary Fig. 5 and Supplementary Data 3). In the model, Au NP **1** is treated as having a single binding site for interaction with **A**, ATP and P. The total concentration of this binding site is equal to the SSC of **A** on Au NP **1** ([Au

NP **1**]$_{tot}$ = 3.7 µM). Starting point of the time course is an equilibrium system between **A** and Au NP **1** ([Au NP **1**•**A**] = 3.47 µM, [**A**]$_0$ = [Au NP **1**]$_0$ = 0.23 µM) to which ATP is added at $t = 0$. This takes into account that a small amount of **A** is not bound when **A** is present at the SSC (see the binding curve in Supplementary Fig. 1). Three species interact with Au NP **1**: probe **A**, ATP and P, the latter representing the combined products of ATP hydrolysis (AMP + 2$P_i$). It is assumed that all associative processes are very fast ($k_a = 10^7$) and that differences in thermodynamic stabilities arise from variations in the dissociation rate constants ($k_d$). Thus, the different affinities of **A**, ATP and P for Au NP **1** are reflected by different dissociation constants ($k_{d,A}$, $k_{d,ATP}$, $k_{d,P}$, respectively). The value for $k_{d,A}$ was arbitrarily set to a value of 2.5 to define binding of **A** to Au NP **1** under saturation conditions. This value is also sufficiently high to ensure a rapid release of probe **A** from Au NP **1** on the addition of ATP, as observed experimentally. The value of 2.5 for $k_{d,A}$ also fixes the dissociation rate constants for ATP ($k_{d,ATP}$) and P ($k_{d,P}$), because the relative affinities of the species for Au NP **1** are known from the abovementioned displacement studies. Michaelis–Menten enzyme kinetics is implemented through a reversible interaction of ATP with the enzyme characterized by the association constant, $k_a$, and the dissociation constant, $k_{d,ATP•enz}$. Formation of the ATP–enzyme complex is followed by a conversion of ATP into products P ($k_{cat}$), which then dissociate from the enzyme (defined by $k_{d,P•enz}$). Development of the model showed that enzyme inhibition by product P is essential for a correct description of the kinetic profile. On the other hand, the introduction of ADP as reaction intermediate is not relevant from a kinetic point of view. The concentration of enzyme was set to an arbitrary value since its actual concentration is not known as the commercial solution is defined by enzyme activity, that is, the amount of ATP (nmol) that is consumed per min per mg. This implies that the values obtained for $k_{cat}$ have no quantitative relevance. The low value of 0.01 µM for [$E$]$_0$ (compared with Au NP **1**) ensures that the enzyme cannot accommodate large quantities of ATP. The FI is obtained by multiplying the concentration of free **A** with a factor X.

**Catalytic transphosphorylation of HPNPP by Au NP 1.** The catalytic activity of Au NP **1** was evaluated by measuring the increase in absorbance at 400 nm (resulting from the formation of *p*-nitrophenolate) at different concentrations of HPNPP. the obtained kinetic data were elaborated with the method of initial rates. Initial rates ($v_0$) were then plotted against the concentration of HPNPP. The resulting enzyme-like saturation profile was fitted with the Michaelis–Menten model, from which the $k_{cat}$ and $K_m$ values are determined (See Supplementary Fig. 7).

## References

1. Ricard, J. Information and communication in living systems. *Emergent Collective Properties, Networks and Information in Biology* **40**, 83–108 (2006).
2. Krauss, G. *Biochemistry of Signal Transduction and Regulation* (Wiley-VCH Verlag GmbH & Co, 2008).
3. Behar, M. & Hoffmann, A. Understanding the temporal codes of intra-cellular signals. *Curr. Opin. Genet. Dev.* **20**, 684–693 (2010).
4. Valeur, B. & Leray, I. Design principles of fluorescent molecular sensors for cation recognition. *Coord. Chem. Rev.* **205**, 3–40 (2000).
5. Beer, P. D. & Gale, P. A. Anion recognition and sensing: The state of the art and future perspectives. *Angew. Chem. Int. Ed.* **40**, 486–516 (2001).
6. Rogers, C. W. & Wolf, M. O. Luminescent molecular sensors based on analyte coordination to transition-metal complexes. *Coord. Chem. Rev.* **233**, 341–350 (2002).
7. Anslyn, E. V. Supramolecular analytical chemistry. *J. Org. Chem.* **72**, 687–699 (2007).
8. Wu, J., Liu, W., Ge, J., Zhang, H. & Wang, P. New sensing mechanisms for design of fluorescent chemosensors emerging in recent years. *Chem. Soc. Rev.* **40**, 3483–3495 (2011).
9. Nguyen, B. T. & Anslyn, E. V. Indicator-displacement assays. *Coord. Chem. Rev.* **250**, 3118–3127 (2006).
10. Wright, A. T. & Anslyn, E. V. Differential receptor arrays and assays for solution-based molecular recognition. *Chem. Soc. Rev.* **35**, 14–28 (2006).
11. Grzybowski, B. A., Wilmer, C. E., Kim, J., Browne, K. P. & Bishop, K. J. M. Self-assembly: from crystals to cells. *Soft Matter* **5**, 1110–1128 (2009).
12. Stuart, M. A. C. *et al.* Emerging applications of stimuli-responsive polymer materials. *Nat. Mater.* **9**, 101–113 (2010).
13. Fialkowski, M. *et al.* Principles and implementations of dissipative (dynamic) self-assembly. *J. Phys. Chem. B* **110**, 2482–2496 (2006).
14. Boekhoven, J. *et al.* Dissipative self-assembly of a molecular gelator by using a chemical fuel. *Angew. Chem. Int. Ed.* **49**, 4825–4828 (2010).
15. Debnath, S., Roy, S. & Ulijn, R. V. Peptide nanofibers with dynamic instability through nonequilibrium biocatalytic assembly. *J. Am. Chem. Soc.* **135**, 16789–16792 (2013).
16. Fanlo-Virgos, H., Alba, A.-N. R., Hamieh, S., Colomb-Delsuc, M. & Otto, S. Transient substrate-induced catalyst formation in a dynamic molecular network. *Angew. Chem. Int. Ed.* **53**, 11346–11350 (2014).

17. Dambenieks, A. K., Vu, P. H. Q. & Fyles, T. M. Dissipative assembly of a membrane transport system. *Chem. Sci.* **5**, 3396–3403 (2014).

18. Fletcher, S. P., Dumur, F., Pollard, M. M. & Feringa, B. L. A reversible, unidirectional molecular rotary motor driven by chemical energy. *Science* **310**, 80–82 (2005).

19. Cheng, C. *et al.* Energetically demanding transport in a supramolecular assembly. *J. Am. Chem. Soc.* **136**, 14702–14705 (2014).

20. Ragazzon, G., Baroncini, M., Silvi, S., Venturi, M. & Credi, A. Light-powered autonomous and directional molecular motion of a dissipative self-assembling system. *Nat. Nanotechnol.* **10**, 70–75 (2015).

21. Ghale, G. & Nau, W. M. Dynamically analyte-responsive macrocyclic host-fluorophore systems. *Acc. Chem. Res.* **47**, 2150–2159 (2014).

22. White, H. D. Kinetics of tryptophan fluorescence enhancement in myofibrils during ATP hydrolysis. *J. Biol. Chem.* **260**, 982–986 (1985).

23. Thomas, III S. W. *et al.* Infochemistry and infofuses for the chemical storage and transmission of coded information. *Proc. Natl Acad. Sci. USA* **106**, 9147–9150 (2009).

24. Pieters, G., Cazzolaro, A., Bonomi, R. & Prins, L. J. Self-assembly and selective exchange of oligoanions on the surface of monolayer protected Au nanoparticles in water. *Chem. Commun.* **48**, 1916–1918 (2012).

25. Bonomi, R., Cazzolaro, A., Sansone, A., Scrimin, P. & Prins, L. J. Detection of enzyme activity through catalytic signal amplification with functionalized gold nanoparticles. *Angew. Chem. Int. Ed.* **50**, 2307–2312 (2011).

26. Pieters, G., Pezzato, C. & Prins, L. J. Controlling supramolecular complex formation on the surface of a monolayer-protected gold nanoparticle in water. *Langmuir* **29**, 7180–7185 (2013).

27. Sapsford, K. E., Berti, L. & Medintz, I. L. Materials for fluorescence resonance energy transfer analysis: beyond traditional donor-acceptor combinations. *Angew. Chem. Int. Ed.* **45**, 4562–4588 (2006).

28. Giljohann, D. A. *et al.* Gold nanoparticles for biology and medicine. *Angew. Chem. Int. Ed.* **49**, 3280–3294 (2010).

29. Bunz, U. H. F. & Rotello, V. M. Gold nanoparticle-fluorophore complexes: sensitive and discerning "noses" for biosystems sensing. *Angew. Chem. Int. Ed.* **49**, 3268–3279 (2010).

30. Saha, K., Agasti, S. S., Kim, C., Li, X. N. & Rotello, V. M. Gold nanoparticles in chemical and biological sensing. *Chem. Rev.* **112**, 2739–2779 (2012).

31. Pezzato, C., Lee, B., Severin, K. & Prins, L. J. Pattern-based sensing of nucleotides with functionalized gold nanoparticles. *Chem. Commun.* **49**, 469–471 (2013).

32. Maiti, S., Pezzato, C., Martin, S. G. & Prins, L. J. Multivalent interactions regulate signal transduction in a self-assembled hg2 + sensor. *J. Am. Chem. Soc.* **136**, 11288–11291 (2014).

33. Molnar, J. & Lorand, L. Studies on apyrases. *Arch. Biochem. Biophys.* **93**, 353–363 (1961).

34. McIntosh, V. J. & Lasley, R. D. Adenosine receptor-mediated cardioprotection: are all 4 subtypes required or redundant? *J. Cardiovasc. Pharmacol. Ther.* **17**, 21–33 (2012).

35. Junger, W. G. Immune cell regulation by autocrine purinergic signalling. *Nat. Rev. Immunol.* **11**, 201–212 (2011).

36. Abbracchio, M. P., Burnstock, G., Verkhratsky, A. & Zimmermann, H. Purinergic signalling in the nervous system: an overview. *Trends Neurosci.* **32**, 19–29 (2009).

37. Jarvis, M. F. & Khakh, B. S. ATP-gated P2X cation-channels. *Neuropharmacology* **56**, 208–215 (2009).

38. Manea, F., Houillon, F. B., Pasquato, L. & Scrimin, P. Nanozymes: gold-nanoparticle-based transphosphorylation catalysts. *Angew. Chem. Int. Ed.* **43**, 6165–6169 (2004).

39. Zaupa, G., Mora, C., Bonomi, R., Prins, L. J. & Scrimin, P. Catalytic self-assembled monolayers on Au nanoparticles: the source of catalysis of a transphosphorylation reaction. *Chem. Eur. J.* **17**, 4879–4889 (2011).

40. Franceschini, C., Scrimin, P. & Prins, L. J. Light-triggered thiol-exchange on gold nanoparticles at low micromolar concentrations in water. *Langmuir* **30**, 13831–13836 (2014).

41. Brown, D. M. & Usher, D. A. Hydrolysis of hydroxyalkyl phosphate esters: effect of changing ester group. *J. Chem. Soc.* 6558–6564 (1965).

42. Tsang, J. S., Neverov, A. A. & Brown, R. S. La(3 + )-Catalyzed methanolysis of hydroxypropyl-p-nitrophenyl phosphate as a model for the RNA transesterification reaction. *J. Am. Chem. Soc.* **125**, 1559–1566 (2003).

43. Yao, Z., Feng, X., Hong, W., Li, C. & Shi, G. A simple approach for the discrimination of nucleotides based on a water-soluble polythiophene derivative. *Chem. Commun.* **31**, 4696–4698 (2009).

44. Webb, M. R. & Corrie, J. E. Fluorescent coumarin-labeled nucleotides to measure ADP release from actomyosin. *Biophys. J.* **81**, 1562–1569 (2001).

## Acknowledgements

Financial support from the ERC (StG-239898) and the University of Padova (CPDA138148) is acknowledged.

## Author contributions

C.P. and L.J.P designed the experiments and interpreted the data; C.P. performed the experiments; L.J.P. developed the models. L.J.P. wrote the manuscript and C.P. commented on it.

# 16

# Tunable solid-state fluorescent materials for supramolecular encryption

Xisen Hou[1,*], Chenfeng Ke[1,*], Carson J. Bruns[1], Paul R. McGonigal[1], Roger B. Pettman[2] & J. Fraser Stoddart[1]

Tunable solid-state fluorescent materials are ideal for applications in security printing technologies. A document possesses a high level of security if its encrypted information can be authenticated without being decoded, while also being resistant to counterfeiting. Herein, we describe a heterorotaxane with tunable solid-state fluorescent emissions enabled through reversible manipulation of its aggregation by supramolecular encapsulation. The dynamic nature of this fluorescent material is based on a complex set of equilibria, whose fluorescence output depends non-linearly on the chemical inputs and the composition of the paper. By applying this system in fluorescent security inks, the information encoded in polychromic images can be protected in such a way that it is close to impossible to reverse engineer, as well as being easy to verify. This system constitutes a unique application of responsive complex equilibria in the form of a cryptographic algorithm that protects valuable information printed using tunable solid-state fluorescent materials.

[1] Department of Chemistry, Northwestern University, Evanston, Illinois 60208-3113, USA. [2] Cycladex, c/o Innovation and New Ventures Office, Northwestern University, 1800 Sherman Avenue, Suite 504, Evanston, Illinois 60201-3789, USA. * These authors contributed equally to this work. Correspondence and requests for materials should be addressed to J.F.S. (email: stoddart@northwestern.edu).

Photoluminescent solid-state materials have been widely applied in dye lasers[1], organic light emitting diodes (OLEDs)[2–4], data recording and storage[5,6], and security printing[7,8]. These materials can be easily applied inexpensively to different surfaces and have been implemented widely as security inks to protect high-value merchandise, government documents and banknotes[9]. Materials with static luminescent outputs, however, are familiar to counterfeiters. In contrast, stimuli-responsive photoluminescent materials, which change their optical outputs in response to external stimuli, possess extra security features that are difficult to mimic, making them suitable for the next generation of security printing. In recent years, a series of stimuli-responsive photoluminescent materials, including those that are thermochromic[10–13], photochromic[14–16], mechanochromic[17–19], solvochromic[20–23] and electrochromic[24,25], have been developed. These smart materials respond to external stimuli with reversible changes to their chemical constitutions or superstructures in the solid state, causing them to emit different luminescent colours. Coding these fluorescent colours in one or two dimensions –a practice which is similar to digital coding (for example, barcodes and QR codes) in computer science– has been suggested[26] as a potential strategy to prevent tampering or counterfeiting. Current stimuli-responsive photoluminescent materials, however, can only provide a small matrix of colours as optical codes. Developing wide-spectrum tunable photoluminescent solid-state materials with multiple fluorescent emissions (a large matrix) remains a major challenge.

Herein, we report the unexpected discovery, during synthesis by a cooperative capture strategy[27–29], of a hetero[4]rotaxane **R4•4Cl**, which contains pyrene stoppers and a diazaperopyrenium unit, derived from **1•Cl** and **2•2Cl**, respectively. Its fluorescent emission in the solid state can be fine-tuned rapidly and reversibly over a wide ($\sim 100$ nm) range of wavelengths as a result of stimuli-responsive aggregation and de-aggregation processes that are governed by a network of supramolecular equilibria. The unique features of this heterorotaxane include (i) the widely tunable colour of its fluorescent emission, providing a large analogue matrix of optical outputs, (ii) variations in fluorescent emission colour that occur when the heterorotaxane is deposited on different types of paper and (iii) the non-linear variation in its colour profile in response to simple chemical stimuli. These properties have allowed us to demonstrate a system of chemical encryption, in which the heterorotaxane is employed as a fluorescent security ink and its complex supramolecular equilibria serve as an encryption algorithm. By encrypting graphical information using these inks, printed images are nigh impossible to mimic, counterfeit and reverse engineer, yet can be easily verified on application of an appropriate authentication reagent, without revealing the original information.

## Results

**Synthesis of hetero[n]rotaxanes.** The key compound in our investigations is the heterorotaxane **R4•4Cl**, which was isolated (Supplementary Figs 1 – 10) as an unexpected side product during the synthesis (Fig. 1a) of the heterorotaxane **R3•4Cl** (Supplementary Figs 11 – 14) from cucurbit[6]uril (CB6), γ-cyclodextrin (γ-CD) and two fluorescent precursors, one (**1•Cl**) derived from pyrene and the other (**2•2Cl**) from a diazaperopyrenium (DAPP) dication. In common with many fluorophores, DAPP exhibits a high fluorescence quantum yield[30,31] in solution ($\Phi = 53\%$) but not in the solid state ($\Phi = 0\%$) as a result of aggregation-induced quenching. To 'turn on' the fluorescence of DAPP in the solid state, we attempted to de-aggregate[32] DAPP by encapsulating **2•2Cl** with γ-CD. No complexation, however,

between **2•2Cl** and γ-CD was observed (Fig. 1b) in aqueous solution (Supplementary Figs 15 – 17). We anticipated that a more effective strategy would be to fix two bulky CB6 rings at the periphery of the DAPP dye as part of a mechanically interlocked molecule, formed by means of the highly efficient and rapid cooperative capture synthesis[27–29] in aqueous solution. To our surprise, the reaction not only afforded the anticipated heterorotaxane **R3•4Cl** (yield = 80%), but also another heterorotaxane **R4•4Cl** (yield = 9%). The formation of **R4•4Cl**, in which γ-CD encircles the DAPP unit, seems to be at odds with the observation that γ-CD does not bind **2•2Cl**. The unfavourable energetics of threading γ-CD onto the dumbbell of the heterorotaxane are outweighed by the positive contribution from the hydrogen bonding network formed between γ-CD and the neighbouring CB6, an observation which is supported by molecular mechanics simulations (Supplementary Fig. 43), as well as the release of high-energy water molecules[33] from the macrocycles. Increasing the amount of γ-CD to 10 equiv. in the reaction mixture favours the formation of **R4•4Cl**, which was isolated as the major product in 83% yield in 3 h (Supplementary Table 1). Although both heterorotaxanes dissolve to some extent in water, **R3•4Cl**, which lacks a solubilizing γ-CD ring, is poorly soluble and undergoes aggregation. In contrast, **R4•4Cl** exhibits significant water solubility of up to 3 mM at room temperature. [1]H NMR spectroscopy reveals that, while **R4•4Cl** undergoes aggregation (Fig. 1c and Supplementary Figs 18 – 19) at room temperature, it experiences de-aggregation to its monomeric form (Fig. 1d) on heating to 80 °C. 2D-NOESY and variable temperature NMR experiments indicate (Supplementary Figs 20 – 21) that the γ-CD ring in **R4•4Cl** has a fixed orientation and position, and does not shuttle along the dumbbell rapidly on the [1]H NMR timescale.

**Photophysical studies.** The UV/Vis absorption spectrum (Fig. 2a) of **R4•4Cl** has two characteristic absorption bands at 341 and 443 nm, which can be attributed to electronic transitions in the pyrenyl and DAPP units, respectively. Despite the presence of the CB6 and γ-CD rings, **R4•4Cl** forms aggregates ($\mathbf{R4^{4+}}_{agg}$) in water, as confirmed (Supplementary Fig. 22) by dynamic light scattering experiments. A blue shift (7 nm) of the absorption band near 450 nm is recorded (Fig. 2b) in the concentration-dependent UV/Vis absorption spectrum of **R4•4Cl** on dilution from 500 to 25 µM, while the shoulder evident at around 350 nm, arising from the pyrene stopper, diminishes. The isosbestic points observed (Fig. 2b) at 431 and 492 nm suggest[34,35] that the aggregation is homogenous and non-cooperative[36–38]. By fitting the data to a dimerisation model[39,40], the aggregation constant $K_{agg}$ was determined (Supplementary Figs 23 – 25) to be $1.4 \times 10^4 \mathrm{M}^{-1}$, indicating strong interactions between **R4•4Cl** monomers.

Irradiating a dilute solution (5 µM) of **R4•4Cl** at excitation wavelengths of either 340 or 443 nm results in identical fluorescence emission spectra (Supplementary Fig. 26), with an emission maximum ($\Phi = 52.4\%$, Supplementary Table 2 and Supplementary Figs 27 – 44) at 510 nm. No emission is observed at 390 nm, indicating (Fig. 2a and Supplementary Fig. S41) the transfer of the excited state energy from pyrene to DAPP by a Förster resonance energy transfer (FRET) mechanism with near-quantitative (> 99%) efficiency. This remarkable FRET efficiency can be rationalised by considering the geometric constraints enforced by the rings that impart (i) a close-to-ideal spatial separation ($\sim 1.2$ Å, calculated by molecular mechanics, Supplementary Figs 45 – 47) between the FRET donors (pyrene) and acceptor (DAPP) in **R4•4Cl**, while (ii) limiting conformational flexibility and (iii) preventing aggregation-induced quenching. These observations suggest that cooperative capture strategies

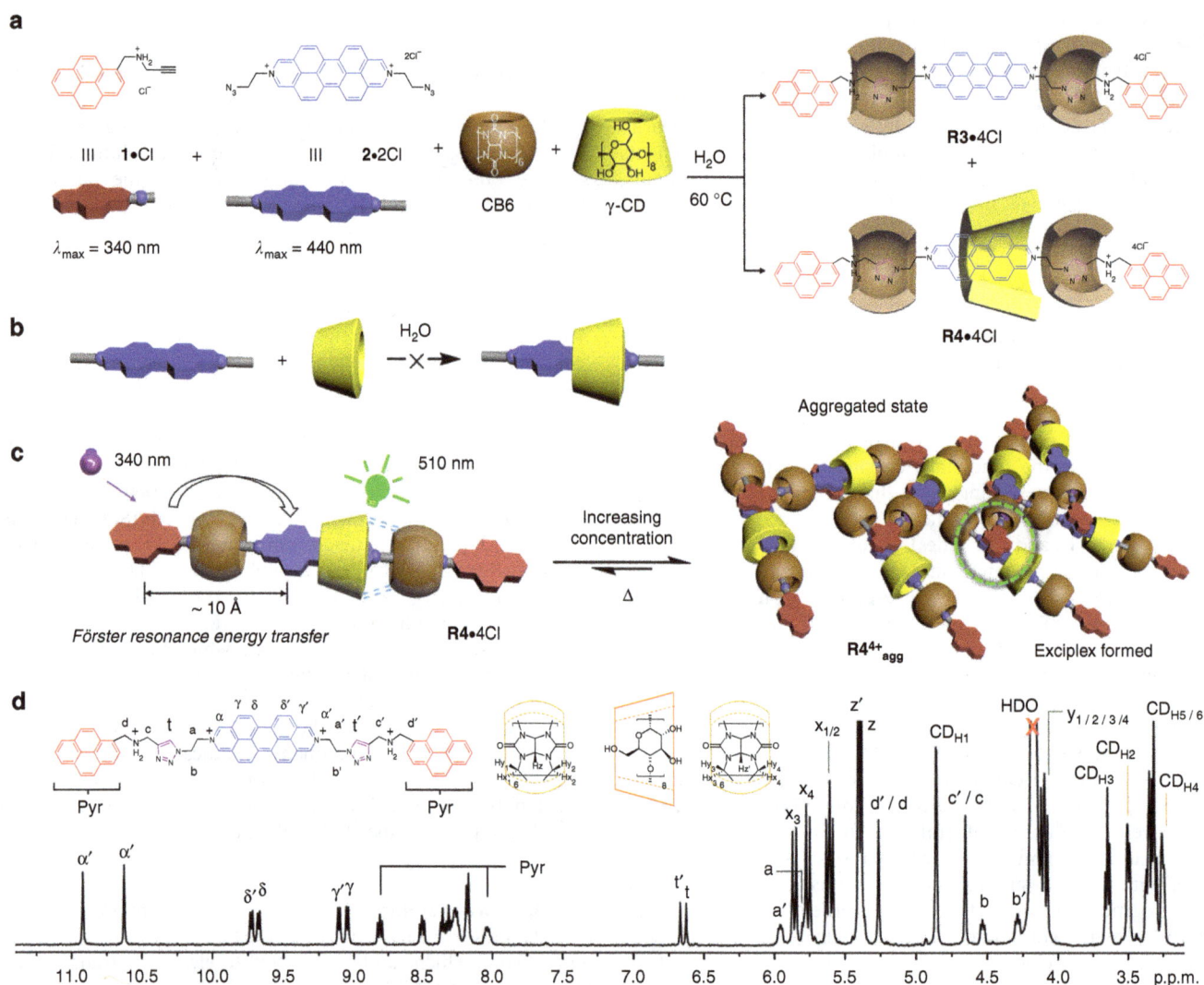

**Figure 1 | Synthesis and characterization of heterorotaxanes.** (**a**) Synthesis of the heterorotaxanes **R3•4Cl** and **R4•4Cl** from the stopper **1•Cl**, the dumbbell precursor **2•2Cl**, CB6 and γ-CD. (**b**) No complexation was observed between **2•2Cl** and γ-CD. (**c**) Graphical representation of the aggregation of **R4**$^{4+}$ monomers in response to changes in concentration or temperature. (**d**) $^1$H NMR spectrum (600 MHz) of **R4•4Cl** (1 mM) recorded in D$_2$O at 80 °C.

have the potential to control the distances between fluorophores for applications such as bio-sensing[41,42].

As **R4•4Cl** undergoes increased aggregation at higher and higher concentrations, its narrow emission band (Fig. 2c) at 510 nm is gradually replaced by a broad, featureless band around 610 nm, implying that either excimers (DAPP homodimers) or exciplexes (pyrenyl–DAPP heterodimers) are being formed in the excited state. Circular dichroism spectra reveal (Fig. 2d) that the aggregation of **R4•4Cl** (200 μM) is temperature dependent. As the temperature is lowered from 80 °C (monomeric state) to 2 °C (aggregated state), the positive induced circular dichroism (ICD) signals diminish, as a negative ICD peak[43–45] attributable to the pyrene stoppers appears at around 350 nm, indicating that these stoppers congregate near the rims of the γ-CD ring. As γ-CD encircles DAPP, it follows that pyrene–DAPP heterodimers (Figs 1c and 2d), and their corresponding exciplexes, are responsible for the observed aggregation and emission behaviour of **R4•4Cl**.

**Dynamic supramolecular equilibria of the hetero[4]rotaxane.** In aqueous solution, the disassembly (Fig. 3a) of aggregated heterorotaxanes (**R4**$^{4+}$$_{agg}$) is promoted by introducing γ-CD,

which encircles the pyrene moieties of **R4•4Cl** and prevents aggregation. The stepwise encapsulation process, which occurs via the formation of a **R4**$^{4+}$ ⊂ CD intermediate, favours a monomeric **R4**$^{4+}$ ⊂ CD$_2$ complex in the presence of excess of γ-CD, with an averaged equilibrium constant $K_{CD} = (K_1 \cdot K_2)^{1/2}$. As a result of the complex equilibria in solution, it is not easy to measure $K_{CD}$ directly. To obtain a good estimate of the binding affinities between the pyrene moieties of **R4•4Cl** and γ-CD, a reference heterorotaxane **SR4•4Cl** was synthesised and the averaged binding affinity ($0.9 \times 10^4$ M$^{-1}$) between its pyrene moiety and γ-CD was measured (Supplementary Figs 48 – 55) in D$_2$O. This encapsulation process can be reversed (Fig. 3a) by introducing a competitive binding agent (**CBA**), which competes for γ-CD in solution with an association constant, $K_{CBA}$. As the aggregation constant $K_{agg}$ and the encapsulation constant $K_{CD}$ are of the same order of magnitude, addition of even a weak **CBA**, for example, 2-adamantylamine hydrochloride (Ad•Cl, $K_{CBA} =$ 90 M$^{-1}$, Supplementary Figs 56 – 59), will efficiently perturb the equilibria between **R4**$^{4+}$$_{agg}$, **R4**$^{4+}$ ⊂ CD$_2$ and **R4**$^{4+}$.

The dynamic nature of **R4**$^{4+}$ in aqueous solution affords us the opportunity to customize an emission profile that remains preserved in the solid state after removal of the solvent. The fluorescent emission spectrum (Fig. 3b and Supplementary

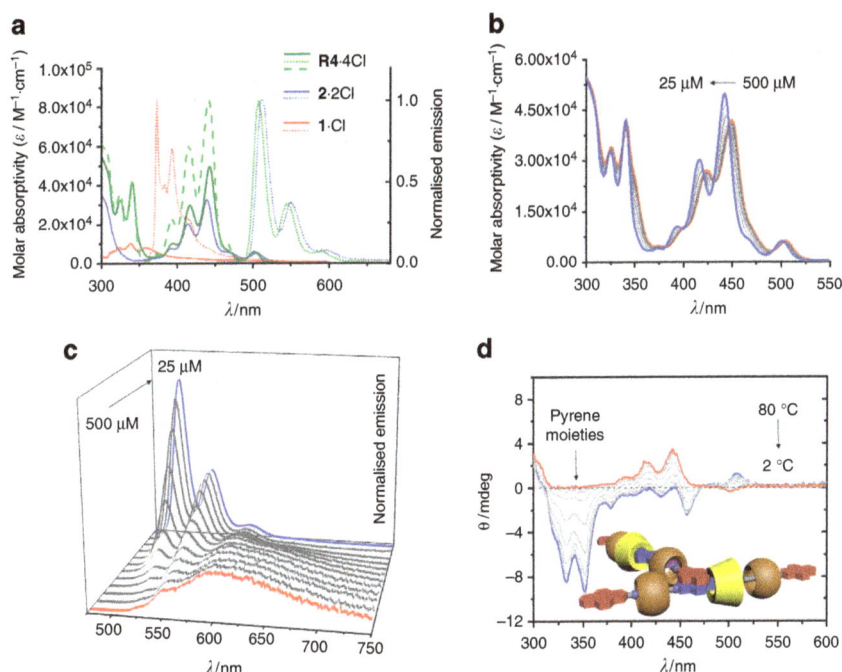

**Figure 2 | Photophysical studies of R4•4Cl.** (**a**) UV/Vis absorption (solid lines) and normalised fluorescence spectra (excitation: dashed lines, emission: dotted lines) of aqueous solutions of **R4•**4Cl (green), stopper **1•**Cl (red) and dumbbell precursor **2•**2Cl (blue). (**b**) Concentration-dependent (25–500 μM) UV/Vis absorption spectra of **R4•**4Cl at 25 °C in water. (**c**) Normalised concentration-dependent (25–500 μM) fluorescence emission spectra ($\lambda_{\text{excitation}} = 341$ nm) of **R4•**4Cl at 25 °C in water. (**d**) Temperature-dependent (2–80 °C) ICD spectra (200 μM) of **R4•**4Cl in water.

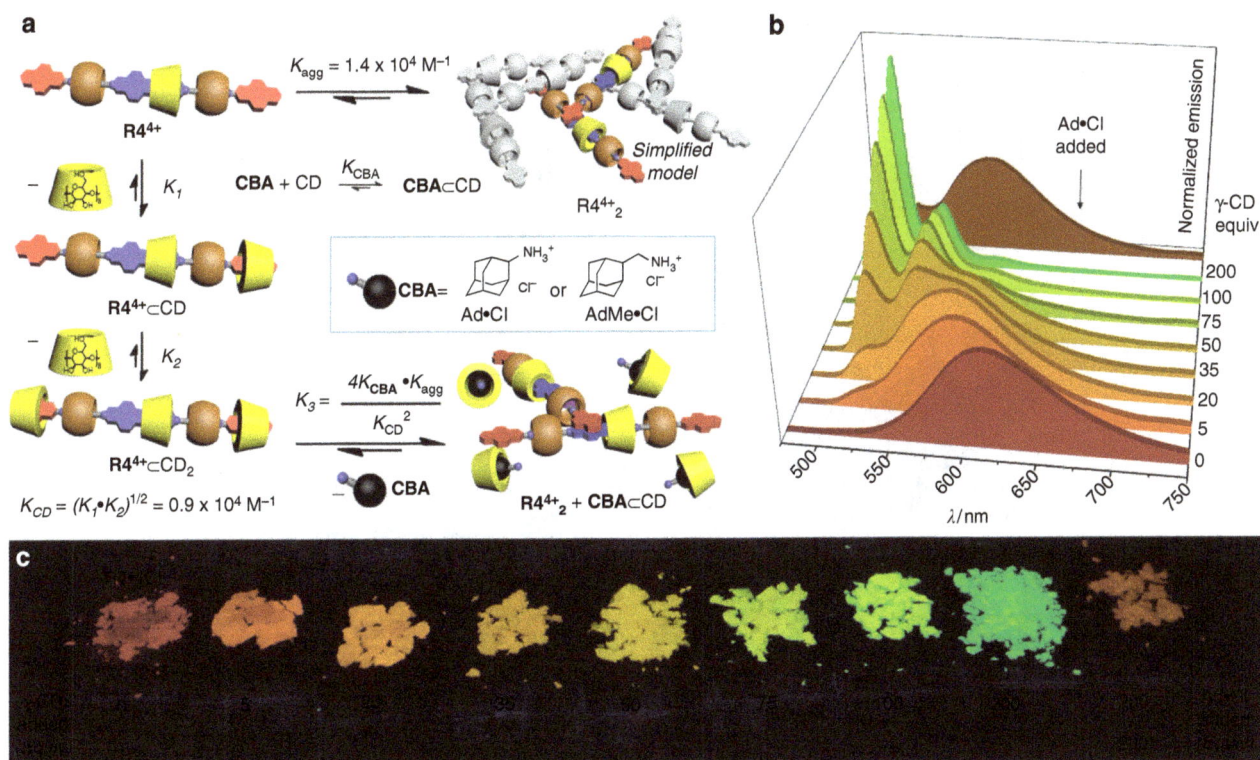

**Figure 3 | Equilibrium network and solid-state fluorescence studies.** (**a**) Graphical representation of the equilibria involving **R4⁴⁺** as its Cl⁻ salt in the presence of γ-CD and **CBA**s. (**b**) Solid-state fluorescence spectra ($\lambda_{\text{excitation}} = 347$ nm) of **R4•**4Cl on adding 0–200 equiv. of γ-CD, followed by 200 equiv. of Ad•Cl. (**c**) Powders obtained from homogeneous mixtures of **R4•**4Cl and varying amounts (0–200 equiv) of γ-CD and Ad•Cl (200 equiv) under UV light.

Figs 60–64) of the amorphous $\mathbf{R4^{4+}}_{\mathbf{agg}}$ ($\lambda_{\text{max}} = 610$ nm, $\Phi = 7.7\%$) is very similar to its emission (Fig. 2c) in a concentrated aqueous solution (Supplementary Figs 65–66).

On the addition of γ-CD, the solid-state fluorescent emission spectra become gradually blue-shifted to 510 nm ($\Phi = 42.5\%$), with the emission colour (Fig. 3c) changing from red to green.

Solid-state emission is also conserved from solution in the presence of a **CBA**. For example, the addition of 200 equiv. of Ad•Cl to a mixture comprising **R4**•4Cl:γ-CD (molar ratio: 1:200) results in a red-shift (Fig. 3b,c) of the emission back to $\lambda_{max} = 580$ nm. Thus, by changing the ratio of **R4**•4Cl, γ-CD and Ad•Cl, the solid-state fluorescence of the material can be tuned reversibly over a wide range of colours from green through to red. This stimulus-responsive tuning of fluorescence spectra in the solid state, over such a wide range of colours and under ambient conditions, is unique and holds promise for applications in security printing technology.

**Applications as fluorescent inks**. By loading $R4^{4+}$-based aqueous solutions (inks) into fountain pens, information can be written (Fig. 4), which is then revealed under UV light. By applying colourless γ-CD and Ad•Cl inks on top of the fluorescent **R4**•4Cl ink (Fig. 4a), additional information can be added ($R4^{4+}_{agg} \rightarrow R4^{4+} \subset CD_2$) or erased ($R4^{4+} \subset CD_2 \rightarrow R4^{4+}_{agg}$) on pre-existing images that is only noticeable (Supplementary Fig. 67 and Supplementary Movie 1) under UV light. During handwriting experiments, we found that the $R4^{4+} \subset CD_2$ ink exhibits an unusual phenomenon, that is, the colour of its emission depends (Fig. 4b) on the type of paper (Supplementary Movie 2). For example, on rag paper, newsprint and banknotes, **R4**•4Cl appears reddish-orange and $R4^{4+} \subset CD_2$ appears green (Fig. 4b) under UV light, which is consistent (Fig. 3c) with the corresponding powders. On different kinds of ordinary white paper (Fig. 4b, Supplementary Figs 68 – 70 and Supplementary Table 3), however, both of

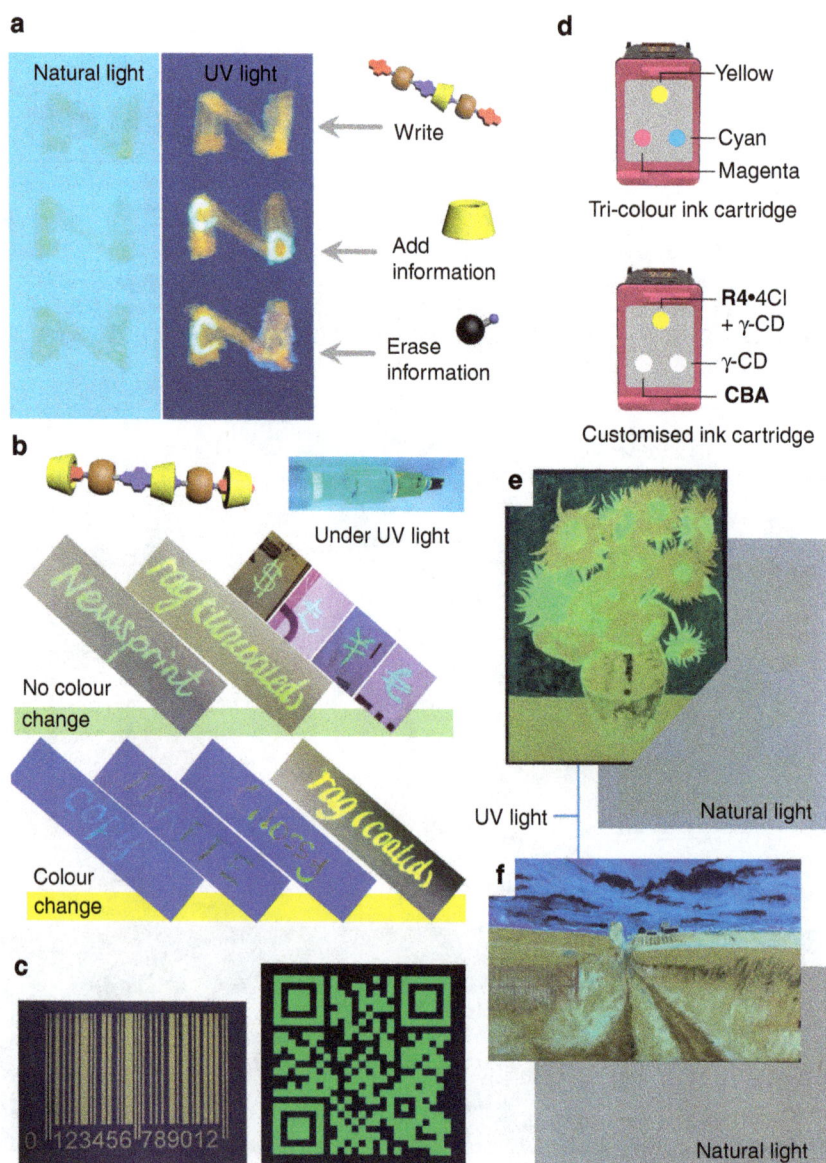

**Figure 4 | Security features of the heterorotaxane $R4^{4+}$- and its complex $R4^{4+} \subset CD_2$-based fluorescent inks. (a)** Reversibly adding and erasing information on the fluorescent ink with γ-CD and Ad•Cl aqueous solution. **(b)** Surface-dependent fluorescence of $R4^{4+} \subset CD_2$ ink on different paper media (newsprint, coated and uncoated rag paper, banknotes, copy, matte and glossy white paper) under UV light. **(c)** A UV barcode and a QR code under UV light printed using a customized black inkjet cartridge filled with $R4^{4+}$ and $R4^{4+} \subset CD_2$ ink, respectively. **(d)** Graphical representations of a customized tri-colour inkjet cartridge, in which aqueous solutions of **R4**•4Cl/γ-CD (**R4**•4Cl: 1 mM, γ-CD: 200 mM), a **CBA** and γ-CD occupy the yellow, magenta and cyan colour channels, respectively. **(e)** Fluorescent replica of van Gogh's 'Sunflowers' on rag paper printed using the customized tri-colour inkjet cartridge under UV and natural light. **(f)** Fluorescent image printed using an inkjet cartridge under UV and natural light, in which the cyan channel was loaded with γ-CD and PyMe•Cl.

these inks appear reddish-orange. The change of colour is most likely a result of noncovalent bonding interactions with papers of different compositions.

$R4^{4+}$-based inks are also compatible with inkjet printing technology. A monochromic barcode and a QR code printed (Fig. 4c) on paper from an inkjet cartridge contains information that, although invisible under natural light, can be read (Supplementary Figs 71 – 72 and Supplementary Movie 3) on a smartphone under UV light. The supramolecular encapsulation/competition between $R4^{4+}$, $\gamma$-CD and Ad•Cl is established rapidly (milliseconds) before the inks dry during the printing process, making it possible to print polychromic fluorescent images. By loading aqueous solutions of $R4^{4+} \subset CD_2$ ($R4^{4+}$: $\gamma$-CD = 1:50), $\gamma$-CD and Ad•Cl into a tri-colour inkjet cartridge (Fig. 4d and Supplementary Fig. 73), we have printed a fluorescent reproduction (Fig. 4e) of van Gogh's 'Sunflowers' with good colour resolution. The colour range of the fluorescent inks can be expanded to accommodate RGB printing (Fig. 4f and Supplementary Figs 74-76) by choosing a fluorescent **CBA** with blue emission, such as the terminal fragment of **R4•4Cl**, 1-pyrenemethylamine hydrochloride (PyMe•Cl). It is worth noting that, by reducing the amount of **R4•4Cl** applied to the paper, the images produced are invisible to the naked eye (Fig. 4f, right) under natural light.

**Supramolecular encryption and authentication theory**. At a fundamental level, the $R4^{4+}$-based fluorescent inks provide an extensive fluorescent colour matrix for encryption coding. More importantly, the nonlinear dependence of this system's output on the concentrations of components and their equilibrium constants (Fig. 3a) points towards a general concept whereby complex supramolecular equilibria in aqueous solutions can be used as a chemical encryption method (Fig. 5). In principle, the colour of a dot printed by the customized tri-colour ink cartridge reflects a complex supramolecular equilibrium in the solution state, which can be simplified as

$$3R4^{4+} + 3CD + CBA \xrightleftharpoons{4K_{agg}\bullet K_{CD}^2\bullet K_{CBA}} R4_2^{4+} + R4^{4+} \subset CD_2$$
$$+ CBA \subset CD \quad (1)$$

Since $K_{agg}$ and $K_{CD}$ are fixed, three key parameters control the supramolecular equilibria and the subsequent fluorescent colour (Fig. 5 and Supplementary Fig. 77) under UV light for a given dot after printing: (1) $[R4^{4+}]_0$, reflecting the absolute amount of

$R4^{4+}$ ink applied on paper, (2) $[CD]_0$ and $[CBA]_0$, reflecting the ratio of $R4^{4+}_{agg}$ and $R4^{4+} \subset CD_2$ on paper and (3) the chemical composition of **CBA**, reflecting a different $K_{CBA}$ in the supramolecular equilibrium. These variables constitute the encryption settings, which can be defined by the user in charge of security printing. By simply varying (i) the sequence (Fig. 5) of inks in Channels ①, ②, ③, (ii) the chemical composition of the **CBA** and (iii) the concentration of the inks, it is possible to generate a large number of fluorescent colour combinations. It is also worth noting that, a combination of different **CBA**s could be loaded in channels ② and ③ simultaneously, thus introducing even more variables to the supramolecular equilibria. In this manner, it would be challenging for counterfeiters to reproduce a printed colour palette even if they had access to the **R4•4Cl** ink, as they would also require a complete knowledge of (i) the **CBA**, of which there could be a large number of possibilities, as well as (ii) the paper media, (iii) channel assignments and (iv) initial ink concentrations. Even relatively small errors in initial concentrations can lead to obvious differences in the colour palette, by a margin which cannot be easily reverse engineered on account of the non-linearity of the equilibrium equations,

$$\begin{aligned} I_\lambda = {} & F^\circ_{R4^{4+}} + F^\circ_{R4^{4+} \subset CD}\bullet K_{CD}([CD]_0 - 3m) \\ & + F^\circ_{R4^{4+} \subset CD_2}\bullet K_{CD}^2\bullet([CD]_0 - 3m)^2\bullet([R4^{4+}]_0 - 3m) \\ & + F^\circ_{R4_2^{4+}}\bullet 4K_{agg}\bullet([R4^{4+}]_0 - 3m)^2 \\ & + F^\circ_{CBA}\bullet([CBA]_0 - m) + F^\circ_{CBA \subset CD}\bullet m \end{aligned}$$
$$(2)$$

where the fluorescence intensity $I_\lambda$ at a given wavelength $\lambda$ is the sum of the emission intensities of each component in an aqueous solution containing **R4•4Cl**, $\gamma$-CD and **CBA**. $F^\circ$ is the molar fluorescence coefficient at wavelength $\lambda$, and $m$ is the molar concentration of **CBA** being encapsulated by $\gamma$-CD. See Supplementary Discussion for detailed derivations.

The dynamic nature of the inks also makes them amenable to a variety of fraud detection mechanisms. In principle, applying a layer of authentication agent(s) can re-establish the supramolecular equilibria in solution and shift the colour outputs (Fig. 5) on paper non-linearly, as described by equations (1) and (2). As the colour-changing process is dynamic and depends on the amount of the authentication agent(s) that has been applied, it is close to impossible that the dynamic colour-changing process could be precisely mimicked. Apart from applying authentication agent(s),

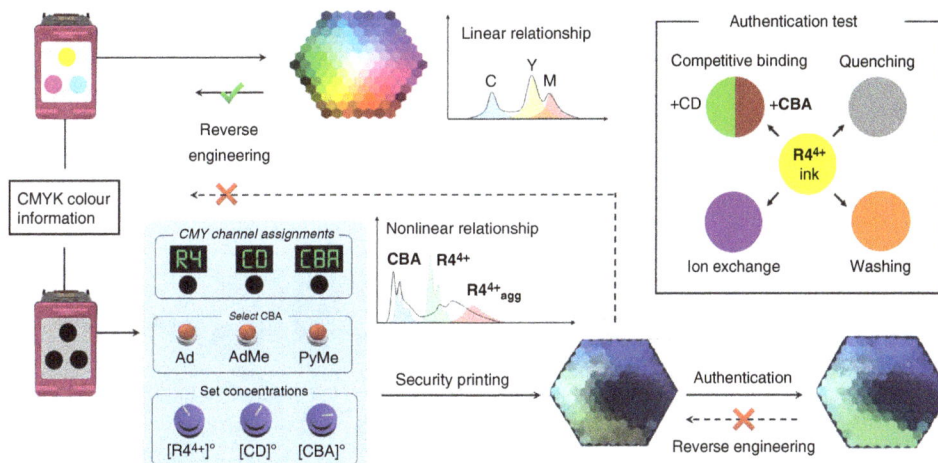

**Figure 5 | Supramolecular encryption and fraud detection using the heterorotaxane-based fluorescent security inks.** A comparison between conventional cyan-magenta-yellow-black (CMYK) printing and supramolecular encrypted printing technology. Inset: possible mechanisms to verify the authenticity of the protected colour document.

which could induce a supramolecular equilibrium shift, mechanisms such as counterion exchange and fluorescence quenching can also be utilized (Fig. 5 inset) to change emission colours in real time.

**Demonstration of the supramolecular encryption theory.** In an attempt to demonstrate the non-linear nature of this system, we have printed fluorescent colour palettes (Fig. 6a – f) using various ink concentrations and different **CBA**s. A broad range of colours from green to red can be printed (Fig. 6b) when Ad•Cl is used as the competitor. The colour palette is sensitive to the association strength of the competitor as predicted and evidenced by the differences between images printed from equally concentrated solutions of Ad•Cl (Fig. 6b) and a stronger-binding competitor (Fig. 6c), namely, 1-adamantanemethylamine hydrochloride (AdMe•Cl, $K_{AdMe} = 127 \, M^{-1}$, Supplementary Figs 58 – 59). Reducing the concentrations of either **CBA** (Fig. 6c,d) or γ-CD (Fig. 6c,e) redistributes the colour spectrum in the yellowish-red region or greenish yellow region, respectively. The colour spectrum is expanded (Fig. 6f) by choosing a blue fluorescent **CBA**, for example, PyMe•Cl. The ability to exchange fluorescent and non-fluorescent **CBA**s in a modular and user-controlled manner elevates the anti-counterfeiting features possessed intrinsically by this new security ink.

**Demonstration of authentication methods.** Authentication mechanisms have also been demonstrated. For example, exposure of an encrypted image (Fig. 6g, centre) to aqueous solutions of non-fluorescent AdMe•Cl or γ-CD changes the existing colour gradient of the image, while printing fluorescent PyMe•Cl aqueous solution creates new colours, such as blue and purple by shifting the complex equilibria. Printing 1,3,6,8-pyrenetetrasulfonic acid tetrasodium (PTSA•4Na) solution gives rise to new colours (Fig. 6g) as a result of counterion exchange. A characteristic colour change can also be brought about through the application of a quencher, such as tryptophan, or even by simply soaking the printed image in water for as little as 1 – 2 min, during which time γ-CD and the **CBA** are washed away to some extent, thus shifting the equilibria. These authentication agents can discriminate, not only between images produced by the $R4^{4+}$-based ink and other fluorescent dyes, but also between images produced using different $R4^{4+}$/**CBA**/γ-CD ink formulations. In a blind test, blocks of a given fluorescence colour (Fig. 6h, top), which appear almost identical to one another under UV light, but are formulated differently, were found to result in noticeably different colours (Fig. 6h bottom) after the application of the same amount of authentication agents. For further details, see Supplementary Fig. 78. As hundreds of chemicals meet the criteria to be **CBA**s, an extremely large library of different ink systems and authentication tests using this supramolecular encryption method can, in principle, be generated.

## Discussion

In summary, we have developed a stimulus-responsive solid-state fluorescent heterorotaxane, which is easily prepared from simple

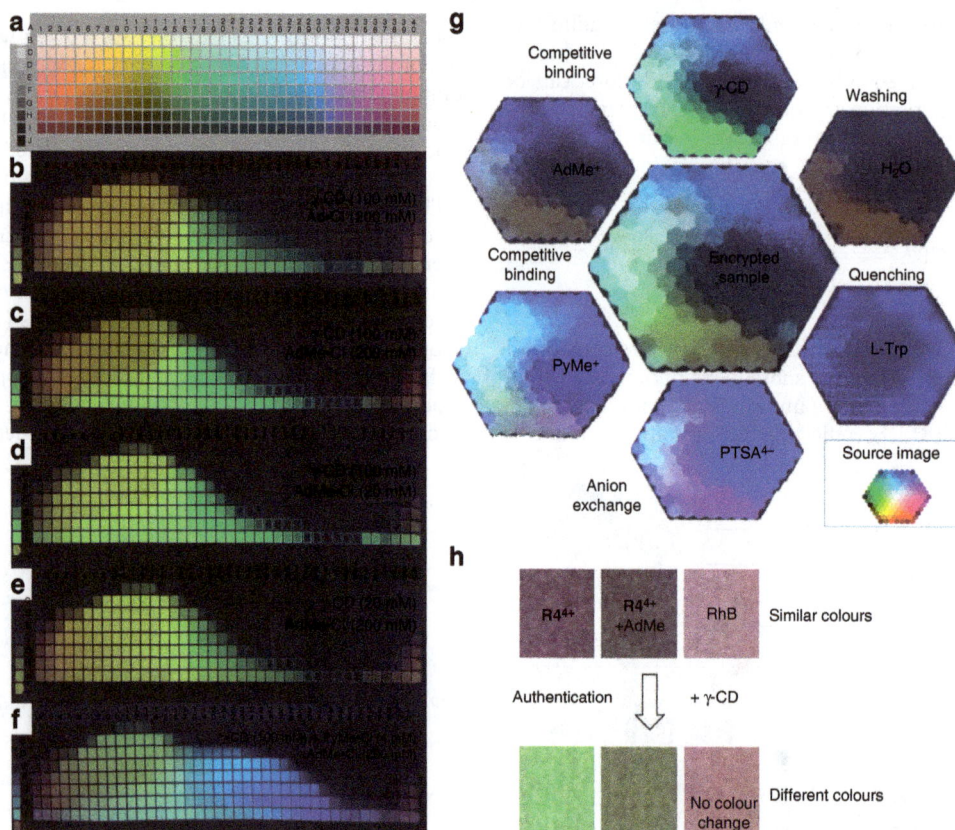

**Figure 6 | Demonstration of the supramolecular encryption and authentication using the heterorotaxane-based fluorescent security inks.**
(**a**) A standard colour palette. (**b-f**) Colour palette images produced using the customized tri-colour inkjet cartridge with (**b**) Ad•Cl (200 mM), (**c**) AdMe•Cl (200 mM) and (**d**) AdMe•Cl (20 mM) in channel ③, (**e**) γ-CD (20 mM) and (**f**) γ-CD (100 mM) with PyMe•Cl (4 mM) in channel ②, respectively. **R4**•4Cl + γ-CD (**R4**•4Cl: 1 mM, γ-CD: 40 mM) solution was loaded in channel ① in the tri-colour inkjet cartridge. (**g**) Encrypted polychromic colour palette samples produced by the customized inkjet cartridge (centre) and its derivatives (around the periphery, after printing a layer of authentication reagents) under UV light. (**h**) Similar colours produced by **R4**$^{4+}$-based security inks have composition-dependent response after chemical authentication. No distinguishable colour change is observed after chemical authentication when rhodamine B (RhB) is applied as the fluorescent ink.

starting materials in high yield by a cooperative capture method. We have applied it as a component of fluorescent security inks with built-in supramolecular encryption. The inks are well-placed for assimilation into a commercial setting on account of the simple and high-yielding synthesis of the heterorotaxane from commodity chemicals. The solid-state emission of these security inks can be fine-tuned over a wide emission range ($\sim 100$ nm) with rapid response (milliseconds) to chemical stimuli. Printed information is encrypted in a chemical language based on a nonlinear equation that describes the dynamic equilibrium network. A potentially enormous library of different fluorescent colour combinations can be generated. In contrast with conventional dyes, the encrypted information printed using the heterorotaxane inks can be verified by chemical authentication methods without revealing the original colour image information. The interplay of fluorescence output with dynamic supramolecular equilibria that we observed quite fortuitously could be a general phenomenon that is not exclusive to the heterorotaxane, or even to mechanically interlocked molecules. It opens up a new way to encrypt and protect information in a manner that is far from easy to mimic.

## Methods

### Synthesis and characterization of the heterorotaxane R4•4Cl.
Stopper precursor 1•Cl (67 mg, 0.22 mmol), dumbbell precursor 2•2Cl (54 mg, 0.10 mmol) and γ-CD (1287 mg, 1.00 mmol) were mixed in $H_2O$ (35 ml) and stirred at 60 °C for 10 min before CB6 (250 mg, 0.25 mmol) was added. The reaction mixture was stirred at 60 °C for 3 h. Insoluble residues were filtered off from the reaction mixture. The filtrate was loaded directly onto a reverse phase C18 column (150 gram, RediSep Rf Gold C18Aq) on an automatic column chromatographic system (Combiflash Rf200, Teledyne Isco) and chromatographed in $H_2O$/MeCN/ 0.1% TFA with a gradient from 0 to 60% MeCN over 40 min at a flow rate of 85 ml min$^{-1}$. Fractions containing $R4^{4+}$ were collected and the counterions of $R4^{4+}$ were exchanged to $PF_6^-$ on addition of an excess of aqueous $NH_4PF_6$. The product R4•4PF$_6$, which precipitates on removal of MeCN from these fractions under reduced pressure, was collected by vacuum filtration and washed extensively with $H_2O$. The hetero[4]rotaxane R4•4Cl was obtained after a second counterion exchange by precipitation from an MeCN solution of R4•4PF$_6$ with an excess of tetrabutylammonium chloride. The yellow precipitate was collected by vacuum filtration, washed with excess MeCN and dried under vacuum to afford R4•4Cl (371 mg, 83%) as an orange powder. $^1$H NMR (600 MHz, $D_2O$, 353 K): δ = 10.92 (s, 2H), 10.63 (s, 2H), 9.73 (d, $J$ = 9.5 Hz, 2H), 9.68 (d, $J$ = 9.4 Hz, 2H), 9.11 (d, $J$ = 9.3 Hz, 2H), 9.05 (d, $J$ = 9.3 Hz, 2H), 8.81 (dd, $J$ = 12.4, 9.3 Hz, 2H), 8.50 (dd, $J$ = 11.5, 7.9 Hz, 2H), 8.40–8.22 (m, 8H), 8.18 (d, $J$ = 8.9 Hz, 4H), 8.08–7.98 (m, 2H), 6.67 (s, 1H), 6.63 (s, 1H), 5.96 (t, $J$ = 8.8 Hz, 2H), 5.86 (d, $J$ = 15.3 Hz, 6H), 5.80 (m, 2H), 5.76 (d, $J$ = 15.5 Hz, 6H), 5.62 (d, $J$ = 15.5, 6H), 5.60 (d, $J$ = 15.5, 6H), 5.40 (s, 12H), 5.39 (s, 12H), 5.26 (s, 4H), 4.86 (t, $J$ = 3.8 Hz, 8H), 4.66 (s, 4H), 4.53 (t, $J$ = 9.0 Hz, 2H), 4.29 (t, $J$ = 8.9 Hz, 2H), 4.10 (m, 24H), 3.65 (t, $J$ = 9.5 Hz, 8H), 3.50 (dd, $J$ = 10.0, 3.8 Hz, 8H), 3.41–3.20 (m, 32H). HR-ESI-MS: calcd for $[M - 4Cl]^{4+}$ $m/z$ = 1,074.8630, found $m/z$ = 1,074.8623; $[M - H - 4Cl]^{3+}$ $m/z$ = 1,432.8149, found $m/z$ = 1,432.8123; $[M - 3Cl]^{3+}$ $m/z$ = 1,445.1406, found $m/z$ = 1,445.1369.

Detailed synthesis and characterisation of stopper and rod precursors, rotaxanes R3•4Cl and SR4•4Cl are available in Supplementary Methods.

### Photophysical studies of the heterorotaxane R4•4Cl.
The UV/Vis spectra of the sample solutions were measured on a Shimadzu UV/Vis/NIR spectrometer (UV 3600 model) with a cell temperature controller. Quartz cuvettes with 1 or 10 mm pathway were used to record the UV/Vis spectra. The fluorescence excitation and emission spectra of the sample solutions were recorded on a HORIBA fluorometer (fluoroMax-4 model). Circular dichroism spectra were recorded on a JASCO circular dichroism spectrophotometer (J-815 model) with a temperature controller. Solid-state UV/Vis spectra were recorded on a Perkin Elmer UV/Vis/NIR spectrometer (LAMBDA 1050 model) equipped with an integrating sphere. Solid-state fluorescence spectra were recorded on an ISS fluorometer (PC1 model) equipped with a variable-angle, front surface sample compartment.

### Ink writing tests.
Four types of inks for pen writing were prepared using R4•4Cl (0.5 mM), Ad•Cl (100 mM), γ-CD (100 mM) and R4 ⊂ CD$_2$ (R4•4Cl = 0.5 mM, γ-CD = 100 mM) solutions, respectively. Typically, 0.5 ml of the ink was loaded in a fountain pen for writing tests. A wide selection of paper-based printing media has been tested, including copy papers (various brands and models), matte presentation paper (HP), glossy presentation paper (HP), resume paper (25 and 100%

cotton), newsprint paper, rag paper (100% cotton, without optical brightener) and cigarette rolling paper. Banknote identification tests were performed on genuine banknotes of US dollars, British pounds sterling, Euros, Chinese Yuan and Japanese Yen. In these tests, the corresponding currency symbols (\$, £, € and ¥) were drawn on the testing banknotes using the fountain pen filled with R4 ⊂ CD$_2$ ink.

### Ink printing tests.
Printing tests were performed on an HP inkjet printer (Photosmart CP4780 model) and an HP colour laser printer (CP1025nw model) with customized ink cartridges and original toners, respectively. Rag paper (100% cotton, without optical brightener, no surface coating side) was chosen for most printing tests based on the ink writing test results.

Ink cartridges for printing tests were customized from HP black and tri-colour cartridges (HP60 model). The filled inks were removed from the cartridge, which was washed extensively with $H_2O$ and EtOH. Aqueous solutions of R4•4Cl (4 mL, 0.5 mM) and R4 ⊂ CD$_2$ (4 ml, R4•4Cl = 0.5 mM, γ-CD = 100 mM) were loaded in two empty, clean black ink cartridges, respectively, to perform the monochromic printing tests. In the polychromic printing tests, aqueous solutions of Ad•Cl (2 ml, 100 mM), R4/γ-CD (2 ml, R4•4Cl = 0.5 mM, γ-CD = 25 mM) and γ-CD (2 ml, 100 mM) were loaded in the magenta, yellow and cyan channels of the cleaned tri-colour ink cartridge, respectively. Fluorescent colour under UV light was tuned by controlling the proportion of the three inks in the customized tri-colour ink cartridge.

## References
1. Hide, F. et al. Semiconducting polymers: a new class of solid-state laser materials. Science 273, 1833–1836 (1996).
2. Zhu, X. H., Peng, J. B., Caoa, Y. & Roncali, J. Solution-processable single-material molecular emitters for organic light-emitting devices. Chem. Soc. Rev. 40, 3509–3524 (2011).
3. Santra, M. et al. Dramatic substituent effects on the photoluminescence of boron complexes of 2-(benzothiazol-2-yl)phenols. Chem. Eur. J. 18, 9886–9893 (2012).
4. Sasabe, H. et al. 3,3′-Bicarbazole-based host materials for high-efficiency blue phosphorescent OLEDs with extremely low driving voltage. Adv. Mater. 24, 3212–3217 (2012).
5. Kumar, K. et al. Printing colour at the optical diffraction limit. Nat. Nanotechnol. 7, 557–561 (2012).
6. Lu, Y. Q. et al. Tunable lifetime multiplexing using luminescent nanocrystals. Nat. Photonics 8, 33–37 (2014).
7. Deisingh, A. K. Pharmaceutical counterfeiting. Analyst 130, 271–279 (2005).
8. Yoon, B. et al. Recent functional material based approaches to prevent and detect counterfeiting. J. Mater. Chem. C 1, 2388–2403 (2013).
9. Prime, E. L. & Solomon, D. H. Australia's plastic banknotes: fighting counterfeit currency. Angew. Chem. Int. Ed. 49, 3726–3736 (2010).
10. Kishimura, A., Yamashita, T., Yamaguchi, K. & Aida, T. Rewritable phosphorescent paper by the control of competing kinetic and thermodynamic self-assembling events. Nat. Mater. 4, 546–549 (2005).
11. Mutai, T., Satou, H. & Araki, K. Reproducible on-off switching of solid-state luminescence by controlling molecular packing through heat-mode interconversion. Nat. Mater. 4, 685–687 (2005).
12. Perruchas, S. et al. Mechanochromic and thermochromic luminescence of a copper iodide cluster. J. Am. Chem. Soc. 132, 10967–10969 (2010).
13. Yan, D. P. et al. Reversibly thermochromic, fluorescent ultrathin films with a supramolecular architecture. Angew. Chem. Int. Ed. 50, 720–723 (2011).
14. Wu, Y. et al. Quantitative photoswitching in bis(dithiazole)ethene enables modulation of light for encoding optical signals. Angew. Chem. Int. Ed. 53, 2090–2094 (2014).
15. Li, Y. et al. Reversible photochromic system based on rhodamine B salicylaldehyde hydrazone metal complex. J. Am. Chem. Soc. 136, 1643–1649 (2014).
16. Tian, H. & Yang, S. J. Recent progresses on diarylethene based photochromic switches. Chem. Soc. Rev. 33, 85–97 (2004).
17. Dong, Y. et al. Piezochromic luminescence based on the molecular aggregation of 9,10-bis((E)-2-(pyrid-2-yl)vinyl)anthracene. Angew. Chem. Int. Ed. 51, 10782–10785 (2012).
18. Sagara, Y. & Kato, T. Mechanically induced luminescence changes in molecular assemblies. Nat. Chem. 1, 605–610 (2009).
19. Sagara, Y. & Kato, T. Brightly tricolored mechanochromic luminescence from a single-luminophore liquid crystal: reversible writing and erasing of images. Angew. Chem. Int. Ed. 50, 9128–9132 (2011).
20. Dias, H. V. R., Diyabalanage, H. V. K., Rawashdeh-Omary, M. A., Franzman, M. A. & Omary, M. A. Bright phosphorescence of a trinuclear copper(I) complex: Luminescence thermochromism, solvatochromism, and concentration luminochromism. J. Am. Chem. Soc. 125, 12072–12073 (2003).
21. Liu, Y., Wang, K. R., Guo, D. S. & Jiang, B. P. Supramolecular assembly of perylene bisimide with beta-cyclodextrin grafts as a solid-state fluorescence sensor for vapor detection. Adv. Funct. Mater. 19, 2230–2235 (2009).

22. Yoon, S. J. *et al.* Multistimuli two-color luminescence switching via different slip-stacking of highly fluorescent molecular sheets. *J. Am. Chem. Soc.* **132,** 13675–13683 (2010).

23. Ni, J., Zhang, X., Wu, Y. H., Zhang, L. Y. & Chen, Z. N. Vapor- and mechanical-grinding-triggered color and luminescence switches for bis(sigma-fluorophenylacetylide) platinum(II) complexes. *Chem. Eur. J.* **17,** 1171–1183 (2011).

24. Liou, G. S., Hsiao, S. H. & Su, T. H. Synthesis, luminescence and electrochromism of aromatic poly(amine-amide)s with pendent triphenylamine moieties. *J. Mater. Chem.* **15,** 1812–1820 (2005).

25. Sun, H. B. *et al.* Smart responsive phosphorescent materials for data recording and security protection. *Nat. Commun.* **5,** 3601 (2014).

26. Diaz, R., Palleau, E., Poirot, D., Sangeetha, N. M. & Ressier, L. High-throughput fabrication of anticounterfeiting colloid-based photoluminescent microtags using electrical nanoimprint lithography. *Nanotechnology* **25,** 345302 (2014).

27. Ke, C. *et al.* Quantitative emergence of hetero[4]rotaxanes by template-directed click chemistry. *Angew. Chem. Int. Ed.* **52,** 381–387 (2013).

28. Ke, C. *et al.* Pillar[5]arene as a co-factor in templating rotaxane formation. *J. Am. Chem. Soc.* **135,** 17019–17030 (2013).

29. Hou, X. *et al.* Efficient syntheses of pillar[6] arene-based hetero[4]rotaxanes using a cooperative capture strategy. *Chem. Commun.* **50,** 6196–6199 (2014).

30. Slamaschwok, A. *et al.* Interactions of the dimethyldiazaperopyrenium dication with nucleic-acids. 1. Binding to nucleic-acid components and to single-stranded polynucleotides and photocleavage of single-stranded oligonucleotides. *Biochemistry* **28,** 3227–3234 (1989).

31. Slamaschwok, A. *et al.* Interactions of the dimethyldiazaperopyrenium dication with nucleic-acids. 2. Binding to double-stranded polynucleotides. *Biochemistry* **28,** 3234–3242 (1989).

32. Biedermann, F., Elmalem, E., Ghosh, I., Nau, W. M. & Scherman, O. A. Strongly fluorescent, switchable perylene bis(diimide) host-guest complexes with cucurbit[8]uril in water. *Angew. Chem. Int. Ed.* **51,** 7739–7743 (2012).

33. Biedermann, F., Uzunova, V. D., Scherman, O. A., Nau, W. M. & De Simone, A. Release of high-energy water as an essential driving force for the high-affinity binding of Cucurbit[n]urils. *J. Am. Chem. Soc.* **134,** 15318–15323 (2012).

34. Seibt, J. *et al.* On the geometry dependence of molecular dimer spectra with an application to aggregates of perylene bisimide. *Chem. Phys.* **328,** 354–362 (2006).

35. Shao, C. Z., Grune, M., Stolte, M. & Würthner, F. Perylene bisimide dimer aggregates: fundamental insights into self-assembly by NMR and UV/Vis spectroscopy. *Chem. Eur. J.* **18,** 13665–13677 (2012).

36. Whitty, A. Cooperativity and biological complexity. *Nat. Chem. Biol.* **4,** 435–439 (2008).

37. Hunter, C. A. & Anderson, H. L. What is cooperativity? *Angew. Chem. Int. Ed.* **48,** 7488–7499 (2009).

38. Ercolani, G. & Schiaffino, L. Allosteric, chelate, and interannular cooperativity: a mise au point. *Angew. Chem. Int. Ed.* **50,** 1762–1768 (2011).

39. Zhang, X., Rehm, S., Safont-Sempere, M. M. & Würthner, F. Vesicular perylene dye nanocapsules as supramolecular fluorescent pH sensor systems. *Nat. Chem.* **1,** 623–629 (2009).

40. Fennel, F. *et al.* Biphasic self-assembly pathways and size-dependent photophysical properties of perylene bisimide dye aggregates. *J. Am. Chem. Soc.* **135,** 18722–18725 (2013).

41. Medintz, I. L. *et al.* Self-assembled nanoscale biosensors based on quantum dot FRET donors. *Nat. Mater.* **2,** 630–638 (2003).

42. Rizzo, M. A., Springer, G. H., Granada, B. & Piston, D. W. An improved cyan fluorescent protein variant useful for FRET. *Nat. Biotechnol.* **22,** 445–449 (2004).

43. Kajtar, M., Horvathtoro, C., Kuthi, E. & Szejtli, J. A simple rule for predicting circular-dichroism induced in aromatic guests by cyclodextrin hosts in inclusion complexes. *Acta. Chim. Acad. Sci. Hung.* **110,** 327–355 (1982).

44. Kodaka, M. A general rule for circular-dichroism induced by a chiral macrocycle. *J. Am. Chem. Soc.* **115,** 3702–3705 (1993).

45. Allenmark, S. Induced circular dichroism by chiral molecular interaction. *Chirality* **15,** 409–422 (2003).

## Acknowledgements

The authors dedicate this manuscript to Professor Yoshihisa Inoue on the occasion of his retirement. We thank Professor Yoshihisa Inoue from Osaka University and Professor Frank Würthner from Universität Würzburg for their useful suggestions. We acknowledge financial support from Northwestern University (NU). X.H. gratefully acknowledges support from the Ryan Fellowship and the Northwestern University International Institute of Nanotechnology.

## Author contributions

C.K. and X.H. conceived the project, X.H. and C.K. performed the experiments and analysed the data under the direction of J.F.S., C.J.B. and P.R.M. gave suggestions to optimize the system, R.B.P. suggested the potential application as ink, and all authors contributed in the manuscript preparation.

# Discovery and enantiocontrol of axially chiral urazoles via organocatalytic tyrosine click reaction

Ji-Wei Zhang[1], Jin-Hui Xu[1], Dao-Juan Cheng[1], Chuan Shi[1], Xin-Yuan Liu[1] & Bin Tan[1]

Axially chiral compounds play an important role in areas such as asymmetric catalysis. The tyrosine click-like reaction is an efficient approach for synthesis of urazoles with potential applications in pharmaceutical and asymmetric catalysis. Here we discover a class of urazole with axial chirality by restricted rotation around an N–Ar bond. By using bifunctional organocatalyst, we successfully develop an organocatalytic asymmetric tyrosine click-like reaction in high yields with excellent enantioselectivity under mild reaction conditions. The excellent remote enantiocontrol of the strategy originates from the efficient discrimination of the two reactive sites in the triazoledione and transferring the stereochemical information of the catalyst into the axial chirality of urazoles at the remote position far from the reactive site.

---

[1] Department of Chemistry, South University of Science and Technology of China, Shenzhen 518055, China. Correspondence and requests for materials should be addressed to X.-Y.L. (email: liuxy3@sustc.edu.cn) or to B.T. (email: tanb@sustc.edu.cn).

Urazoles are important heterocyclic compounds with potential pharmaceutical applications and valuable utilities in the area of protein modification chemistry due to the simplicity of chemical synthesis and ease of optimization of reaction conditions[1-4]. In addition, oxidation of urazoles gives rise to a very useful class of persistent cyclic hydrazyl radicals for versatile transformations[5,6]. Consequently, there is a large demand for easy access to a broad variety of these compounds. In this regard, the tyrosine click reaction provides a straightforward strategy to access such compounds under mild conditions as illustrated in Fig. 1a, in which a class of cyclic diazodicarboxamides (triazodiones) reacted selectively and rapidly with the phenol side chain of tyrosine as first developed by the Barbas group for the application in bioconjugate chemistry[7,8]. Although the development of other methodologies towards the synthesis of these compounds has also been reported[9-11], to the best of our knowledge, there is no any report involving the direct construction of chiral urazoles in a catalytic enantioselective manner. Inspired by a developing research field on atropisomeric compounds possessing an N–Ar chiral axis[12], we envisioned that urazoles directly obtained from tyrosine click-like reaction could be recognized as a type of axially chiral skeleton containing an N–Ar chiral axis because of the presence of two N–Ar bonds in arylurazoles.

After discovery of the axially chiral urazoles (Fig. 1b, compound **D**), we turned our attention to construct the chiral urazoles in an atroposelective approach via tyrosine click-like reaction. In this scenario, three major challenges would be encountered: (1) the selection of suitable catalyst to interact with the substrates in high efficiency to inhibit the very strong background reaction; (2) the choice of an appropriate chiral catalyst prompt to efficiently induce remote axial enantiocontrol at the distant position via organocatalytic desymmetrization strategy[13-19]; (3) the use of mild reaction conditions to circumvent the axial rotation. Recently, some strategies have been successfully developed for the organocatalytic synthesis of axially chiral compounds[20-33]. Although the task of controlling the remote axial chirality under the current reaction system is a formidable challenge, the success of the above results provides strong evidence that organocatalysis can be performed in the control of axial chirality by using rationally designed substrate or catalyst. It is well known that bifunctional organocatalysts have made a great contribution to the field of asymmetric catalysis[34-37]. In such catalysts, the acidic and basic centres acting as both hydrogen-bonding donors and acceptors, respectively, thus activating the nucleophile and electrophile at the same time in an appropriate spatial configuration. As shown in Fig. 1c, we speculated that the utility of bifunctional organocatalysts could be expected by distinguishing the two nonequivalent reactive nitrogen centres (*a* and *b*) in the triazoledione and transferring the central chirality of the catalyst into the axial chirality far from the reaction site. As part of our continued interest in the area of synthesis of axially chiral compounds[38] and asymmetric catalysis[39], herein, we would like to exhibit the remote control of the axial chirality of arylurazoles by using a desymmetrization strategy via organocatalytic tyrosine click reaction of 4-aryl-1,2,4-triazole-3,5-dione (ATAD). The key feature of our strategy is the ability of a bifunctional organocatalyst to transfer its

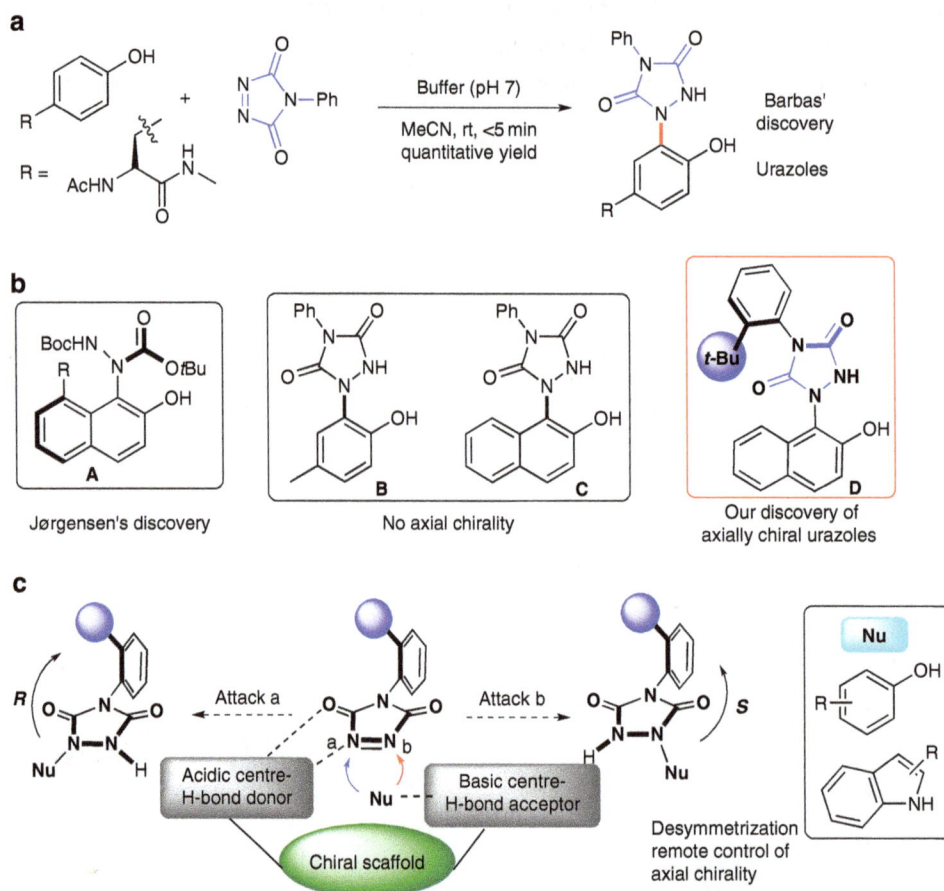

**Figure 1 | Synthesis of urazoles via tyrosine click reaction and discovery of axial chirality and strategy for remote enantiocontrol. (a)** Synthesis of urazoles via tyrosine click reaction (Barbas' discovery). **(b)** Discovery of urazoles with axial chirality. **(c)** Our strategy for remote enantiocontrol of axial chirality of urazoles.

stereochemical information to a remote position and thereafter efficiently control its axial chirality.

## Results

**Discovery of urazoles with axial chirality.** In 2006, the Jørgensen group discovered a new class of axially chiral skeleton **A** via asymmetric amination of 8-amine-2-naphthol with azodi-carboxylates (Fig. 1b)[40,41]. Motivated by this pioneering discovery, we synthesized the compounds **B** and **C** through tyrosine click reaction and imagined that such compounds should have axial chirality due to the significant restricted rotation between nitrogen atom and the directly attached phenol ring or naphthol ring (Fig. 1b). Disappointedly, they did not display axial

chirality based on the chiral stationary high-performance liquid chromatography (HPLC) analysis presumably because of the relatively low rotational barrier of the N–Ar bond. To further screen different aryl substituents of triazodiones, we are pleased to find that urazole **D** with a steric bulky substituent (*t*-butyl group) in the *ortho* position of the phenyl ring shows apparently axial chirality. As such, a class of urazoles with axial chirality was discovered (Fig. 1b).

**Optimization of reaction conditions involving naphthols.** To investigate the feasibility of our hypothesis, we initiated to conduct the tyrosine click reaction of naphthol (**1a**) with 4-(2-*tert*-butylphenyl)-3*H*-1,2,4-triazole-3,5-dione (**2a**) by using

---

**Table 1 | Optimization of the organocatalytic enantioselective tyrosine click reaction*.**

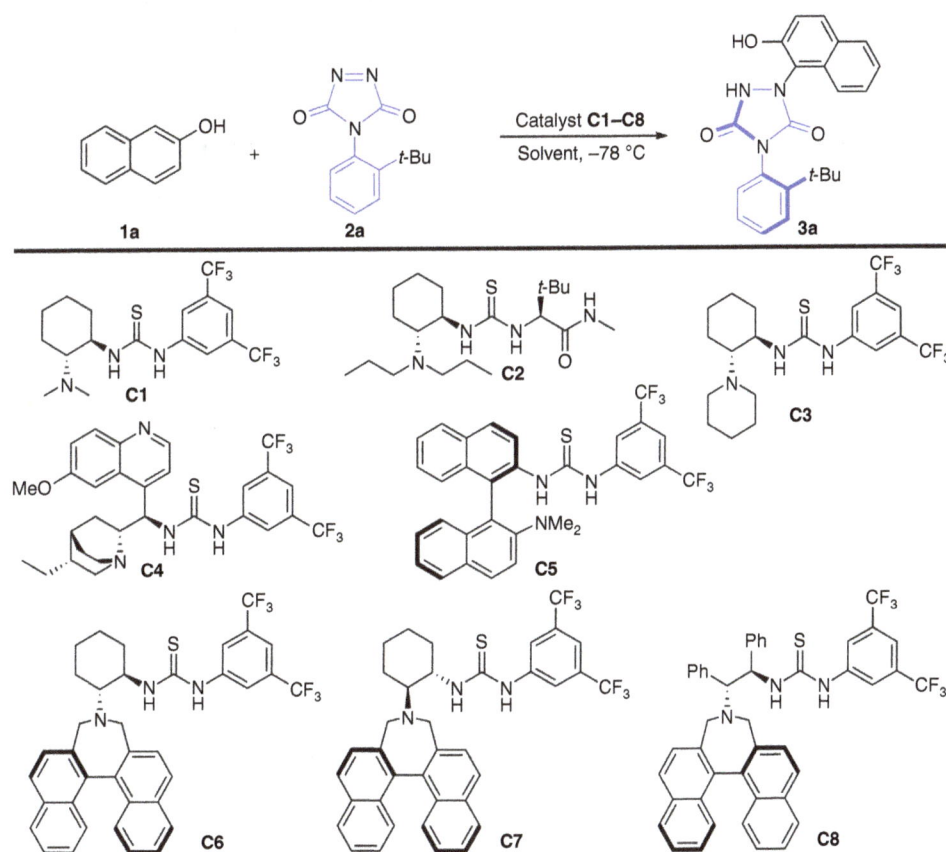

| Entry | Solvent | Catalyst | Time (min) | Yield (%)[†] | ee (%)[‡] |
|---|---|---|---|---|---|
| 1 | DCM | C1 | <5 | 57 | 25 |
| 2 | DCM | C2 | <5 | 59 | 5 |
| 3 | DCM | C3 | <5 | 68 | 75 |
| 4 | DCM | C4 | <5 | 65 | 11 |
| 5 | DCM | C5 | <5 | 63 | − 9 |
| 6 | DCM | C6 | <5 | 61 | − 45 |
| 7 | DCM | C7 | <5 | 68 | 91 |
| 8 | DCM | C8 | <5 | 66 | − 43 |
| 9 | Toluene | C7 | 90 | 65 | 90 |
| 10 | Et₂O | C7 | 25 | 73 | 97 |
| 11[§] | Et₂O | C7 | 30 | 82 | 99 |
| 12[‖] | Et₂O | C7 | 90 | 70 | 98 |

DCM, dichloromethane; HPLC, high-performance liquid chromatography.
*Reactions were performed with **1a** (0.10 mmol), **2a** (0.12 mmol) and catalyst (10 mol%) in 2.0 ml solvent.
[†]Isolated yield.
[‡]Determined by HPLC analysis on a chiral stationary phase.
[§]Reaction was conducted with 5 mol% catalyst.
[‖]3 mol% catalyst was used.

Takemoto catalyst (**C1**)[42] in dichloromethane (DCM) at room temperature. To our delight, the desired product **3a** was obtained in almost quantitative yield in less than 5 min, albeit without any enantioselectivity. Using the analysis of chiral HPLC, the urazole compound **3a** was confirmed to be atropisomeric and two peaks corresponding to the enantiomers were observed on the chiral HPLC at room temperature without any change during the analysis timescale. In the absence of organocatalyst, the reaction also proceeded very smoothly (less than 5 min for the model reaction) in quantitative yield, indicating that the strong background reaction might be the major challenge for efficiently realizing enantioselective transformation. With these initial results in hand and to improve the enantioselectivity, we turned our attention to decrease the reaction temperature to −78 °C. Gratifyingly, the reaction proceeded completely within just 5 min and the desired product was obtained in 57% isolated yield with 25% enantioselectivity excess (ee). We next investigated different bifunctional thiourea-tertiary amine catalysts (Table 1, entries 2–5). Among the tested catalysts, Takemoto catalyst **C3** with a cyclic tertiary amine proved to be very promising, with the ee value up to 75%. Considering that the additional aromatic stacking interaction might be involved in the transition states, catalysts **C6** and **C7** with an axial binaphthyl moiety were tested[43]. Catalyst **C7** displayed an excellent enantiocontrol (entry 7), while catalyst **C6** with opposite configuration of diamine gave rise to poor enantioselectivity (entry 6). As shown in entry 8, the

diamine skeleton in the catalyst had a great influence on the asymmetric induction. Of the solvents tested for the reaction catalysed by **C7**, diethyl ether proved optimal with respect to the enantioselectivity (Table 1, entry 10). It is noteworthy that the reaction proceeded smoothly without having any affect on enantioselectivity (99% ee) and with an improved chemical yield up to 82% when 5 mol% of catalyst was used (entry 11).

**Substrate scope.** After the optimal reaction condition being established, we set out to explore the substrate scope with respect to various phenols and 2-naphthols as reactants (Table 2). All of the investigated reactions were complete within 60 min and gave products in moderate to good yields (51–85%) and with excellent enantioselectivities (90–99% ee). As regarding the use of a variety of 2-naphthols, bearing electron-withdrawing (Table 2, products **3b–3f**) and electron-donating (Table 2, products **3g–3h**) groups, the reaction of these 2-naphthols with **2a** gave the expected products with very high stereoselectivities. These results indicated that there was only limited influence on stereoselectivity regardless of the electronic properties of the substituents at the different positions on the aromatic ring. It is noteworthy that the use of 4-substituted phenol, such as 4-*tert*-butyl-phenol and 4-phenyl-phenol, also afforded the desired products **3i** and **3j** in excellent stereocontrol with a modified reaction conditions, respectively, demonstrating that the substrate scope could not be only limited to naphthols.

---

**Table 2 | Substrate scope of naphthols or phenols*.**

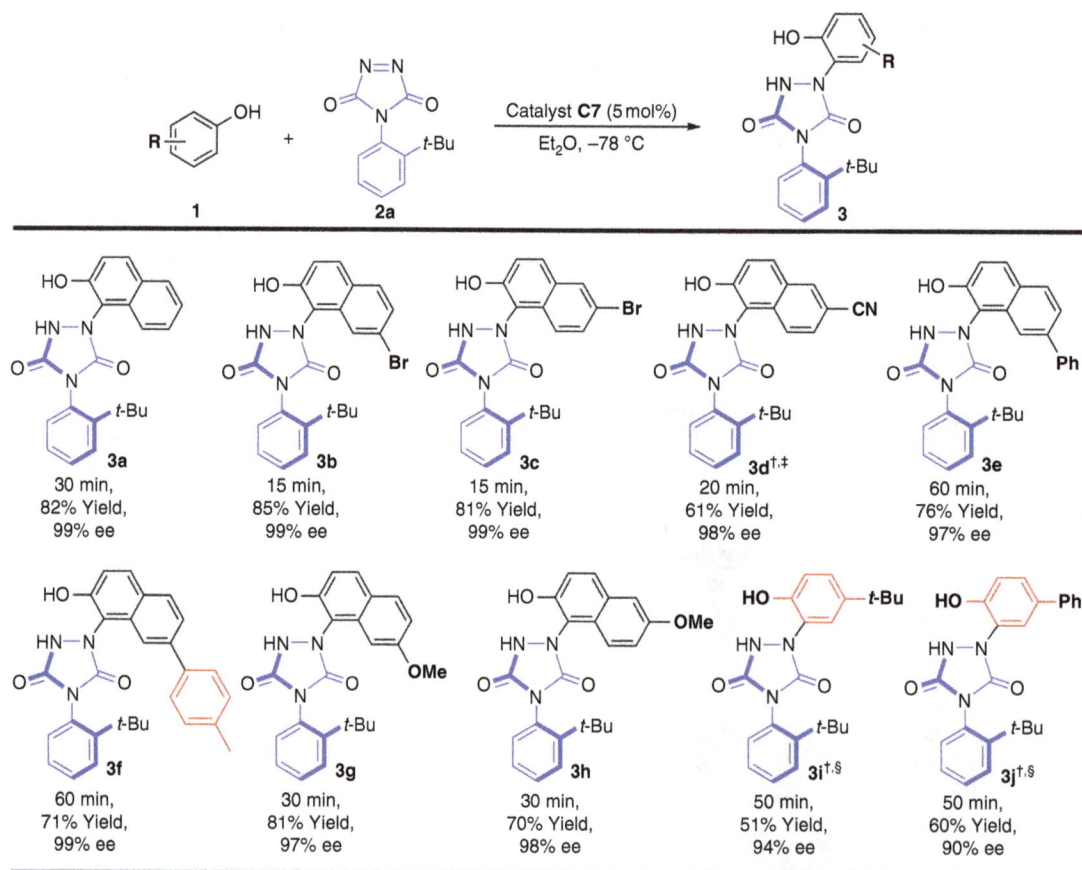

| 3a | 3b | 3c | 3d[†,‡] | 3e |
| --- | --- | --- | --- | --- |
| 30 min, 82% Yield, 99% ee | 15 min, 85% Yield, 99% ee | 15 min, 81% Yield, 99% ee | 20 min, 61% Yield, 98% ee | 60 min, 76% Yield, 97% ee |

| 3f | 3g | 3h | 3i[†,§] | 3j[†,§] |
| --- | --- | --- | --- | --- |
| 60 min, 71% Yield, 99% ee | 30 min, 81% Yield, 97% ee | 30 min, 70% Yield, 98% ee | 50 min, 51% Yield, 94% ee | 50 min, 60% Yield, 90% ee |

DCM, dichloromethane; HPLC, high-performance liquid chromatography.
*Reactions were performed with **1** (0.1 mmol), **2a** (0.12 mmol) and catalyst **C7** (5 mol%) in 2.0 ml Et₂O. Isolated yields and the ee values were determined with HPLC analysis using the chiral stationary phase.
†Reactions were performed with 20 mol% catalyst **C7** in 2.0 ml solvent.
‡In DCM at −78 °C.
§In toluene at −40 °C.

Next, we explored the generality of the reaction with regard to variation of ATADs. A broad range of ATADs containing different substituents at the aromatic ring reacted smoothly with 2-naphthol **1a** to produce the corresponding axially chiral urazoles with high efficiency and excellent entantiocontrol (Table 3). The electronic and position properties of the aromatic ring substituents did not affect the selectivities of the tyrosine click reactions. It should be pointed out that the ortho group is not only restricted to *tert*-butyl group or iodo, and the bromo or phenyl group at the ortho position could also be obtained with excellent enantioselectivities (**3p** and **3q**). It should be emphasized that the presence of I or Br is very convinient to do the further transformation for diversity-oriented synthesis and drug discovery due to the high reactivity in many transition metal-catalysed reactions[44]. Experiments on the configurational stability of the product were carried out by heating a solution of **3a** in toluene or MeCN at 80 °C for 12 h. Chiral HPLC analysis showed that the ee value of **3a** did not have any effect. Therefore, the obtained axially chiral compounds may have potential wide applications as asymmetric organocatalysts/ligands.

**Optimization of reaction conditions involving indoles.** To expand the synthetic utility of this methodology and further develop the application of the very reactive ATAD, we next

focused our attention on more challenging nucleophiles. Although much progress has been made in the development of organocatalytic asymmetric intermolecular transformation by using indoles as nucleophiles[45,46], to the best of our knowledge, only few examples involving 2-substituted indoles as nucleophile have been reported with good enantiocontrol, which is probably ascribed to the interrupted interaction between the substrates and the organocatalyst[47]. We envisaged that the very reactive and multifunctional electrophile ATAD might provide new possibility to proceed such a remote control process with good stereoselectivity with bifunctional organocatalysts. To our delight, by using the standard reaction conditions (Table 4, entry 1), we found that the reaction of 2-phenylindole **4a** with 4-(2-*tert*-butylphenyl)-3H-1,2,4-triazole-3,5-dione (**2a**) proceeded smoothly by simply using the catalyst **C6**, giving the desired product **5a** in 74% yield with 15% ee. However, after making great efforts on investigation of the optimized reaction conditions, we could not improve the enantioselectivity by using thiourea-tertiary amine organocatalyst (see Supplementary Table 1 for details). On the basis of these findings and own comprehension on the phosphoric acid catalysis[48-50], we envisioned that phosphoric acid might perform bifunctional action to activate indole and ATAD simultaneously and control the enantioselectivity[51-54]. As shown in Table 4, phosphoric acid

**Table 3 | The reaction substrate scope of 4-aryl-1,2,4-triazoline-3,5-diones\*.**

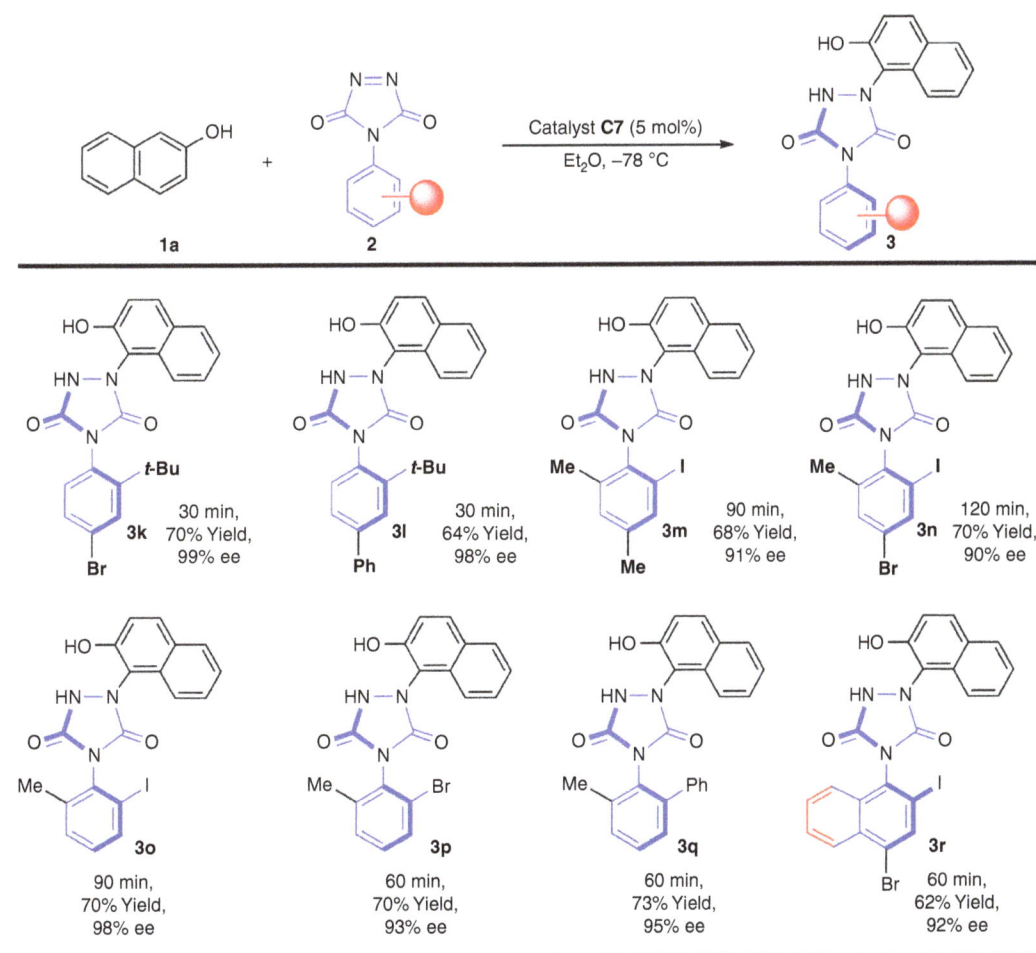

HPLC, high-performance liquid chromatography.
\*Reactions were performed with **1a** (0.1 mmol), **2** (0.12 mmol) and catalyst **C7** (5 mol%) in 2.0 ml Et₂O. Isolated product and the ee values were determined by HPLC analysis using a chiral stationary phase.

**Table 4 | Optimization of the asymmetric tyrosine click-like reaction involving indoles as nucleophiles*.**

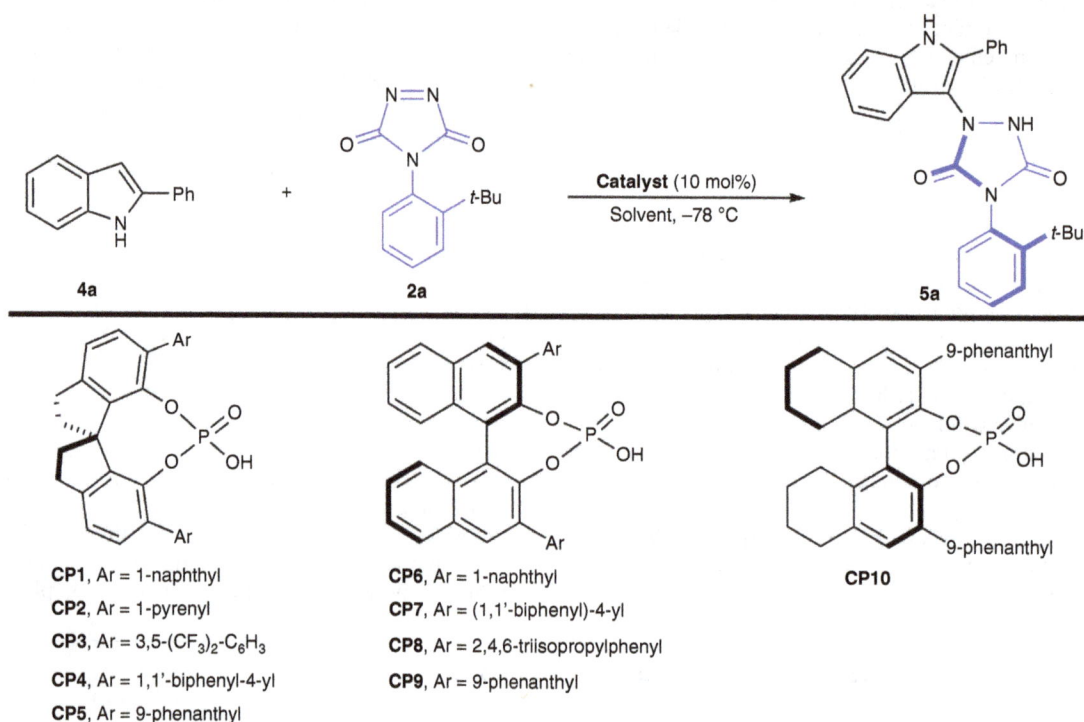

Catalyst (10 mol%)
Solvent, −78 °C

**CP1**, Ar = 1-naphthyl
**CP2**, Ar = 1-pyrenyl
**CP3**, Ar = 3,5-(CF$_3$)$_2$-C$_6$H$_3$
**CP4**, Ar = 1,1'-biphenyl-4-yl
**CP5**, Ar = 9-phenanthyl

**CP6**, Ar = 1-naphthyl
**CP7**, Ar = (1,1'-biphenyl)-4-yl
**CP8**, Ar = 2,4,6-triisopropylphenyl
**CP9**, Ar = 9-phenanthyl

**CP10**

| Entry | Solvent | Catalyst (10%) | Time | Yield (%)[†] | ee (%)[‡] |
|---|---|---|---|---|---|
| 1 | Et$_2$O | **C6** | 48 h | 69 | 5 |
| 2 | Et$_2$O | **C7** | 24 h | 74 | −15 |
| 3 | Et$_2$O | **CP1** | 10 h | 96 | 89 |
| 4 | DCM | **CP1** | <5 min | 99 | 68 |
| 5 | Toluene | **CP1** | 60 min | 94 | 60 |
| 6 | DCM/Et$_2$O (1/1) | **CP1** | 10 min | 99 | 95 |
| 7 | DCM/Et$_2$O (1/2) | **CP1** | 20 min | 98 | 95 |
| 8 | DCM/Et$_2$O (1/1) | **CP2** | 10 min | 99 | 95 |
| 9 | DCM/Et$_2$O (1/1) | **CP3** | 10 min | 98 | 37 |
| 10 | DCM/Et$_2$O (1/1) | **CP4** | 10 min | 99 | 85 |
| 11 | DCM/Et$_2$O (1/1) | **CP5** | 10 min | 99 | 97 |
| 12 | DCM/Et$_2$O (1/1) | **CP6** | 10 min | 96 | −77 |
| 13 | DCM/Et$_2$O (1/1) | **CP7** | 10 min | 96 | −47 |
| 14 | DCM/Et$_2$O (1/1) | **CP8** | 10 min | 97 | −79 |
| 15 | DCM/Et$_2$O (1/1) | **CP9** | 10 min | 98 | −95 |
| 16 | DCM/Et$_2$O (1/1) | **CP10** | 10 min | 99 | −89 |
| 17 | DCM/Et$_2$O (1/1) | **CP5** (5%) | 10 min | 99 | 97 |
| 18 | DCM/Et$_2$O (1/1) | **CP5** (3%) | 10 min | 99 | 95 |
| 19[§] | DCM/Et$_2$O (1/2) | **CP5** (1%) | 40 min | 99 | 95 |

DCM, dichloromethane; HPLC, high-performance liquid chromatography.
*Reactions were performed with **4a** (0.1 mmol), **2a** (0.12 mmol) and 10 mol% **catalyst** (entries 1-15) in 2.0 ml solvent.
[†]Determined by $^1$H NMR analysis using CH$_2$Br$_2$ as an internal standard.
[‡]Determined by HPLC analysis on a chiral stationary phase.
[§]1 mol% Catalyst **CP5**, solvent: DCM/Et$_2$O = 1/2.

catalyst proved to be a suitable organocatalyst for this tranformation. On optimizing the reaction conditions through variation of the phosphoric acid catalysts, solvents and catalyst loadings (Table 4, entries 3–15), the following protocol was proved to be optimal: reaction of **4a** and **2a** with the molar ratio of 1.0:1.2 by using phosphoric acid **CP5** (5 mol%) as catalyst in DCM/Et$_2$O (1/1) at −78 °C for 10 min, **5a** was obtained in exellent yield with 97% ee (Table 4, entry 15). It should be noted that the chiral spiro-phosphoric acid catalyst displayed better enantioselectivity than the BINOL-derived catalyst if the substituent in the 3 and 3′ positions is the same (entries 6 and 12; entries 10 and 13; entries 11 and 15).

**Substrate scope with indoles as nucleophiles**. Having identified the optimized reaction conditions, the reaction was extended to include various 2-substituted indoles and triazoledione compounds with catalyst **CP5**. As shown in Table 5, the reaction proceeded smoothly to give the desired product **5a–5m** in very high yield (86–96%) and excellent enantioselectivity (84–97% ee). It should be noted that the electronic nature, bulkiness or positions of the substituents on the cyclic diazo compounds and substituted indoles have only minimal effect on efficiencies and enantioselectivities. In addition to aromatic groups, alkyl substituents on indole were used to acquire the desired products (**5l** and **5m**) with excellent yields and

**Table 5 | The substrate scope by using indoles as nucleophiles*.**

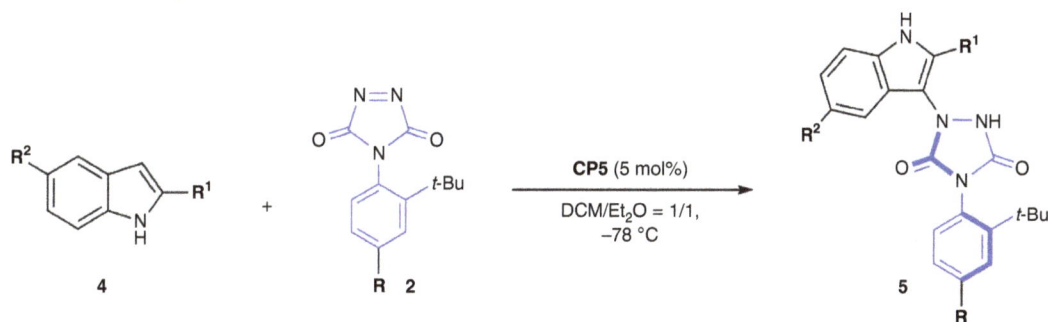

| Entry | R | R¹ | R² | Time (min) | 5 | Yield (%)[†] | ee (%)[‡] |
|---|---|---|---|---|---|---|---|
| 1 | H | Ph | H | 10 | 5a | 96 | 97 |
| 2 | H | 4-F-Ph | H | 15 | 5b | 95 | 96 |
| 3 | H | 4-Cl-Ph | H | 20 | 5c | 95 | 92 |
| 4 | H | Ph | Br | 15 | 5d | 94 | 94 |
| 5 | H | Ph | Ph | 15 | 5e | 90 | 96 |
| 6 | Br | Ph | H | 10 | 5f | 92 | 93 |
| 7 | Br | 4-F-Ph | H | 15 | 5g | 92 | 93 |
| 8 | Br | 4-Cl-Ph | H | 20 | 5h | 93 | 91 |
| 9 | Ph | Ph | H | 10 | 5i | 92 | 94 |
| 10 | Ph | 4-F-Ph | H | 20 | 5j | 91 | 95 |
| 11 | Ph | 4-Cl-Ph | H | 20 | 5k | 86 | 92 |
| 12 | H | Isopropyl | H | <5 | 5l | 95 | 90 |
| 13 | H | Methyl | H | <5 | 5m | 95 | 84 |

DCM, dichloromethane; HPLC, high-performance liquid chromatography.
*Reactions were performed with **2** (0.12 mmol), **4** (0.10 mmol) and catalyst **CP5** (5% mmol) in 2.0 ml solvent (DCM/Et₂O = 1/1).
†Isolated yield.
‡The ee values were determined by HPLC analysis using a chiral stationary phase.

good stereoselectivities in just less than 5 min (Table 5, entries 12–13).

**Preliminary evaluation as chiral ligands.** To verify the stability of such axial compounds, we heated the obtained product **5a** in MeCN at 80 °C for 12 h and no ee erosion was observed. Thus, this kind of axially chiral compounds displayed a high rotation energy about the N–Ar bond, indicating that the chiral urazoles may have potential applications in the field of asymmetric organocatalysts and Lewis acid catalysis. To really investigate the potential application of the resultant axially chiral urazoles in the field of asymmetric catalysis, we chose the addition of N-methylindole (**8**) to N-methylisatin (**9**) as a model reaction and evaluated the potential application in the asymmetric catalysis (see Supplementary Table 2). Gratifyingly, the reaction proceeded completely within 8 h at 5 °C and the desired product (**10**) was obtained in 96% yield with 62% ee (Fig. 2a), demonstrating that the newly developed axially chiral urazoles have the potential application in asymmetric synthesis. Further work encompassing the application of axially chiral urazoles as ligands or catalysts for enantioselective reactions is currently in progress in our laboratory.

**Gram-scale synthesis of enantiopure urazoles.** To further demonstrate the utility of the tyrosine click-like reaction, gram-scale syntheses of products **3a** and **5a** were carried out. As displayed in Fig. 2b, there was almost no change in reactivity and stereoselectivity, suggesting that this method should have the potential for large-scale chemical production (also see Supplementary Note 3). It should be worth highlighting that the reaction by using 2-phenyl indole as nucleophile was proceeded

very smoothly, with only 1 mol% of phosphoric acid catalyst **CP5**. The absolute configuration of **3p** was attributed to be aS and **5f** was assigned to be aR using X-ray diffraction analysis of their methylation derivatives **6p** and **7f** (Fig. 2c, see also Supplementary Fig. 1).

**Discussion**

We have successfully developed an organocatalytic asymmetric tyrosine click-like reaction in high yields with excellent enantioselectivity under mild reaction conditions in an excellent remote enantiocontrol manner. The reaction represents a very convenient approach to an interesting class of axially chiral urazole derivatives, with potential biological activities and potential application as effective chiral organocatalysts/ligands. The excellent remote enantiocontrol of the process stems from the efficient discrimination of the two reactive sites in the triazoledione-involving phenols or indoles as nucleophile and transferring the chirality of the catalyst into the axial chirality of urazoles at the remote position far from the reactive site. The application of this strategy to a broader substrate scope and mechanistic investigations of the desymmetrization strategy are currently underway in our group.

**Methods**

**General information.** Reagents were purchased at the highest commercial quality and used without further purification, unless otherwise stated. Analytical thin layer chromatography (TLC) was performed on precoated silica gel 60 F254 plates. Flash column chromatography was performed using Tsingdao silica gel (60, particle size 0.040–0.063 mm). Visualization on TLC was achieved by the use of ultraviolet light (254 nm). NMR spectra were recorded on a Bruker DPX 400 spectrometer at 400 MHz for ¹H NMR, 100 MHz for ¹³C NMR and 376 MHz for ¹⁹F NMR in CDCl₃ or acetone-d₆ with tetramethylsilane as internal standard. Chemical shifts are reported in p.p.m., and coupling constants are given in Hz.

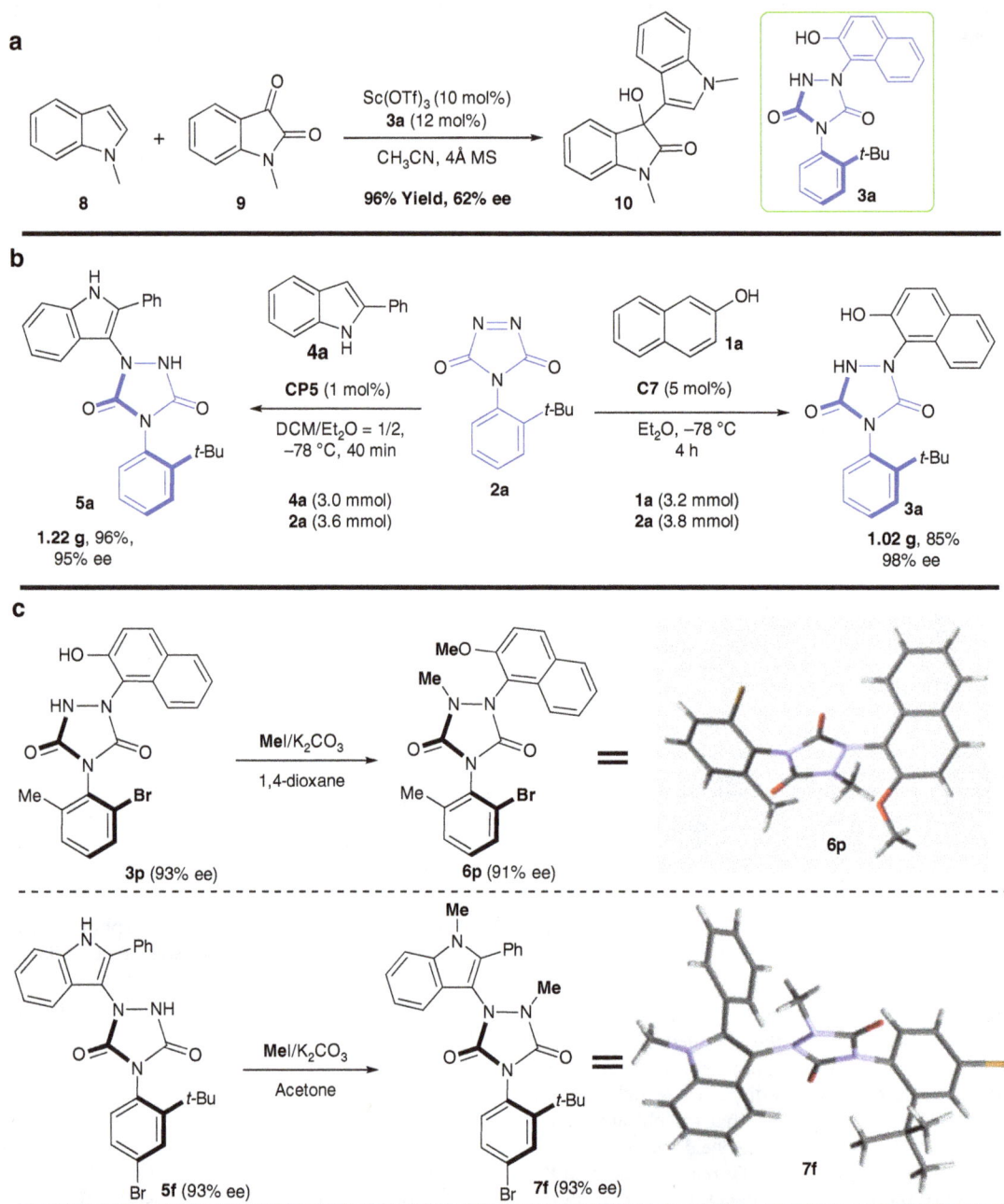

**Figure 2 | Application in asymmetric catalysis and gram-scale synthesis of 3a/5a and further transformation. (a)** Potential application of catalytic asymmetric synthesis of substituted 3-hydroxy-2-oxindole. **(b)** Gram-scale synthesis of axially chiral urazoles via tyrosine click reaction. **(c)** Further transformation for confirmation of absolute configuration.

Data for $^1$H NMR are recorded as follows: chemical shift (p.p.m.), multiplicity (s, singlet; d, doublet; t, triplet; q, quartet; m, multiplet), coupling constant (Hz) and integration. Data for $^{13}$C NMR are reported in terms of chemical shift ($\delta$, p.p.m.). High-resolution mass spectra were recorded on a LC-TOF spectrometer (Micromass). Enantiomeric excess was determined on Agilent HPLC using the DAICEL CHIRAL column. For preparation of 4-aryl-1,2,4-triazoline-3,5-diones, see Supplementary Note 1.

Racemic compounds were obtained without catalyst.

**General procedure for synthesis of axially chiral urazoles 3.** In a Schlenk tube, 4-aryl-1,2,4-triazoline-3,5-diones **2** (0.12 mmol) and catalyst **C7** (5 mol%, 0.005 mmol) were dissolved in Et$_2$O (2 ml; also see Supplementary Note 2). The solution was stirred for 10 min at $-78\,°C$ before 2-naphthols and phenols **1** (0.10 mmol) were added. The resulting solution was stirred at $-78\,°C$ until the red colour disappeared. After monitored with TLC, the reaction mixture was acidified with 6 N HCl and concentrated. Then, the obtained crude material was purified using silica gel column chromatography (CH$_2$Cl$_2$ to CH$_2$Cl$_2$/Acetone = 10/1) to afford the pure products **3**. In some cases, reactions were performed with 20 mol%

of catalyst **C7** in 2.0 ml solvent, for **3d** in DCM at $-78\,°C$; **3i** and **3j** in dry toluene at $-40\,°C$.

**General procedure for synthesis of axially chiral urazoles 5.** In a Schlenk tube, 4-aryl-1,2,4-triazoline-3,5-diones **2** (0.12 mmol) and catalyst **CP5** (5 mol%, 0.005 mmol) were dissolved in DCM/Et$_2$O = 1/1 (2 ml; also see Supplementary Note 2). The solution was stirred for 10 min at $-78\,°C$ before 2-substituted indole **4** (0.10 mmol) was added. The resulting solution was stirred under this condition until the purple colour disappeared. After being monitored with TLC, the reaction mixture was concentrated, and then purified using silica gel column chromatography (CH$_2$Cl$_2$/Acetone = 20/1) to afford the pure products **5**.

## References

1. Hall, I. H., Wong, O. T., Simlot, S., Miller, III. M. C. & Izydore, R. A. Antineoplastic activities and cytotoxicity of 1-acyl and 1,2-diacyl-1,2,4-triazolidine-3,5-diones in murine and human tissue culture cells. *Anticancer Res.* **12**, 1355–1362 (1992).

2. Martinez, A. *et al.* SAR and 3D-QSAR studies on thiadiazolidinone derivatives: exploration of structural requirements for glycogen synthase Kinase 3 inhibitors. *J. Med. Chem.* **48**, 7103–7112 (2005).

3. Adibia, H., Abirib, R., Mallakpourc, S., Zolfigold, M. A. & Majnoonie, M. B. Evaluation of *in vitro* antimicrobial and antioxidant activities of 4-substituted-1,2,4-triazolidine-3,5-dione derivatives. *J. Rep. Pharm. Sci.* **1**, 87–93 (2012).

4. Saluja, P., Khurana, J. M., Nikhilb, K. & Royb, P. Task-specific ionic liquid catalyzed synthesis of novel naphthoquinone–urazole hybrids and evaluation of their antioxidant and *in vitro* anticancer activity. *RSC Adv.* **4**, 34594–34603 (2014).

5. Pirkle, W. H. & Gravel, P. L. Persistent cyclic diacylhydrazyl radicals from urazoles and pyrazolidine-3,5-diones. *J. Org. Chem.* **43**, 808–815 (1978).

6. Breton, G. W. & Hoke, K. R. Application of radical cation spin density maps toward the prediction of photochemical reactivity between N-methyl-1,2,4-triazoline-3,5-dione and substituted benzenes. *J. Org. Chem.* **78**, 4697–4707 (2013).

7. Ban, H., Gavrilyuk, J. & Barbas, C. F. III Tyrosine bioconjugation through aqueous ene-type reactions: a click-like reaction for tyrosine. *J. Am. Chem. Soc.* **132**, 1523–1525 (2010).

8. Ban, H. *et al.* Facile and stabile linkages through tyrosine: bioconjugation strategies with the tyrosine-click reaction. *Bioconjug. Chem.* **24**, 520–532 (2013).

9. Alajarin, M., Cabrera, J., Sanchez-Andrada, P., Orenes, R. & Pastor, A. 4-Alkenyl-2-aminothiazoles: smart dienes for polar [4 + 2] cycloadditions. *Eur. J. Org. Chem.* **2013**, 474–489 (2013).

10. Breton, G. W. Acid-catalyzed reactions of N-methyl-1,2,4-triazoline-3,5-dione (METAD) with some polyaromatic hydrocarbons. *Adv. Chem. Lett.* **1**, 68–73 (2013).

11. Breton, G. W., Hughes, J. S., Pitchko, T. J., Martin, K. L. & Hardcastle, K. Unexpected σ bond rupture during the reaction of N-methyl-1,2,4-triazoline-3,5-dione with acenaphthylene and indene. *J. Org. Chem.* **79**, 8212–8220 (2014).

12. Takahashi, I., Suzuki, Y. & Kitagawa, O. Asymmetric synthesis of atropisomeric compounds with an N-C chiral axis. *Org. Prep. Proced. Int.* **46**, 1–23 (2014).

13. Díaz de Villegas, M. D., Gálvez, J. A., Etayo, P., Badorrey, R. & López-Ram-de-Víu, P. Recent advances in enantioselective organocatalyzed anhydride desymmetrization and its application to the synthesis of valuable enantiopure compounds. *Chem. Soc. Rev.* **40**, 5564–5587 (2011).

14. Rubush, D. M., Morges, M. A., Rose, B. J., Thamm, D. H. & Rovis, T. An asymmetric synthesis of 1,2,4-trioxane anticancer agents via desymmetrization of peroxyquinols through a brønsted acid catalysis cascade. *J. Am. Chem. Soc.* **134**, 13554–13557 (2012).

15. Wang, Z., Chen, Z. & Sun, J. Catalytic enantioselective intermolecular desymmetrization of 3-substituted oxetanes. *Angew. Chem. Int. Ed.* **52**, 6685–6688 (2013).

16. Wang, Z. *et al.* Catalytic enantioselective intermolecular desymmetrization of azetidines. *J. Am. Chem. Soc.* **137**, 5895–5898 (2015).

17. Meng, S.-S. *et al.* Chiral phosphoric acid catalyzed highly enantioselective desymmetrization of 2-substituted and 2,2-disubstituted 1,3-diols via oxidative cleavage of benzylidene acetals. *J. Am. Chem. Soc.* **136**, 12249–12252 (2014).

18. Gualtierotti, J.-B., Pasche, D., Wang, Q. & Zhu, J. Phosphoric acid catalyzed desymmetrization of bicyclic bislactones bearing an all-carbon stereogenic center: total syntheses of ( − )-Rhazinilam and ( − )-Leucomidine B. *Angew. Chem. Int. Ed.* **53**, 9926–9930 (2014).

19. Iorio, N. D. *et al.* Remote control of axial chirality: aminocatalytic desymmetrization of N-arylmaleimides via vinylogous Michael addition. *J. Am. Chem. Soc.* **136**, 10250–10253 (2014).

20. Miyaji, R., Asano, K. & Matsubara, S. Bifunctional organocatalysts for the enantioselective synthesis of axially chiral isoquinoline N-Oxides. *J. Am. Chem. Soc.* **137**, 6766–6769 (2015).

21. Armstrong, R. J. & Smith, M. D. Catalytic enantioselective synthesis of atropisomeric biaryls: a cation-directed nucleophilic aromatic substitution reaction. *Angew. Chem. Int. Ed.* **53**, 12822–12826 (2014).

22. Li, G.-Q. *et al.* Organocatalytic aryl–aryl bond formation: an atroposelective [3,3]-rearrangement approach to BINAM derivatives. *J. Am. Chem. Soc.* **135**, 7414–7417 (2013).

23. De, C. K., Pesciaioli, F. & List, B. Catalytic asymmetric benzidine rearrangement. *Angew. Chem. Int. Ed.* **52**, 9293–9295 (2013).

24. Gustafson, J. L., Lim, D. & Miller, S. J. Dynamic kinetic resolution of biaryl atropisomers via peptide-catalyzed asymmetric bromination. *Science* **328**, 1251–1255 (2010).

25. Barrett, K. T. & Miller, S. J. Enantioselective synthesis of atropisomeric benzamides through peptide-catalyzed bromination. *J. Am. Chem. Soc.* **135**, 2963–2966 (2013).

26. Barrett, K. T., Metrano, A. J., Rablen, P. R. & Miller, S. J. Spontaneous transfer of chirality in an atropisomerically enriched two-axis system. *Nature* **509**, 71–75 (2014).

27. Cozzi, P. G., Emer, E. & Gualandi, A. Atroposelective organocatalysis. *Angew. Chem. Int. Ed.* **50**, 3847–3849 (2011).

28. Shirakawa, S., Liu, K. & Maruoka, K. Catalytic asymmetric synthesis of axially chiral o-iodoanilides by phase-transfer catalyzed alkylations. *J. Am. Chem. Soc.* **134**, 916–919 (2012).

29. Shirakawa, S., Wu, X. & Maruoka, K. Kinetic resolution of axially chiral 2-amino-1,1'-biaryls by phase-transfer-catalyzed N-allylation. *Angew. Chem. Int. Ed.* **52**, 14200–14203 (2013).

30. Lu, S., Poh, S. B. & Zhao, Y. Kinetic resolution of 1,1'-biaryl-2,2'-diols and amino alcohols through NHC-catalyzed atroposelective acylation. *Angew. Chem. Int. Ed.* **53**, 11041–11045 (2014).

31. Ma, G., Deng, J. & Sibi, M. P. Fluxionally chiral DMAP catalysts: kinetic resolution of axially chiral biaryl compounds. *Angew. Chem. Int. Ed.* **53**, 11818–11821 (2014).

32. Mori, K. *et al.* Enantioselective synthesis of multisubstituted biaryl skeleton by chiral phosphoric acid catalyzed desymmetrization/kinetic resolution sequence. *J. Am. Chem. Soc.* **135**, 3964–3970 (2013).

33. Link, A. & Sparr, C. Organocatalytic atroposelective aldol condensation: synthesis of axially chiral biaryls by arene formation. *Angew. Chem. Int. Ed.* **53**, 5458–5461 (2014).

34. Akiyama., T. Stronger Brønsted acids. *Chem. Rev.* **107**, 5744–5758 (2007).

35. Okino, T., Hoashi, Y., Furukawa, T., Xu, X. & Takemoto, Y. Enantio- and diastereoselective Michael reaction of 1,3-dicarbonyl compounds to nitroolefins catalyzed by a bifunctional thiourea. *J. Am. Chem. Soc.* **127**, 119–125 (2005).

36. Siau, W.-Y. & Wang, J. Asymmetric organocatalytic reactions by bifunctional amine-thioureas. *Catal. Sci. Technol.* **1**, 1298–1310 (2011).

37. Chauhan, P., Mahajan, S., Kaya, U., Hack, D. & Enders, D. Bifunctional amine-squaramides: powerful hydrogen-bonding organocatalysts for asymmetric domino/cascade reactions. *Adv. Synth. Catal.* **357**, 253–281 (2015).

38. Cheng, D.-J. *et al.* Highly enantioselective kinetic resolution of axially chiral BINAM derivatives catalyzed by a Brønsted acid. *Angew. Chem. Int. Ed. Engl.* **53**, 3684–3687 (2014).

39. Lin, J.-S. *et al.* Brønsted acid catalyzed asymmetric hydroamination of alkenes: synthesis of pyrrolidines bearing a tetrasubstituted carbon stereocenter. *Angew. Chem. Int. Ed. Engl.* **54**, 7847–7851 (2015).

40. Brandes, S., Bella, M., Kjærsgaard, A. & Jørgensen, K. A. Chirally aminated 2-naphthols-organocatalytic synthesis of non-biaryl atropisomers by asymmetric Friedel–Crafts amination. *Angew. Chem. Int. Ed.* **45**, 1147–1151 (2006).

41. Brandes, S. *et al.* Non-biaryl atropisomers in organocatalysis. *Chem. Eur. J.* **12**, 6039 (2006).

42. Okino, T., Hoashi, Y. & Takemoto, Y. Enantioselective Michael reaction of malonates to nitroolefins catalyzed by bifunctional organocatalysts. *J. Am. Chem. Soc.* **125**, 12672–12673 (2003).

43. Peng, F. *et al.* Organocatalytic enantioselective Michael addition of 2,4-pentandione to nitroalkenes promoted by bifunctional thioureas with central and axial chiral elements. *J. Org. Chem.* **73**, 5202 (2008).

44. Miyaura, N. & Suzuki, A. Palladium-catalyzed cross-coupling reactions of organoboron compounds. *Chem. Rev.* **95**, 2457–2483 (1995).

45. Bartoli, G., Bencivenni, G. & Dalpozzo, R. Organocatalytic strategies for the asymmetric functionalization of indoles. *Chem. Soc. Rev.* **39**, 4449–4465 (2010).

46. Dalpozzo, R. Strategies for the asymmetric functionalization of indoles: an update. *Chem. Soc. Rev.* **44**, 742–778 (2015).

47. Qiao, Z. *et al.* An organocatalytic, δ-regioselective, and highly enantioselective nucleophilic substitution of cyclic Morita–Baylis–Hillman alcohols with indoles. *Angew. Chem. Int. Ed.* **49**, 7294–7298 (2010).

48. Akiyama, T., Itoh, J., Yokota, K. & Fuchibe, K. Enantioselective Mannich-type reaction catalyzed by a chiral Brønsted acid. *Angew. Chem. Int. Ed.* **43**, 1566–1568 (2004).

49. Uraguchi & Terada, M. Chiral Brønsted acid-catalyzed direct Mannich reactions via electrophilic activation. *J. Am. Chem. Soc.* **126**, 5356–5357 (2004).

50. Parmar, D., Sugiono, E., Raja, S. & Rueping, M. Complete field guide to asymmetric BINOL-phosphate derived Brønsted acid and metal catalysis: history and classification by mode of activation; Brønsted acidity, hydrogen bonding, ion pairing, and metal phosphates. *Chem. Rev.* **114**, 9047–9153 (2014).

51. Zhang, Z. & Antilla, J. Enantioselective construction of pyrroloindolines catalyzed by chiral phosphoric acids: total synthesis of ( − )-debromoflustramine B. *Angew. Chem. Int. Ed.* **51**, 11778–11782 (2012).

52. Wang, S.-G., Yin, Q., Zhuo, C.-X. & You, S.-L. Asymmetric dearomatization of β-naphthols through an amination reaction catalyzed by a chiral phosphoric acid. *Angew. Chem. Int. Ed.* **54**, 647–650 (2015).

53. Kang, Q., Zhao, Z.-A. & You, S.-L. Highly enantioselective Friedel – Crafts reaction of indoles with imines by a chiral phosphoric acid. *J. Am. Chem. Soc.* **129**, 1484–1485 (2007).

54. You, S.-L., Cai, Q. & Zeng, M. Chiral Brønsted acid catalyzed Friedel–Crafts alkylation reactions. *Chem. Soc. Rev.* **38**, 2190–2201 (2009).

## Acknowledgements

We greatly appreciate the financial support from the National Natural Science Foundation of China (Nos 21572095 and 21572096). B.T. thanks the Thousand Young Talents Program for financial support. We sincerely dedicated this paper to professor Carlos F. Barbas III for a deep memory.

## Author contributions

J.-W.Z. performed experiments. D.-J.C. took part in the initial reaction development. J.-H.X. and C.S. helped with characterizing all new compounds. B.T. and X.-Y.L. conceived and directed the project and wrote the paper.

# 18

# Fluorescence microscopy as an alternative to electron microscopy for microscale dispersion evaluation of organic–inorganic composites

Weijiang Guan[1], Si Wang[1], Chao Lu[1] & Ben Zhong Tang[2]

Inorganic dispersion is of great importance for actual implementation of advanced properties of organic–inorganic composites. Currently, electron microscopy is the most conventional approach for observing dispersion of inorganic fillers from ultrathin sections of organic–inorganic composites at the nanoscale by professional technicians. However, direct visualization of macrodispersion of inorganic fillers in organic–inorganic composites using high-contrast fluorescent imaging method is hampered. Here we design and synthesize a unique fluorescent surfactant, which combines the properties of the aggregation-induced emission (AIE) and amphiphilicity, to image macrodispersion of montmorillonite and layered double hydroxide fillers in polymer matrix. The proposed fluorescence imaging provides a number of important advantages over electron microscope imaging, and opens a new avenue in the development of direct three-dimensional observation of inorganic filler macrodispersion in organic–inorganic composites.

[1] State Key Laboratory of Chemical Resource Engineering, Beijing University of Chemical Technology, 15 Beisanhuan East Road, PO Box 98, Beijing 100029, China. [2] Department of Chemistry, Hong Kong Branch of Chinese National Engineering Research Center for Tissue Restoration and Reconstruction, Hong Kong University of Science and Technology, Clear Water Bay, Hong Kong 999077, China. Correspondence and requests for materials should be addressed to C.L. (email: luchao@mail.buct.edu.cn) or to B.Z.T. (email: tangbenz@ust.hk).

Since the birth of organic–inorganic composites, they have undoubtedly been one of the most important and active scientific fields[1-5]. For such composite materials, the dispersion state of inorganic fillers in organic matrix plays a vital role for attainable improvements of their properties[6-9]. The dispersion–property relationship can be understood through the direct observation of the spatial distribution of inorganic fillers. The conventional observation method for inorganic filler microdispersion in organic matrix is performed by electron microscopy techniques, such as transmission electron microscopy (TEM) and three-dimensional (3D)-TEM tomography[6-14]. Notwithstanding the impressive amount of their data on microdispersion of inorganic fillers, they suffer from their own intrinsic limitations. First, the sample preparation is very complex and time-consuming, requiring professional technicians to cut and thin with special care; second, it is only suitable for evaluating nanodispersion scale in a small area window, and thus the obtained results could not be truly representative of macrodispersion of inorganic fillers; in addition, heavy-element staining is required for some poor-contrast components[10-14]. Therefore, it appears as a crucial need to develop a preparation-free, operation-simple and high-contrast method for visualization of macrodispersion of inorganic fillers.

The ultrafast and non-invasive 3D visualization is one of native functionalities for confocal fluorescence microscopy (CFM)[15-17]. In principle, CFM would be a powerful method for analyses of inorganic filler macrodispersion and spatial distribution in organic matrix with a large-enough window if the inorganic fillers could emit light. However, the fluorescence quenching usually occurs upon the formation of the fluorophore aggregates in composite materials via $\pi$–$\pi$ stacking interactions[18,19]. Therefore, the direct visualization of macrodispersion of inorganic fillers in organic–inorganic composites using high-contrast 3D fluorescent imaging method is hampered. Such a limitation of fluorescent labelling should be overcome by means of the aggregation-induced emission (AIE)-active fluorophores, which are generally non-fluorescent in solution but induced emission highly in the aggregated state or solid state[20-24].

Herein, we choose a typical organic–inorganic composite, montmorillonite (MMT) polymer composite, to achieve this possibility. In general, the naturally occurring MMT is hydrophilic, and requires organic modification by intercalating cationic surfactants into the interlayer space through ion exchange to form organically compatible[25-28]. The organo-modified MMT can be well dispersed in polymer matrix with remarkable improvement of material properties, such as increased strength and heat resistance, decreased gas permeability and flammability,

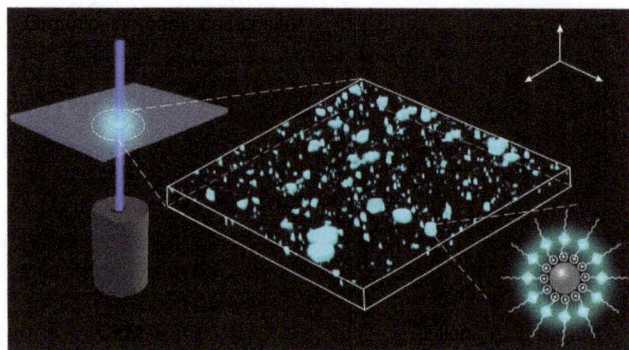

and increased biodegradability[29-33]. It is anticipated that organo-modified MMT can emit light if the intercalated cationic surfactants are attached with an AIE-active fluorophore. In this work, we first synthesize tetraphenylethene (TPE)-cored dodecyltrimethylammonium bromide cationic surfactant (denoted as TPE-DTAB) by incorporating a typical AIE-active TPE luminophore[34-36] into DTAB. Furthermore, we demonstrate the feasibility of TPE-DTAB as a novel probe for visualization of MMT macrodispersion and spatial distribution in polyvinyl chloride (PVC) matrix with some unique advantages, such as preparation-free, ultrafast and non-invasive (Fig. 1). The generality of the present strategy has also been verified by direct visualization of macrodispersion of layered double hydroxide (LDH) fillers in PVC matrix. This work also serves to demonstrate the general potential in the use of suitable AIE-active luminophores for the visualization of the filler macrodispersion in other organic–inorganic composites.

## Results

**Characterizations of TPE-DTAB.** As depicted in Supplementary Fig. 1, in our synthetic strategy for TPE-DTAB surfactant, we first prepared a fluorescent TPE core with two hydroxyl groups (1, TPE-2OH) to allow access to alkyl chain and alkyl bromide, respectively[37]. Then TPE-2OH was treated with equivalent moles of NaH to activate a hydroxyl group of TPE-2OH, followed by introducing n-octane in one side of TPE-2OH to yield compound 2 (ref. 38). After removal of the unreacted TPE-2OH, the reaction between 1,4-dibromobutane and compound 2 was performed in the presence of alkaline $K_2CO_3$ to generate bromo-functionalized compound 3 (ref. 39). The molecular structures of the above intermediate compounds were characterized and verified by $^1H$ nuclear magnetic resonance (NMR) spectroscopy (Supplementary Figs 2–4). Next, the alkyl bromide of compound 3 was converted to the alkyl trimethylamine bromide by adding excess amount of trimethylamine. Finally, TPE-DTAB (4) was successfully synthesized[40]. From the $^1H$ NMR spectrum of the purified TPE-DTAB (Fig. 2a), the characteristic peaks of TPE were clearly

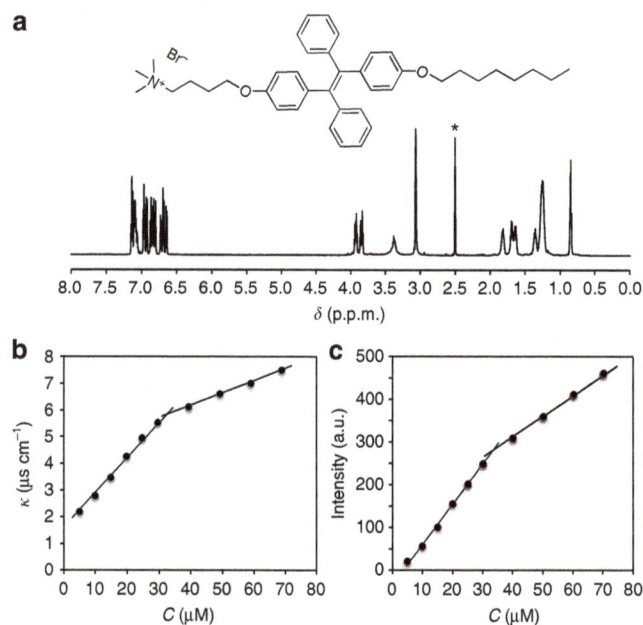

**a**

**b**

**c**

**Figure 2 | Molecular structure and characterization of TPE-DTAB.**
(**a**) $^1H$ NMR spectrum of TPE-DTAB in [$D_6$]dimethyl sulfoxide (the solvent peak is marked with asterisk). Plots of conductivity (**b**) and fluorescence intensity at 490 nm (**c**) versus the concentration of TPE-DTAB. $\lambda_{ex} = 325$ nm.

**Figure 1 | Schematic representation of visualization of 3D macrodispersion of fillers in organic–inorganic composites.** The inorganic fillers modified and bound with AIE molecules are dispersed inside the organic matrix, and then directly visualized by CFM.

visible, and the other peaks corresponded nicely to the alkyl protons of DTAB. In addition, in the positive-ion mode mass spectrum (MS; Supplementary Fig. 5), a mass-to-charge ($m/z$) ratio of 590.3996 was found to be consistent with the exact mass of TPE-DTAB without bromide ion. Finally, in combination with the measurement of the $^{13}$C NMR spectrum (Supplementary Fig. 6), we confirmed the structure of the cationic TPE-DTAB.

The critical micelle concentration (CMC) value of the cationic TPE-DTAB surfactant solution was determined by electrical conductivity method[41–43]. The conductivities ($\kappa$) of the TPE-DTAB at different concentrations ($C$) were plotted in Fig. 2b. At low concentrations, the solution conductivity increased linearly with an increase in the concentration of TPE-DTAB up to 32 μM. However, when the TPE-DTAB concentrations were above 32 μM, the straight line was observed with a decreased slope. The change of the slope was ascribed to the fact that the ionic micelles have less charge per unit mass than their unimers[41–43]. The breakpoint at $\sim$32 μM is generally considered to be the CMC of the TPE-DTAB surfactant.

The optical properties of the TPE-DTAB surfactant solution were investigated by measuring its ultraviolet − visible absorption and fluorescence spectra. As shown in Supplementary Fig. 7, two absorption peaks appeared at about 251 and 320 nm, which were attributed to the absorption of benzene and TPE unit[44]. On the other hand, the fluorescence emission wavelength remained at 490 nm, indicative of a typical TPE pattern of light[34–36,44]. The fluorescence intensity of the TPE-DTAB surfactant solution at 490 nm was plotted as a function of the TPE-DTAB concentration (Fig. 2c). Two straight lines with different slopes in the intensity–concentration plot were found with an inflection point at $\sim$32 μM, indicating the changes of the TPE-DTAB morphology[45]. Moreover, ultraviolet absorption and photoluminescence excitation spectra of TPE-DTAB were recorded below and above CMC (Supplementary Fig. 8). It can be seen that absorbance is simply proportional to photoluminescence species. However, inner filter effect could happen if absorbance is higher than 0.05, causing disproportional increasing of photoluminescence intensity with the increase of the concentration of photoluminescence species[46]. Experimentally, we obtained the linear relationship between TPE-DTAB concentration and photoluminescence intensity. The result may be ascribed to the combined effects, including inner filter effect, AIE effect and some other uncertain effects. The resulting CMC value by fluorescence measurement matched well with that obtained from the conductivity measurement, demonstrating that the synthesized TPE-DTAB is a kind of cationic fluorescent surfactant.

**Characterizations of TPE-DTAB-modified MMT.** It is known that Na$^+$-MMT is hydrophilic and well dispersed in polar solvent like water; however, the agglomeration of Na$^+$-MMT particles is inevitably occurred when they are mixed with organic polymer matrix to fabricate organic–inorganic composites. To obtain hydrophobic MMT, the TPE-DTAB was intercalated into the interlayer space of Na$^+$-MMT via ion exchange method[25–28]. The X-ray diffraction patterns of Na$^+$-MMT and TPE-DTAB-modified MMT were measured to study the structure change of MMT. It is known that $2\theta$ change in the range of 2°–10° for layered silicates indicated the formation of new ordered intercalated cationic layers[47–49]. As shown in Fig. 3a, the (001) diffraction peak of Na$^+$-MMT occurred at 7.00°, reflecting a d-spacing ($d_{001}$) of 12.6 Å. Interestingly, the (001) peak of the TPE-DTAB-modified MMT was shifted to 4.61° and 2.27°, indicating an enlarged d-spacing of 19.2 and 39 ± 1 Å, respectively, suggesting that the MMT interlayers might be intercalated with TPE-DTA$^+$ ions in different arrangements (Supplementary Fig. 9)[47–49].

Fourier transform infrared spectra of the TPE-DTAB, the Na$^+$-MMT and the TPE-DTAB-modified MMT further confirmed the combination of the TPE-DTA$^+$ ions and the MMT particles (Fig. 3b). In comparison with the Na$^+$-MMT, the TPE-DTAB-modified MMT exhibited two new absorption bands at 2,926 and 2,855 cm$^{-1}$, which were ascribed to the asymmetrical and symmetrical stretching vibrations of methylene groups in the alkyl chains of TPE-DTAB, respectively[50]. Moreover, the absorption intensity at 3,624 cm$^{-1}$ of the TPE-DTAB-modified MMT was lower than that of the Na$^+$-MMT because of the improved hydrophobicity[51]. In conclusions, the TPE-DTA$^+$ ions can be inserted into the interlayers of MMT.

On the other hand, the inorganic cations at the external surfaces of MMT particles would be also replaced by TPE-DTA$^+$ ions as a result of the ion exchange reaction between TPE-DTAB and Na$^+$-MMT. It is known that the adsorption of ionic surfactants at the particle surface usually has a great impact on the $\zeta$ potential of the particle[52–55]. Therefore, to clarify the configuration of TPE-DTAB at the MMT surfaces, we measured the $\zeta$ potential of the TPE-DTAB, Na$^+$-MMT and TPE-DTAB-modified MMT, respectively (Fig. 3c). The TPE-DTAB showed a positive $\zeta$ potential of 43.5 mV, whereas the Na$^+$-MMT particles had the negatively charged surfaces with a negative $\zeta$ potential of − 28.8 mV. For the TPE-DTAB-modified MMT, its $\zeta$ potential was changed to approximately zero, indicating that TPE-DTA$^+$ ions were adsorbed at the MMT surface via electrostatic attraction to expose their hydrophobic tails to the aqueous environment[55]. In addition, the Na$^+$-MMT particles dispersed well in water; however, the TPE-DTAB-modified MMT particles seemed to be very swollen and hydrophobic. In comparison, the hydrophobic property of TPE-DTAB-modified MMT particles were further confirmed via their good dispersion in petroleum ether (PE) for 24 h (inset of Fig. 3c).

It is essential to study the fluorescence property of TPE-DTAB-modified MMT in the solid state. Figure 3d showed the fluorescence spectra of the Na$^+$-MMT powder and the TPE-DTAB-modified MMT powder. The TPE-DTAB-modified MMT powder could emit strong blue–green fluorescence, whereas no fluorescence emissions appeared for the Na$^+$-MMT powder. Moreover, the fluorescence quantum yield was significantly enhanced from 5.79% (TPE-DTAB solution) to 42.10% (TPE-DTAB-modified MMT powder), demonstrating that the intramolecular motions of the AIE-active TPE-DTA$^+$ ions were tightly restrained by the rigid framework of MMT layers to suppress the non-radiative decay[56]. Therefore, the as-prepared TPE-DTAB-modified MMT powder was endowed not only a hydrophobic property but also an excellent fluorescence-enhanced performance, distinguishing from conventional fluorophores (aggregation-caused quenching effect).

**Macrodispersion of MMT in PVC/MMT composite.** The PVC/TPE-DTAB-modified MMT (5 wt%) composite was prepared to investigate the macrodispersion of MMT fillers in PVC matrix by TEM and CFM, respectively. TEM image was obtained from the lateral slice of the PVC/TPE-DTAB-modified MMT composite for the morphological characterization and microdispersion of TPE-DTAB-modified MMT particles in PVC matrix (Fig. 4a). However, TEM observation area was not large enough, and the analysis of microdispersion state was usually dependent on the cut cross-section. In contrast, the as-prepared PVC/MMT composite can be directly observed through CFM by means of the strong AIE-active fluorescence emissions of TPE-DTAB-modified MMT particles. As shown in Fig. 4b, the dark background (600 × 600 μm$^2$) was dotted with hundreds of luminescent particles to provide a fluorescence distribution map,

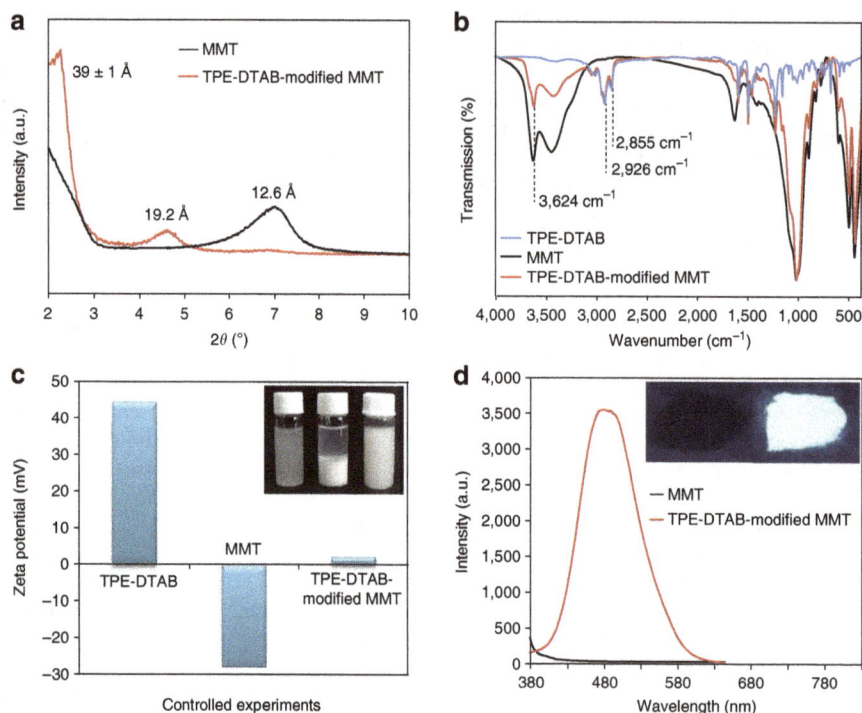

**Figure 3 | Characterizations of TPE-DTAB-modified MMT.** (**a**) Powder X-ray diffraction patterns of Na$^+$-MMT and TPE-DTAB-modified MMT. (**b**) Fourier transform infrared spectra of TPE-DTAB, Na$^+$-MMT and TPE-DTAB-modified MMT. (**c**) $\zeta$ potential measurements of TPE-DTAB, Na$^+$-MMT and TPE-DTAB-modified MMT; the inset showed the photographs of Na$^+$-MMT in water, TPE-DTAB-modified MMT in water and TPE-DTAB-modified MMT in petroleum ether for 24 h, respectively. (**d**) Fluorescence spectra of Na$^+$-MMT and TPE-DTAB-modified MMT; the inset showed the photographs of Na$^+$-MMT powder (left) and TPE-DTAB-modified MMT powder (right) under ultraviolet irradiation at 365 nm.

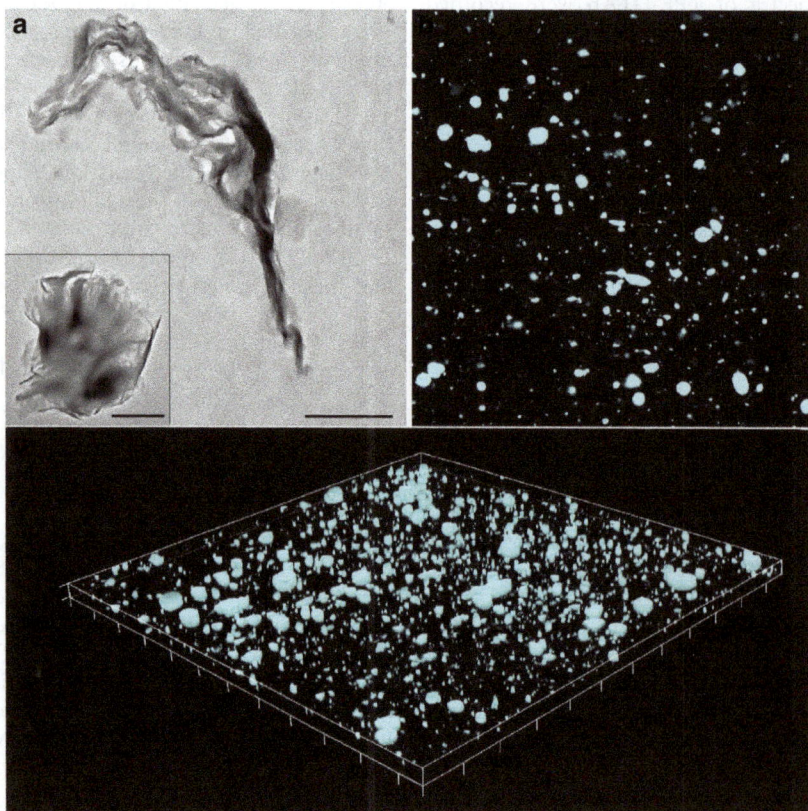

**Figure 4 | Macrodispersion of organo-modified MMT in PVC/MMT composite.** (**a**) Cross-sectional TEM micrograph of PVC/TPE-DTAB-modified MMT (5 wt%) composite; the inset showed TEM image of TPE-DTAB-modified MMT. Scale bar, 1 μm. (**b**) Fluorescence microscopy image (600 × 600 μm$^2$) of PVC/TPE-DTAB-modified MMT (5 wt%) composite. (**c**) 3D representation of TPE-DTAB-modified MMT dispersion (cyan parts) in PVC matrix. All fluorescence microscopy images were taken with a 405-nm laser.

clearly demonstrating the macrodispersion state (*XY* plane) of the TPE-DTAB-modified MMT particles in PVC. Furthermore, a total of 30 images in the *XY* plane were collected at different depths using Z-scan technique. As a result, the spatial distribution of TPE-DTAB-modified MMT particles in PVC matrix was achieved by combining these obtained *XY*-plane images (Fig. 4c). A large area can be observed in such images, and thus the imaging of the real dispersed state of MMT was captured, facilitating an impartial judgement of the overall dispersion state of the composite material. More importantly, this preparation-free imaging method is very convenient and time saving ($\sim$5 s for one picture), similar to an ultrafast and non-invasive computed tomography scan to macrodispersion of the composite material.

On the other hand, the dispersion of TPE-DTAB-modified MMT was compared with that of traditional organoclays. Herein, we incorporated TPE-DTAB as a small percentage of guests in traditional organo-MMT (that is, cetyltrimethyl ammonium bromide (CTAB)-modified MMT). Then, the PVC/CTAB/TPE-DTAB-modified MMT (5 wt%) composite was prepared for dispersion evaluation by CFM. As shown in Supplementary Fig. 10, with the help of the fluorescent TPE-DTAB, we could also directly observe the spatial distribution of CTAB/TPE-DTAB-modified MMT by CFM. In comparison with PVC/TPE-DTAB-modified MMT composite, PVC/CTAB/TPE-DTAB-modified MMT composite exhibited a similar dispersion state, although its fluorescence intensity was much lower as a result of small amounts of TPE-DTAB.

**Macrodispersion of LDH in PVC/LDH composite.** In addition, another popular organic–inorganic composite, PVC/LDH composite[57,58], was also investigated to verify the generality of the developed visualization strategy. Similarly, the positively charged LDHs were organically modified by an AIE-active anionic surfactant (TPE-SDS) to possess both organic compatibility and AIE characteristic. Supplementary Fig. 11 showed the TEM image of the LDH microdispersion in PVC matrix. On the other hand, the spatial distribution of LDHs were also directly visualized by CFM (Supplementary Fig. 12). These results demonstrated that the proposed visualization strategy exhibited the generality for studying the spatial distribution of inorganic fillers in organic matrix.

## Discussion

In conclusion, we have synthesized a new fluorescent surfactant with AIE effect through attachment of TPE units onto cationic surfactants. The resulting cationic surfactant TPE-DTAB facilitates the construction of organo-modified MMT in order to be compatible with polymer matrix. Furthermore, the synthesis of TPE-DTAB surfactant enables us to develop a simple and versatile fluorescence imaging platform for direct observation of macrodispersion of MMT fillers in polymer matrix. The generality of this strategy further highlights 3D visualization of LDH filler macrodispersion in PVC matrix. Note that scanning electron microscopy could give microscale imaging with simple sample preparation. However, scanning electron microscopy is usually used for investigating surface topography. With the help of penetration depth of electrons, it might also observe dispersion of particles in slight interior of samples. In comparison to the nanoscale electron microscopy techniques (for example, TEM and 3D-TEM tomography), the AIE-active CFM technology offers the advantages of a wide-view dispersion image with simplicity, sensitivity and high-contrast, but a limited spatial resolution of fine structure by light diffraction ($\sim$200 nm in the lateral dimension and 500 nm in the axial dimension)[15,17]. In contrast, super-resolution microscopy imaging techniques, such

as stimulated emission depletion microscopy and stochastic optical reconstruction microscopy, can obtain a higher resolution than the diffraction limit. For example, stimulated emission depletion microscopy can achieve an image resolution of 20–40 nm in the lateral dimensions and 30–50 nm in the axial dimension[17]; while stochastic optical reconstruction microscopy can reach 20–30 nm and 50–60 nm for the lateral resolution and axial resolution, respectively[59]. It is worth mentioning that super-resolution microscopy is potentially applicable to the proposed AIE method if we can synthesize some AIE molecules suitable for such technique. On the other hand, cathodoluminescence by the interaction between electron beams and solid luminescent materials is a similar idea for the proposed AIE method, which has been widely used to investigate the nanoscale properties of solid samples[60]. Specially, the defect structure of solid samples could be visualized by cathodoluminescence microscopy[61], which is difficult for the proposed AIE method. In addition, the proposed AIE molecules in our study are designed for charged inorganic and organic substances, which is inapplicable to non-charged inorganic and organic substances. However, it has been reported that non-charged inorganics could be easily modified by luminescent polymers through physisorption[62]. With these merits in the current protocol, this facile imaging platform may open viable opportunities and inspirations for macrodispersion of other inorganic fillers in organic–inorganic composites, especially for dispersion of composite materials difficult to be distinguished using electron microscope imaging.

## Methods

**Materials.** Zinc dust and 4-hydroxyl benzophenone were purchased from Sigma-Aldrich Chemical Co. Sodium hydride (NaH), $TiCl_4$, poly(vinyl chloride) with Mn = 67,750 (polydispersity index (PDI) = 2.32), dioctyl phthalate (DOP), CTAB, calcium stearate, zinc stearate, NaCl and $K_2CO_3$ were purchased from J&K Chemical Ltd. Anhydrous dimethyl formamide (DMF), tetrahydrofuran (THF) and sodium sulfate were purchased from Alfa Aesar. 1-Bromooctane, 1,4-dibromobutane and trimethylamine in THF (100 ml, 1.0 M) were purchased from Tokyo Chemical Industry. Ethanol, PE, acetone, ethyl acetate (EA) and dichloromethane were purchased from Beijing Chemical Reagent Company. Sodium montmorillonite (Na⁺-MMT) with cation exchange capacity values of 145 meq per 100 g (from Nanocor, PGW grades) was used without further purification. Carbonate intercalated Mg-Al LDHs (with a molar ratio of 2:1 between $Mg^{2+}$ and $Al^{3+}$) were synthesized and characterized according to the literature[54]. TPE-SDS was synthesized and characterized according to our previous work[36]. All reagents were of analytical grade and used without further purification. Water was purified with a Milli-Q purification system (Milli-Q).

**Synthesis of compound 1.** In a 250-ml, two-necked, round-bottom flask equipped with a condenser, zinc dust (2.9 g, 44 mmol) and 4-hydroxybenzophenone (2.0 g, 10 mmol) were dissolved in 100 ml dry THF under nitrogen. The mixture was cooled to −78 °C and $TiCl_4$ (2.5 ml, 22 mmol) was added dropwise. After the addition, the mixture was allowed to warm to room temperature in 0.5 h, and then was heated to reflux for overnight. The reaction was quenched with 10% aqueous $K_2CO_3$ solution. The mixture was extracted with diethyl ether for three times and the combined organic layer was washed with brine twice and dried over sodium sulfate. After solvent evaporation, the crude product was separated through silica-gel chromatography flushed with PE/EA (v/v 1:1). 1.56 g product was obtained as a white solid with a yield of 86%. ¹H NMR (600 MHz, $CDCl_3$, δ): 6.94 − 7.05 (m, 10H), 6.76 − 6.81 (m, 4H), 6.47 − 6.52 (m, 4H).

**Synthesis of compound 2.** Under $N_2$ atmosphere, sodium hydride (0.022 g, 0.55 mmol) was added to the solution of **1** (0.182 g, 0.50 mmol) in dry DMF (10 ml) and the mixture was stirred for extra 30 min at room temperature. Then 1-bromooctane (0.145 g, 0.75 mmol) was added and the reaction mixture was stirred at 60 °C for 8.0 h. When the reaction completed, the solvent DMF was removed and the residue was redissolved with EA, and the resulting solution was washed with water for three times and dried over sodium sulfate. The solution was concentrated and the residue was purified through silica gel chromatography flushed with PE/EA (7:1 v/v), compound **2** (0.121 g, 0.25 mmol) as a yellow liquid was obtained with a yield of 51%. ¹H NMR (600 MHz, $CDCl_3$, δ): 6.98 − 7.11 (m, 10H), 6.83 − 6.92 (m, 4H), 6.52 − 6.63 (m, 4H), 3.82 − 3.88 (m, 2H), 1.67 − 1.76 (m, 2H), 1.37 − 1.41 (m, 2H), 1.24 − 1.29 (m, 8H), 0.85 − 0.88 (t, 3H).

**Synthesis of compound 3.** The compound **2** (0.174 g, 0.50 mmol) and $K_2CO_3$ (0.076 g, 0.55 mmol) were mixed in 20 ml acetone. After stirred for 1 h, 1,4-dibromobutane (0.118 g, 0.55 mmol) was added and the resulting mixture was stirred for 24 h at 60 °C. After evaporation of acetone, the obtained solids were first redispersed in EA and then filtered to remove the insoluble $K_2CO_3$. The purified product was obtained in 74% yield after purification separation by silica gel chromatography using PE/EA (10:1 (v/v)). $^1H$ NMR (600 MHz, $CDCl_3$, $\delta$): 7.00 − 7.12 (m, 10H), 6.88 − 6.95 (m, 4H), 6.59 − 6.65 (m, 4H), 4.20 − 4.25 (m, 2H), 3.84 − 3.93 (m, 4H), 2.43 − 2.46 (m, 2H), 1.81 − 1.86 (m, 4H), 1.70 − 1.76 (m, 2H), 1.38 − 1.43 (m, 2H), 1.26 − 1.32 (m, 8H), 0.87 − 0.89 (t, 3H).

**Synthesis of TPE-DTAB (compound 4).** A 100-ml flask with a magnetic spin bar was charged with **3** (0.3 g, 0.50 mmol) dissolved in 20 ml of THF. To this solution, trimethylamine (1.0 M, 5 ml) was added. The mixture was heated to reflux and stirred for 3 days. During this period, 5 ml of trimethylamine in THF was added at several intervals. After THF and extra trimethylamine were evaporated, the residue was washed with chloroform and acetone and then dried overnight *in vacuo* at 60 °C. A yellowish product was obtained in 88% yield. $^1H$ NMR (600 MHz, DMSO-$d_6$, $\delta$): 7.07 − 7.15 (m, 6H), 6.92 − 6.98 (m, 4H), 6.80 − 6.88 (m, 4H), 6.64 − 6.73 (m, 4H), 3.91 − 3.95 (m, 2H), 3.82 − 3.87 (m, 2H), 3.34 − 3.39 (m, 2H), 3.07 − 3.09 (t, 9H), 1.79 − 1.85 (m, 2H), 1.61 − 1.72 (m, 4H), 1.34 − 1.36 (m, 2H), 1.24 − 1.27 (m, 8H), 0.84 − 0.86 (t, 3H). $^{13}C$ NMR (600 MHz, DMSO-$d_6$, $\delta$): 157.54, 157.29, 144.27, 144.20, 139.82, 139.64, 136.22, 132.41, 132.36, 131.19, 128.28, 128.17, 126.79, 126.76, 114.23, 114.19, 114.14, 114.09, 67.67, 66.97, 65.40, 52.60, 31.69, 29.20, 29.16, 29.12, 26.11, 26.02, 25.99, 22.55, 19.72, 14.43. MS: $m/z$: 590.3996 ($[M-Br]^+$, calculated for $C_{41}H_{52}NO_2$, 590.3993).

**Synthesis of TPE-DTAB-modified MMT and TPE-SDS-modified LDH.**
TPE-DTAB-modified MMT was prepared from $Na^+$-MMT by ion exchange method. Typically, a 0.5 g portion of $Na^+$-MMT was mixed with 50 ml of deionized water. Then, 0.5 g of TPE-DTAB was added in the MMT solution. The ion exchange was carried out under stirring for 1 h at 60 °C. Then, the reaction solution was centrifuged at 5,000 r.p.m. for 5 min, and the precipitate was washed with distilled water to remove the physically adsorbed TPE-DTAB. The obtained TPE-DTAB-modified MMT was dried under vacuum at 60 °C, and then finely powdered in an agate mortar for further use. TPE-SDS-modified LDH powder was prepared according to the same procedure.

**Synthesis of CTAB/TPE-DTAB-modified MMT.** $Na^+$-MMT (0.5 g) was dispersed in 50 ml of deionized water. Then, a mixture of TPE-DTAB (0.05 g) and CTAB (0.25 g) was added to the MMT dispersion and stirred for 1 h at 60 °C. After centrifugation and washing, the obtained CTAB/TPE-DTAB-modified MMT solid was dried under vacuum at 60 °C and ground into powder.

**Preparation of PVC/MMT composite.** The PVC/TPE-DTAB-modified MMT composite, containing 10.0 g PVC powder, 5.0 g DOP and 0.5 g TPE-DTAB-modified MMT powder, was prepared by blending in a heated double-roller mixer for 5 min at 140 °C. The resulting composites were molded at 120 °C and then cooled at room temperature to give thin films with a thickness of 1 mm. The PVC/CTAB/TPE-DTAB-modified MMT composite was prepared according to the same procedure.

**Preparation of PVC/LDH composite.** The PVC/TPE-SDS-modified LDH composite, containing 10.0 g PVC powder, 5.0 g DOP, 0.2 g TPE-SDS-modified LDH powder, 0.23 g calcium stearate and 0.1 g zinc stearate, was prepared by blending in a heated double-roller mixer for 5 min at 140 °C. The resulting composite was molded at 120 °C and then cooled at room temperature to give thin films with a thickness of 1 mm.

**TEM sample preparation.** The composite films were ultrathin-sectioned with a diamond knife at − 120 °C using a Leica EM UC6 ultramicrotome. The obtained ultrathin sections were then collected in a trough filled with deionized water and placed on 200-mesh copper grids.

**Characterization.** Proton and carbon-13 nuclear magnetic resonance ($^1H$ NMR and $^{13}C$ NMR) spectra were recorded at room temperature with a 600-MHz Bruker spectrometer (Bruker). MS was carried out with Quattro microtriple quadrupole mass spectrometer (Waters). Electrical conductivity measurements were performed using a EC 215 conductivity meter (Shanghai Jingmi Instrumental Co.). TEM photographs were performed on a Tecnai G220 TEM (FEI Company) at an accelerating voltage of 200 kV. Ultraviolet–visible spectra were measured on a USB 4000 miniature fibre optic spectrometer in absorbance mode with a DH-2000 deuterium and tungsten halogen light source (Ocean Optics). Fluorescence spectra were obtained using a F-7000 fluorescence spectrophotometer at a slit of 5.0 nm with a scanning rate of 1,200 nm min$^{-1}$. X-ray diffraction measurements of MMT and TPE-DTAB-modified MMT were performed with a Brucker D8 ADVANCE X-ray diffractometer (Bruker) equipped with graphite–monochromatized Cu/Kα

radiation ($\lambda = 0.1541$ nm). The samples as unoriented powders were step-scanned in steps of 0.02° ($2\theta$) in the range of 2–10°. Fourier transform infrared spectroscopy experiments were carried out on Nicolet 380 system (Thermo) containing a controlled environment chamber equipped with $CaF_2$ windows. Zeta potential was determined using a Malvern Zetasizer 3000HS nano-granularity analyzer. The quantum yield values were obtained from the reconvolution fit analysis (Edinburgh F980 analysis software) equipped with an integrating sphere. Fluorescence microscope images were recorded on a confocal laser scanning microscope (Leica, TCS SP8).

## References

1. Ziolo, R. F. *et al.* Matrix-mediated synthesis of nanocrystalline γ-$Fe_2O_3$: a new optically transparent magnetic material. *Science* **257**, 219–223 (1992).
2. Podsiadlo, P. *et al.* Ultrastrong and stiff layered polymer nanocomposites. *Science* **318**, 80–83 (2007).
3. Capadona, J. R., Shanmuganathan, K., Tyler, D. J., Rowan, S. J. & Weder, C. Stimuli-responsive polymer nanocomposites inspired by the sea cucumber dermis. *Science* **319**, 1370–1374 (2008).
4. Manias, E. Stiffer by design. *Nat. Mater.* **6**, 9–11 (2007).
5. Li, Q. *et al.* Flexible high-temperature dielectric materials from polymer nanocomposites. *Nature* **523**, 576–579 (2015).
6. Rittigstein, P., Priestley, R. D., Broadbelt, L. J. & Torkelson, J. M. Model polymer nanocomposites provide an understanding of confinement effects in real nanocomposites. *Nat. Mater.* **6**, 278–282 (2007).
7. Chandran, S., Begam, N., Padmanabhan, V. & Basu, J. K. Confinement enhances dispersion in nanoparticle–polymer blend films. *Nat. Commun.* **5**, 3697–3706 (2014).
8. Mangal, R., Srivastava, S. & Archer, L. A. Phase stability and dynamics of entangled polymer–nanoparticle composites. *Nat. Commun.* **6**, 7198–7207 (2015).
9. Suter, J. L., Groen, D. & Coveney, P. V. Chemically specific multiscale modeling of clay–polymer nanocomposites reveals intercalation dynamics, tactoid self-assembly and emergent materials properties. *Adv. Mater.* **27**, 966–984 (2015).
10. Giannelis, E. P. Polymer layered silicate nanocomposites. *Adv. Mater.* **8**, 29–35 (1996).
11. Balazs, A. C., Emrick, T. & Russell, T. P. Nanoparticle polymer composites: where two small worlds meet. *Science* **314**, 1107–1110 (2006).
12. Wang, J., Lin, L., Cheng, Q. & Jiang, L. A strong bio-inspired layered PNIPAM–clay nanocomposite hydrogel. *Angew. Chem. Int. Ed.* **51**, 4676–4680 (2012).
13. Ojijo, V. & Ray, S. S. Nano-biocomposites based on synthetic aliphatic polyesters and nanoclay. *Prog. Mater. Sci.* **62**, 1–57 (2014).
14. Ray, S. S. Recent trends and future outlooks in the field of clay-containing polymer nanocomposites. *Macromol. Chem. Phys.* **215**, 1162–1179 (2014).
15. Weckhuysen, B. M. Chemical imaging of spatial heterogeneities in catalytic solids at different length and time scales. *Angew. Chem. Int. Ed.* **48**, 4910–4943 (2009).
16. Ameloot, R. *et al.* Three-dimensional visualization of defects formed during the synthesis of metal–organic frameworks: a fluorescence microscopy study. *Angew. Chem. Int. Ed.* **52**, 401–405 (2013).
17. Hell, S. W. Nanoscopy with focused light (Nobel Lecture). *Angew. Chem. Int. Ed.* **54**, 8054–8066 (2015).
18. Levitsky, I. & Krivoshlykov, S. G. Rational design of a Nile Red/polymer composite film for fluorescence sensing of organophosphonate vapors using hydrogen bond acidic polymers. *Anal. Chem.* **73**, 3441–3448 (2001).
19. Cao, X. D., Li, C. M., Bao, H. F., Bao, Q. L. & Dong, H. Fabrication of strongly fluorescent quantum dot–polymer composite in aqueous solution. *Chem. Mater.* **19**, 3773–3779 (2007).
20. Hong, Y. N., Lam, J. W. Y. & Tang, B. Z. Aggregation-induced emission. *Chem. Soc. Rev.* **40**, 5361–5388 (2011).
21. Hu, R. R., Leung, N. L. C. & Tang, B. Z. AIE macromolecules: syntheses, structures and functionalities. *Chem. Soc. Rev.* **43**, 4494–4562 (2014).
22. Mei, J. *et al.* Aggregation-induced emission: the whole is more brilliant than the parts. *Adv. Mater.* **26**, 5429–5479 (2014).
23. Kwok, R. T. K., Leung, C. W. T., Lam, J. W. Y. & Tang, B. Z. Biosensing by luminogens with aggregation-induced emission characteristics. *Chem. Soc. Rev.* **44**, 4228–4238 (2015).
24. Mei, J., Leung, N. L. C., Kwok, R. T. K., Lam, J. W. Y. & Tang, B. Z. Aggregation-induced emission: together we shine, united we soar. *Chem. Rev.* **115**, 11718–11940 (2015).
25. Schmidt, D. F., Clément, F. & Giannelis, E. P. On the origins of silicate dispersion in polysiloxane/layered-silicate nanocomposites. *Adv. Funct. Mater.* **16**, 417–425 (2006).

26. Iijima, M., Kobayakawa, M., Yamazaki, M., Ohta, Y. & Kamiya, H. Anionic surfactant with hydrophobic and hydrophilic chains for nanoparticle dispersion and shape memory polymer nanocomposites. *J. Am. Chem. Soc.* **131**, 16342–16343 (2009).

27. Maheshwari, S. *et al.* Layer structure preservation during swelling, pillaring, and exfoliation of a zeolite precursor. *J. Am. Chem. Soc.* **130**, 1507–1516 (2008).

28. Schmidt, D. F. & Giannelis, E. P. Silicate dispersion and mechanical reinforcement in polysiloxane/layered silicate nanocomposites. *Chem. Mater.* **22**, 167–174 (2010).

29. Mark, J. E. Some novel polymeric nanocomposites. *Acc. Chem. Res.* **39**, 881–888 (2006).

30. Gilman, J. W. *et al.* Polymer/layered silicate nanocomposites from thermally stable trialkylimidazolium-treated montmorillonite. *Chem. Mater.* **14**, 3776–3785 (2002).

31. Möller, M. W. *et al.* UV-cured, flexible, and transparent nanocomposite coating with remarkable oxygen barrier. *Adv. Mater.* **24**, 2142–2147 (2012).

32. Dang, Z.-M., Yuan, J.-K., Yao, S.-H. & Liao, R.-J. Flexible nanodielectric materials with high permittivity for power energy storage. *Adv. Mater.* **25**, 6334–6365 (2013).

33. Dawson, J. I. & Oreffo, R. O. C. Clay: new opportunities for tissue regeneration and biomaterial design. *Adv. Mater.* **25**, 4069–4086 (2013).

34. Wang, J. *et al.* Click synthesis, aggregation-induced emission, E/Z isomerization, self-organization, and multiple chromisms of pure stereoisomers of a tetraphenylethene-cored luminogen. *J. Am. Chem. Soc.* **134**, 9956–9966 (2012).

35. Li, J. W. *et al.* An aggregation-induced-emission platform for direct visualization of interfacial dynamic self-assembly. *Angew. Chem. Int. Ed.* **53**, 13518–13522 (2014).

36. Guan, W. J., Zhou, W. J., Lu, C. & Tang, B. Z. Synthesis and design of aggregation-induced emission surfactants: direct observation of micelle transitions and microemulsion droplets. *Angew. Chem. Int. Ed.* **54**, 15160–15165 (2015).

37. Yu, C. M. *et al.* Hyperbranched polyester-based fluorescent probe for histone deacetylase via aggregation-induced emission. *Biomacromolecules* **14**, 4507–4514 (2013).

38. Gu, X. Polymorphism-dependent emission for di(p-methoxylphenyl) dibenzofulvene and analogues: optical waveguide/amplified spontaneous emission behaviors. *Adv. Funct. Mater.* **22**, 4862–4872 (2012).

39. Xu, X. *et al.* Functionalization of graphene by tetraphenylethylene using nitrene chemistry. *RSC Adv.* **2**, 7042–7047 (2012).

40. Jiang, B.-P., Guo, D.-S., Liu, Y.-C., Wang, K.-P. & Liu, Y. Photomodulated fluorescence of supramolecular assemblies of sulfonatocalixarenes and tetraphenylethene. *ACS Nano* **8**, 1609–1618 (2014).

41. Rosen, M. J. & Kunjappu, J. T. *Surfactants and Interfacial Phenomena* (Wiley, 2012).

42. Holmberg, K., Jönsson, B., Kronberg, B. & Lindman, B. *Surfactants and Polymers in Aqueous Solution* (Wiley, 2002).

43. Shi, L., Lundberg, D., Musaev, D. G. & Menger, F. M. [12]Annulene gemini surfactants: structure and self-assembly. *Angew. Chem. Int. Ed.* **46**, 5889–5891 (2007).

44. Tong, H. *et al.* Protein detection and quantitation by tetraphenylethene-based fluorescent probes with aggregation-induced emission characteristics. *J. Phys. Chem. B* **111**, 11817–11823 (2007).

45. Xia, Y. J. *et al.* Water-soluble nano-fluorogens fabricated by self-assembly of bolaamphiphiles bearing AIE moieties: towards application in cell imaging. *J. Mater. Chem. B* **3**, 491–497 (2015).

46. Lakowicz, J. R. *Principles of Fluorescence Spectroscopy* (Spriger, 2006).

47. Mariott, W. R. & Chen, E. Y.-X. Stereochemically controlled PMMA-exfoliated silicate nanocomposites using intergallery-anchored metallocenium cations. *J. Am. Chem. Soc.* **125**, 15726–15727 (2003).

48. Xu, L. *et al.* Interfacial modification of magnetic montmorillonite (MMT) using covalently assembled LbL multilayers. *J. Phys. Chem. C* **118**, 20357–20362 (2014).

49. Pirillo, S., Luna, C. R., López-Corral, I., Juan, A. & Avena, M. J. Geometrical and electronic properties of hydrated sodium montmorillonite and tetracycline montmorillonite from DFT calculations. *J. Phys. Chem. C* **119**, 16082–16088 (2015).

50. Liyanage, A. U., Ikhuoria, E. U., Adenuga, A. A., Remcho, V. T. & Lerner, M. M. Synthesis and characterization of low-generation polyamidoamine (PAMAM) dendrimer–sodium montmorillonite (Na-MMT) clay nanocomposites. *Inorg. Chem.* **52**, 4603–4610 (2013).

51. Gu, Z., Gao, M., Lu, L., Liu, Y. & Yang, S. Montmorillonite functionalized with zwitterionic surfactant as a highly efficient adsorbent for herbicides. *Ind. Eng. Chem. Res.* **54**, 4947–4955 (2015).

52. Lotya, M. *et al.* Liquid phase production of graphene by exfoliation of graphite in surfactant/water solutions. *J. Am. Chem. Soc.* **131**, 3611–3620 (2009).

53. Lin, S. C., Shih, C.-J., Strano, M. S. & Blankschtein, D. Molecular insights into the surface morphology, layering structure, and aggregation kinetics of surfactant-stabilized graphene dispersions. *J. Am. Chem. Soc.* **133**, 12810–12823 (2011).

54. Guan, W. J., Zhou, W. J., Huang, Q. W. & Lu, C. Chemiluminescence as a novel indicator for interactions of surfactant–polymer mixtures at the surface of layered double hydroxides. *J. Phys. Chem. C* **118**, 2792–2798 (2014).

55. Chen, S., Zhou, W. J., Cao, Y. Q., Xue, C. C. & Lu, C. Organo-modified montmorillonite enhanced chemiluminescence via inactivation of halide counterions in a micellar solution. *J. Phys. Chem. C* **118**, 2851–2856 (2014).

56. Guan, W. J., Lu, J., Zhou, W. J. & Lu, C. Aggregation-induced emission molecules in layered matrices for two-color luminescence films. *Chem. Commun.* **50**, 11895–11898 (2014).

57. Liu, J., Chen, G. M. & Yang, J. P. Preparation and characterization of poly(vinyl chloride)/layered double hydroxide nanocomposites with enhanced thermal stability. *Polymer* **49**, 3923–3927 (2008).

58. Awad, W. H. *et al.* Material properties of nanoclay PVC composites. *Polymer* **50**, 1857–1867 (2009).

59. Huang, B., Wang, W., Bates, M. & Zhuang, X. W. Three-dimensional super-resolution imaging by stochastic optical reconstruction microscopy. *Science* **319**, 810–813 (2008).

60. Bischak, C. G. *et al.* Cathodoluminescence-activated nanoimaging: noninvasive near-field optical microscopy in an electron microscope. *Nano Lett.* **15**, 3383–3390 (2015).

61. Götze, J. Application of cathodoluminescence microscopy and spectroscopy in geosciences. *Microsc. Microanal.* **18**, 1270–1284 (2012).

62. Nakao, A. & Fujiki, M. Visualizing spontaneous physisorption of non-charged π-conjugated polymers onto neutral surfaces of spherical silica in nonpolar solvents. *Polym. J.* **47**, 434–442 (2015).

## Acknowledgements

This work was supported by National Basic Research Program of China (973 Program, 2014CB932103), the National Natural Science Foundation of China (21575010 and 21375006), Innovation and Promotion Project of Beijing University of Chemical Technology and the Innovation and Technology Commission (ITC-CNERC14SC01).

## Author contributions

W.J.G. and C.L. conceived the experiments. W.J.G. and S.W. carried out the experiments. W.J.G., C.L. and B.Z.T. contributed to data analysis and writing of this manuscript.

# Putting pressure on aromaticity along with *in situ* experimental electron density of a molecular crystal

Nicola Casati[1], Annette Kleppe[2], Andrew P. Jephcoat[3] & Piero Macchi[4]

When pressure is applied, the molecules inside a crystal undergo significant changes of their stereoelectronic properties. The most interesting are those enhancing the reactivity of systems that would be otherwise rather inert at ambient conditions. Before a reaction can occur, however, a molecule must be activated, which means destabilized. In aromatic compounds, molecular stability originates from the resonance between two electronic configurations. Here we show how the resonance energy can be decreased in molecular crystals on application of pressure. The focus is on *syn*-1,6:8,13-Biscarbonyl[14]annulene, an aromatic compound at ambient conditions that gradually localizes one of the resonant configurations on compression. This phenomenon is evident from the molecular geometries measured at several pressures and from the experimentally determined electron density distribution at 7.7 GPa; the observations presented in this work are validated by periodic DFT calculations.

[1] Paul Scherrer Institute, WLGA/229, CH-5232 Villigen, Switzerland. [2] Diamond light source Ltd., Harwell Science and innovation Campus, Didcot OX11ODE, UK. [3] Institute for Study of the Earth's interior, Okayama University, Yamada 827, Misasa, Tottori 682-0193, Japan. [4] Department of Chemistry and Biochemistry, University of Bern, Freiestrasse 3, Bern CH-3012, Switzerland. Correspondence and requests for materials should be addressed to P.M. (email: piero.macchi@dcb.unibe.ch) or to N.C. (email: nicola.casati@psi.ch).

One of the most important and famous classes of chemical compounds is that of aromatic molecules. According to Hückel[1,2], the conjugation of an odd number of electron pairs in a ring is stabilized by the resonance between two equivalent electronic configurations. The implications for the chemistry of these compounds are enormous: the extra stability of aromatic hydrocarbons implies more severe conditions to induce reactions, compared with non-conjugated poly-olefins. Aromaticity affects also the structure of a molecule and its response to external magnetic fields, in particular, the nuclear resonant frequency. In general, a planar geometry formed by equally distant C atoms, an induced diamagnetic current in the ring and a scarce tendency to react are the main clues of aromaticity[3]. Nevertheless, an unbiased and universal criterion for quantifying aromaticity remains elusive. Over the years, chemists have challenged the very concept of aromaticity at times, synthesizing ever more exotic molecular systems and using different investigating methods (diffraction, NMR, reactivity tests, molecular orbital calculations and so on), with the aim of finding universal criteria based on structural, energetic and magnetic parameters[4–8]. Insight often came by comparing different molecular species, mimicking a continuous variation of the more relevant parameters, to solve the aromatic riddle[3,9]. In this work, instead, we adopt a different strategy to investigate aromaticity, which is probing its variation in a single species, while modifying continuously the molecular geometry through compression.

Reducing the aromaticity of a species requires significant external energy, to stabilize one of the two electronic configurations over the other, breaking the resonance. Typically, chemists make use of heat, light or electrochemical potential to attack an aromatic molecule. An alternative source of external energy is pressure, which rises the internal energy and the enthalpy of a system. This may perturb the molecular conformation and/or the electronic state stable at the ambient conditions, in favour of an otherwise inaccessible configuration.

Recent research in high-pressure (HP) solid-state chemistry led to the discovery of very peculiar phenomena, such as the polymerization of molecules like $N_2$ (ref. 10), CO (ref. 11) and $CO_2$ (ref. 12), or the transformation of metals into non-metals, for example, Na (ref. 13). Studies on organic crystals are less abundant and have appeared only more recently[14,15]. Benzene—that is, the prototype of aromaticity—is an exception, because the first, seminal investigations of its HP forms date back to 1960s (ref. 16). While its phase diagram remains controversial, there is consensus on the occurrence of an irreversible polymerization above 24 GPa, but the structure of this phase is known only from theoretical predictions[17], without experimental confirmation yet. Ciabini et al.[18] estimated that below a critical intermolecular distance (C–C ≈ 2.6 Å) lattice phonons are able to bring atoms of neighbouring molecules sufficiently close to induce an intermolecular addition reaction, which eventually leads to one or more polymeric products[19]. On crystals of s-triazine, a progressive destabilization of the π bonding orbitals has been reported, based on two-photon induced fluorescence[20], and the enhanced reactivity of the species was ascribed to this process. In both cases, however, details are missing for the pressure-induced distortions of the molecular geometry that could favour reactivity. In fact, the HP crystalline phases of benzene that anticipate the polymerization[21] are not known with sufficient precision to enable fine speculations and the high-pressure structures of triazine are poorly characterized. It is also worth noting that most of related studies are conducted in non-hydrostatic conditions.

To obtain relevant information on the aromaticity of species under pressure, the structural, electronic and energetic changes should be monitored. The elective method for investigating molecular geometries is single crystal X-ray diffraction, which maps the electron density (ED, $\rho(\mathbf{r})$) distribution in a crystal. Measurements of particular accuracy and completeness are able to reveal not only the maxima of $\rho(\mathbf{r})$ (coinciding with the nuclear sites) but also the smaller fraction of electrons present in between the atoms and responsible of the chemical bonding[22]. The ED mapping from X-ray diffraction is nowadays a well-established technique, but it requires quenching the atomic motion at low temperature for a sufficient deconvolution of the ED from the atomic displacements. In addition, a more sophisticated modelling based on atomic multipolar expansion[23] is necessary to extract the desired information from the diffraction data. In this respect, benzene may not be the perfect test case, because its solidification must occur in the HP apparatus and the crystal sample cannot be of the highest quality. Experimental ED determinations of molecular crystals at HP are not known so far. A few examples reported ED maps of simple inorganic compounds[24] or pure elements[25], obtained by maximum entropy method. Models using multipoles restricted to theoretical values have been tested against X-ray diffraction data for propionamide[26] and piperazinium hydrogen oxalate[27], but no full refinement was so far reported.

The main obstacles to ED mapping from HP X-ray diffraction are due to the pressure apparatus (the Diamond Anvil Cell, DAC), which reduces drastically the resolution, completeness and quality of available data[28]. However, such pitfalls may be overcome by a combination of higher pressure (which significantly attenuates thermal motion even at ambient temperature), modern synchrotron sources (easily providing high-intensity and very short wavelength radiation) and careful experimental strategies; these are discussed in details below.

A complementary approach to study the HP forms of molecules in crystals is first principle calculations, in particular density functional theory (DFT) with periodic boundary conditions[29]. This not only allows to validate the experimental observations and predict the occurrence of new phases, but also to calculate quantities otherwise not available or too difficult to measure, such as electron correlation, current density and electronic energy of a system. Moreover, a theoretical analysis enables extending the pressure range achievable with experiments.

In the following, we report on the experimental and theoretical investigations of an aromatic molecule in its crystal form and we analyse how its aromatic character is reduced by the pressure-induced modifications of the molecular geometry.

## Results

**HP single crystal X-ray diffraction.** Our investigation focused on a doubly bridged annulene, namely syn-1,6:8,13-Biscarbonyl[14] annulene (BCA), Fig. 1, for which high-quality crystals are available. Initially studied within the debate on aromaticity of annulenes[30], BCA was also the subject of a very detailed ED study at ambient P and low T (19 K) (ref. 31), which is an excellent benchmark. Due to the strain in the ring, the aromaticity of BCA is quite smaller than for benzene, therefore, a stronger response to perturbation is expected.

Single crystals of BCA were investigated with multi-temperature and multi-pressure X-ray diffraction, to determine experimentally the structural changes on varying the thermo-dynamic conditions. HP diffraction experiments were carried out at the I15 beamline of Diamond light source using a pinhole defined monochromatic beam with 0.31 Å wavelength and an Atlas CCD detector. Six simple data collections, based on perpendicular scans ($\omega$ and $\varphi$ at $\chi = 90$) with frontal detector (resulting in a resolution up to 0.8 Å), were performed in the

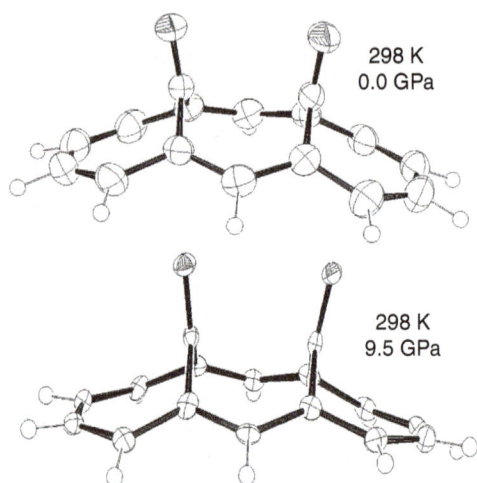

**Figure 1 | BCA molecule.** The molecular geometry and atomic displacement parameters of BCA at ambient pressure and at 9.5 GPa.

range from 0.0001 to 9.5 GPa. These were intended to determine the main structural changes occurring to the molecule as a function of pressure, up to the hydrostatic limit of the pressure-transmitting medium that we adopted (methanol:ethanol 4:1 mixture; *ca.* 10 GPa). The data collection at 7.7 GPa (experiment EE7741-1) was more extensive, involving several $\omega$ and $\varphi$ scans at four different $\chi$ positions (0°, 30°, 60° and 90°), using different $\theta$ positions for the detector, as it was intended to measure with the best accuracy the diffraction intensities and to reach the highest resolution ($d = 0.5$ Å). Accuracy here means: (a) high data completeness (that is, the portion of reciprocal space that is measurable), obtained using two crystals in the DAC; (b) redundant measures of the diffracted intensities (with every $\omega$ and $\varphi$ scan re-performed with a 2° offset of $\varphi$ and $\omega$, respectively) to minimize random errors and therefore enable a multipolar expansion of the ED. The high-energy radiation chosen reduces absorption effects and allows collection of higher-order reflections. A beam smaller than the crystal was selected by a 30-μm pinhole, to probe one crystal at a time and to maximize the sample/diamond diffraction intensity ratio. Periodic and molecular DFT calculations were used to simulate the structures, validate the experimental results, compute the geometry at pressures above the experimentally available range and compute those properties that are not directly accessible through experiments, like the current densities and the electron delocalization indices. Details of all experiments and calculations are in Methods section and in Supplementary Methods.

**Theoretical calculations and molecular geometries.** With its 14 carbon atom ring, BCA is a $4n + 2$ Hückel system, therefore potentially aromatic. However, the two carbonyl bridges distort the planarity of the ring, reducing the conjugation of C–C bonds and therefore decreasing the aromaticity of the system. The C–C distances are quite important indicators: in an aromatic molecule, the ring skeleton bonds should have homogeneous distances, quite shorter than single bonds but longer than double bonds. In the gas phase of a molecule like benzene, all C–C distances are equal by symmetry (1.389 Å), whereas in BCA the lower molecular symmetry ($C_{2v}$) does not imply equalization of C–C distances and the two bridges produce an heterogeneous distribution in the range 1.37–1.41 Å. Four C–C bonds are symmetry independent, Fig. 2. Gas phase DFT calculations on the isolated molecule (at B3LYP/6–31 + G(2d,2p) level) indicate that C1–C14, C2–C3 and all their symmetry equivalents, are shorter whereas C1–C2, C3–C4 and all their symmetry equivalents, are

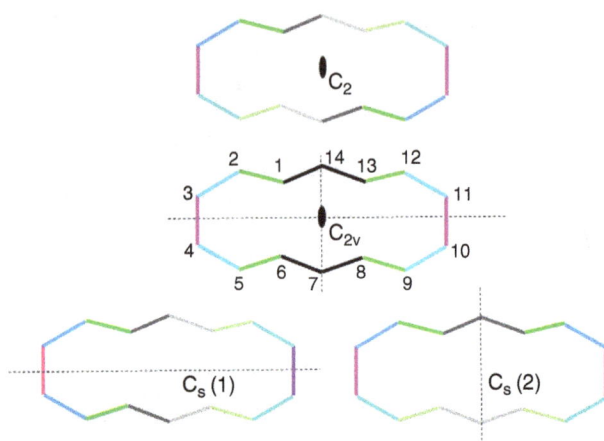

**Figure 2 | The possible configurations of BCA.** BCA symmetries (only the annulene skeleton is drawn): $C_{2v}$ is the most symmetric isomer, stable in the gas phase and very close to the molecule in the crystal at ambient pressure; $C_s(1)$ is the sub-symmetry which is close to the configuration found at HP; $C_2$ and $C_s(2)$ are the other two possible sub-symmetries. Note that only $C_s(1)$ implies breaking the resonance between $\psi_1$ and $\psi_2$.

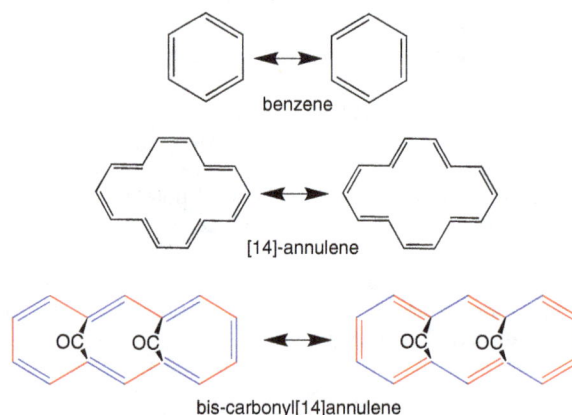

**Figure 3 | Resonant configurations in aromatic molecules.** Skeletal formulae of the resonant configuration in selected aromatic molecules.

longer. Nevertheless, a $C_{2v}$ symmetry still implies a perfect resonance between the two electronic configurations $\psi_1$ and $\psi_2$ (Figs 2 and 3). In the crystal phase, the BCA molecule sits on a general position (therefore, without any intramolecular symmetry element), although its geometry remains close to the gas phase $C_{2v}$ isomer, as confirmed by X-ray diffraction. The small root mean square deviation from $C_{2v}$ progressively increases on lowering the temperature (from 0.031 Å at 298 K to 0.036 Å at 19 K). It is worth noting that, even in the absence of carbonyl bridges, a ring strain significantly affects the aromaticity of the 'parent', unbridged [14]-annulene (Fig. 3), which is in fact not planar[32].

The aromaticity is not only reflected by the C–C distances, hence by the position of the $\rho(\mathbf{r})$ maxima, but also by the amount of ED in the chemical bonds, correlated with the bond strength. In this respect, the detailed experimental study by Destro and Merati[31] on BCA went beyond a routine geometrical analysis, providing also an accurate determination of $\rho(\mathbf{r})$ at 19 K. At the bond critical points, (that is, saddle points of the three-dimensional ED function, according to Bader's QTAIM[33]), $\rho(\mathbf{r})$ parallels the bond distances and confirms the perfect resonance between the two electronic configurations. Interestingly, this analysis revealed an unexpected bond critical point interconnecting the two bridging carbons (C15–C16 in Supplementary Fig. 1). Although the ED at the critical point is small, the feature is not anticipated from a simple Lewis structural formula of BCA. Moreover, the C–C distance of 2.593 Å is more than 1 Å longer than a typical covalent bond. A Møller-Plesset[34] perturbation calculation on the isolated molecule, indicates a small population of two virtual molecular orbitals containing in-phase combinations of the $\pi^*$-type $C=O$ orbitals. Nevertheless, this interaction is principally of closed-shell type.

As anticipated, the aromaticity also affects the magnetically induced current density in the ring. This modifies substantially the interaction of an external magnetic field with the spin active nuclei, as it can be monitored by nuclear magnetic resonance. However, the atomic chemical shift may depend on the position of the spin active nuclei (most often the protons, revealed by $H^1$ NMR). Therefore, geometries of cyclic molecules that differ for conformation may introduce biases because of shielding/deshielding effects not directly related to the aromatic behaviour. For this reason, a nuclear-independent chemical shift (NICS) indicator has been introduced[35] to provide an unbiased indication of the ring current. NICS is available only from theoretical simulations, because it requires calculating the shielding due to the ring current in the centre of the ring, assuming a virtual atom in that site. The NICS is the negative of the shielding tensor, for sake of consistency with traditional NMR chemical shifts. We calculated a negative NICS at the centre of the BCA ring, which implies the diatropic ring current produced by aromaticity[3].

As mentioned in the introduction, our goal was observing how all parameters, correlated with aromaticity, vary when the molecule is compressed. The HP experiments revealed a quite large compressibility of the crystal, which exceeds 25% from ambient pressure up to 9.5 GPa, in keeping with theoretical predictions (Fig. 4): the experimental $K_0$, calculated using a fourth order Birch–Murnaghan equation, is only 7.4 GPa (compared, for example, to 37.1 GPa for quartz)[36]. As the crystal shrinks, the electric field experienced by a molecule and generated by all other molecules in the crystal, increases significantly (Fig. 5). Because the molecule sits on a general position, inside the monoclinic unit cell, atoms which would be equivalent under the ideal $C_{2v}$ symmetry of the molecule, experience a different crystal electric field and this asymmetry increases with pressure. Therefore, an external stress gradually, but not uniformly, perturbs all the covalent bonds of the annulene skeleton.

Good indicators of this phenomenon are the average C–C distances of the hypothetical double bonds for each of the two resonant configurations (Figs 3 and 6). At ambient $P$, the observed and calculated pseudo-$C_{2v}$ symmetry implies that the two sets of bonds have coincident average distances (1.396 Å). As $P$ increases, however, one of the two resonant configurations of Fig. 3 ($\psi 1$) becomes progressively dominant. This is evident, because all the double bonds of this configuration shorten with respect to the ambient pressure geometry, whereas all the hypothetical double bonds of the alternative configuration ($\psi_2$) are unaltered or even slightly elongated (see also Supplementary

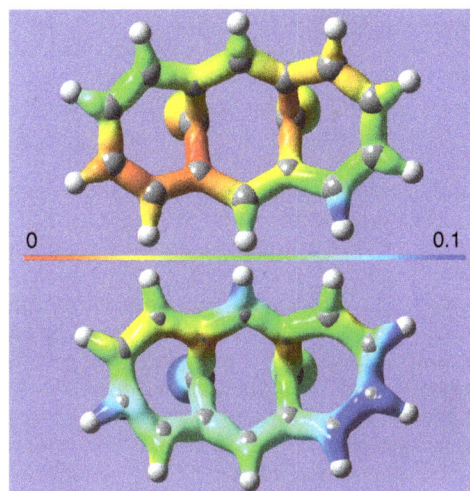

**Figure 5 | BCA electric field.** The magnitude of the electric field acting on a BCA molecule, plotted on a electron density isovalue surface of 0.2 a.u. of BCA at 0.0 GPa (top) and 7.7 GPa (bottom). The colour scale for the electric field is in atomic units ($= 5.1 \times 10^{11}$ Vm$^{-1}$, blue). a.u., arbitrary units.

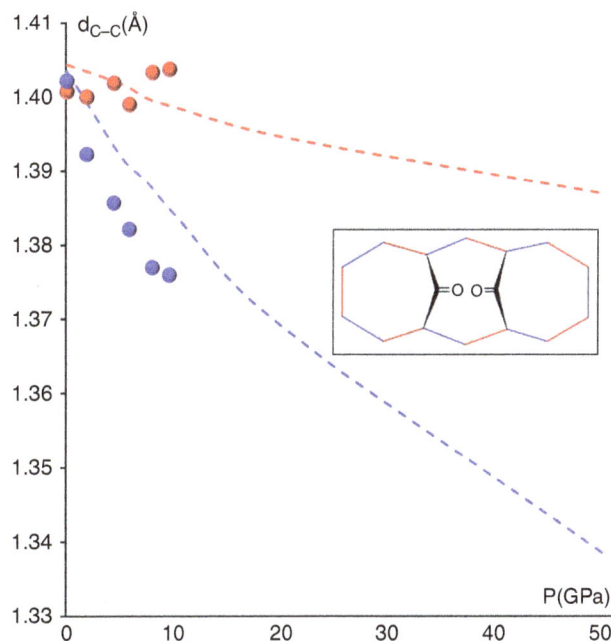

**Figure 6 | Resonant configurations bond evolution.** The average bond distances for hypothetical C–C double bonds of $\psi_1$ and $\psi_2$ (blue and red, following Fig. 3) as a function of pressure from the single crystal measurements (circles) or the theoretical calculations (dashed lines). Experimental data are corrected for thermal libration effects.

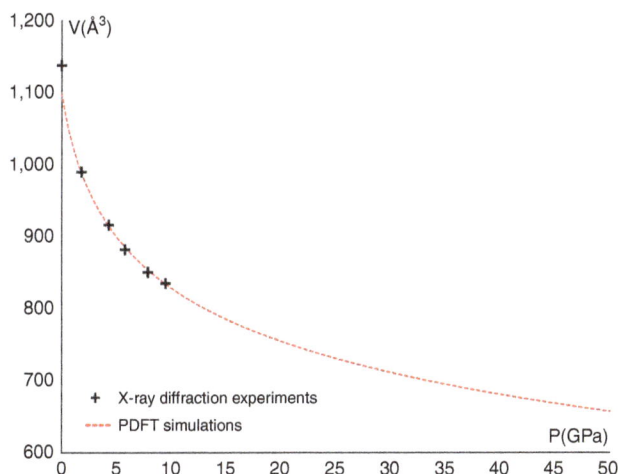

**Figure 4 | BCA compressibility.** The unit cell volume as a function of pressure, from single crystal diffraction experiments and from periodic DFT calculations.

Fig. 7 and Table 4). From the experimental structure at 9.5 GPa, the average distances of the two sets of bonds are 1.375 and 1.405 Å, respectively. The gap is even larger at 50 GPa, from the theoretical simulations (1.338 vs 1.387 Å, see also Supplementary Fig. 8) This shows that BCA clearly distorts from $C_{2v}$ symmetry, but the C–C bonds in the annulene ring respect one of the two mirror symmetries, namely the $C_s(1)$ configuration represented in Fig. 2. This distortion destabilizes the molecule (by ca. 65 kcal mol$^{-1}$) not only because of breaking the aromaticity, but also because of other geometrical distortions imposed by the reduced volume available for each molecule, including a significant bending of the bridging CO's, Fig. 1. The distance between them shortens from 2.591 (ambient pressure) to 2.525 Å (at 9.5 GPa, from experiment) or even 2.46 Å (at 50 GPa, from calculations), at the expense of the C=O bonds that experience a small but continuous elongation under pressure. This observation confirms the weak bonding proposed by Destro and Merati[31], and it suggests a slight strengthening with pressure.

**Electron and current density distribution.** The partial localization of one electronic configuration should be visible also from the analysis of the ED distribution. For this reason, we determined the accurate $\rho(\mathbf{r})$ from the X-ray diffraction intensities measured at 7.7 GPa. We used two methods to reconstruct $\rho(\mathbf{r})$: the traditional multipolar refinement[22] and the X-ray constrained wave function[37]. Details of the model refinements are in SupplementaryMethods, whereas the technical details on the necessities and pitfalls of ED determinations at HP will be subject of a forthcoming paper. Here we report only on the most relevant results of the $\rho(\mathbf{r})$ analysis. To validate the experimental models, the ED was also computed at various pressure points with periodic DFT calculations.

In Fig. 7, we plot $\rho(\mathbf{r})$ at isosurface values (0.305 arbitrary units) chosen to visualize the ED level of the mixed single–double bonds of the molecular skeleton. At 0.0001 GPa, the theoretical calculations and the experimental model[31] show an almost perfectly symmetric distribution. The bonds C1–C14 and C2–C3 (and all pseudosymmetry related ones, Fig. 2) display larger amount of ED, in agreement with their shorter distances, whereas C1–C2 and C3–C4 and all pseudosymmetry equivalents, have lower density. This scenario respects the resonance scheme, that,

however, partially breaks at 7.7 GPa. In fact, both the experimental models and the periodic DFT calculations demonstrate that bonds C2–C3, C4–C5 and C6–C7 gain ED whereas C1–C2, C3–C4 and C5–C6 loose it (Fig. 7). In the left part of the molecule, the scenario is now closer to a localized configuration because the ED accumulations are clearly associated only with the hypothetical double bonds of $\psi_1$ in Fig. 3. On the other hand, in the right part of the molecule a larger delocalization persists. To explain this, one should consider that localizing $\psi_1$ implies strengthening bonds C8–C9, C10–C11 and C12–C13 (originally weaker at ambient conditions) at the expense of C7–C8, C9–C10, C11–C12 and C13–C14 (originally stronger). Anyway, C8–C9, C10–C11 and C12–C13 have increased the ED amount compared with their pseudo-symmetric counter parts C5–C6, C3–C4 and C1–C2. A complete localization of $\psi_1$ would eventually occur at 50 GPa, where only periodic DFT calculations are available without confirmation from the experiment. At this pressure, all the double bonds of $\psi_1$ are associated with larger amounts of ED peaks, compared with all single bonds. The molecule has therefore become closer to a cyclic non-aromatic poly-ene.

The qualitative agreement between the experimental models (multipole or X-ray constrained wave function) and the first principle calculations (periodic DFT) at 7.7 GPa is remarkable, which excludes potential biases in the analysis and it allows to thrust the theoretical values computed at 50 GPa.

From the multipolar model or the X-ray constrained wave function, we could also determine the molecular graph (Fig. 8), that is, the set of lines of maximum ED (bond paths) that interconnect bonded atoms, following QTAIM. The partial localization of one of the two resonant configurations is visible also from the increased ED at the critical points (saddle points along the bond paths). On average, $\rho(\mathbf{r})$ is 2.29 and 2.15 e Å$^{-3}$ in the partial double or single bonds, respectively, which varies from what observed at ambient pressure at 19 K (on average, 2.05 e Å$^{-3}$ for both configurations).

As observed at the ambient $P$ and low $T$ by Destro and Merati[31], a bond path links the two carbonyl carbons, a feature confirmed also by the periodic DFT calculations.

Having an experimental wave function available, we can compute other quantities, otherwise not available from the

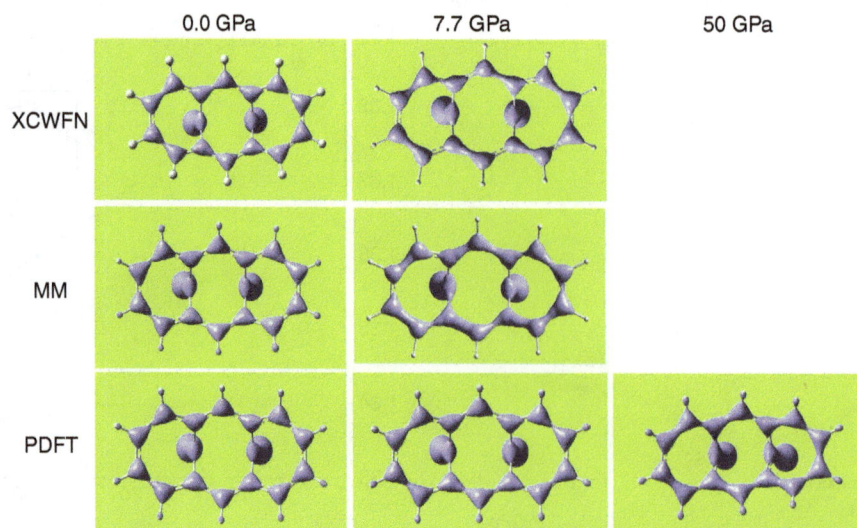

**Figure 7 | ED distribution of BCA.** Plots are shown at various pressure and from various sources: PDFT are periodic DFT calculations at B3LYP level of theory; XCWFN are X-ray constrained wave functions computed at Hartree-Fock level, but constrained against the experimentally measured diffraction intensities; MM is the electron density derived from multipolar expansion, with coefficients refined against experimentally measured intensities. Experimental data at ambient pressure are taken from Destro and Merati[31], collected at 19 K; the 7.7 GPa data are from this work.

multipolar model only; in particular, the delocalization indexes ($\delta$) (ref. 38) that address the amount of electron pairs shared between two atoms. While the theoretical calculations for the $C_{2v}$ symmetric geometry give on average $\delta_{C-C} = 1.34e$ for both configurations, at 7.7 GPa the partially localized double bonds have $\delta_{C-C} = 1.42e$ against $\delta_{C-C} = 1.28e$ of the partially localized single bonds. For the theoretically calculated structure at 50 GPa, $\delta_{C-C} = 1.47e$ and $1.22e$, respectively, in keeping with the further shortening of the double bonds and lengthening of the single bonds. Noteworthy, the non-planar and highly strained geometry of BCA would hamper a full localization of double bonds and the ideal $\delta_{C-C} = 2.0e$.

As anticipated above, the aromaticity of a molecule also affects the current density in the ring and the shielding experienced by the atomic nuclei. Current density can only be calculated, with gauge invariant orbitals for molecules, after geometry optimizations in the crystal at various pressure points. These results are illustrated in Fig. 9, where one can easily see the modified orientation of the current density vectors at 50 GPa. The NICS can be also used to analyse the pressure-induced changes. Recent works suggest to scan the out of plane components along a direction perpendicular to the ring[39]. In BCA, we can scan only along the direction opposed to the two CO bridges, otherwise the shielding of $C = O$ would severely interfere. In agreement with standard aromatic systems, NICS is a bit larger (in absolute value) at *ca.* 0.5–1.0 Å out of the plane, and then it decreases (Fig. 9c). NICS decreases from the ambient pressure to the 50 GPa structure. This is a further proof, based on magnetic criteria, that the aromaticity of BCA decreases as a function of pressure, in keeping with the structural, electronic and energetic criteria above discussed.

## Discussion

We have analysed the pressure-induced loss of aromaticity of a carbonyl annulene. The molecule activates by partially localizing one of the two resonant configurations. This mechanism may occur also in other aromatic systems under pressure and could be representative of the steps that anticipate addition reactions leading to polymerization, like in benzene.

A fortunate circumstance for BCA is that the activation mechanism is induced by a smaller pressure and it does not damage the crystal quality, thus the molecular geometries as well as the ED could be determined with sufficient accuracy to elucidate many important details.

This observation of progressive aromatic loss in BCA is extremely useful to test how different indicators of aromaticity respond to drastic changes of the molecular geometry. Criteria based on C–C distances appear to be extremely sensitive. Moreover, they are quite easy to determine whether single crystal X-ray diffraction is available. Deviations from a uniform distribution of C–C bond distances directly correlate with the raise of molecular energy that can be determined with theoretical methods but not experimentally. While magnetic criteria are also sensitive to an increased or decreased aromaticity, NICS are only available from theoretical calculations, whereas the [1]H NMR chemical shifts, although measurable even at HP, may not reveal so directly the ongoing changes of electronic configuration. In fact, at each pressure point, the protons would probe the ring current in different positions, due to the geometrical distortions that involve themselves as well. Therefore, they would be unable to reveal the actual modifications of the current density due only to the breaking of aromaticity.

In this study, we have also presented a way to determine experimentally the charge density of a molecular crystal under pressure, which provides additional indicators of the aromaticity based on the topological analysis of the ED. We believe the use of a synchrotron is presently mandatory to achieve the necessary quality and quantity of unique reflections; potential improvements on our setup include the use of poly-nanocrystalline diamonds, which should eliminate the problem of diamond dips (though introducing a significant background), the use of photon counting detectors (possibly with a high Z material such as GaAs), which would reduce the noise and an improved usage of panoramic cells, presently limited in the literature. The ongoing research on new liquid jet microsources of relatively high energy may also enable in the near future such experiments with laboratory sources, nevertheless the present limitations are difficult to overcome and limit the possibilities of such studies, though not their potential. As we have shown for BCA at 7.7 GPa, it was possible to successfully refine a full multipolar model and an X-ray constrained wave function. This enabled us to visualize the partial localization of double and single bonds, which eventually becomes more complete at higher pressure, according to theoretical predictions. ED criteria are seamlessly replicating the geometric criteria that, in this case, provide easily accessible and sufficiently reliable indicators of even minor changes of aromaticity. This cross-validation opens up the possibility of using this method for far more complex observation of ED localization in molecular systems.

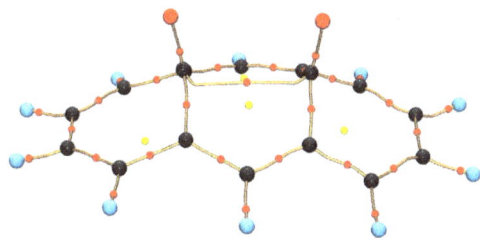

**Figure 8 | BCA bonding scheme.** Molecular graph for BCA at 7.7 GPa from the experimentally refined multipolar model. Atoms are represented by black (carbon), red (oxygen) and blue (hydrogen) spheres, critical points are shown as small red dots. Noteworthy, depending on the refinement model, a bond path interconnecting the two O atoms can also be localized.

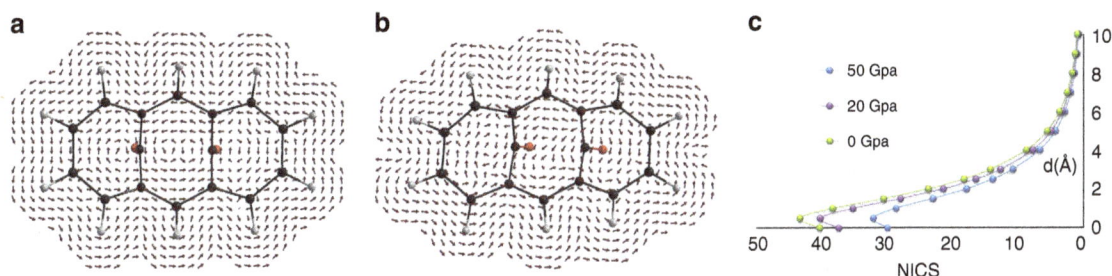

**Figure 9 | BCA ring current.** The calculated current density $J(r)$ for the molecule of BCA at 0.0001 GPa (**a**) and 50 GPa (**b**). (**c**) NICS scan perpendicularly to the average plane of the BCA ring (anti with respect to the bridging carbonyls).

## Methods

**HP diffraction measurements.** The two separate kinds of experiments were conducted using a Betsa and an own University of Oxford screw driven-type DAC equipped with 0.5 mm culet diamonds. In the first setup a single crystal of BCA was loaded, using a methanol;ethanol 4:1 mixture as pressure medium and ruby fluorescence for pressure measurement. At $P = 7.7$ GPa, a second setup was adopted, using two crystals pre-oriented with crystallographic axis almost normal to each. A sufficiently high redundancy was also sought to correct for problems such as diamond dips, which were also separately identified by recording transmission scans through the cell using a diode immediately after the cell itself. With this setup, the overall data redundancy was 4.7 (6.0 for data up to 0.8 Å resolution) and 70% of the unique reflections were measured (88% for data up to 0.8 Å resolution). This guarantees a sufficient sampling of the reciprocal space, especially in the region where the valence electrons are mostly scattering.

For both setups, a monochromatic radiation of 40 keV was focused down to about $90 \times 90 \, \mu m^2$ and then collimated by pinhole of 30 μm in diameter. In the 7.7 GPa experiment the full beam size was always probing only one of the two crystals, which were separately centred and measured. The beam was notably smaller than the crystal, which means all the beam was effectively used for diffraction by the crystal itself. In the case of beams larger than the crystal, problems may arise from the significant diffraction of diamonds, saturating the detector and of the other elements of the cell, which contribute to background. Data were treated using the dedicated HP routines present in the package of CrysalisPro[40] for shading areas, carefully assigning a well-describing vector and opening angle to the cell. Diamond reflections were individually masked in a similar way to already described procedures[41], which proved also important in obtaining a smooth background subtraction from the programme itself. Suspiciously badly fitting reflections were investigated and manually rejected when: (a) their intensity was significantly lower than their equivalents and on the border of masks, (b) their intensity was significantly higher than their equivalents and on the tails of a diamond reflection and (c) they were on images collected at angles were an obvious diamond dip occurred, as revealed by the mentioned transmission scans. No change of space group was detected up to the maximum pressure; therefore, standard refinements were carried on the known structure using ShelXL[42] as included in the WingX package[43]. The non-standard P2₁/n space group was used for all determinations for sake of consistency with previous studies[31].

### Experimental determination and modelling of the ED distribution at 7.7 GPa.

To obtain an accurate ED mapping, it is necessary to collect accurately and extensively the X-ray data, up to a sufficient resolution. High resolution is necessary because the large number of parameters of a multipolar model requires many more intensities to match a sufficient observation/parameter ratio. The data set should be sufficiently complete to avoid systematic effects in the refinement. This is particularly cogent for the low angle data, because the valence electrons, which are mainly responsible for aspherical scattering, are not contributing to reflection intensities at higher resolution. To reach the goal of determination of the ED in a crystal, it is also necessary a very accurate measurement of the reflection intensities, which means minimizing the effect of experimental errors such as absorption by the sample, extinction and, very important here, absorption by diamonds and metal gasket. Apart from accurate correction of the data, the modern area detector technologies offer exceptional possibility to improve the precision of the measurement by repeated collection of the same intensities.

The data set collected at 7.7 GPa was an ideal candidate to attempt a determination of accurate ED in BCA, because the pressure is sufficient to reduce the thermal parameters by a factor of *ca.* 4 (making the atomic displacement parameters comparable to those measured at 120 K) and because both crystals were still sufficiently free from damages. Below this pressure, the atomic motion is still too large and above it the hydrostaticity of the medium decreases and therefore some damage could occur to the samples, which does not affect a conventional structure determination but hampers any accurate ED mapping. Data from the two difference crystals were linearly scaled and merged without any weighting scheme using the appropriate routine in WingX.

A full multipolar model could be refined based on the 7.7 GPa data, expanding each C and O atom up to an octupole level and each H atom up to a dipole level (refining only the bond directed dipole). H positions were fixed at values calculated from the periodic DFT calculations. The final R factor is larger than what one could obtain from low temperature experiments at ambient pressure (*ca.* 2% versus 6%), but an accurate analysis of the residuals reveal that they obey a normal distribution and the larger peaks do not occur in important regions of the molecules. This implies that, despite being noisy, the data do not contain systematic errors and the model is therefore not significantly biased. The program XD2006[44] was used for the multipolar modelling and for the calculation of the ED and the molecular graph. MolCoolQT[45] and Gaussview[46] were used for plotting isosurfaces.

**Theoretical calculations.** Calculations on molecules (structural optimization, ED, NICS and wing current) were carried out with Gaussian09 (ref. 47), using B3LYP/6-31(2d,2p) level of theory. AimAll[48] was used to compute and visualize the ED, the theoretical molecular graph and the ring current. Calculations on periodic systems (geometry optimization and ED at various pressure points)

were carried out at the same level of theory, including empirical corrections for dispersion effects. The program CRYSTAL14 (ref. 29) was used.

X-ray constrained wave function calculations were carried out using the program TONTO[49]. This method is a calculation of molecular orbitals by means of a modified variational approach, which minimizes a function that couples the Hartree–Fock energy of a molecule and the experimentally observed structure factors of its crystal. The same basis set of previous calculations was used. X-ray constrained wave functions enable to exploit the experimental information, though avoiding a dangerous over-fitting of noisy data that could occur using the multipolar model. This is especially cogent when data are not of the highest quality, such as the X-ray diffraction measured from crystal in a DAC.

## References

1. Hückel, E. *Grundzüge der Theorie ungesättigter und aromatischer Verbindungen* 71–85 (Verlag Chemie, 1938).
2. Von, W., Doering, E. & Detert, F. L. Cycloheptatrienylium Oxide. *J. Am. Chem. Soc.* **73**, 876–877 (1951).
3. Stanger, A. What is aromaticity: a critique of the concept of aromaticity—can it really be defined? *Chem. Commun.* 1939–1947 (2009).
4. Krygowski, T. M. & Cyranski, M. K. Structural aspects of aromaticity. *Chem. Rev.* **101**, 1385–1419 (2009).
5. Mitchell, R. H. Measuring aromaticity by NMR. *Chem. Rev.* **101**, 1301–1315 (2001).
6. Gomes, J. A. N. F. & Mallion, R. B. Aromaticity and ring currents. *Chem. Rev.* **101**, 1349–1383 (2001).
7. Bürgi, H. B. Getting more out of crystal-structure analyses. *Helv. Chim. Acta* **86**, 1625–1640 (2003).
8. Abersfelder, K., White, A. J. P., Rzepa, H. S. & Scheschkewitz, D. A tricyclic aromatic isomer of hexasilabenzene. *Science* **327**, 564–566 (2010).
9. Hey, J. *et al.* Heteroaromaticity approached by charge density investigations and electronic structure calculations. *Phys. Chem. Chem. Phys.* **15**, 20600 (2013).
10. Eremets, M. I., Gavriliuk, A. G., Trojan, I. A., Dzivenko, D. A. & Boehler, R. Single-bonded cubic form of nitrogen. *Nat. Mater.* **3**, 558–563 (2004).
11. Lipp, M. J., Evans, W. J., Baer, B. J. & Yoo, C. S. High-energy density extended CO solid. *Nat. Mater.* **4**, 211–215 (2005).
12. Santoro, M. & Gorelli, F. A. High pressure solid state chemistry of carbon dioxide. *Chem. Soc. Rev.* **35**, 918–931 (2006).
13. Ma, Y. *et al.* Transparent dense sodium. *Nature* **458**, 182–185 (2009).
14. Boldyreva, E. High-pressure diffraction studies of molecular organic solids. A personal view. *Acta Crystallogr. A* **64**, 218–231 (2008).
15. Katrusiak, A. High-pressure crystallography. *Acta Crystallogr. A* **A64**, 131–148 (2007).
16. Piermarini, G. J., Mighell, A. D., Weir, C. E. & Block, S. Crystal structure of benzene II at 25 kilobars. *Science* **165**, 1250–1255 (1969).
17. Wen, X.-D., Hoffmann, R-. & Achcroft, N. W. Benzene under High Pressure: a Story of Molecular Crystals Transforming to Saturated Networks, with a Possible Intermediate Metallic Phase. *J. Am. Chem. Soc.* **133**, 9023–9035 (2011).
18. Ciabini, L. *et al.* Triggering dynamics of the high-pressure benzene amorphization. *Nat. Mater.* **6**, 39–43 (2006).
19. Fitzgibbons, T. C. *et al.* Benzene-derived carbon nanothreads. *Nat. Mater.* **15**, 43–47 (2015).
20. Fanetti, S., Citroni, M. & Bini, R. Tuning the aromaticity of s-triazine in the crystal phase by pressure. *J. Phys. Chem. C* **118**, 13764–13768 (2014).
21. Budzianowski, A. & Katrusiak, A. Pressure-frozen benzene I revisited. *Acta Crystallogr. A* **B62**, 94–101 (2006).
22. Coppens, P. *X-ray Charge Density and Chemical Bonding* (Oxford University, 1997).
23. Hansen, N. K. & Coppens, P. Testing aspherical atom refinements on small-molecule data sets. *Acta Crystallogr. A* **A34**, 909–921 (1978).
24. Yamanaka, T., Okada, T. & Nakamoto, Y. Electron density distribution and static dipole moment of KNbO3 at high pressure. *Phys. Rev. B* **80**, 094108 (2009).
25. Tse, J. S., Klug, D. D., Patchkovskii, S. & Dewhurst, J. K. Chemical bonding, electron-phonon coupling, and structural transformations in high-pressure phases of Si. *J. Phys. Chem. B* **110**, 3721–3726 (2006).
26. Fabbiani, F. P. A., Dittrich, B., Pulham, C. R. & Warren, J. E. Towards charge-density analysis of high-pressure molecular crystal structures. *Acta Crystallogr. A.* **67**, C376 (2011).
27. Macchi, P. & Casati, N. Strong hydrogen bonds in crystals under high pressure. *Acta Crystallogr. A* **67**, C163–C164 (2011).
28. Katrusiak, A. High-pressure crystallography. *Acta Crystallogr. A* **A64**, 135–148 (2008).
29. Dovesi, R. *et al.* CRYSTAL14: a program for the ab initio investigation of crystalline solids. *Int. J. Quantum Chem.* **114**, 1287–1317 (2014).
30. Destro, R. & Simonetta, M. Syn-1,6 : 8,13-Bisearbonyl[14]annulene. *Acta Crystallogr. A* **B33**, 3219–3221 (1977).
31. Destro, R. & Merati, F. Bond lengths, and beyond. *Acta Crystallogr. A* **B51**, 559–570 (1995).
32. Chiang, C. C. & Paul, I. C. Crystal and molecular structure of [14]annulene. *J. Am. Chem. Soc.* **94**, 4741–4743 (1972).

33. Bader, R. W. F. *A Quantum Theory* (Oxford University Press, 1990).
34. Møller, C. & Plesset, M. S. Note on an approximation treatment for many-electron systems. *Phys. Rev.* **46,** 618–622 (1934).
35. Merino, G., Heine, T. h. & Seifert, G. The induced magnetic field in cyclic molecules. *Chemistry* **10,** 4367–4371 (2004).
36. Angel, R. J., Allan, D. R., Miletich, R. & Finger, W. The use of quartz as an internal pressure standard in high-pressure crystallography. *J. Appl. Crystallogr.* **30,** 461–466 (1997).
37. Jayatilaka, D. & Grimwood, D. J. Wavefunctions derived from experiment. I. Motivation and theory. *Acta Crystallogr. A* **A57,** 76–86 (2001).
38. Bader, R. F. W. & Stephens, M. E. Spatial localization of the electronic pair and number distributions in molecules. *J. Am. Chem. Soc.* **97,** 7391–7399 (1975).
39. Stanger, A. Nucleus-independent chemical shifts (NICS): distance dependence and revised criteria for aromaticity and antiaromaticity. *J. Org. Chem.* **71,** 883–893 (2006).
40. CrysAlis PRO. Agilent Technologies UK Ltd, Yarnton, England, 2014.
41. Casati, N., Macchi, P. & Sironi, A. Improving the quality of diamond anvil cell data collected on an area detector by shading individual diamond overlay. *J. Appl. Crystallogr.* **40,** 628–630 (2007).
42. Sheldrick, G. M. A short history of SHELX. *Acta Crystallogr. A* **64,** 112–122 (2008).
43. Farrugia, L. J. WinGX suite for small-molecule single-crystal crystallography. *J. Appl. Crystallogr* **32,** 837–838 (1999).
44. Volkov, A. *et al.* XD2006 - A Computer Program Package for Multipole Refinement, Topological Analysis of Charge Densities and Evaluation of Intermolecular Energies from Experimental and Theoretical Structure Factors (2006).
45. Hübschle, C. B. & Dittrich, B. MoleCoolQt - a molecule viewer for charge-density research. *J. Appl. Cryst.* **44,** 238–240 (2011).
46. Dennington, R., Keith, T. & Millam, J. GaussView, Version 5, Semichem Inc. (Shawnee Mission, KS, USA, 2009).
47. Frisch, M. J. *et al.* Gaussian 09, Revision D.01, Gaussian, Inc. (Wallingford, CT, USA, 2009).
48. Keith, T. A. TK Gristmill Software, AIMAll (Version 14.11.23) Software (Overland Park, KS, USA, 2014) (aim.tkgristmill.com).
49. Jayatilaka, D. & Grimwood, D. J. Tonto: a fortran based object-oriented system for quantum chemistry and crystallography. *Lect. Notes Comput. Sci.* **2660,** 142–151 (2003).

## Acknowledgements

We thank the Swiss National Science foundation (project 144534 and 162861) for financial support. We thank Dr Anna Krawczuk, Dr Shaun Evans and Dr Heribert Whilelm for assistance during one of the experiments at Diamond Light Source, Professor Riccardo Destro for providing the crystal samples.

## Author contributions

N.C. and P.M. conceived the project. N.C., A.P.J. and A.K. set-up and carried out the HP experiments and N.C. analysed the data refined the models. P.M. carried out the multi-temperature experiments, the multipolar refinements and the theoretical calculations. N.C. and P.M. wrote the paper.

# Practical carbon–carbon bond formation from olefins through nickel-catalyzed reductive olefin hydrocarbonation

Xi Lu[1,2], Bin Xiao[1], Zhenqi Zhang[1], Tianjun Gong[1], Wei Su[1], Jun Yi[1], Yao Fu[1] & Lei Liu[2]

New carbon–carbon bond formation reactions expand our horizon of retrosynthetic analysis for the synthesis of complex organic molecules. Although many methods are now available for the formation of $C(sp^2)$-$C(sp^3)$ and $C(sp^3)$-$C(sp^3)$ bonds via transition metal-catalyzed cross-coupling of alkyl organometallic reagents, direct use of readily available olefins in a formal fashion of hydrocarbonation to make $C(sp^2)$-$C(sp^3)$ and $C(sp^3)$-$C(sp^3)$ bonds remains to be developed. Here we report the discovery of a general process for the intermolecular reductive coupling of unactivated olefins with alkyl or aryl electrophiles under the promotion of a simple nickel catalyst system. This new reaction presents a conceptually unique and practical strategy for the construction of $C(sp^2)$-$C(sp^3)$ and $C(sp^3)$-$C(sp^3)$ bonds without using any organometallic reagent. The reductive olefin hydrocarbonation also exhibits excellent compatibility with varieties of synthetically important functional groups and therefore, provides a straightforward approach for modification of complex organic molecules containing olefin groups.

[1] Hefei National Laboratory for Physical Sciences at the Microscale, iChEM, CAS Key Laboratory of Urban Pollutant Conversion, Anhui Province Key Laboratory of Biomass Clean Energy, University of Science and Technology of China, Hefei 230026, China. [2] Department of Chemistry, Tsinghua University, Beijing 100084, China. Correspondence and requests for materials should be addressed to Y.F. (email: fuyao@ustc.edu.cn) or to L.L. (email: lliu@mail.tsinghua.edu.cn).

Olefins are important synthons in organic chemistry[1,2]. They are readily available as stable and inexpensive compounds with great diversity. Simple olefins are both raw materials and products in petrochemical industry. For example, ethylene is produced mostly through steam cracking. They are converted to higher olefins, polyethylene materials and various commodity chemicals[3]. On the other hand, olefin groups are also widely represented in natural products with complex structures and many functional groups. Not only the extensive source but also the unique chemical reactivity of olefins attracts chemists, as the olefin moieties are resistant to a good number of synthetic transformations. Some unique transition metal catalyst systems can activate the olefin double bonds leading to highly elegant as well as useful reactions. Famous examples include the Wacker process[4], olefin metathesis[1,5], olefin hydroformylation[6] and Heck reaction[7] that have been extensively used in the preparation of complex organic molecules both in laboratory and in industry. These reactions establish the central role of olefins in modern synthetic organic chemistry as well as fine chemical industry[8].

More recently, unactivated olefins have been used directly as chemical input in some novel cross-coupling reactions (for example, carbon-heteroatom coupling reactions[9–15] and few examples of carbon–carbon coupling reactions[16–20]). These findings suggest that olefins can be recognized as nuclephilic radical equivalents[9,10,12,15,17,20] or alkylmetallics equivalents[11,13,14,18,19,21–23] from a novel perspective. In some of these emerging methods that involve transition metal catalysts (Cu (ref. 22), Fe (refs 9,20), Co (ref. 15), Mn (ref. 24) and so on), silanes were used as hydride source as well as reductant. New reactions that use olefins as chemical input are expected to bring new opportunities to organic synthesis. For instance, use of olefins to replace alkylmetallic reagents in traditional cross-coupling reaction fashion[25] (for example, Kumada coupling reaction) with aryl/alkyl electrophiles would have appealing advantages such as better functional group compatibility and broader substrate availability.

We now report the discovery of a new catalytic reaction of olefins, namely, Ni-catalyzed intermolecular reductive olefin hydrocarbonation between olefins and alkyl/aryl halides in an anti-Markovnikov fashion. This reaction provides an efficient strategy for the construction of carbon–carbon bonds[26,27] from more stable and less expensive substrates as compared with the existing methods using organometallic reagents[25,28–30]. In terms

---

**Table 1 | Optimization of reaction conditions for reductive olefin hydrocarbonation reaction.**

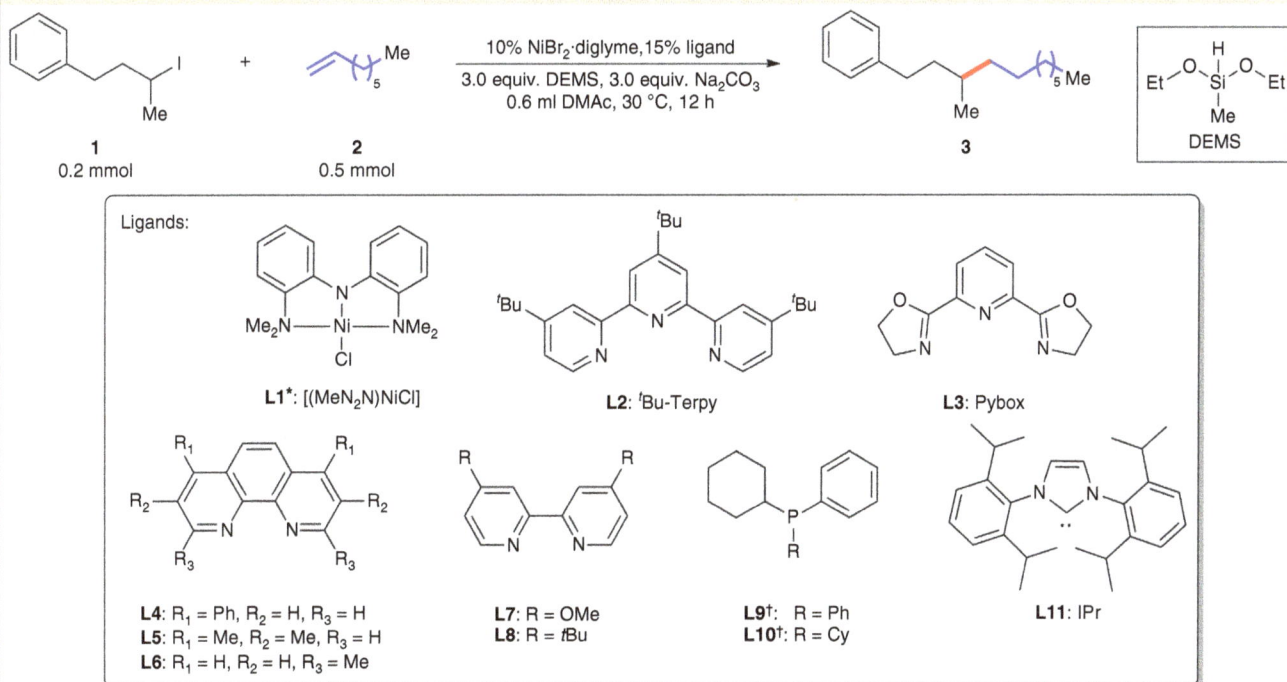

*For **L1**, NiBr$_2$·diglyme was not added, †For **L9** or **L10**, 20% Ligand was added. DEMS, Diethoxymethylsilane; DMAc, N,N-dimethylacetamide; Diglyme, diethylene glycol dimethyl ether.

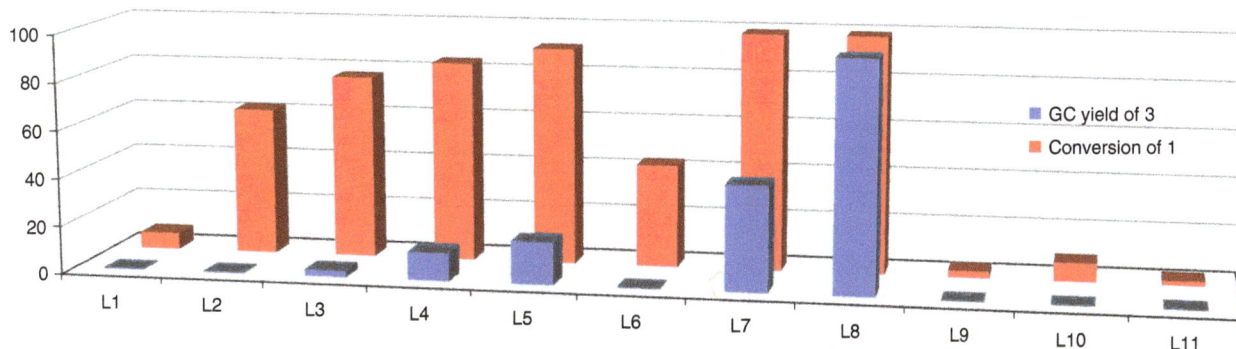

**Table 2 | Substrates scope for reductive olefin hydrocarbonation reaction.**

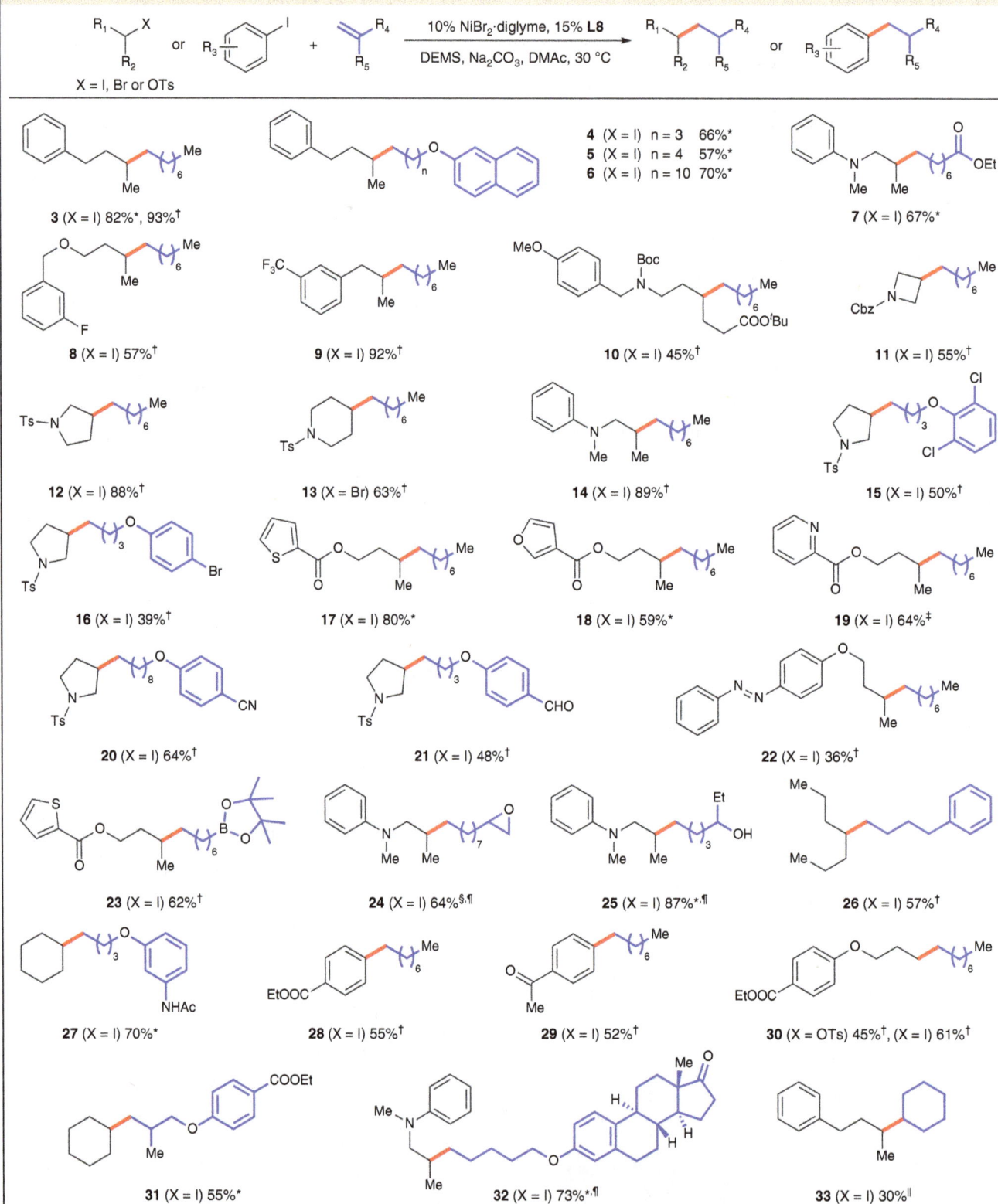

Reaction scheme: R1(R2)–X or R3–I (aryl iodide) + CH2=C(R4)(R5) → with 10% NiBr2·diglyme, 15% L8, DEMS, Na2CO3, DMAc, 30 °C → R1(R2)–CH2–C(R4)(R5) or R3–CH(R4)(R5). X = I, Br or OTs

**3** (X = I) 82%*, 93%†

**4** (X = I) n = 3  66%*
**5** (X = I) n = 4  57%*
**6** (X = I) n = 10  70%*

**7** (X = I) 67%*

**8** (X = I) 57%†

**9** (X = I) 92%†

**10** (X = I) 45%†

**11** (X = I) 55%†

**12** (X = I) 88%†

**13** (X = Br) 63%†

**14** (X = I) 89%†

**15** (X = I) 50%†

**16** (X = I) 39%†

**17** (X = I) 80%*

**18** (X = I) 59%*

**19** (X = I) 64%‡

**20** (X = I) 64%†

**21** (X = I) 48%†

**22** (X = I) 36%†

**23** (X = I) 62%†

**24** (X = I) 64%§,¶

**25** (X = I) 87%*,¶

**26** (X = I) 57%†

**27** (X = I) 70%*

**28** (X = I) 55%†

**29** (X = I) 52%†

**30** (X = OTs) 45%†, (X = I) 61%†

**31** (X = I) 55%*

**32** (X = I) 73%*,¶

**33** (X = I) 30%‖

Boc, t-butyloxycarbonyl; Cbz, benzyloxycarbonyl; Ts, tosyl. The reactions were conducted in 0.2 mmol scale at 30 °C. Isolated yields. *1.5 equiv. alkene, 2 equiv. DEMS, 2 equiv. Na₂CO₃ and 0.6 ml DMAc were used, 12 h. †2.5 equiv. alkene, 3 equiv. DEMS, 3 equiv. Na₂CO₃ and 0.6 ml DMAc were used, 12 h. ‡2.5 equiv. alkene, 3 equiv. DEMS, 3 equiv. KHCO₃, 0.6 ml DMAc were used, 24 h. §1.5 equiv. alkene, 2 equiv. DEMS, 2 equiv. Cs₂CO₃, 0.6 ml DMAc were used, 12 h. ‖4 equiv. cyclohexene, 6 equiv. DEMS, 6 equiv. KHCO₃ and 0.6 ml DMAc were used, 24 h. ¶Diastereomeric mixture.

of practicality, the reaction shows high levels of 'chemo'- and 'regio'-selectivity, so that a wide range of sensitive functional groups can be tolerated (for example, epoxide, aldehyde and alcohol) in the transformation with minimal substrate protection necessary[31]. As Ni-catalyzed carbon–carbon bond formation processes have enjoyed great success in modern synthesis[32], the present reaction is expected to find important applications in organic chemistry.

**a**

**b**

Kumada coupling reaction

Reductive olefin hydrocarbonation--*This work*

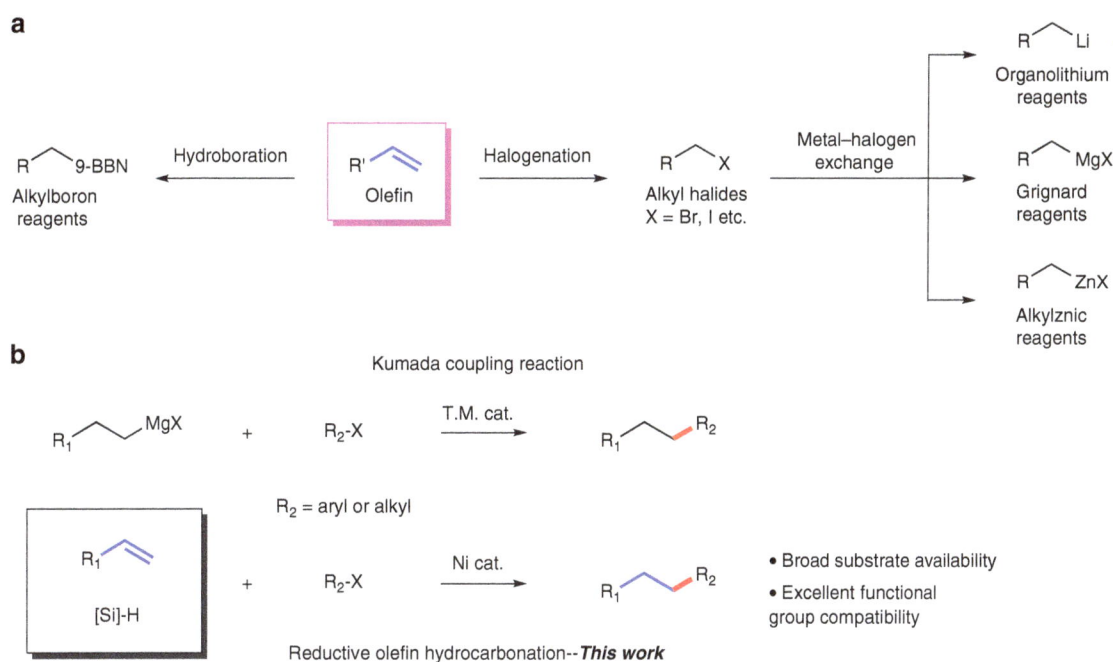

**Figure 1 | Carbon–carbon bonds formation from olefins.** (**a**) Alkyl organometallic reagents used in cross-coupling reactions. Alkylboron reagents[47-49] are usually made through alkene hydroboration. Grignard[50,51], organolithium[52,53] and alkylznic reagents[42,54] are generally obtained through insertion of metals into alkyl halides. However, an often ignored problem is that most terminal alkyl halides are converted from olefins[55]. (**b**) Comparison of reductive olefin hydrocarbonation reaction with transition metal-catalyzed Kumada-coupling reaction. From a viewpoint of synthetic chemistry, the combination of olefins with silanes could be recognized as equivalent to alkyl organometallic reagents. 9-BBN = 9-borabicyclo[3.3.l]nonane.

## Results

**Reaction discovery.** We screened various Ni catalysts, base, silane and solvents for the reductive olefin hydrocarbonation reaction of 1-octene with **1** (see Table 1 and Supplementary Tables 1–5). The pincer complex **L1** was tested first, but only trace amount of desired product was obtained with large amount of alkyl iodide recovered. We then tested the terpyridine ligand **L2** and pybox ligand **L3**. Higher conversion of alkyl iodide was observed but the yield was only slightly improved. To our delight, we observed significant formation of the desired product with the phenanthroline family ligands **L4** and **L5**. We then tested **L6** bearing an *ortho*-methyl group but **L6** was inferior. On the other hand, a bipyridine ligand **L7** exhibited much better reactivity. Remarkably, when 4,4'-di-*tert*-butyl-2,2'-bipyridine (**L8**) was used, the GC yield increased to 96% with an isolated yield of 93% for the desired product. We also tested some monodentate phosphine ligands (**L9** & **L10**) and carbene ligand (**L11**), but they were not effective.

**Substrates scope.** The substrate scope of the reductive olefin hydrocarbonation reaction was shown in Table 2. A variety of carbon electrophiles and olefins with different functional groups could be readily converted to the desired products with modest to excellent yields (30–93%). Not only alkyl iodides (for example, **3**), bromides (for example, **13**) and tosylates (for example, **30**) were good substrates, but also aryl iodides[33,34] (for example, **28**) could be transformed successfully. With respect to olefins, both mono-substituted (for example, **26**) and 1,1-di-substituted alkenes (for example, **31**) could be used. Because of the mild reaction conditions, a wide range of synthetically relevant functional groups could survive the transformation. For instance, ether (**4–6**), ester (**7**), fluoride (**8**), trifluoromethyl (**9**), carbamate (**10–11**), sulfonamide (**12–13**), amine (**14**), aryl choride (**15**) and bromide (**16**) were well tolerated. Heterocycles such as thiophene (**17**), furan (**18**), and pyridine (**19**) could also be used in the

reaction. Several base-sensitive groups, such as nitrile (**20**) and ketone (**29**, **32**) posed no problem. Even more active groups, such as unprotected benzaldehyde (**21**) and azo groups (**22**), were compatible with the reaction. As an interesting substrate, **23** containing a pinacol boronate ester[35] could selectively undergo the reductive olefin hydrocarbonation reaction with its carbon–boron bond intact. To our surprise, the reaction could even be conducted in the presence of an epoxide group[36] (**24**) or an unprotected OH group (**25**). Noteworthily, an internal alkene[11,14] (for example, **33**) could have been converted in the reaction, although further ligand optimization was needed to improve the yields.

**Modification of complex molecules.** To further demonstrate the high degree of functional group compatibility of the reductive olefin hydrocarbonation reaction (Fig. 1), we exploited its use as a novel tool for the modification of complex biologically interesting molecules (Fig. 2). As an example, a cholesterol derivative (**34**) could react with **35** to produce **36** without affecting either the internal alkene or alcohol groups (Fig. 2a). Hecogenin derivative (**37**), which contained both ketal and ketone groups, was also a good substrate for the modification process (Fig. 2b). Furthermore, calciferol (**40**) was converted to **42** selectively in the presence of the hydroxyl, internal alkene and even 1,3-diene groups (Fig. 2c).

Modification of a cinchonidine derivative (**43**) resulted in the 'chemo'-selective formation of **44**, while tolerating both the amino group and quinoline structure (Fig. 2d, left). Single-crystal XRD analysis of **44** confirmed that the skeleton of cinchonidine was fully maintained during the modification process. In addition, the coupling of quinine (**46**) and a fructose derivative (**45**) enabled the production of highly complex molecules in a convergent fashion (Fig. 2d, right).

The reductive olefin hydrocarbonation reaction of sclareol (**48**) proceeded smoothly in the presence of different two tertiary

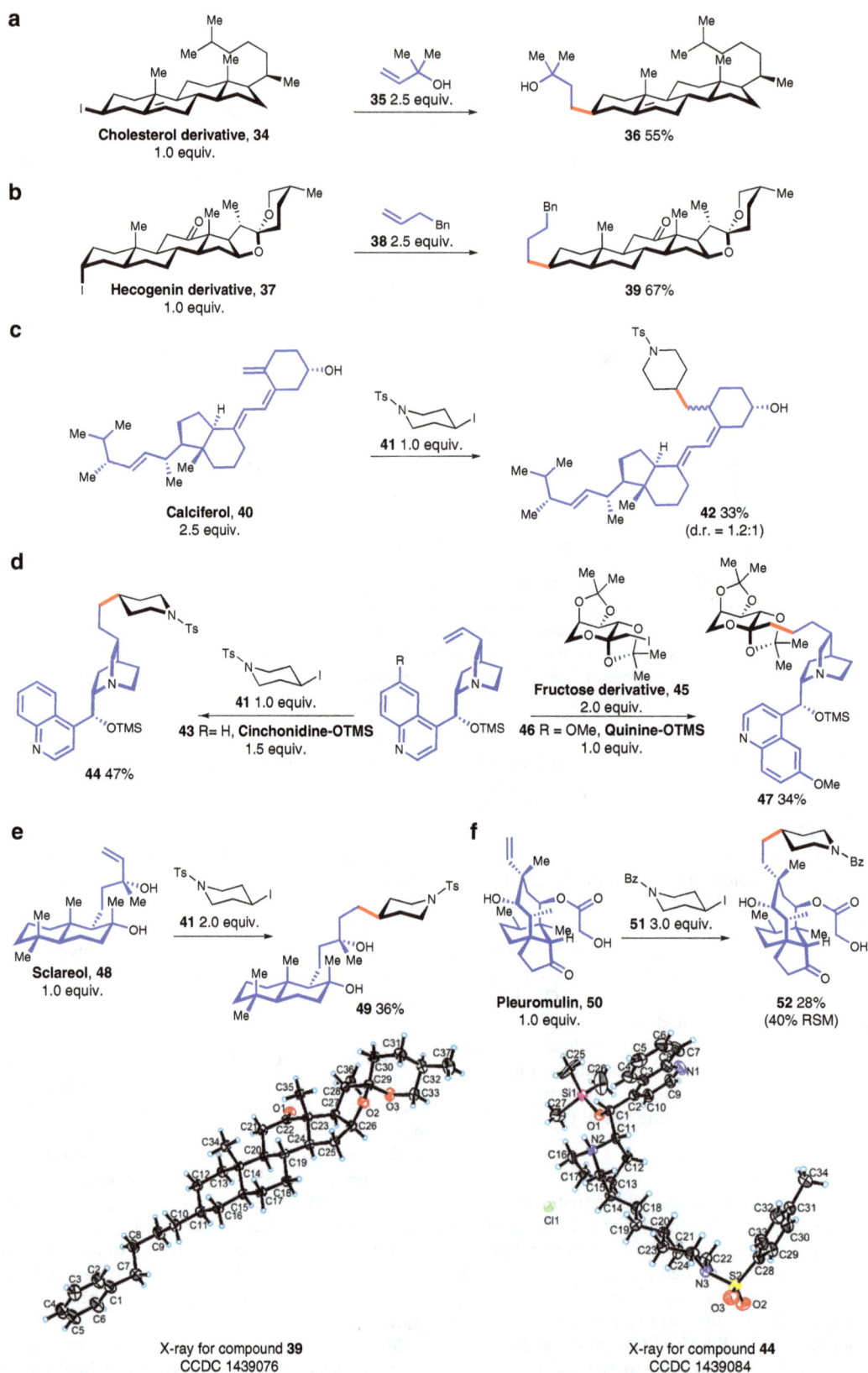

**Figure 2 | Modification of complex molecules. (a)** 10% NiBr₂.diglyme, 15% **L8**, 3.0 equiv. DEMS, 3.0 equiv. Na₂CO₃, 2 ml DMAc, 30 °C, 12 h.
**(b)** The same conditions as in **a** the newly formed carbon–carbon bond was between C10 and C11. **(c)** 20% NiBr₂.diglyme, 30% **L8**, 3.0 equiv. DEMS,
3.0 equiv. Na₂CO₃, 2 ml DMAc, 30 °C, 12 h. **(d)** conditions for compound **44**: 20% NiBr₂.diglyme, 30% **L8**, 2.0 equiv. DEMS, 2.0 equiv. Na₂CO₃,
2 ml THF/DMAc (v/v = 1/3), 30 °C, 12 h, the newly formed carbon–carbon bond was between C19 and C20; conditions for compound **47**: same
conditions as in **c**. **(e)** Same conditions as in **c**. **(f)** 20% NiBr₂.diglyme, 30% **L8**, 4.0 equiv. DEMS, 4.0 equiv. Na₂CO₃, 2 ml DMAc, 30 °C, 12 h.
Bn, benzyl; Bz, benzoyl; TMS, trimethylsilyl.

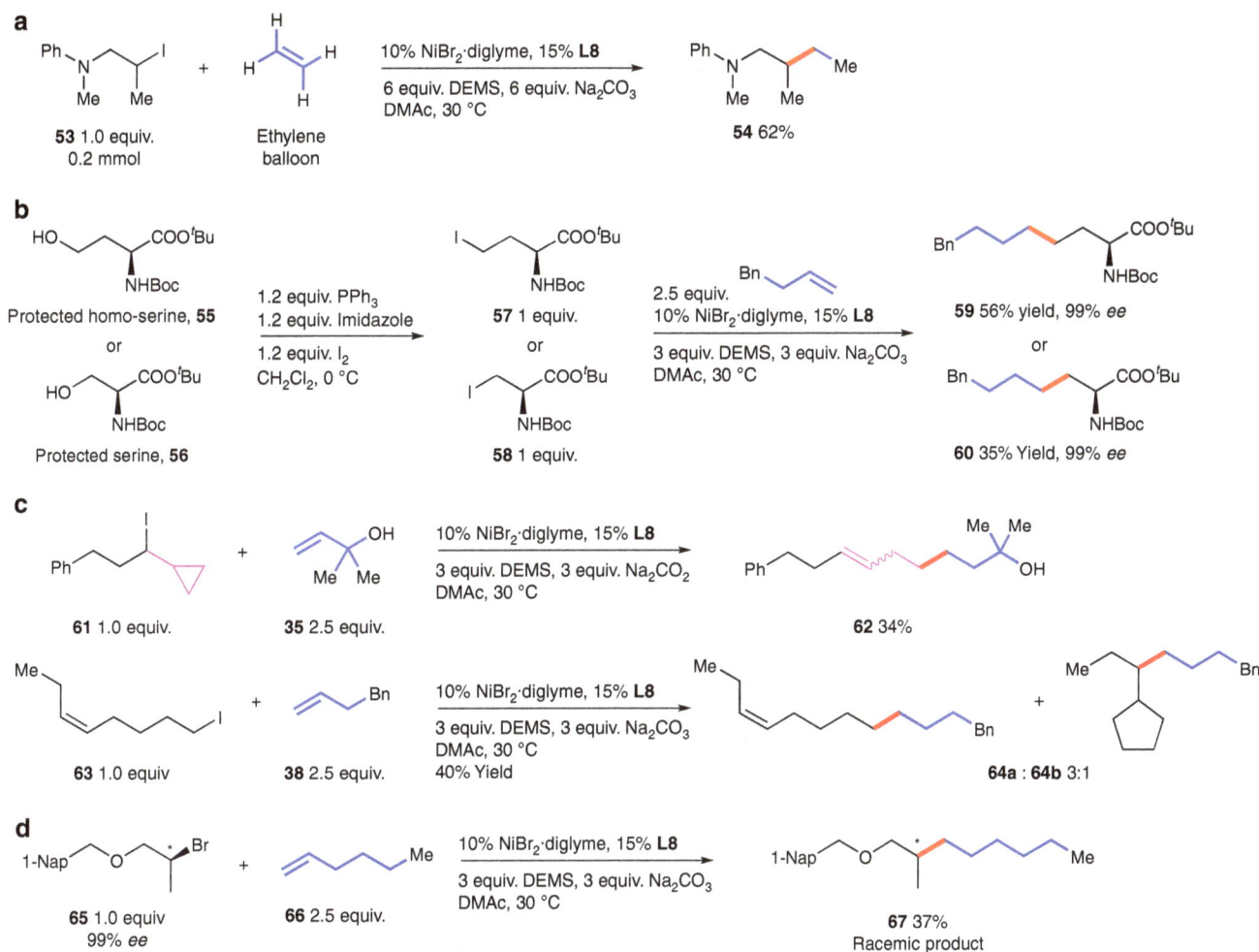

**Figure 3 | Other applications of reductive olefin hydrocarbonation reaction. (a)** Conversion of ethylene. **(b)** Synthesis of non-natural amino acids. **(c)** Radical clock experiments. **(d)** Stereochemistry of reductive olefin hydrocarbonation reaction. Nap, naphthyl.

alcohol groups (Fig. 2e). In a more complex example with pleuromulin (**50**) (Fig. 2f), we obtained the desired product (**52**) in 28% yield (with 40% recovery of starting material) despite the presence of carbamate, ester, keton, unprotected primary and secondary alcohol groups in the reactant. Therefore, the reductive olefin hydrocarbonation reaction presents attractive opportunities for the modification of natural products or other complex molecules.

**Other applications.** Ethylene, as the simplest and most abundant olefin, has been attracting increasing attentions in synthetic organic chemistry[3]. We were delighted that ethylene as C2 source was indeed a good substrate in the reductive olefin hydrocarbonation reaction (Fig. 3a). Compound **54** was obtained in 62% isolated yield.

The reductive olefin hydrocarbonation reaction was useful for the synthesis of non-natural amino acids[37] (Fig. 3b). As an example, homoserine-derived iodide **57** could be converted to **59** with a yield of 56%. More interestingly, the reaction of a racemization-prone serine derived iodide **58** was also successful affording **60** in 99% ee. This finding was surprising because in our previous study[38] on the Ni-catalyzed reaction of **58** we observed significant racemization of the amino acid.

To gain more insights into the reaction mechanism, radical clock experiments were carried out (Fig. 3c). Compound **61** containing a cyclopropyl ring was used as radical clock substrate

(Fig. 3c, top). In this coupling reaction, we obtained only the ring-opened product **62** in 34% isolated yield.[39] We also tested the reaction with (Z)-8-Iodooct-3-ene (**63**) (Fig. 3c, bottom). A mixture of linear coupling product (**64a**) and ring-cyclized product (**64b**) was obtained with a ratio of 3:1. The formation of ring-cyclized product (**64b**) revealed that this reaction proceeds through a radical cyclization process[40].

Finally, we took advantage of optical pure secondary alkyl bromide (**65**) to study the stereochemistry of this reductive olefin hydrocarbonation reaction (Fig. 3d). When (S)-**65** was alkylated with 1-hexene, we obtained a racemic product (**67**) in 37% yield[41]. Furthermore, radical inhibiting experiment using TEMPO (2,2,6,6-tetramethylpiperidinooxy) as a radical trap was carried out (see Supplementary Discussion). The reaction was largely inhibited when 0.2 equiv. TEMPO was added, indicating a radical type reaction mechanism[32]. Nonetheless, details for the mechanism of this reaction are not clear at present[42–46]. Further investigations are ongoing in our lab.

In summary, we have developed a practical and user-friendly method for the formation of carbon–carbon bonds through Ni-catalyzed intermolecular coupling of aryl or alkyl electrophiles with olefins under reductive conditions. This newly developed reductive olefin hydrocarbonation reaction provides a useful and general approach for the construction of carbon–carbon bonds by directly using olefins as nucleophile precursors. This reaction exhibited excellent compatibility with varieties of synthetically important functional groups and therefore, provided an efficient new approach for the modification of complex molecules. Our next challenge was the development of asymmetric version of this new carbon–carbon bond forming reaction and its extension to internal olefins.

## Methods

**Materials.** For NMR and high-performance liquid chromatography spectra of compounds in this manuscript, see Supplementary Figs 1–119. For details of the synthetic procedures, see Supplementary Methods. For X-ray data see Supplementary Data 1.

**Procedure.** NiBr$_2$·diglyme (7.0 mg, 0.02 mmol, 10 mol%), 4,4'-di-tert-butyl-2, 2'-bipyridine (8.0 mg, 0.03 mmol, 15 mol%) and Na$_2$CO$_3$ (42.4 mg, 0.4 mmol, 2.0 equiv.) were added to a Schlenk tube equipped with a stir bar. The vessel was evacuated and filled with argon (three cycles). To these solids, 0.6 ml DMAc was added under argon atmosphere. The reaction mixture was stirred at room temperature for 30 s. To the reaction mixture, electrophile (0.2 mmol, 1.0 equiv.), alkene (0.3 mmol, 1.5 equiv.) and DEMS (0.4 mmol, 2.0 equiv.) were added under a positive flow of argon. The reaction mixture was stirred at 30 °C for 12 h. To remove the DMAc, the reaction mixture was poured into 50 ml of ice water and the resulting mixture was extracted with ethyl acetate (4 × 30 ml). The combined organic layer was dried over Na$_2$SO$_4$, filtered, concentrated in vacuum and purified by column chromatography on silica gel.

## References

1. Hoveyda, A. H. & Zhugralin, A. R. The remarkable metal-catalysed olefin metathesis reaction. *Nature* **450**, 243–251 (2007).
2. Williams, J. M. J. *Preparation of Alkenes: A Practical Approach* (Oxford University Press, 1996).
3. Saini, V., Stokes, B. J. & Sigman, M. S. Transition-metal-catalyzed laboratory-scale carbon-carbon bond-forming reactions of ethylene. *Angew. Chem. Int. Ed. Engl.* **52**, 11206–11220 (2013).
4. Takacs, J. M. & Jiang, X. The Wacker reaction and related alkene oxidation reactions. *Curr. Org. Chem.* **7**, 369–396 (2003).
5. Connon, S. J. & Blechert, S. Recent developments in olefin cross-metathesis. *Angew. Chem. Int. Ed. Engl.* **42**, 1900–1923 (2003).
6. Evans, P. A. *Modern Rhodium-Catalyzed Organic Reactions* 93–110 (Wiley-VCH, 2005).
7. Meijere, A. D. & Diederich, F. *Metal-Catalyzed Cross-Coupling Reactions* 2nd edn 217–315 (Wiley-VCH, 2008).
8. Corey, E. J. & Chen, X.-M. *The Logic of Chemical Synthesis* (John Wiley & Sons, Inc., 1995).
9. Leggans, E. K., Barker, T. J., Duncan, K. K. & Boger, D. L. Iron(III)/NaBH$_4$-mediated additions to unactivated alkenes: synthesis of novel 20'-Vinblastine analogues. *Org. Lett.* **14**, 1428–1431 (2012).
10. Gaspar, B. & Carreira, E. M. Catalytic hydrochlorination of unactivated olefins with para-toluenesulfonyl chloride. *Angew. Chem. Int. Ed. Engl.* **47**, 5758–5760 (2008).
11. Yang, Y., Shi, S.-L., Niu, D., Liu, P. & Buchwald, S. L. Catalytic asymmetric hydroamination of unactivated internal olefins to aliphatic amines. *Science* **349**, 62–66 (2015).
12. Gui, J. *et al.* Practical olefin hydroamination with nitroarenes. *Science* **348**, 886–891 (2015).
13. Sakae, R., Hirano, K. & Miura, M. Ligand-controlled regiodivergent Cu-catalyzed aminoboration of unactivated terminal alkenes. *J. Am. Chem. Soc.* **137**, 6460–6463 (2015).
14. Xi, Y., Butcher, T. W., Zhang, J. & Hartwig, J. F. Regioselective, asymmetric formal hydroamination of unactivated internal alkenes. *Angew. Chem. Int. Ed. Engl.* **55**, 776–780 (2016).
15. Waser, J. & Carreira, E. M. Convenient synthesis of akylhydrazides by the Cobalt-catalyzed hydrohydrazination reaction of olefins and azodicarboxylates. *J. Am. Chem. Soc.* **126**, 5676–5677 (2004).
16. Gaspar, B. & Carreira, E. M. Mild cobalt-catalyzed hydrocyanation of olefins with tosyl cyanide. *Angew. Chem. Int. Ed. Engl.* **46**, 4519–4522 (2007).
17. Lo, J. C., Gui, J., Yabe, Y., Pan, C. M. & Baran, P. S. Functionalized olefin cross-coupling to construct carbon-carbon bonds. *Nature* **516**, 343–348 (2014).
18. Wang, Y.-M., Bruno, N. C., Placeres, Á. L., Zhu, S. & Buchwald, S. L. Enantioselective synthesis of carbo- and heterocycles through a CuH-catalyzed hydroalkylation approach. *J. Am. Chem. Soc.* **137**, 10524–10527 (2015).
19. Su, W. *et al.* Ligand-controlled regiodivergent copper-catalyzed alkylboration of alkenes. *Angew. Chem. Int. Ed. Engl.* **54**, 12957–12961 (2015).
20. Lo, J. C., Yabe, Y. & Baran, P. S. A practical and catalytic reductive olefin coupling. *J. Am. Chem. Soc.* **136**, 1304–1307 (2014).
21. Maksymowicz, R. M., Roth, P. M. C. & Fletcher, S. P. Catalytic asymmetric carbon-carbon bond formation using alkenes as alkylmetal equivalents. *Nat. Chem.* **4**, 649–654 (2012).
22. Miki, Y., Hirano, K., Satoh, T. & Miura, M. Copper-catalyzed intermolecular regioselective hydroamination of styrenes with polymethylhydrosiloxane and hydroxylamines. *Angew. Chem. Int. Ed. Engl.* **52**, 10830–10834 (2013).
23. Miki, Y., Hirano, K., Satoh, T. & Miura, M. Copper-catalyzed enantioselective formal hydroamination of oxa- and azabicyclic alkenes with hydrosilanes and hydroxylamines. *Org. Lett.* **16**, 1498–1501 (2014).
24. Waser, J. & Carreira, E. M. Catalytic hydrohydrazination of a wide range of alkenes with a simple Mn complex. *Angew. Chem. Int. Ed. Engl.* **43**, 4099–4102 (2004).
25. Jana, R., Pathak, T. P. & Sigman, M. S. Advances in transition metal (Pd,Ni,Fe)-catalyzed cross-coupling reactions using alkyl-organometallics as reaction partners. *Chem. Rev.* **111**, 1417–1492 (2011).
26. Roughley, S. D. & Jordan, A. M. The medicinal chemist's toolbox: an analysis of reactions used in the pursuit of drug candidates. *J. Med. Chem.* **54**, 3451–3479 (2011).
27. Geist, E., Kirschning, A. & Schmidt, T. $sp^3$-$sp^3$ Coupling reactions in the synthesis of natural products and biologically active molecules. *Nat. Prod. Rep.* **31**, 441–448 (2014).
28. Chemler, S. R., Trauner, D. & Danishefsky, S. J. The B-alkyl Suzuki-Miyaura cross-coupling reaction: development, mechanistic study, and applications in natural product synthesis. *Angew. Chem. Int. Ed. Engl.* **40**, 4544–4568 (2001).
29. Negishi, E.-i. Magical power of transition metals: past, present, and future (Nobel Lecture). *Angew. Chem. Int. Ed. Engl.* **50**, 6738–6764 (2011).
30. Li, L., Wang, C.-Y., Huang, R. & Biscoe, M. R. Stereoretentive Pd-catalysed Stille cross-coupling reactions of secondary alkyl azastannatranes and aryl halides. *Nat. Chem.* **5**, 607–612 (2013).
31. Young, I. S. & Baran, P. S. Protecting-group-free synthesis as an opportunity for invention. *Nat. Chem.* **1**, 193–205 (2009).
32. Tasker, S. Z., Standley, E. A. & Jamison, T. F. Recent advances in homogeneous nickel catalysis. *Nature* **509**, 299–309 (2014).
33. Bair, J. S. *et al.* Linear-selective hydroarylation of unactivated terminal and internal olefins with trifluoromethyl-substituted arenes. *J. Am. Chem. Soc.* **136**, 13098–13101 (2014).
34. Everson, D. A., Shrestha, R. & Weix, D. J. Nickel-catalyzed reductive cross-coupling of aryl halides with alkyl halides. *J. Am. Chem. Soc.* **132**, 920–921 (2010).
35. Yang, C.-T. *et al.* Alkylboronic esters from copper-catalyzed borylation of primary and secondary alkyl halides and pseudohalides. *Angew. Chem. Int. Ed. Engl.* **51**, 528–532 (2012).
36. Zhao, Y. & Weix, D. J. Nickel-catalyzed regiodivergent opening of epoxides with aryl halides: co-catalysis controls regioselectivity. *J. Am. Chem. Soc.* **136**, 48–51 (2014).
37. Hughes, A. B. *Origins and Synthesis of Amino Acids* Vol. 1 (Wiley-VCH, 2009).
38. Lu, X. *et al.* Expedient synthesis of chiral α-amino acids through nickel-catalyzed reductive cross-coupling. *Chem. Eur. J.* **20**, 15339–15343 (2014).
39. Monks, B. M. & Cook, S. P. Palladium-catalyzed intramolecular iodine-transfer reactions in the presence of β-hydrogen atoms. *Angew. Chem. Int. Ed. Engl.* **52**, 14214–14218 (2013).
40. Cheung, C. W., Zhurkin, F. E. & Hu, X. Z-selective olefin synthesis via iron-catalyzed reductive coupling of alkyl halides with terminal arylalkynes. *J. Am. Chem. Soc.* **137**, 4932–4935 (2015).
41. Yi, J. *et al.* Alkylboronic esters from palladium- and nickel-catalyzed borylation of primary and secondary alkyl bromides. *Adv. Synth. Catal.* **354**, 1685–1691 (2012).
42. Vettel, S., Vaupel, A. & Knochel, P. Nickel-catalyzed preparations of functionalized organozincs. *J. Org. Chem.* **61**, 7473–7481 (1996).
43. Breitenfeld, J., Scopelliti, R. & Hu, X. Synthesis, reactivity, and catalytic application of a Nickel pincer hydride complex. *Organometallics* **31**, 2128–2136 (2012).
44. Tang, S., Liu, K., Liu, C. & Lei, A. Olefinic C-H functionalization through radical alkenylation. *Chem. Soc. Rev.* **44**, 1070–1082 (2015).
45. Luo, S. *et al.* Fe-promoted cross coupling of homobenzylic methyl ethers with Grignard reagents via $sp^3$ C-O bond cleavage. *Chem. Commun.* **49**, 7794–7796 (2013).
46. Li, Z. & Liu, L. Recent advances in mechanistic studies on Ni catalyzed cross-coupling reactions. *Chin. J. Catal.* **36**, 3–14 (2015).
47. Suzuki, A. Cross-coupling reactions of organoboranes: an easy way to construct C-C bonds (Nobel lecture). *Angew. Chem. Int. Ed. Engl.* **50**, 6722–6737 (2011).
48. Miyaura, N. *Metal-Catalyzed Cross-Coupling Reactions* 2th edn 41–123 (Wiley-VCH, 2008).
49. Kotha, S., Lahiri, K. & Kashinath, D. Recent applications of the Suzuki-Miyaura cross-coupling reaction in organic synthesis. *Tetrahedron* **58**, 9633–9695 (2002).
50. Knochel, P. *et al.* Highly functionalized organomagnesium reagents prepared through halogen-metal exchange. *Angew. Chem. Int. Ed. Engl.* **42**, 4302–4320 (2003).

51. Faràdy, L., Bencze, L. & Markó, L. Transition-metal alkyls and hydrides: III. Alkyl-olefin exchange reaction of Grignard reagents catalyzed by nickel chloride. *J. Organomet. Chem.* **10,** 505–510 (1967).

52. Wakefield, B. J. *The Chemistry of Organolithium Compounds* 1st edn (Elsevier, 1974).

53. Giannerini, M., Fañanás-Mastral, M. & Feringa, B. L. Direct catalytic cross-coupling of organolithium compounds. *Nat. Chem.* **5,** 667–672 (2013).

54. Knochel, P. & Singer, R. D. Preparation and reactions of polyfunctional organozinc reagents in organic synthesis. *Chem. Rev.* **93,** 2117–2188 (1993).

55. Mo, F. & Dong, G. Regioselective ketone α-alkylation with simple olefins via dual activation. *Science* **345,** 68–72 (2014).

## Acknowledgements

We thank the financial supports by the National Basic Research Program of China (973 program; No. 2012CB215306, 2013CB932800), NSFC (21325208, 21361140372, 21532004, 21572212), IPDFHCPST (2014FXCX006), CAS (KFJ-EW-STS-051), FRFCU and PCSIRT.

## Author contributions

X.L. and B.X. contributed equally to this work. X.L. and B.X. designed and carried out the experimental work. Z.Z., T.G., W.S. and J.Y. helped to complete the experimental work. Y.F. and L.L. directed the project and wrote the manuscript.

# Atom-economic catalytic amide synthesis from amines and carboxylic acids activated *in situ* with acetylenes

Thilo Krause[1], Sabrina Baader[1], Benjamin Erb[1] & Lukas J. Gooßen[1]

Amide bond-forming reactions are of tremendous significance in synthetic chemistry. Methodological research has, in the past, focused on efficiency and selectivity, and these have reached impressive levels. However, the unacceptable amounts of waste produced have led the ACS GCI Roundtable to label 'amide bond formation avoiding poor atom economy' as the most pressing target for sustainable synthetic method development. In response to this acute demand, we herein disclose an efficient one-pot amide coupling protocol that is based on simple alkynes as coupling reagents: in the presence of a dichloro[(2,6,10-dodecatriene)-1,12-diyl]ruthenium catalyst, carboxylate salts of primary or secondary amines react with acetylene or ethoxyacetylene to vinyl ester intermediates, which undergo aminolysis to give the corresponding amides along only with volatile acetaldehyde or ethyl acetate, respectively. The new amide synthesis is broadly applicable to the synthesis of structurally diverse amides, including dipeptides.

[1] FB Chemie-Organische Chemie, Technische Universität Kaiserslautern, Erwin Schrödinger Strasse Geb. 54, 67663 Kaiserslautern, Germany. Correspondence and requests for materials should be addressed to L.J.G. (email: goossen@chemie.uni-kl.de).

A mide bond formation is one of the most frequently used transformations in organic chemistry[1-4]. The most desirable amide synthesis, a direct condensation of carboxylic acids with amines, is hindered by the intrinsic acid–base reactivity of the starting materials. The thermal amide bond formation from the ammonium carboxylate salts requires high temperatures[5-7], which can be lowered by Lewis acids or boronic acid derivatives. However, even the best known systems are limited to a narrow range of amines and require scavenging the reaction water, for example, by large amounts of molecular sieves. (Fig. 1, left)[8-13]. Therefore, amides are usually synthesized by aminolysis of activated carboxylic acid derivatives, such as halides, anhydrides, azides, or activated esters, that are mostly generated in an extra step with aggressive, expensive or waste-intensive reagents[14-20]. The other main strategy for amide bond formation involves the in situ activation of carboxylic acids by peptide coupling reagents, such as carbodiimides or phosphonium salts[21-31]. Such amide syntheses are highly optimized and provide access to almost any amide structure in near quantitative yields. In modern protein synthesis, they are complemented by efficient chemical and enzymatic peptide ligation methods[32-37]. However, the atom economy of all these processes is low, and the cumulative waste generated during amide synthesis is unacceptable. As a result, the ACS GCI Roundtable has identified 'amide bond formation avoiding poor atom economy' as the most pressing target for sustainable synthetic method development[38].

Over the last years, some elegant strategies for waste-minimized amide synthesis have been devised (Fig. 1), for example, dehydrogenative couplings of alcohols, aldehydes or alkynes with amines, or additions of alcohols to nitriles[39-51]. However, for most synthetic organic chemists, carboxylic acids and amines are still the optimal substrate base for amide synthesis.

To address the central issue of atom economy in the synthesis of amides from ammonium carboxylates, we looked for an activator with minimal molecular weight and low intrinsic reactivity that would scavenge the reaction water in a catalytic condensation process. We envisioned that a hydroacyloxylation catalyst with unprecedented activity might enable the generation of vinyl esters from ammonium carboxylates and gaseous acetylene. Aminolysis of these intermediates would furnish the desired amides along with volatile acetaldehyde.

Ru[II], Ag[I] and Au[I] complexes efficiently promote the addition of carboxylic acids to alkynes under mild conditions, as reported by Mitsudo, Dixneuf, Bruneau and others[52-59]. The aminolysis of enol esters takes place under similarly mild conditions[60-63]. However, for all known catalysts, the two reaction steps are incompatible. As a result, this technology appeared limited to

two-step procedures with isolation of sensitive enol esters. For example, Kita et al. reported an amide synthesis via isolated ketene acetal intermediates[64], and Breinbauer et al. synthesized polypeptides via a Ru-catalysed hydroacyloxylation of alkynes followed by enzymatic aminolysis[65]. These reactions demonstrate the potential of this concept, giving access to amides in high yields under mild conditions, as demanded especially by peptide chemists. However, this approach can reach synthetic maturity only through a catalytic one-pot process that overcomes all its associated problems, for example, carboxylate salt formation with basic amines which hinders catalytic hydroacyloxylation, the control of hydroamination as a side reaction, and the challenging activation of gaseous acetylene, which state-of-the-art catalysts have not been extending to[66].

We disclose herein an amidation protocol which allows the use of low-molecular acetylene and its more activated homologue ethoxyacetylene as a sustainable alternative for state-of-the-art coupling agents. These procedures are convincing in terms of the amount, toxicity and separation of the formed byproducts, yet, broadly applicable, convenient and comparable cheap.

## Results

**Development of a one-pot amide synthesis.** Evaluation of state-of-the-art catalysts, for example, [Ru(methallyl)$_2$dppb] or [RuCl$_2$(PPh$_3$)($p$-cymene)][58,67-69], in the reaction between benzoic acid (**1a**) and 1-hexyne confirmed that they give high yields only in the absence of benzylamine. None of them catalysed the reaction of **1a** with acetylene to give vinyl benzoate (**3a**; Supplementary Tables 4 and 5).

However, we were pleased to find that simple RuCl$_3$ catalyses the conversion of benzylammonium benzoate (**6aa**) to the desired $N$-benzyl benzamide in up to 75% yield at 80 °C under acetylene at 1.7 bar, which is its usual tank pressure (Table 1, entry 1). Systematic evaluation of Ru[III] and Ru[IV] precursors revealed that **Ru-1** was most effective (entries 2 and 3). Phosphine and nitrogen ligands adversely affected the yield (Supplementary Tables 1 and 2). This is surprising, because the only known Ru[IV] hydroacyloxylation catalyst is the triphenylphosphine complex reported by Cadierno et al.[70]

Dioxane was found to be the best solvent, but the reaction also works well in toluene, THF and ethyl acetate (entries 4 – 8). The reaction is surprisingly tolerant to oxygen and water up to a certain threshold (Supplementary Table 1).

Under optimal conditions, that is, 2 mol% **Ru-1** or RuCl$_3$ in dioxane at 80 °C, the amide forms in near quantitative yield within 6 h with acetylene as the carboxylate activator (Table 1, entry 9, Supplementary Table 1). Higher alkynes are inactive as activators, but with ethoxyacetylene and **Ru-1** as

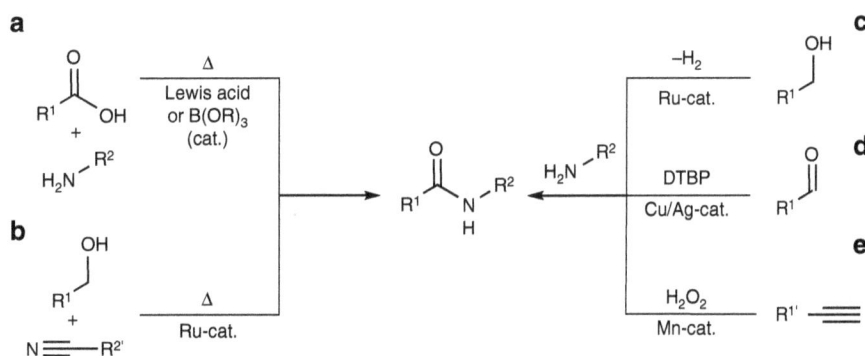

**Figure 1 | Atom-efficient approaches to amide bond formation. (a)** Thermal or Lewis acid-mediated dehydration of ammonium carboxylates. **(b)** Catalytic addition of alcohols to nitriles. **(c)** Dehydrogenative coupling of alcohols with amines. **(d)** Oxidative coupling of aldehydes and amines. **(e)** Oxidative coupling of alkynes and amines.

## Table 1 | One-pot activation and amidation of carboxylic acids with acetylene*.

| Entry | Solvent | [Ru] (mol%) | Yield (%) |
|---|---|---|---|
| 1 | 1,4-dioxane | RuCl$_3$•3H$_2$O (1) | 75 |
| 2 | 1,4-dioxane | **Ru-2** (0.5) | 73 |
| 3 | 1,4-dioxane | **Ru-1** (1) | 81 |
| 4 | Ethyl acetate | **Ru-1** (1) | 66 |
| 5 | Toluene | **Ru-1** (1) | 64 |
| 6 | Tetrahydrofuran | **Ru-1** (1) | 62 |
| 7 | Acetonitrile | **Ru-1** (1) | 25 |
| 8 | Water | **Ru-1** (1) | 0 |
| 9 | 1,4-dioxane | **Ru-1** (2) | 94 (93) |
| 10 | 1,4-dioxane | **Ru-1** (3) | 84 |
| 11 | 1,4-dioxane | **Ru-1** (5) | 71 |
| 12[†] | 1,4-dioxane | – | 0 |
| 13[‡] | 1,4-dioxane | **Ru-1** (2) | 87 |
| 14[§] | 1,4-dioxane | **Ru-1** (2) | 1 |
| 15[‡,||] | NMP | **Ru-1** (1.5) | 99 (99) |

NMP, N-methyl-2-pyrrolidone; **Ru-1**, dichloro[(2,6,10-dodecatriene)-1,12-diyl]ruthenium.
*Reaction conditions: 0.5 mmol **6aa**, 0.25 mmol **4a**, 1.7 bar acetylene, Ru-catalyst, 0.5 ml solvent, 80 °C, 6 h. Yields were determined by GC analysis using n-tetradecane as internal standard; isolated yields in parentheses.
[†]without Ru-catalyst.
[‡]1.5 mmol **2b** instead of **2a**.
[§]1.5 mmol 1-hexyne instead of **2a**.
[||]1 ml solvent, 40 °C, 4 h.

## Table 2 | Scope of the amidation with acetylene as the activating agent*.

*Reaction conditions: 1.0 mmol **6**, 0.5 mmol **4**, 1.7 bar acetylene, 2 mol% **Ru-1**, 1 ml 1,4-dioxane, 80 °C, 6 h. Isolated yields.

catalyst, full conversion was observed already at 40 °C within 4 h (entries 13 – 15). Under identical conditions, RuCl$_3$ gives only unsatisfactory yields for this activator (Supplementary Table 2). The advantages of the somewhat less atom-economic ethoxyacetylene are that it is more easily handled on small scales than gaseous acetylene, and that inert ethyl acetate rather than acetaldehyde is released.

Both new protocols were compared with two-step procedures using established catalysts[64], in which the enol esters are formed in a separate step, with consecutive addition of the amine either in the same solvent or after solvent exchange. With acetylene as the activator, no conversion could be achieved, and with ethoxyacetylene, the yields obtained in these two-step syntheses were much lower than those obtained with our convenient one-step protocols (Supplementary Tables 1 and 2).

**Applicability of the developed processes.** The scope of the ecologically and economically beneficial acetylene protocol is illustrated in Table 2. Aliphatic, aromatic and heteroaromatic carboxylates were successfully coupled with primary amines. Unfortunately, the substrate scope of this protocol is limited by the solubility of the alkylammonium carboxylates in dioxane, the optimal solvent for acetylene gas.

Such restrictions do not apply to the ethoxyacetylene protocol in the solvent N-methyl-2-pyrrolidone, which is applicable to a remarkably wide range of substrates (Table 3). Aromatic, heteroaromatic and aliphatic carboxylic acids reacted with benzylamine to give high yields of the corresponding amides. Diverse functionalities including halo, ether, amide, aldehyde, ester and even-free OH groups were tolerated. Other primary and secondary amines were successfully converted to the

**Table 3 | Scope of the amidation with ethoxyacetylene as activating agent*.**

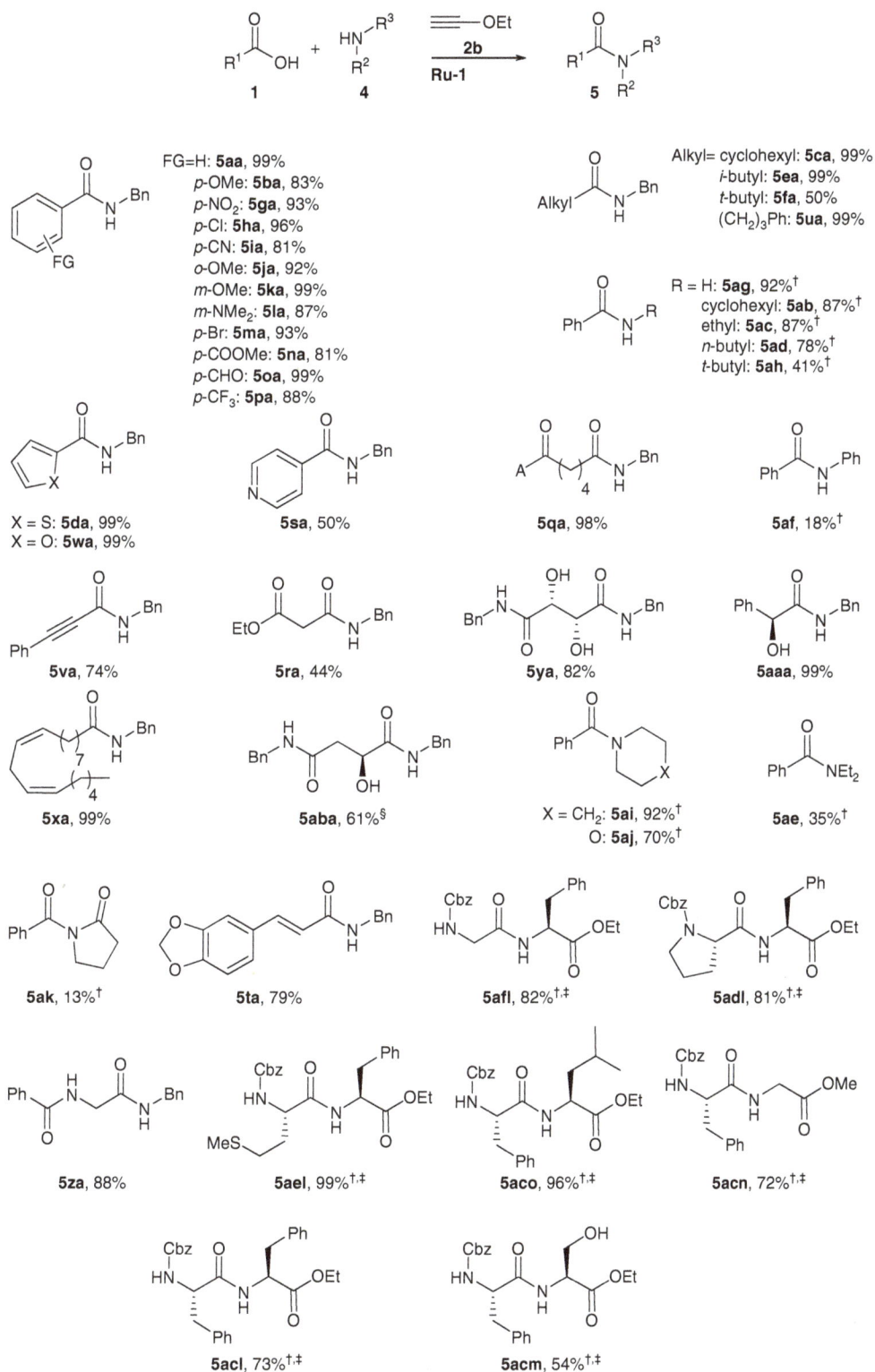

FG=H: **5aa**, 99%
p-OMe: **5ba**, 83%
p-NO$_2$: **5ga**, 93%
p-Cl: **5ha**, 96%
p-CN: **5ia**, 81%
o-OMe: **5ja**, 92%
m-OMe: **5ka**, 99%
m-NMe$_2$: **5la**, 87%
p-Br: **5ma**, 93%
p-COOMe: **5na**, 81%
p-CHO: **5oa**, 99%
p-CF$_3$: **5pa**, 88%

Alkyl= cyclohexyl: **5ca**, 99%
i-butyl: **5ea**, 99%
t-butyl: **5fa**, 50%
(CH$_2$)$_3$Ph: **5ua**, 99%

R = H: **5ag**, 92%[†]
cyclohexyl: **5ab**, 87%[†]
ethyl: **5ac**, 87%[†]
n-butyl: **5ad**, 78%[†]
t-butyl: **5ah**, 41%[†]

X = S: **5da**, 99%
X = O: **5wa**, 99%

**5sa**, 50%

**5qa**, 98%

**5af**, 18%[†]

**5va**, 74%

**5ra**, 44%

**5ya**, 82%

**5aaa**, 99%

**5xa**, 99%

**5aba**, 61%[§]

X = CH$_2$: **5ai**, 92%[†]
O: **5aj**, 70%[†]

**5ae**, 35%[†]

**5ak**, 13%[†]

**5ta**, 79%

**5afl**, 82%[†,‡]

**5adl**, 81%[†,‡]

**5za**, 88%

**5ael**, 99%[†,‡]

**5aco**, 96%[†,‡]

**5acn**, 72%[†,‡]

**5acl**, 73%[†,‡]

**5acm**, 54%[†,‡]

*Reaction conditions: 1.0 mmol **1**, 1.5 mmol **4**, 1.5 mmol **2b**, 1.5 mol% **Ru-1**, 2 ml NMP, 40 °C, 4 h. Isolated yields.
[†]80 °C, 6 h.
[‡]2 ml of toluene instead of NMP.
[§]3 mmol **2b** and **4a**.

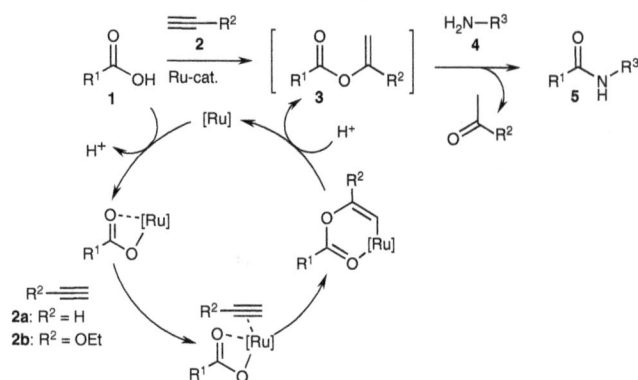

**Figure 2 | Catalytic amide condensation via enol esters.** The proposed catalytic cycle starts with the coordination of a carboxylate and an alkyne to the ruthenium catalyst, followed by an addition of the carboxylate to the alkyne. After protonolysis, the enol ester intermediate is released, which then acts as an acylating agent for the amine, yielding the desired amide along with the carbonyl-byproduct.

corresponding benzamides in good to excellent yields when increasing the temperature to 80 °C to ensure full conversion (Supplementary Table 3). Remarkably, the coupling of less nucleophilic compounds such as amides, aniline and diethylamine with benzoic acid also gave the desired products, albeit in low yields. Other oxygen- or sulfur-based nucleophiles could not be converted.

The synthetic concept may also be used for peptide couplings. Various N-protected amino acids were successfully coupled with amino acid esters. Without additives, racemization could not fully be suppressed but remained below 10%, which is a good basis for dedicated optimization.

**Mechanistic considerations.** The reaction mechanism was investigated by *in situ* nuclear magnetic resonance spectroscopy. The experiments confirmed the intermediacy of enol esters, which formed within minutes and were consumed in the course of the reaction (see Supplementary Table 6 and Supplementary Fig. 1 respectively). We thus conclude that as outlined in Fig. 2, the reaction proceeds via a Ru-catalysed hydroacyloxylation via a standard catalytic cycle[67,71] followed by aminolysis. In ESI MS investigations of the reaction mixture, species with $m/z$ values of 754 and 647 were dominant. These were identified as [RuCl$_2$(benzyl amine)$_3$(ethoxyacetylene)$_2$ (benzoate)]$^+$ and [RuCl$_2$(benzyl amine)$_2$(ethoxyacetylene)$_2$ (benzoate)]$^+$. In tandem mass spectrometry (MS) experiments, these adducts fragmented with loss of benzyl amine ligands and formation of the six-coordinate [RuCl$_2$(benzyl amine)$_1$ (ethoxyacetylene)$_2$(benzoate)]$^+$ complex, which we believe to be the catalyst resting state. It is reasonable to assume that it is a Ru(IV)-complex, since it bears three anions and is still positively charged. These investigations suggest the intermediacy of high-valent Ru-species, which explains why Ru$^{IV}$ pecursors have a higher activity than the Ru$^{II}$ and Ru$^0$ precursors employed in other catalytic additions. For the details of the spectroscopic investigation, see Supplementary Figs 2–5. In-depth, studies are required to clarify whether the carboxylate addition proceeds via Ru-complexes with $\eta^2$-coordinated alkynes or via Ru-alkylidene complexes.

In conclusion, the feasibility of catalytic amidation reactions with minimal waste production has been demonstrated. Even though extensive optimization is still required, this reaction concept could become an important factor in meeting one of the key challenges of Green Chemistry.

## Methods
For analytical data and preparation methods of the compounds in this article, see Supplementary Figs 6–111 and Supplementary Methods.

**General techniques.** All reactions were performed in oven-dried glassware containing a Teflon-coated stirring bar and dry septum under a nitrogen atmosphere. For the exclusion of atmospheric oxygen from the reaction media, solvents were degassed by argon sparge and purified by standard procedures before use. Non-aqueous amines were distilled before use. All reactions were monitored by gas chromatography (GC) using *n*-tetradecane as an internal standard or by high-performance liquid chromatography using anisole as an internal standard. Response factors of the products with regard to *n*-tetradecane/anisole were obtained experimentally by analysing known quantities of the substances. GC analyses were carried out using an HP-5 capillary column (Phenyl Methyl Siloxane 30 m × 320 × 0.25, 100/2.3-30-300/3) and a temperature programme beginning with 2 min at 60 °C followed by 30 °C/min ramp to 300 °C, then 3 min at this temp. High-performance liquid chromatography analyses were carried out using a Shimadzu LC-2010A. The stationary phase was a reversed phase column LiChroCart PAH C-18 from *Merck KGaA* with acetonitrile and water as eluents at 60 °C and the following solvent programme: starting from 10 vol% acetonitrile for 1 min, followed by increasing acetonitrile to 70 vol% during 23 min, then decreasing again to 10 vol% rapidly and maintaining this value for the next 2 min. Column chromatography was performed using a Combi Flash Companion-Chromatography-System (Isco-Systems) and Redi*Sep* packed columns (12 g). nuclear magnetic resonance spectra were obtained on Bruker AMX 400 or on Bruker Avance 600 systems using DMSO-$d_6$, Chloroform-$d_3$ or Toluene-$d_8$ as solvent, with proton and carbon resonances at 400/600 MHz and 101/151 MHz, respectively. Mass spectral data were acquired on a GC-MS Saturn 2,100 T (Varian). Infrared spectra were recorded on Perkin Elmer Spectrum 100 FT-IR Spectrometer with Universal ATR Sampling Accessory. Melting points are uncorrected and were measured on a Mettler FP 61. ESI MS data were acquired on a Bruker Esquire 6,000 and evaluated with mMass software. Sample solutions at concentrations of $\sim 1 \times 10^{-4}$ M were continuously infused into the ESI chamber at a flow rate of 2 µl min$^{-1}$ using a syringe pump. We use nitrogen as drying gas at a flow rate of 3.0–4.0 l min$^{-1}$ at 300 °C and spray the solutions at a nebulizer pressure of 4 psi with the electrospray needle held at 4.5 kV. CHN-elemental analyses were performed with a Hanau Elemental Analyzer vario Micro cube and HRMS with a Waters GCT Premier.

**Synthesis of amides using acetylene as activator.** An oven-dried headspace vial with Teflon-coated stirring bar was charged with the corresponding ammonium carboxylate (1 mmol) and dichloro[(2,6,10-dodecatriene)-1,12-diyl]ruthenium (5.06 mg, 20 µmol). The atmosphere was changed three times with nitrogen, then N-methylpyrrolidone (1 ml) and the corresponding amine (0.5 mmol) were added. The vial was placed in an autoclave reactor, the atmosphere was changed twice with acetylene, and a pressure of 1.7 bar was set. The mixture was then heated to 80 °C for 6 h. After cooling down to room temperature, the mixture was diluted with 20 ml of ethyl acetate and washed with each 20 ml of saturated NaHCO$_3$ solution, water and brine. The organic layer was dried with MgSO$_4$, the solvent removed under reduced pressure and the residue purified by column chromatography (SiOH, ethyl acetate/cyclohexane gradient).

**Synthesis of amides using ethoxyacetylene as activator.** An oven-dried headspace vial with Teflon-coated stirring bar was charged with the corresponding carboxylic acid (1 mmol) and dichloro[(2,6,10-dodecatriene)-1,12-diyl]ruthenium (5.06 mg, 15.0 µmol). The atmosphere was changed three times with nitrogen, then N-methylpyrrolidone (2 ml), benzyl amine (164 mg, 167 µl, 1.5 mmol) and ethoxyacetylene (40 wt%-solution in hexane; 210 mg, 299 µl and 1.5 mmol) were added in this order. The mixture was then heated to 40 °C for 4 h. After cooling down to room temperature, the mixture was diluted with 20 ml of ethyl acetate and washed with each 20 ml of sat. NaHCO$_3$ solution, water and brine. The organic layer was dried with MgSO$_4$, the solvent removed under reduced pressure and the residue purified by column chromatography (SiOH, ethyl acetate/ cyclohexane gradient).

## References
1. Montalbetti, C. A. G. N. & Falque, V. Amide bond formation and peptide coupling. *Tetrahedron* **61**, 10827–10852 (2005).
2. Lanigan, R. M. & Sheppard, T. D. Recent developments in amide synthesis: direct amidation of carboxylic acids and transamidation reactions. *Eur. J. Org. Chem.* **2013**, 7453–7465 (2013).
3. Pattabiraman, V. R. & Bode, J. W. Rethinking amide bond synthesis. *Nature* **480**, 471–479 (2011).

4. Lundberg, H., Tinnis, F., Selander, N. & Adolfsson, H. Catalytic amide formation from non-activated carboxylic acids and amines. *Chem. Soc. Rev.* **43**, 2714–2742 (2014).

5. Gooßen, L. J., Ohlmann, D. M. & Lange, P. P. The thermal amidation of carboxylic acids revisited. *Synthesis* **2009**, 160–164 (2009).

6. Mitchell, J. A. & Reid, E. E. The preparation of aliphatic amides. *J. Am. Chem. Soc.* **53**, 1879–1883 (1931).

7. Allen, C. L., Chhatwal, A. R. & Williams, J. M. J. Direct amide formation from unactivated carboxylic acids and amines. *Chem. Commun.* **48**, 666–668 (2012).

8. Nelson, P. & Pelter, A. 954. Trisdialkylaminoboranes: new reagents for the synthesis of enamines and amides. *J. Chem. Soc.* 5142–5144 (1965).

9. Ishihara, K., Ohara, S. & Yamamoto, H. 3, 4, 5-Trifluorobenzeneboronic acid as an extremely active amidation catalyst. *J. Org. Chem.* **61**, 4196–4197 (1996).

10. Tinnis, F., Lundberg, H. & Adolfsson, H. Direct catalytic formation of primary and tertiary amides from non-activated carboxylic acids, employing carbamates as amine source. *Adv. Synth. Catal.* **354**, 2531–2536 (2012).

11. Allen, C. L. & Williams, J. M. J. Metal-catalysed approaches to amide bond formation. *Chem. Soc. Rev.* **40**, 3405–3415 (2011).

12. Mohy El Dine, T., Erb, W., Berhault, Y., Rouden, J. & Blanchet, J. Catalytic chemical amide synthesis at room temperature: One more step toward peptide synthesis. *J. Org. Chem.* **80**, 4532–4544 (2015).

13. Lundberg, H. & Adolfsson, H. Hafnium-Catalyzed direct amide formation at room temperature. *ACS Catal.* **5**, 3271–3277 (2015).

14. Villeneuve, G. B. & Chan, T. H. A rapid, mild and acid-free procedure for the preparation of acyl chlorides including formyl chloride. *Tetrahedron Lett.* **38**, 6489–6492 (1997).

15. Lal, G. S., Pez, G. P., Pesaresi, R. J., Prozonic, F. M. & Cheng, H. Bis (2-methoxyethyl)aminosulfur Trifluoride: a new broad-spectrum deoxofluorinating agent with enhanced thermal stability. *J. Org. Chem.* **64**, 7048–7054 (1999).

16. Shioiri, T., Ninomiya, K. & Yamada, S. Diphenylphosphoryl azide. new convenient reagent for a modified Curtius reaction and for peptide synthesis. *J. Am. Chem. Soc.* **94**, 6203–6205 (1972).

17. Carpino, L. A., Beyermann, M., Wenschuh, H. & Bienert, M. Peptide synthesis via amino acid halides. *Acc. Chem. Res.* **29**, 268–274 (1996).

18. Lee, J. B. Preparation of acyl halides under very mild conditions. *J. Am. Chem. Soc.* **88**, 3440–3441 (1966).

19. Olah, G. A., Nojima, M. & Kerekes, I. Synthetic methods and reactions; IV. 1 Fluorination of carboxylic acids with cyanuric fluoride. *Synthesis* **1973**, 487–488 (1973).

20. Carpino, L. A. & El-Faham, A. Tetramethylfluoroformamidinium hexafluorophosphate: a rapid-acting peptide coupling reagent for solution and solid phase peptide synthesis. *J. Am. Chem. Soc.* **117**, 5401–5402 (1995).

21. Windridge, G. & Jorgensen, E. C. 1-Hydroxybenzotriazole as a racemization-suppressing reagent for the incorporation of im-benzyl-L-histidine into peptides. *J. Am. Chem. Soc.* **93**, 6318–6319 (1971).

22. Kisfaludy, L., Schön, I., Szirtes, T., Nyéki, O. & Löw, M. A novel and rapid peptide synthesis. *Tetrahedron Lett.* **15**, 1785–1786 (1974).

23. König, W. & Geiger, R. Eine neue methode zur synthese von peptiden: aktivierung der carboxylgruppe mit dicyclohexylcarbodiimid unter zusatz von 1-hydroxy-benzotriazolen. *Chem. Ber.* **103**, 788–798 (1970).

24. Kisfaludy, L. & Schön, I. Preparation and applications of pentafluorophenyl esters of 9-fluorenylmethyloxycarbonyl amino acids for peptide synthesis. *Synthesis* **1983**, 325–327 (1983).

25. Mikozlajczyk, M. & Kiezlbasiński, P. Recent developments in the carbodiimide chemistry. *Tetrahedron* **37**, 233–284 (1981).

26. Carpino, L. A. & El-Faham, A. The diisopropylcarbodiimide/ 1-hydroxy-7-azabenzotriazole system: segment coupling and stepwise peptide assembly. *Tetrahedron* **55**, 6813–6830 (1999).

27. Guinó, M. & Kuok (Mimi), H. K. Wang-aldehyde resin as a recyclable support for the synthesis of α,α-disubstituted amino acid derivatives. *Org. Biomol. Chem.* **3**, 3188–3193 (2005).

28. Coste, J., Frérot, E., Jouin, P. & Castro, B. Oxybenzotriazole free peptide coupling reagents for N-methylated amino acids. *Tetrahedron Lett.* **32**, 1967–1970 (1991).

29. Han, S.-Y. & Kim, Y.-A. Recent development of peptide coupling reagents in organic synthesis. *Tetrahedron* **60**, 2447–2467 (2004).

30. Valeur, E. & Bradley, M. Amide bond formation: beyond the myth of coupling reagents. *Chem. Soc. Rev.* **38**, 606–631 (2009).

31. Gabriel, C. M., Keener, M., Gallou, F. & Lipshutz, B. H. Amide and peptide bond formation in water at room temperature. *Org. Lett.* **17**, 3968–3971 (2015).

32. Dawson, P., Muir, T., Clark-Lewis, I. & Kent, S. Synthesis of proteins by native chemicalligation. *Science* **266**, 776–779 (1994).

33. Bode, J. W., Fox, R. M. & Baucom, K. D. Chemoselective amide ligations by decarboxylative condensations of N-alkylhydroxylamines and α-ketoacids. *Angew. Chem. Int. Ed.* **45**, 1248–1252 (2006).

34. Zhang, Y., Xu, C., Lam, H. Y., Lee, C. L. & Li, X. Protein chemical synthesis by serine and threonine ligation. *Proc. Natl Acad. Sci. USA* **110**, 6657–6662 (2013).

35. Nilsson, B. L., Kiessling, L. L. & Raines, R. T. Staudinger ligation: a peptide from a thioester and azide. *Org. Lett.* **2**, 1939–1941 (2000).

36. Noda, H., Erős, G. & Bode, J. W. Rapid ligations with equimolar reactants in water with the potassium acyltrifluoroborate amide formation. *J. Am. Chem. Soc.* **136**, 5611–5614 (2014).

37. Fouché, M., Masse, F. & Roth, H.-J. Hydroxymethyl salicylaldehyde auxiliary for a glycine-dependent amide-forming ligation. *Org. Lett.* **17**, 4936–4939 (2015).

38. Constable, D. J. C. et al. Key green chemistry research areas: a perspective from pharmaceutical manufacturers. *Green Chem.* **9**, 411–420 (2007).

39. Tamaru, Y., Yamada, Y. & Yoshida, Z. Direct oxidative transformation of aldehydes to amides by palladium catalysis. *Synthesis* **1983**, 474–476 (1983).

40. Tillack, A., Rudloff, I. & Beller, M. Catalytic amination of aldehydes to amides. *Eur. J. Org. Chem.* **2001**, 523–528 (2001).

41. Yoo, W.-J. & Li, C.-J. Highly efficient oxidative amidation of aldehydes with amine hydrochloride salts. *J. Am. Chem. Soc.* **128**, 13064–13065 (2006).

42. Gunanathan, C., Ben-David, Y. & Milstein, D. Direct synthesis of amides from alcohols and amines with liberation of H2. *Science* **317**, 790–792 (2007).

43. Ekoue-Kovi, K. & Wolf, C. One-pot oxidative esterification and amidation of aldehydes. *Chem. Eur. J.* **14**, 6302–6315 (2008).

44. Dobereiner, G. E. & Crabtree, R. H. Dehydrogenation as a substrate-activating strategy in homogeneous transition-metal catalysis. *Chem. Rev.* **110**, 681–703 (2010).

45. De Sarkar, S. & Studer, A. Oxidative amidation and azidation of aldehydes by NHC catalysis. *Org. Lett.* **12**, 1992–1995 (2010).

46. Chen, C. & Hong, S. H. Oxidative amide synthesis directly from alcohols with amines. *Org. Biomol. Chem.* **9**, 20–26 (2011).

47. Zhang, L. et al. Aerobic oxidative coupling of alcohols and amines over Au–Pd/ resin in water: Au/Pd molar ratios switch the reaction pathways to amides or imines. *Green Chem.* **15**, 2680–2684 (2013).

48. Kang, B., Fu, Z. & Hong, S. H. Ruthenium-catalyzed redox-neutral and single-step amide synthesis from alcohol and nitrile with complete atom economy. *J. Am. Chem. Soc.* **135**, 11704–11707 (2013).

49. Li, F., Ma, J., Lu, L., Bao, X. & Tang, W. Combination of gold and iridium catalysts for the synthesis of N-alkylated amides from nitriles and alcohols. *Catal. Sci. Technol.* **5**, 1953–1960 (2015).

50. Miyamura, H., Min, H., Soulé, J.-F. & Kobayashi, S. Size of gold nanoparticles driving selective amide synthesis through aerobic condensation of aldehydes and amines. *Angew. Chem. Int. Ed.* **54**, 7564–7567 (2015).

51. Owston, N. A., Parker, A. J. & Williams, J. M. J. Iridium-catalyzed conversion of alcohols into amides via oximes. *Org. Lett.* **9**, 73–75 (2007).

52. Bruneau, C. in *Hydrofunctionalization* (eds Ananikov, V. P. & Tanaka) M.**43**, 203–230 (Springer Heidelberg 2011).

53. Ishino, Y., Nishiguchi, I., Nakao, S. & Hirashima, T. Novel synthesis of enol esters through silver-catalyzed reaction of acetylenic compounds with carboxylic acids. *Chem. Lett.* **5**, 641–644 (1981).

54. Chary, B. C. & Kim, S. Gold(I)-catalyzed addition of carboxylic acids to alkynes. *J. Org. Chem.* **75**, 7928–7931 (2010).

55. Rotem, M. & Shvo, Y. Addition of carboxylic acids to alkynes catalyzed by ruthenium complexes. Vinyl ester formation. *Organometallics* **2**, 1689–1691 (1983).

56. Mitsudo, T., Hori, Y. & Watanabe, Y. Selective addition of unsaturated carboxylic acids to terminal acetylenes catalyzed by bis(.eta.5-cyclooctadienyl)ruthenium(II)-tri-n-butylphosphine. A novel synthesis of enol esters. *J. Org. Chem.* **50**, 1566–1568 (1985).

57. Ruppin, C., Lecolier, S. & Dixneuf, P. H. Regioselective synthesis of isopropenyl esters by ruthenium catalysed addition of N-protected amino-acids to propyne. *Tetrahedron Lett.* **29**, 5365–5368 (1988).

58. Bruneau, C., Neveux, M., Kabouche, Z., Ruppin, C. & Dixneuf, P. H. Ruthenium-catalysed additions to alkynes: synthesis of activated esters and their use in acylation reactions. *Synlett* **1991**, 755–763 (1991).

59. Gooßen, L. J., Paetzold, J. & Koley, D. Regiocontrolled ru-catalyzed addition of carboxylic acids to alkynes: practical protocols for the synthesis of vinyl esters. *Chem. Commun.* 706–707 (2003).

60. Kita, Y. et al. Facile and efficient syntheses of carboxylic anhydrides and amides using (trimethylsilyl)ethoxyacetylene. *J. Org. Chem.* **51**, 4150–4158 (1986).

61. Kabouche, Z., Bruneau, C. & Dixneuf, P. H. Enol esters as intermediates for the facile conversion of amino acids into amides and dipeptides. *Tetrahedron Lett.* **32**, 5359–5362 (1991).

62. Neveux, M., Bruneau, C., Lécolier, S. & Dixneuf, P. H. Novel syntheses of oxamides, oxamates and oxalates from diisopropenyl oxalate. *Tetrahedron* **49**, 2629–2640 (1993).

63. Bruneau, C. & Dixneuf, P. H. Selective transformations of alkynes with ruthenium catalysts. *Chem. Commun.* **6,** 507–512 (1997).

64. Kita, Y., Maeda, H., Omori, K., Okuno, T. & Tamura, Y. Novel efficient synthesis of 1-ethoxyvinyl esters using ruthenium catalysts and their use in acylation of amines and alcohols: synthesis of hydrophilic 3′-N-acylated oxaunomycin derivatives. *J. Chem. Soc. Perkin Trans.* **1,** 2999–3005 (1993).

65. Schröder, H. *et al.* Racemization-free chemoenzymatic peptide synthesis enabled by the ruthenium-catalyzed synthesis of peptide enol esters via alkyne-addition and subsequent conversion using alcalase-cross-linked enzyme aggregates. *Adv. Synth. Catal.* **355,** 1799–1807 (2013).

66. Ashton Acton, Q. *Benzoic Acids—Advances in Research and Application* (Scholarly Editions, 2013).

67. Doucet, H., Martin-Vaca, B., Bruneau, C. & Dixneuf, P. H. General synthesis of (Z)-alk-1-en-1-yl esters via ruthenium-catalyzed anti-Markovnikov trans-addition of carboxylic acids to terminal alkynes. *J. Org. Chem.* **60,** 7247–7255 (1995).

68. Doucet, H., Höfer, J., Bruneau, C. & Dixneuf, P. H. Stereoselective synthesis of Z-enol esters catalysed by [bis(diphenylphosphino)alkane]bis (2-methylpropenyl)ruthenium complexes. *J. Chem. Soc. Chem. Commun.* **850,** 850–851 (1993).

69. Gooßen, L. J., Salih, K. S. M. & Blanchot, M. Synthesis of secondary enamides by ruthenium-catalyzed selective addition of amides to terminal alkynes. *Angew. Chem. Int. Ed.* **47,** 8492–8495 (2008).

70. Cadierno, V., Francos, J. & Gimeno, J. Ruthenium(IV)-catalyzed Markovnikov addition of carboxylic acids to terminal alkynes in aqueous medium. *Organometallics* **30,** 852–862 (2011).

71. Alonso, F., Beletskaya, I. P. & Yus, M. Transition-metal-catalyzed addition of heteroatom – hydrogen bonds to alkynes. *Chem. Rev.* **104,** 3079–3160 (2004).

## Acknowledgements
We thank Astra Zeneca, DFG (SFB/TRR-88, '3MET') and Deutsche Bundesstiftung Umwelt (fellowship to S.B.) for financial support, Umicore AG for the donation of chemicals and Johannes Lang for technical assistance performing ESI MS measurements.

## Author contributions
L.J.G. planned and supervised the research; T.K. conceived and performed most experiments together with S.B.; B.E. performed additional experiments; T.K. and S.B. isolated and characterized the products; T.K. and L.J.G. co-wrote the manuscript.

# 22

# Organic–inorganic supramolecular solid catalyst boosts organic reactions in water

Pilar García-García[1], José María Moreno[1], Urbano Díaz[1], Marta Bruix[2] & Avelino Corma[1,3]

Coordination polymers and metal-organic frameworks are appealing as synthetic hosts for mediating chemical reactions. Here we report the preparation of a mesoscopic metal-organic structure based on single-layer assembly of aluminium chains and organic alkylaryl spacers. The material markedly accelerates condensation reactions in water in the absence of acid or base catalyst, as well as organocatalytic Michael-type reactions that also show superior enantioselectivity when comparing with the host-free transformation. The mesoscopic phase of the solid allows for easy diffusion of products and the catalytic solid is recycled and reused. Saturation transfer difference and two-dimensional $^1H$ nuclear Overhauser effect NOESY NMR spectroscopy show that non-covalent interactions are operative in these host–guest systems that account for substrate activation. The mesoscopic character of the host, its hydrophobicity and chemical stability in water, launch this material as a highly attractive supramolecular catalyst to facilitate (asymmetric) transformations under more environmentally friendly conditions.

[1] Instituto de Tecnología Química, UPV-CSIC, Universidad Politécnica de Valencia, Avenida de los Naranjos s/n, E-46022 Valencia, Spain. [2] Instituto de Química Física Rocasolano, CSIC, Serrano 119, 28006 Madrid, Spain. [3] King Fahd University of Petroleum and Minerals, PO Box 989, 31261 Dhahran, Saudi Arabia. Correspondence and requests for materials should be addressed to A.C. (email: acorma@itq.upv.es).

Metal-organic frameworks (MOFs) emerged as a new type of porous crystalline solids with tailorable structures and functions. The high versatility and possibilities that these materials offer are making them very attractive for specific applications such as gas storage[1-3], separations[4,5], luminescent[6], conducting[7], and magnetic materials[8], drug delivering systems[9] and catalysis[10].

MOF-type metal-organic structures are frequently obtained as 3D frameworks based on the coordination binding between metallic nodes and rigid bi- or multi-podal linkers[11,12]. Preparation of modifiable layered metal-organic materials formed by individual sheets, spatially organized through electrostatic interactions is, however, not described in the literature, as far as we know. Only incipient studies related with the formation of MOF nanosheets by modulation of growth kinetics have been reported[13]. Nevertheless, the existence of 3D conventional MOFs which exhibited lamellar structural tendency in the spatial distribution of their structural building units could be indicative of the possibility to obtain metal-organic structures by assembly of lamellar units. In particular, from a topological point of view, this type of MOFs are based on infinite inorganic M-O-M chains, formed by only discrete metal clusters, separated by aryl dicarboxylate linkers perpendicularly located to individual metallic nodes alignment[14,15]. This topology can be observed in different well-known MOFs such as MIL-53(Al) (ref. 16), DUT-4, DUT-5 (ref. 17), DUT-8 (refs 18,19), MOF-Zn-DABCO (ref. 20), Cu-(tpa)[21] or MIL-68(Al) (ref. 22), in which it is possible to distinguish the 2D lamellar inorganic sub-networks.

Notably, many enzymes can promote chemical reactions via hydrophobic cavities wherein the cumulative influence of not only the active site in the enzyme but also the non-covalent bonds interactions account for substrate activation. Several synthetic host–guest systems have been used that copy these remarkable properties such as cyclodextrins[23-25], cucurbiturils[26,27], (metalla)crown ethers[28], calixarenes[29], carcerands[30] and zeolites[31]. Solvation effects[32,33] have also been studied and proved to account for rate acceleration in chemical reactions in such a way that approach enzyme-like catalysis. Lately, self-assembled hosts[34-37] have demonstrated to function as 'molecular flasks' to entail unusual reactions, or unique

chemical phenomena. However, the development and engineering of larger and more sophisticated host structures that can incorporate several reaction components is still a challenge. An appealing strategy for supramolecular catalysis involves the design of a host system able to accommodate a given (ideally any) homogeneous catalyst and also the two reacting components, facilitating the reaction of the two partners even further and/or allowing for more environmentally friendly reaction conditions, such as lower reaction temperature, lower homogeneous catalyst loading or the use of water as a green solvent. Here we report the initial steps towards this design plan. Taking into consideration the lamellar tendency of MIL-53 structures, we prepare a novel metal-organic-type material, Al-ITQ-HB, which is based on ordered individual aluminium clusters-type sheets, similar to the sub-networks detectable in MIL-53-type materials. Key in our design is the use of a specific organic spacer, 4-heptylbenzoic acid (HB), with only one reactive carboxylate group that interacts with inorganic metallic nodes through stable coordination bonds. This particular organic spacer contains hydrocarbonated tails, which control the separation between metallic nanosheets, inhibit the 3D growth observed in conventional MOFs, favouring the formation of mesoscopic non-ordered phases with well-defined metal-organic monolayers[38]. We use this material as a supramolecular hybrid structure in cases where the hydrophobic part acts to concentrate reactants and acts as hydrophobic pockets wherein molecules are activated when performing reactions even in aqueous media. The material markedly accelerates condensation reactions in water in the absence of acid or base catalyst, as well as organocatalytic Michael-type reactions that also show superior enantioselectivity when comparing with the host-free transformation.

## Results

### Synthesis and characterization of the metal-organic material.
We started with a MIL-53 (Al)-type material and decided to synthesize 2D nanosheets (see Fig. 1). This was achieved by inhibiting the 3D growth in the MIL-53 (Al) crystals and favouring the formation of mesoscopic phases that fitted our

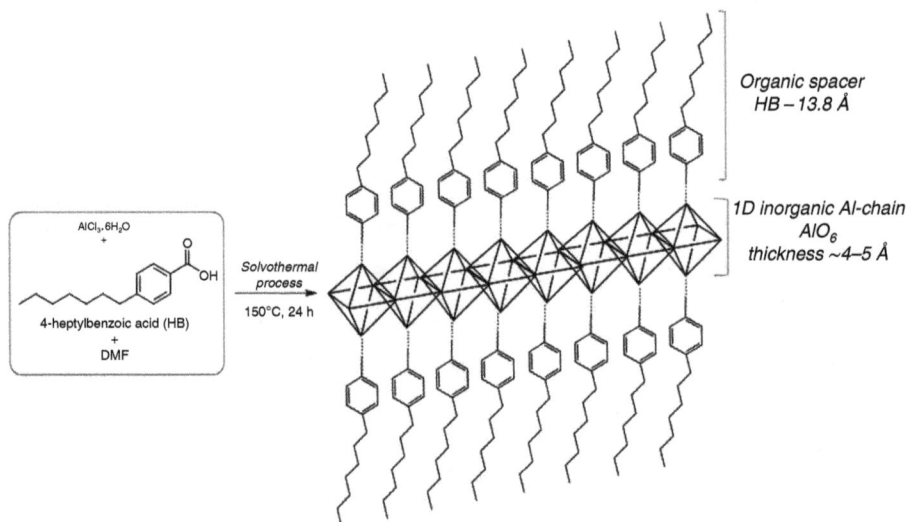

**Figure 1 | Synthetic route followed to obtain mesoscopic-type hybrid material.** The synthetic procedure was replicated >10 times. Representation of Al-ITQ-HB as individual organic-inorganic nanosheet is shown. The individual nanosheets would be formed by consecutive corner-sharing octahedral (AlO$_6$) units, conforming 1D inorganic chains, which are separated by alkyl benzene monocarboxylate ligands, located on both sides of metallic nodes. So, the coordinative association between inorganic chains and organic spacers in the opposite sides of metallic nodes forms each individual organic–inorganic layer which would be the basis of this type of metal-organic mesoscopic structures. DMF, dimethylformamide.

catalytic purposes. In our case, *para*-heptyl benzene mono-carboxylate compound (HB) was used as molecular spacers (Fig. 1). This organic spacer was utilized instead of conventional rigid aryl dicarboxylate linkers, such as terephthalic acid, normally used in the preparation of the major part of MOF-type structures. Solvothermal process in the presence of aluminium chloride, dimethylformamide and HB as inorganic source, solvent and organic spacers, respectively, facilitated the preparation of the mesoscopic hybrid material whose walls were based on metal-organic layers with octahedral aluminium units separated by the hydrocarbonated tails acting as spacers, perpendicularly located to inorganic nodes. Representation of individual organic–inorganic layers is also drawn in Fig. 1, highlighting the role of organic spacers (HB) as effective growing inhibitors of the standard 3D metal-organic structure.

The X-ray diffraction pattern of the as-synthesized hybrid material Al-ITQ-HB (Supplementary Fig. 1), showed that mesoscopic organization could be achieved being clearly observable a *(100)* low-angle diffraction peak ($d_{100} = \sim 35 \text{ Å}$) which is characteristic of mesoporous solids with short-range hexagonal ordering. However, *(100)* diffraction band could also be distinctive of 2D lamellar frameworks formed by individual sheets perpendicularly disposed to axis-*a*, indicating that a certain regularity would exist in the separation (basal space of $\sim 35 \text{ Å}$) between successive piled organic–inorganic nanosheets. In any case, the analysed hybrid solid did not exhibit a very high structural order, as deduced from the low intensity and broadness of the low-angle *(100)* diffraction band together with the absence of repetitive *(h00)* diffraction bands.

The mesoscopic long-order organization was observed from TEM microscopy, being clearly detected the presence of mesoporous which were distributed along plate-like crystals (Fig. 2a,b, Supplementary Fig. 2). The micrographs also showed the existence of areas formed by short sheets which would evidence the dual nature of the Al-ITQ-HB materials, that is, lamellar and mesoporous due to the presence of metal-organic nanolayers, probably assembled around non-ordered

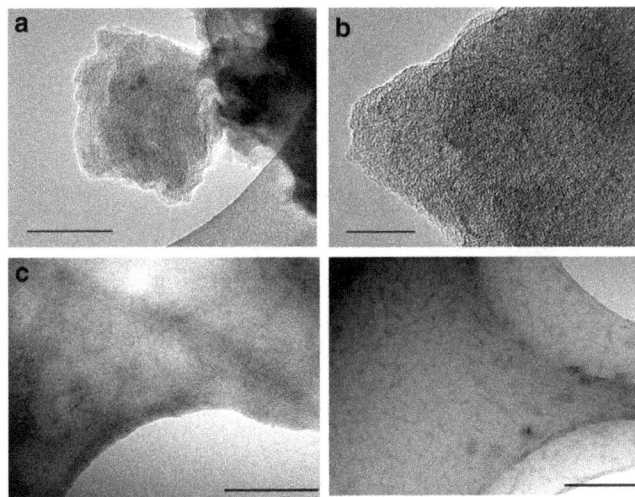

**Figure 2 | HRTEM images of Al-ITQ-HB sample. (a)** As-synthesized material. Scale bar, 100 nm. **(b)** As-synthesized material. Scale bar, 20 nm. Mesoporous cavities are clearly detected. A pore-size distribution around 20 Å in diameter is observed. **(c,d)** material after post-synthesis treatment with dichloromethane generating a stable solution. Scale bars, 200 nm. The high hydrophobicity exhibited by Al-ITQ-HB solid, due to the elevated organic content, facilitated the interlayer penetration of solvent molecules with the consequent delamination effect, being observable individual nanolayers dispersed in the stable solution.

mesocavities. However, the low structural periodicity exhibited by this type of hybrid solids along the channel direction would hinder the easy detection of the lamellar sub-domains from electronic microscopy. Despite that, the lamellar nature of the hybrid sample was confirmed through the exfoliation phenomenon achieved in presence of non-polar solvents (Fig. 2c,d, Supplementary Fig. 3).

From the elemental CHNS analysis (Supplementary Table 1), it was possible to estimate the amount of organic content present in the hybrid material, Al-ITQ-HB. The results indicated that the organic counterpart contribution was around 40 wt%, corresponding to alkylaryl fragments. Furthermore, the practical absence of nitrogen content in the solid showed that most of the dimethylformamide, used as solvent during the solvothermal synthesis, was removed during the successive washing steps.

The weight loss with temperature and the corresponding derivative (TGA and DTA curves; Supplementary Fig. 4) for the mesoscopic metal-organic hybrid material allowed establishing, not only the amount of organic spacers incorporated in the Al-ITQ-HB solid, but also their hydrothermal stability. After elimination of the major part of hydration water and residual dimethylformamide (both detected at around 80–150 °C), it was possible to see a main weight loss, between 450 and 600 °C, assigned to the presence of *para*-alkyl benzene monocarboxylate molecules (HB) used as spacers. It is important to remark the weight loss observed between 250 and 400 °C assigned to the decomposition (dehydration phenomenon) of $AlO_4(OH)_2$ units present in the 1D inorganic chains[39], which would confirm again the presence of lamellar organic–inorganic sub-domains integrated into the mesoscopic framework. In this temperature range, contribution of alkyl chains from incorporated organic linkers HB, would also be included into this weight loss.

The $^{13}C$ CP/MAS NMR spectrum of Al-ITQ-HB (Supplementary Fig. 5B) confirms the total integrity of the organic spacer (HB) after the solvothermal synthesis processes since all carbon atoms were unequivocally assigned. In addition, the presence and integrity of alkyl benzene monocarboxylate units located in the walls of Al-ITQ-HB solid was confirmed by infrared spectroscopy (Supplementary Fig. 6). On the other hand, $^{27}Al$ MAS NMR spectrum (Supplementary Fig. 7) of the Al-ITQ-HB sample, presented a defined peak at 6.7 p.p.m. chemical shift that is characteristic of aluminium (oxo)hydroxide species, showing a regular octahedral environment of metallic atoms. This result would confirm the formation of 1D inorganic chains of $(AlO_6)$ octahedral units in the assembled organic–inorganic nanolayers[40].

Textural properties for Al-ITQ-HB solid were analysed from Argon adsorption isotherm (Supplementary Fig. 8), being estimated a reduced surface area and porous volume ($S_{BET} \sim 35 \text{ m}^2 \text{g}^{-1}$, $V_{TOTAL} \sim 0.11 \text{ cm}^3 \text{g}^{-1}$) due to, probably, the big amount of organic content ($\sim 40 \text{ wt%}$) present in the hybrid material that hinders the correct gas adsorption along the mesocavities. The marked hydrophobic character of the material would reinforce this phenomenon. However, the Hörvath–Kawazoe pore-size distribution (Supplementary Fig. 9) showed that the majority of pores were centred at $\sim 20–25 \text{ Å}$, this value being coincident with that estimated from electronic microscopy, remarking the mesoscopic nature of the Al-ITQ-HB material.

CO adsorption at low temperature (100 K) was monitored by FTIR spectroscopy to gather information relating Lewis acidic sites in both samples, MIL-53 (Al) and Al-ITQ-HB (Supplementary Figs 10 and 11). An absorption band is observed at $2,158 \text{ cm}^{-1}$ that already indicates that the Lewis acidity of these materials is very weak. This being confirmed by the fact that a complete disappearance of this band is observed at 100 K by subsequent treatment of the sample in vacuum.

**Knoevenagel condensation reaction**. The extremely flexible hybrid material obtained could act as a supramolecular catalyst in cases where the hydrocarbonated alkylaryl spacers would constitute hydrophobic pockets when working in aqueous media. Indeed, we found that Al-ITQ-HB efficiently promotes the Knoevenagel condensation of 2-naphthaldehyde (**1a**) with Meldrum's acid (**2**) in $H_2O$:$CHCl_3$ in the absence of base or acid catalysts. Under the optimized conditions (Supplementary Tables 2 and 3), the condensation product (**3a**) was formed in 91% yield after 10 h (Table 1, entry 1). The reaction conditions are very mild (neutral, room temperature) and the reaction proceeds catalytically (30 mol% of Al-ITQ-HB) as the product easily diffuse through the mesoporous channels liberating the host to embrace another pair of reacting components. Without Al-ITQ-HB, **1a** gave only small amount of the **3a** (4%) under otherwise identical conditions. Carrying out the reaction at higher molarity of $CHCl_3$ as the sole solvent in the absence of Al-ITQ-HB, afforded the product in low yield of 10%, (Supplementary Table 4), allowing us to conclude that rate enhancement is not a mere consequence of the concentration rise. Al-ITQ-HB is not simply increasing molarity of reagents but also promotes the condensation in water. Control experiment with conventional MIL-53(Al) showed barely formation of product (17%). The hydrophobic cavity of Al-ITQ-HB is, therefore, crucial for the reaction because the individual MOF components Al(OH)$(C_2H_3O_2)_2$ or HB acid did not promote the transformation substantially (Supplementary Table 4). Micellar catalysis was also tested by using the equivalent amount of sodium 4-heptylbenzoate that resulted in low product formation (19% yield under otherwise identical conditions, Supplementary Table 4). Rate acceleration appeared more obvious when following the Knoevenagel condensation reaction over time (Supplementary Figs 12 and 13). We observe that initial reaction rate in the case of Al-ITQ-HB in the presence of water (Supplementary Fig. 12) is 540 times higher than when no additive is used. Initial reaction rate is also superior for Al-ITQ-HB versus utilization of HB acid (54 times higher), sodium 4-heptylbenzoate (54 times higher) or MIL-53 (Al) (216 times higher). Rise in reaction rate for Al-ITQ-HB was also observed in organic solvents (such as $CHCl_3$, Supplementary Fig. 13) wherein initial rate is 105 times superior when using Al-ITQ-HB versus no additive utilization.

Substrate scope of the Knoevenagel reaction is shown in Table 1 (see also Supplementary Methods). Under the very mild conditions, a variety of aromatic aldehydes underwent efficient condensation when using Al-ITQ-HB. Control experiments without Al-ITQ-HB showed that reaction hardly occurred. Once again, 3D MIL-53 (Al) showed no relevant conversion to the **3a** highlighting the prominent role of the hydrophobic pocket and demonstrating that substrate activation is not a mere consequence of Lewis acidic activation by the framework, or the influence of $\pi$-stacking interaction of the starting aromatic aldehyde with the framework. The method proceeded better with aromatic aldehydes having electron donating groups (Table 1, entry 4), although rate acceleration is observed in all cases.

After completion of the reaction, the supramolecular catalyst Al-ITQ-HB could be recovered by simple centrifugation and can be reused for the next run with a slight deterioration of the catalytic activity (72% conversion). PXRD showed that the recycled Al-ITQ-HB maintain its mesoscopic nature after use, being observable the characteristic low-angle diffraction band (Supplementary Fig. 14).

**Multicomponent reaction**. The activity of Al-ITQ-HB was also tested in the multicomponent reaction of isatin, dimedone and malononitrile for the synthesis of the spirooxindole product (Fig. 3). This transformation proceeds under micellar catalysis in water by means of sodium stearate (10 mol%) (ref. 41). We found out that Al-ITQ-HB efficiently promoted the three-component reaction in an organic solvent such as chloroform, while the product is barely formed when using conventional MIL-53 (Al)

---

**Table 1 | Al-ITQ-HB catalysed Knoevenagel condensation of aldehydes 1 with Meldrum's acid 2\*.**

| Entry | 1 | | Time (h) | Yield[†] | | |
|-------|---|---|----------|-----------|---|---|
| | | | | With Al-ITQ-HB | Without Cat | With MIL-53(Al) |
| 1 | 2-naphthaldehyde (CHO) | 1a | 10 | 91 | 4 | 17 |
| 2 | 1-naphthaldehyde (CHO) | 1b | 24 | 91 | 15 | 15 |
| 3 | benzaldehyde (CHO) | 1c | 8 | 86 | 11 | 15 |
| 4 | 2-OMe benzaldehyde (CHO) | 1d | 8 | 93 | 15 | 21 |
| 5 | 4-Cl benzaldehyde (CHO) | 1e | 24 | 62 | 9 | 30 |

\*Reaction conditions: aldehyde (**1**) (0.1 mmol), Meldrum's acid (**2**) (0.1 mmol) and catalyst (30 mol%) in $H_2O$:$CHCl_3$ 1:1 (0.5 M) at room temperature.
[†]$^1$H NMR yields. Yields are the average of at least two runs.

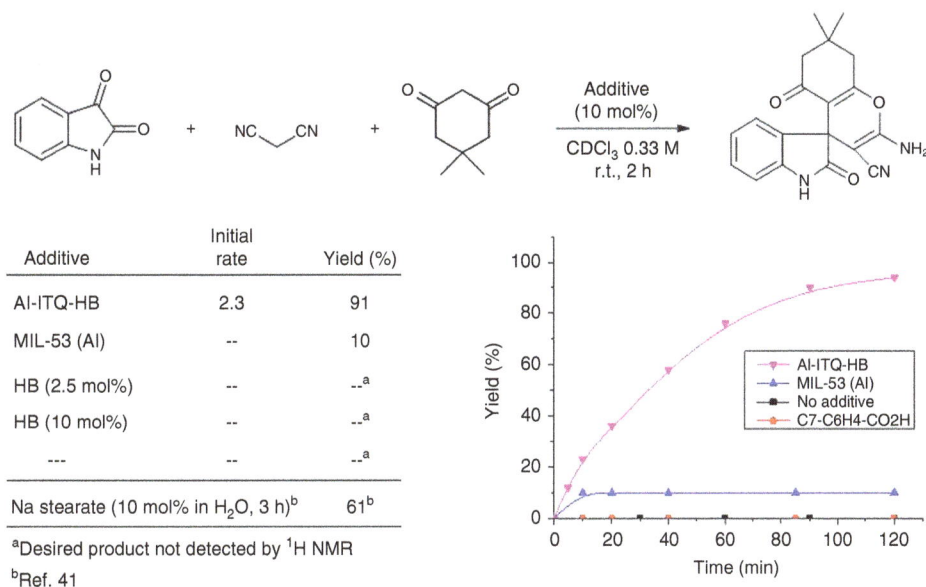

| Additive | Initial rate | Yield (%) |
|---|---|---|
| Al-ITQ-HB | 2.3 | 91 |
| MIL-53 (Al) | -- | 10 |
| HB (2.5 mol%) | -- | --[a] |
| HB (10 mol%) | -- | --[a] |
| --- | -- | --[a] |
| Na stearate (10 mol% in H$_2$O, 3 h)[b] | | 61[b] |

[a]Desired product not detected by $^1$H NMR
[b]Ref. 41

**Figure 3 | Al-ITQ-HB catalysed multicomponent reaction for the synthesis of spirooxindole product.** Reaction conditions: isatin (0.2 mmol), dimedone (0.2 mmol), malononitrile (0.2 mmol) and additive (10 mol% except noted) in CDCl$_3$ (0.33 M) at room temperature. $^1$H NMR yields (d6-DMSO) are reported (internal standard: Ph$_3$CH).

and not detected by $^1$H NMR after 2 h reaction time when using HB acid or no additive at all (Fig. 3).

While micellar catalysis by means of sodium stearate gave comparable results as our Al-ITQ-HB on the bases of initial reaction rates (Supplementary Fig. 16), the attained yield after 2 h reaction time is higher for Al-ITQ-HB. Once again, in the presence of water, Al-ITQ-HB probed rate enhancement as compared with the MOF-free reaction under otherwise similar conditions (Supplementary Fig. 17). Superiority of Al-ITQ-HB is also shown versus utilization of the 3D MIL-53 (Al) or HB acid (Supplementary Fig. 17).

**Organocatalytic Michael-type reactions.** Given the excellent results attained in rate acceleration in the above condensation reactions and the mesoscopic nature of Al-ITQ-HB, we envisioned that this framework could also be applied to facilitate asymmetric organocatalytic reactions. Organocatalysis have become a very productive field or research that well complements metal catalysis and biocatalysis in offering new opportunities for (enantio)selective transformations[42]. Probably, the main limitations in organocatalysis are the fact of low turnover numbers and slow reaction rates often affecting procedures. As an example, Michael addition of isobutyraldehyde to nitrostyrene catalysed by several primary amine bifunctional organocatalysts (similar to **4** in Fig. 4) have shown to require loadings of 20–30 mol%, to achieve good to high product formation after relatively long reaction time of 2 days (refs 43,44). In particular, 10 mol% of bifunctional amino-urea catalyst **4** provided the addition product in 14% yield and high enantioselectivity after 24 h reaction time (see equation 1 in Fig. 4). Remarkably, the combination of organocatalyst **4** with our supramolecular host catalyst Al-ITQ-HB provided substantial rate acceleration while maintaining the same degree of asymmetric induction. By this means, high yield (of 96%) can be achieved with relatively low organocatalyst loading, at room temperature, without the aid of noxious additives and after reasonable reaction time. Further analysis of the Al-ITQ-HB mediated reaction proved the role of the hydrophobic pocket in the host catalyst. Similar trends are observed here as in the case of the above depicted Knoevenagel condensation reaction. The presence of additives such as

1-phenylheptane, HB acid or MIL-53(Al) had no influence on the reaction rate (Supplementary Table 5). While host catalyst, Al-ITQ-HB, also shows rate enhancement in bare organic solvents such as toluene and CH$_2$Cl$_2$, higher product yields are attained in the presence of water evidencing the interplay of the hydrophobic effect that accounts partly for the observed efficiency. It was found that micellar catalysis exerted similar effect as our hydrophobic Al-ITQ-HB (Supplementary Table 5). However, while micellar medium required the so called critical micellar concentration to be operative, with our system the rate enhancement is observed even at low host catalyst load (Supplementary Table 5). The use of brine as reaction media instead of plain water showed advantage in the host-free reaction, although it did not provide further reaction acceleration in the presence of Al-ITQ-HB.

Following the organocatalytic Michael-type reaction over time shows the magnitude of the rate acceleration (Supplementary Fig. 18). We observe an important increase in the reaction rate in the case of Al-ITQ-HB versus utilization of HB acid, sodium 4-heptylbenzoate or no catalyst at all.

The observed efficiency achieved by combining host Al-ITQ-HB with the chiral bifunctional organocatalyst prompted us to evaluate the generality of this effect in other organocatalytic asymmetric transformations. Indeed, analogous rate enhancement is observed in the Michael addition of dimethylmalonate to 4-phenyl-3-buten-2-one (see equation 2 in Fig. 4). What it is more, higher enantioselectivity is observed when performing the reaction in the presence of host Al-ITQ-HB, evidencing that the supramolecular catalyst is indeed hosting the reacting 3-component system with non-covalent bond interactions playing a decisive role in rate acceleration and asymmetric induction. Furthermore, supramolecular host catalyst Al-ITQ-HB showed rate as well as enantioselectivity enhancement in the Michael addition of nitromethane to 4-phenyl-3-buten-2-one (see equation 3 in Fig. 4), establishing a considerable generality for the hydrophobic host system to facilitate organic reactions in aqueous media.

**NMR spectroscopy study of host–guest interactions.** Intrigued by the way of action of solid Al-ITQ-HB in exerting the observed

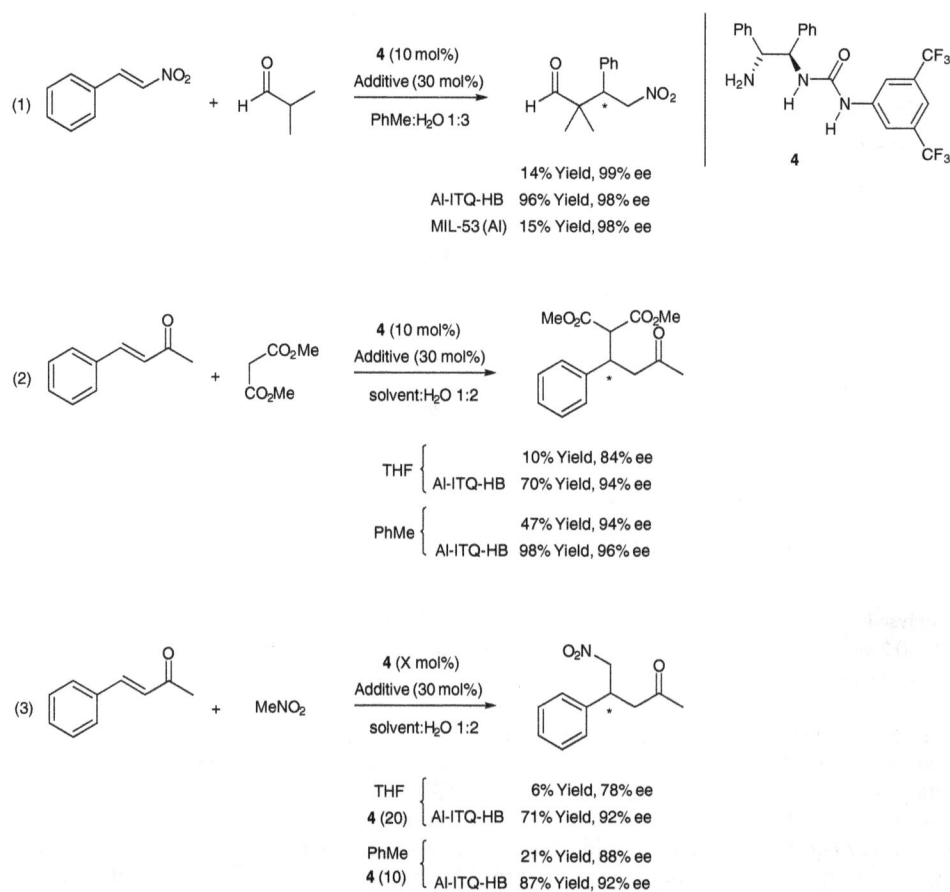

**Figure 4 | Enantioselective organocatalytic Michael reactions promoted by Al-ITQ-HB.** Organocatalyst **4** was used in the amount of 10 mol% in all cases except noted. Additives used were Al-ITQ-HB (results highlighted in blue) or MIL-53 (Al) (results highlighted in red) in the amount of 30 mol%. All the experiments were performed at room temperature for 24 h. Yields are the average of at least two runs.

rate acceleration in the varied reactions shown above, we completed this study with information from techniques that provide a direct and detailed view on the underlying interactions at the molecular level. Solution NMR spectroscopy could provide such a view since intermolecular interactions can be studied through their effect on the chemical shift, relaxation times and translational diffusion coefficients of the various species. In the case of protein receptors and ligand binding, specific NMR approaches are routinely applied to provide information about the binding process and conformations. Due to the high flexibility of our solid Al-ITQ-HB and its high organic content, we wondered whether such solution NMR-based approaches could also be utilized in our heterogeneous system for the *in situ* characterization of the molecular interactions. In fact, 2D NOESY NMR experiments were carried out on suspension mixtures of naphthaldehyde **1a** + Al-ITQ-HB, organocatalyst **4** + Al-ITQ-HB and nitrostyrene + Al-ITQ-HB (Supplementary Figs 19, 20 and 21, respectively). In all three cases, it is shown through space correlation between the aromatic protons of the substrates **1a**, **4** and nitrostyrene with the hydrocarbon chain of Al-ITQ-HB. What it is more, changes in the chemical shifts and an increase in signal linewidths are observed in the $^1$H NMR spectrum of organocatalyst **4** (in CDCl$_3$) when forming the suspension mixture with Al-ITQ-HB (Supplementary Fig. 22) evidencing the interactions between both systems and probably since Al-ITQ-HB may be disrupting the self-aggregation of the organocatalyst.

Saturation transfer difference (STD) NMR spectroscopy is a powerful technique to investigate ligand binding to proteins[45]. Working with an excess of ligand, it consist in saturating some

resonances of the protein target in the on-resonance experiment taking care not to affect the ligand resonances. This selective saturation subsequently spreads through the entire network of dipolar-coupled protons in the protein via spin diffusion. When a ligand binds to the protein, part of the saturation is transferred onto its protons. Since each protein undergoes multiple binding events during the saturation time, a sizable fraction of the ligands is affected, leading to a reduction of the ligand resonance intensity. This is easily characterized by subtracting a reference, the off-resonance experiment wherein saturation is applied outside the frequency range where ligand and protein resonances occur. The difference spectrum, therefore, yields non-zero intensities only for binding ligands. Furthermore, as ligand protons in close contact with the protein receive more saturation than more distant ones, the relative intensity of the ligand resonances can be interpreted in terms of a binding epitope. Despite the broad information that this technique could provide in supramolecular systems, it has been almost exclusively utilized in protein–ligand studies and only a few studies have been performed on synthetic hosts[46,47]. Here we demonstrate the potential of STD NMR for the investigation of interactions between hybrid metal-organic Al-ITQ-HB and several substrates. A suspension mixture of **1a** and Al-ITQ-HB was analysed (Fig. 5a). Saturation was performed at the resonance of the methyl group (0.9 p.p.m.) in the Al-ITQ-HB far away from resonances of aromatic protons of naphthaldehyde, whereas off-resonance frequency was − 150 p.p.m. (Supplementary Methods). Pulse sequences followed were those reported in literature for protein–ligand systems[48]. Figure 5b shows the result

a

b

c

d

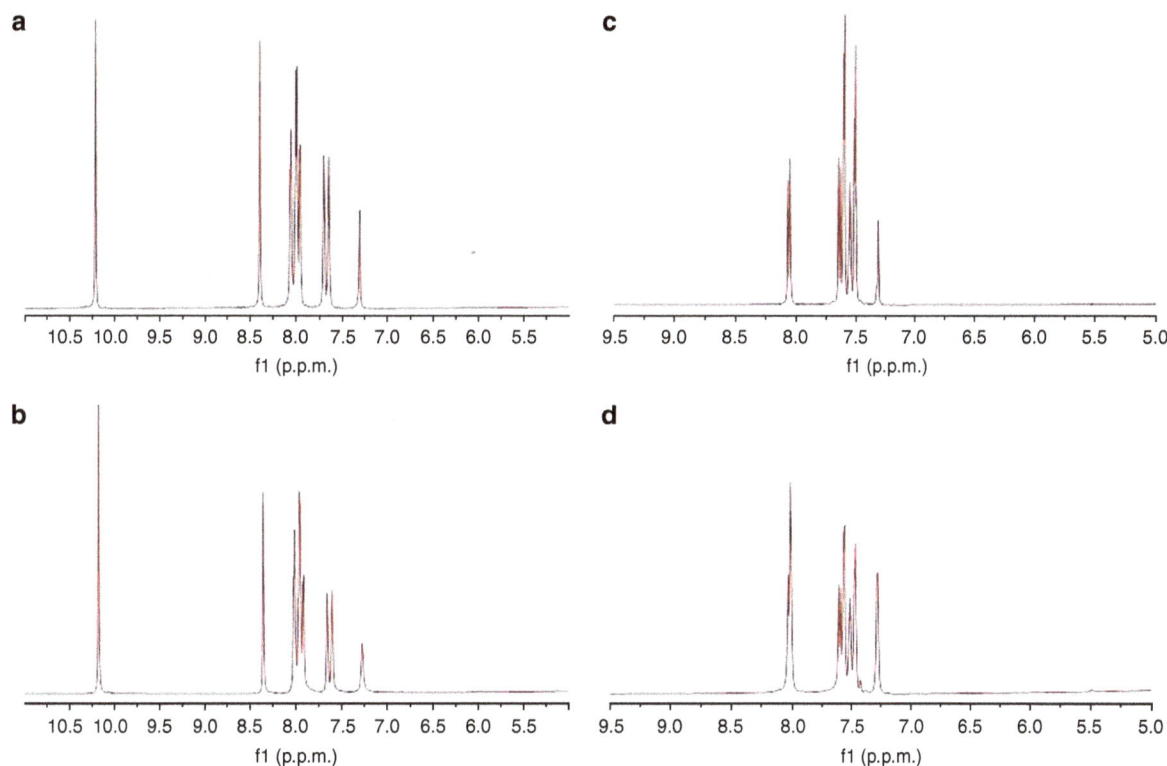

**Figure 5 | STD spectra (STD$_{off}$—STD$_{on}$) of substrate and Al-ITQ-HB suspensions. (a)** $^1$H NMR reference spectrum of naphthaldehyde **1a** and Al-ITQ-HB suspension. **(b)** STD NMR of naphthaldehyde **1a** and Al-ITQ-HB suspension with saturation of methyl signal (0.9 p.p.m.) of Al-ITQ-HB. All of the naphthaldehyde resonances appear in the STD experiment with equal intensity as in the reference $^1$H NMR demonstrating that naphthaldehyde is indeed binding to Al-ITQ-HB. **(c)** $^1$H NMR reference spectrum of nitrostyrene and Al-ITQ-HB suspension. **(d)** STD NMR of nitrostyrene and Al-ITQ-HB suspension with saturation of methyl signal (0.9 p.p.m.) of Al-ITQ-HB. All of the nitrostyrene resonances appear in the STD experiment with equal intensity as in the reference $^1$H NMR demonstrating that nitrostyrene is indeed binding to Al-ITQ-HB.

of the STD experiment to the mixture shown in Fig. 5a. All of the naphthaldehyde resonances appear in the STD experiment with equal intensity as in the reference $^1$H NMR (Fig. 5a) demonstrating that naphthaldehyde is indeed binding to the supramolecular host catalyst. The major difference here with respect to the protein–ligand systems relays on the fact that we do not have a one-molecule site that binds to one protein-binding site, but rather a hydrophobic surface area that is large, with respect to the individual molecules, providing many locations and modes for binding. Similar results were attained when the binding substrate chosen was nitrostyrene (Fig. 5c = reference $^1$H NMR and Fig. 5d = STD NMR experiment). Further details on this study and blank experiments are shown in the Supplementary Figs 23–27. These results show that non-covalent interactions are operative in these host–guest systems that together with the above demonstrated hydrophobic effect and the higher effective molarity account for the observed substrate activation.

## Discussion

We have constructed a mesoscopic and lamellar metal-organic material that was successfully used as a host catalyst for the condensation reactions in water under neutral conditions. Activity displayed appears to emulate enzyme's way of action in natural systems, wherein the hydrophobic pocket and non-covalent interactions account for the extended efficiency. The mesoscopic host material can also shelter a chiral organocatalyst and two reacting components, facilitating asymmetric transformations under more environmentally friendly conditions, meaning lower catalyst loading, room-temperature conditions and water as a green solvent. The mesoscopic phase of

the solid allows for easy diffusion of starting materials and products resulting in procedures promoted by a catalytic amount of the solid. Although the formed lipophilic products prefer the hydrophobic environment in the catalyst system over the aqueous solvent, the obtained products are easily extracted and the solid could be recycled and reused. Furthermore, we have shown that STD NMR is a valuable technique for the *in situ* solution characterization of intermolecular interactions between substrates and dispersed hybrid metal-organic materials, allowing for detection of binding ligands.

## Methods

**Synthesis of the mesoscopic hybrid material Al-ITQ-HB.** General experimental information can be found in the Supplementary Methods. Al-ITQ-HB material was synthesized from equimolar quantities of AlCl$_3$.6H$_2$O (3.1 mmol) and HB acid (3.1 mmol) which were dissolved in two different solutions with 15 ml dimethylformamide in each. The two solutions were mixed and the resulting slurry (pH = 2.5) was introduced into a stainless steel autoclave, being heated at 150 °C for 24 h under autogeneous pressure and static conditions. Once cooled to room temperature, the solution (pH = 5.5) was filtered and the collected powder was washed with distilled water. Then, the sample was stirred in methanol for 24 h to efficiently remove the remaining unreacted linker and dimethylformamide solvent molecules. Finally, the material was isolated and dried under vacuum at room temperature.

**Knoevenagel condensation reaction.** General experimental information can be found in the Supplementary Methods. The mesoscopic hybrid material, Al-ITQ-HB (8.0 mg) was placed in a 1-ml glass vessel. Aldehyde **1a** (0.1 mmol, 15.6 mg) and compound **2** (0.1 mmol, 14.4 mg) were then added. Chloroform (100 µl) and water (100 µl) were subsequently added and the reaction mixture was left to stir vigorously at room temperature for 10 h. The product was extracted with EtOAc (3 × 1 ml), and the solvent evaporated *in vacuo* to give **3a** (Supplementary Methods) in 91% yield as determined by analysis by $^1$H NMR.

**Multicomponent reaction.** The mesoscopic hybrid material, Al-ITQ-HB (5.3 mg, 10 mol%) or the corresponding additive (as indicated in the Supplementary Figs 15–17) was placed in a 1-ml glass vessel. Isatin (29.4 mg, 0.2 mmol), dimedone (28 mg, 0.2 mmol) and Ph$_3$CH (12.2 mg, 0.05 mmol) were then added. Subsequently, a stock solution of malononitrile in CDCl$_3$ was then added (0.6 ml, 0.2 mmol of malononitrile) and the reaction mixture was left to stir vigorously at room temperature. Evolution of the reaction was followed by $^1$H NMR by taking aliquots of the reaction mixture and dissolving the product in d6-DMSO.

**Michael addition of isobutyraldehyde to *trans*-β-nitrostyrene.** General experimental information can be found in the Supplementary Methods. The mesoscopic hybrid material, Al-ITQ-HB (24 mg) and organocatalyst **4** (14.1 mg, 0.03 mmol) were placed in a 2-ml glass vessel. *trans*-β-Nitrostyrene (45.0 mg, 0.3 mmol) was then added. Toluene (150 µl) and water (450 µl) were subsequently added followed by isobutyraldehyde (81 µl, 0.9 mmol) and the reaction mixture was left to stir vigorously at room temperature for 24 h. The product was extracted with EtOAc (5 × 1 ml), and host catalyst separated by centrifugation. Volatile components were then removed under reduced pressure and the crude product was purified by column chromatography using hexane/ethyl acetate (90/10) as eluent to yield addition product (Supplementary Fig. 28 and Supplementary Methods) in 96% yield (63.7 mg, 0.29 mmol). Enantiomeric excess was determined to be 98% using HPLC on a chiral stationary phase. (Kromasil 5-Cellucoat, *n*-hexane/isopropanol = 70:30, 1 ml min$^{-1}$, 220 nm, t (major) = 10.1 min and t (minor) = 13.6 min (Supplementary Fig. 31).

**Michael addition of dimethylmalonate to 4-phenyl-3-buten-2-one.** The mesoscopic hybrid material, Al-ITQ-HB (24 mg, 0.09 mmol, 30 mol%) and organocatalyst **4** (14.1 mg, 0.03 mmol) were placed in a 2-ml glass vessel. 4-Phenyl-3-buten-2-one (43.9 mg, 0.3 mmol) was then added. Toluene (150 µl) and water (300 µl) were subsequently added followed by dimethylmalonate (69 µl, 0.6 mmol) and the reaction mixture was left to stir vigorously at room temperature for 24 h. The product was extracted with EtOAc (5 × 1 ml), and host catalyst separated by centrifugation. Volatile components were then removed under reduced pressure and the crude product was purified by column chromatography using hexane/ethyl acetate as eluent to yield addition product (Supplementary Fig. 30 and Supplementary Methods) in 98% yield (81.8 mg, 0.29 mmol). Enantiomeric excess was determined to be 96% using HPLC on a chiral stationary phase. (Kromasil 5-amycoat, *n*-hexane/isopropanol = 90:10, 1 ml min$^{-1}$, 220 nm, t (minor) = 9.79 min and t (mayor) = 11.3 min (Supplementary Fig. 33).

**Michael addition of nitromethane to 4-phenyl-3-buten-2-one.** The mesoscopic hybrid material, Al-ITQ-HB (24 mg, 0.09 mmol, 30 mol%) and organocatalyst **4** (14.1 mg, 0.03 mmol) were placed in a 2-ml glass vessel. 4-Phenyl-3-buten-2-one (43.9 mg, 0.3 mmol) was then added. Toluene (150 µl) and water (300 µl) were subsequently added followed by nitromethane (150 µl) and the reaction mixture was left to stir vigorously at room temperature for 24 h. The product was extracted with EtOAc (5 × 1 ml), and host catalyst separated by centrifugation. Volatile components were then removed under reduced pressure and the crude product was purified by column chromatography using hexane/ethyl acetate as eluent to yield addition product (Supplementary Fig. 29 and Supplementary Methods) in 87% yield (54.1 mg, 0.26 mmol). Enantiomeric excess was determined to be 92% using HPLC on a chiral stationary phase. (Daicel Chiralcel OJ, *n*-hexane/isopropanol = 40:60, 1.5 ml min$^{-1}$, 220 nm, t (minor) = 9.3 min and t (mayor) = 13.3 min (Supplementary Fig. 32).

## References

1. Li, B. *et al.* A porous metal-organic framework with dynamic pyrimidine groups exhibiting record high methane storage working capacity. *J. Am. Chem. Soc.* **136**, 6207–6210 (2014).
2. Getman, R. B., Bae, Y.-S., Wilmer, C. E. & Snurr, R. Q. Review and analysis of molecular simulations of methane, hydrogen, and acetylene storage in metal–organic frameworks. *Chem. Rev.* **112**, 703–723 (2012).
3. Suh, M. P., Park, H. J., Prasad, T. K. & Lim, D.-W. Hydrogen storage in metal–organic frameworks. *Chem. Rev.* **112**, 782–835 (2012).
4. Li, B., Wen, H.-M., Zhou, W. & Chen, B. Porous metal-organic frameworks for gas storage and separation: what, how, and why? *J. Phys. Chem. Lett.* **5**, 3468–3479 (2014).
5. Li, J.-R., Sculley, J. & Zhou, H.-C. Metal-organic frameworks for separations. *Chem. Rev.* **112**, 869–932 (2012).
6. Cui, Y., Yue, Y., Qian, G. & Chen, B. Luminescent functional metal-organic frameworks. *Chem. Rev.* **112**, 1126–1162 (2012).
7. Yoon, M., Suh, K., Natarajan, S. & Kim, K. Proton conduction in metal–organic frameworks and related modularly built porous solids. *Angew. Chem. Int. Ed.* **52**, 2688–2700 (2013).
8. Kurmoo, M. Magnetic metal-organic frameworks. *Chem. Soc. Rev.* **38**, 1353–1379 (2009).
9. Horcajada, P. *et al.* Metal–organic frameworks in biomedicine. *Chem. Rev.* **112**, 1232–1268 (2012).
10. Liu, J. *et al.* Applications of metal-organic frameworks in heterogeneous supramolecular catalysis. *Chem. Soc. Rev.* **43**, 6011–6061 (2014).
11. Rowsell, J. L. C. & Yaghi, O. M. Metal-organic frameworks: a new class of porous materials. *Micropor. Mesopor. Mat.* **73**, 3–14 (2004).
12. Eubank, J. F. *et al.* The next chapter in MOF pillaring strategies: trigonal heterofunctional ligands to access targeted high-connected three dimensional nets, isoreticular platforms. *J. Am. Chem. Soc.* **133**, 17532–17535 (2011).
13. Rodenas,*T. *et al.* Metal-organic framework nanosheets in polymer composite materials for gas separation. *Nat. Mater.* **14**, 48–55 (2015).
14. Chang, Z. *et al.* Rational construction of 3D pillared metal–organic frameworks: synthesis, structures, and hydrogen adsorption properties. *Inorg. Chem.* **50**, 7555–7562 (2011).
15. Cheetham, A. K., Rao, C. N. R. & Feller, R. K. Structural diversity and chemical trends in hybrid inorganic-organic framework materials. *Chem. Commun.* 4780–4795 (2006).
16. Loiseau, T. *et al.* A rationale for the large breathing of the porous aluminum terephthalate (MIL-53) upon hydration. *Chem. Eur. J.* **10**, 1373–1382 (2004).
17. Senkovska, I. *et al.* New highly porous aluminium based metal-organic frameworks: Al(OH)(ndc) (ndc = 2,6-naphthalene dicarboxylate) and Al (OH) (bpdc) (bpdc = 4,4′-biphenyl dicarboxylate). *Micropor. Mesopor. Mat.* **122**, 93–98 (2009).
18. Klein, N. *et al.* Structural flexibility and intrinsic dynamics in the M2(2,6-ndc)2(dabco) (M = Ni, Cu, Co, Zn) metal-organic frameworks. *J. Mater. Chem.* **22**, 10303–10312 (2012).
19. Hoffmann, H. C. *et al.* High-pressure *in situ* 129Xe NMR spectroscopy and computer simulations of breathing transitions in the metal–organic framework Ni2(2,6-ndc)2(dabco) (DUT-8(Ni)). *J. Am. Chem. Soc.* **133**, 8681–8690 (2011).
20. Gu, J.-M., Kim, W.-S. & Huh, S. Size-dependent catalysis by DABCO-functionalized Zn-MOF with one-dimensional channels. *Dalton Trans.* **40**, 10826–10829 (2011).
21. Carson, C. G. *et al.* Synthesis and structure characterization of copper terephthalate metal–organic frameworks. *Eur. J. Inorg. Chem.* **2009**, 2338–2343 (2009).
22. Yang, Q. *et al.* Probing the adsorption performance of the hybrid porous MIL-68(Al): a synergic combination of experimental and modelling tools. *J. Mater. Chem.* **22**, 10210–10220 (2012).
23. Li, H. *et al.* Visible light-driven water oxidation promoted by host-guest interaction between photosensitizer and catalyst with a high quantum efficiency. *J. Am. Chem. Soc.* **137**, 4332–4335 (2015).
24. Hapiot, F., Bricout, H., Menuel, S., Tilloy, S. & Monflier, E. Recent breakthroughs in aqueous cyclodextrin-assisted supramolecular catalysis. *Catal. Sci. Technol.* **4**, 1899–1908 (2014).
25. Harada, A., Takashima, Y. & Nakahata, M. Supramolecular polymeric materials via cyclodextrin-guest interactions. *Acc. Chem. Res.* **47**, 2128–2140 (2014).
26. Cong, H. *et al.* Substituent effect of substrates on cucurbit[8]uril-catalytic oxidation of aryl alcohols. *J. Mol. Catal. A Chem.* **374-375**, 32–38 (2013).
27. Masson, E., Ling, X., Joseph, R., Kyeremeh-Mensah, L. & Lu, X. Cucurbituril chemistry: a tale of supramolecular success. *RSC Adv.* **2**, 1213–1247 (2012).
28. Song, F.-T., Ouyang, G.-H., Li, Y., He, Y.-M. & Fan, Q.-H. Metallacrown ether catalysts containing phosphine-phosphite polyether ligands for Rh-catalyzed asymmetric hydrogenation—enhancements in activity and enantioselectivity. *Eur. J. Org. Chem.* **2014**, 6713–6719 (2014).
29. Rebilly, J.-N. & Reinaud, O. Calixarenes and resorcinarenes as scaffolds for supramolecular metallo-enzyme mimicry. *Supramol. Chem.* **26**, 454–479 (2014).
30. Ajami, D., Liu, L. & Rebek, Jr J. Soft templates in encapsulation complexes. *Chem. Soc. Rev.* **44**, 490–499 (2015).
31. Corma, A. & Garcia, H. Supramolecular host-guest systems in zeolites prepared by ship-in-a-bottle synthesis. *Eur. J. Inorg. Chem.* **2004**, 1143–1164 (2004).
32. Kemp, D. S., Cox, D. D. & Paul, K. G. Physical organic chemistry of benzisoxazoles. IV. Origins and catalytic nature of the solvent rate acceleration for the decarboxylation of 3-carboxybenzisoxazoles. *J. Am. Chem. Soc.* **97**, 7312–7318 (1975).
33. Thorn, S. N., Daniels, R. G., Auditor, M. T. & Hilvert, D. Large rate accelerations in antibody catalysis by strategic use of haptenic charge. *Nature* **373**, 228–230 (1995).
34. Yoshizawa, M., Klosterman, J. K. & Fujita, M. Functional molecular flasks: new properties and reactions within discrete, self-assembled hosts. *Angew. Chem. Int. Ed.* **48**, 3418–3438 (2009).

35. Yoshizawa, M., Tamura, M. & Fujita, M. Diels-Alder in aqueous molecular hosts: unusual regioselectivity and efficient catalysis. *Science* **312**, 251–254 (2006).

36. Murase, T., Nishijima, Y. & Fujita, M. Cage-catalyzed knoevenagel condensation under neutral conditions in water. *J. Am. Chem. Soc.* **134**, 162–164 (2012).

37. Zhao, C., Toste, F. D., Raymond, K. N. & Bergman, R. G. Nucleophilic substitution catalyzed by a supramolecular cavity proceeds with retention of absolute stereochemistry. *J. Am. Chem. Soc.* **136**, 14409–14412 (2014).

38. Choi, M. *et al.* Stable single-unit-cell nanosheets of zeolite MFI as active and long-lived catalysts. *Nature* **461**, 246–249 (2009).

39. Loiseau, T. *et al.* MIL-96, a porous aluminum trimesate 3D structure constructed from a hexagonal network of 18-membered rings and μ3-Oxo-centered trinuclear units. *J. Am. Chem. Soc.* **128**, 10223–10230 (2006).

40. Bezverkhyy, I. *et al.* MIL-53(Al) under reflux in water: formation of γ-AlO(OH) shell and H2BDC molecules intercalated into the pores. *Micropor. Mesopor. Mat.* **183**, 156–161 (2014).

41. Wang, L.-M. *et al.* Sodium stearate-catalyzed multicomponent reactions for efficient synthesis of spirooxindoles in aqueous micellar media. *Tetrahedron* **66**, 339–343 (2010).

42. List, B. *Science of Synthesis: Asymmetric Organocatalysis 1, Lewis Base and Acid Catalysts* (Georg Thieme Verlag, 2012).

43. He, T., Gu, Q. & Wu, X.-Y. Highly enantioselective Michael addition of isobutyraldehyde to nitroalkenes. *Tetrahedron* **66**, 3195–3198 (2010).

44. Avila, A., Chinchilla, R., Fiser, B., Gómez-Bengoa, E. & Nájera, C. Enantioselective Michael addition of isobutyraldehyde to nitroalkenes organocatalyzed by chiral primary amine-guanidines. *Tetrahedron Asymmetry* **25**, 462–467 (2014).

45. Meyer, B. & Peters, T. NMR spectroscopy techniques for screening and identifying ligand binding to protein receptors. *Angew. Chem. Int. Ed.* **42**, 864–890 (2003).

46. Szczygiel, A., Timmermans, L., Fritzinger, B. & Martins, J. C. Widening the view on dispersant – pigment interactions in colloidal dispersions with saturation transfer difference NMR spectroscopy. *J. Am. Chem. Soc.* **131**, 17756–17758 (2009).

47. Basilio, N., Martín-Pastor, M. & García-Río, L. Insights into the structure of the supramolecular amphiphile formed by a sulfonated calix[6]arene and alkyltrimethylammonium surfactants. *Langmuir* **28**, 6561–6568 (2012).

48. Mayer, M. & Meyer, B. Characterization of ligand binding by saturation transfer difference NMR spectroscopy. *Angew. Chem. Int. Ed.* **38**, 1784–1788 (1999).

## Acknowledgements

This work was funded by ERC-AdG-2014-671093-SynCatMatch and the Generalitat Valenciana (Prometeo). M.B. acknowledges the funding: CTQ2014-52633-P. The Severo Ochoa program (SEV-2012-0267) is thankfully acknowledged.

## Author contributions

A.C. conceived and directed the project. J.M.M. and U.D. synthesized and collected the characterization data for material Al-ITQ-HB. A.C., U.D. and J.M.M. analysed the synthesis and characterization data for material Al-ITQ-HB. A.C. and P.G.-G. planned, conducted and analysed the data for the catalytic experiments. P.G.-G. and M.B. planned, conducted and analysed the data for the STD and NOESY NMR experiments. P.G.-G., U.D. and A.C. wrote the manuscript. All authors discussed the results and commented on the manuscript.

# 23

# Bio-based polycarbonate as synthetic toolbox

O. Hauenstein[1], S. Agarwal[1] & A. Greiner[1]

Completely bio-based poly(limonene carbonate) is a thermoplastic polymer, which can be synthesized by copolymerization of limonene oxide (derived from limonene, which is found in orange peel) and $CO_2$. Poly(limonene carbonate) has one double bond per repeating unit that can be exploited for further chemical modifications. These chemical modifications allow the tuning of the properties of the aliphatic polycarbonate in nearly any direction. Here we show synthetic routes to demonstrate that poly(limonene carbonate) is the perfect green platform polymer, from which many functional materials can be derived. The relevant examples presented in this study are the transformation from an engineering thermoplastic into a rubber, addition of permanent antibacterial activity, hydrophilization and even pH-dependent water solubility of the polycarbonate. Finally, we show a synthetic route to yield the completely saturated counterpart that exhibits improved heat processability due to lower reactivity.

[1] Macromolecular Chemistry II and Center for Colloids and Interfaces, University of Bayreuth, Universitätsstrasse 30, 95440 Bayreuth, Germany. Correspondence and requests for materials should be addressed to A.G. (email: greiner@uni-bayreuth.de).

The petroleum-based plastics industry is facing two major challenges. On the one hand, there is the urgent environmental problem of pollution of the ocean with about five million tonnes of plastic waste per year[1,2]. On the other hand, there is a natural limitation of petroleum resources, which eventually leads to a running out of oil and natural gas within this century[3]. To overcome these limitations, efforts are directed towards the development of degradable polymers[4,5] and the use of bio-based monomers[6–8], respectively. Sometimes, both classes are combined, that is, the polymer is bio-based and biodegradable, for example, poly(lactic acid) or polyhydroxyalkanoates[9–11]. In other instances, the polymer can be assigned either to the class of bio-based non-degradable plastics, such as bio-polyethylene or bio-poly(ethylene terephthalate), or to the class of biodegradable petro-based plastics, for example, poly($\varepsilon$-caprolactone) or poly(butylene adipate-co-terephthalate)[8,12]. However, even for a material that is assigned to both classes—as for poly(lactic acid)—the origin of the bio-based monomers is questionable, as lactic acid is derived from glucose, which is again derived from corn starch. The latter is also an important food resource and, as such, in competition with the use as precursor for the conversion into plastics. In contrast, limonene—a doubly unsaturated terpene—is a bio-based non-food resource, which is mainly derived from the peel of citrus fruits[13–15]. As the major component of orange oil (>90%), it is an abundantly available side product of the orange industry, produced in amounts of roughly 500 kt per year[16]. Its versatility as a monomer is reflected by the great variety of polymers that are derived from limonene[13,17–19]. In 2004, Coates and colleagues[20] reported the elegant metal-catalysed conversion of its oxidation product limonene oxide (LO) with $CO_2$, to give a low-molecular-weight poly(limonene carbonate) (PLimC) (Fig. 1). An Al(III)-based catalyst was recently found to incorporate not only the *trans*- but also the *cis*-isomer of LO, which is an important step towards higher conversions[21].

Inspired by the work of Coates *et al.*[20], we modified the copolymerization to yield high-molecular-weight (>100 kDa) PLimC in kilogram quantities[22] with a glass transition temperature of 130 °C, higher transparency than bisphenol-A polycarbonate (bisphenol A polycarbonate (BPA PC); 94 versus 89%) and improved mechanical properties compared with the petro-based counterpart poly(cyclohexene carbonate) (strain at break of 2 versus 15% for PLimC)[23]. The amorphous thermoplastic still possesses one double bond per repeating unit. This suggests a broad range of modifications to tune the properties in almost any direction. Thus, we consider PLimC a platform, from which countless functional materials can be derived. To support this statement, we give here relevant examples of straightforward addition reactions, that is, thiol-ene click chemistry[24], acid-catalysed electrophilic addition and metal-catalysed hydrogenation. The first two are conducted as polymer-analogous reactions, whereas the latter involves modification of the pre-monomer limonene. The manipulations lead to dramatic

changes in the property profile of the engineering thermoplastic PLimC, including a transformation into a rubbery material, antibacterial activity, increased hydrophilicity or even water solubility and, last but not least, improved melt processability.

## Results

**Modification of unsaturated PLimC.** The valorization of the platform polymer PLimC is illustrated in Fig. 2. When butyl-3-mercaptopropionate (B3MP) is used, an enormous change in mechanical properties is achieved. This leads to a transformation of the high-$T_g$ thermoplastic into rubbery PLimC-B3MP with a nearly three orders of magnitude decreased Young's modulus. Furthermore, this chemistry is also applied to transform PLimC into an antibacterial material, by covalently attaching a tertiary amine to the backbone (PLimC-N) and subsequent quaternization with an aromatic moiety (PLimC-NQ). The antibacterial activity of PLimC-NQ was successfully tested against *Escherichia coli* bacteria. Another aspect is the hypothetical biodegradability of PLimC, which would be expected for an aliphatic PC, as it was shown for others[25–27]. Initial composting studies on PLimC revealed no degradation after prolonged exposure at elevated temperatures. This rather shows the high biostability of the material, which is also desirable for many applications. Here we present two major synthetic routes, to tune the degradation behaviour of the rigid and hydrophobic polymer, that is, either an acid-catalysed electrophilic addition of poly(ethylene glycol)monomethyl ether (PEG-3-OH) resulting in PLimC-PEG or thiol-ene chemistry with mercaptoethanol (ME) to give PLimC-ME or mercaptoacetic acid (MAc) to yield PLimC-MAc, respectively. For the latter, this eventually even leads to pH-dependent water solubility, that is, the material dissolves readily in basic environment. Apart from the above-mentioned addition reactions, the synthetic route to the fully hydrogenated PC poly(menthene carbonate) (PMenC) is also reported for the first time. The saturated PC might be a viable choice for replacing PLimC in thermal processing, as no cross-linking can occur. Thus, starting from the bio-based platform polymer PLimC, we could introduce antibacterial activity (ideal as coating material)[28,29], elastomeric behaviour, hydrophilization, water solubility (both should accelerate biodegradation) and, in fact, inertness by hydrogenation to improve processing.

**The transformation into elastic PLimC.** The enormous versatility of PLimC is reflected in an experiment, where the thiol-functionalized ester B3MP—the pure ester is found in many fruits[30]—is clicked to the double bond in nearly quantitative yield (schematic in Fig. 3 and Supplementary Figs 1–3). The $T_g$ of PLimC lies at 130 °C, rendering it a typical engineering thermoplastic such as polyamide, poly(ethylene terephthalate), BPA PC and so on (yellow region in Ashby plot of Fig. 3 and Supplementary Table 1)[31]. The covalently attached butyl ester B3MP changes the thermal and hence the mechanical properties dramatically, that is, the $T_g$ drops to 5 °C (Table 1, Supplementary Table 2 and Supplementary Fig. 4). Tensile testing of the new material revealed its high maximum elongation $\varepsilon$ combined with a low Young's modulus and tensile strength $\sigma_s$, respectively. PLimC-B3MP with 0–2% residual double bonds in the backbone was prepared by variation of the reaction time (Supplementary Fig. 5). Curing the unsaturated polymers at 100 °C for 5 h renders the cross-linked samples insoluble, whereas the mechanical properties can be adjusted (Supplementary Fig. 6; for a detailed discussion on the curing process, see Supplementary Discussion). These observations combined with the elasticity (Supplementary Fig. 7) suggest a transition from the engineering thermoplastics

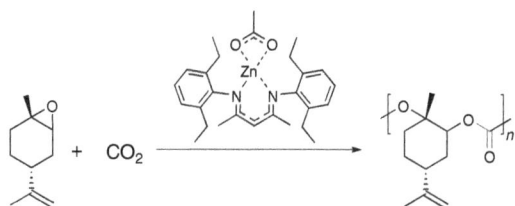

**Figure 1 | Synthetic route towards PLimC.** The copolymerization of limonene oxide and $CO_2$ in the presence of a $\beta$-diiminate zinc catalyst was discovered by Coates *et al.*[20] and optimized in our group to yield high-molecular-weight PLimC.

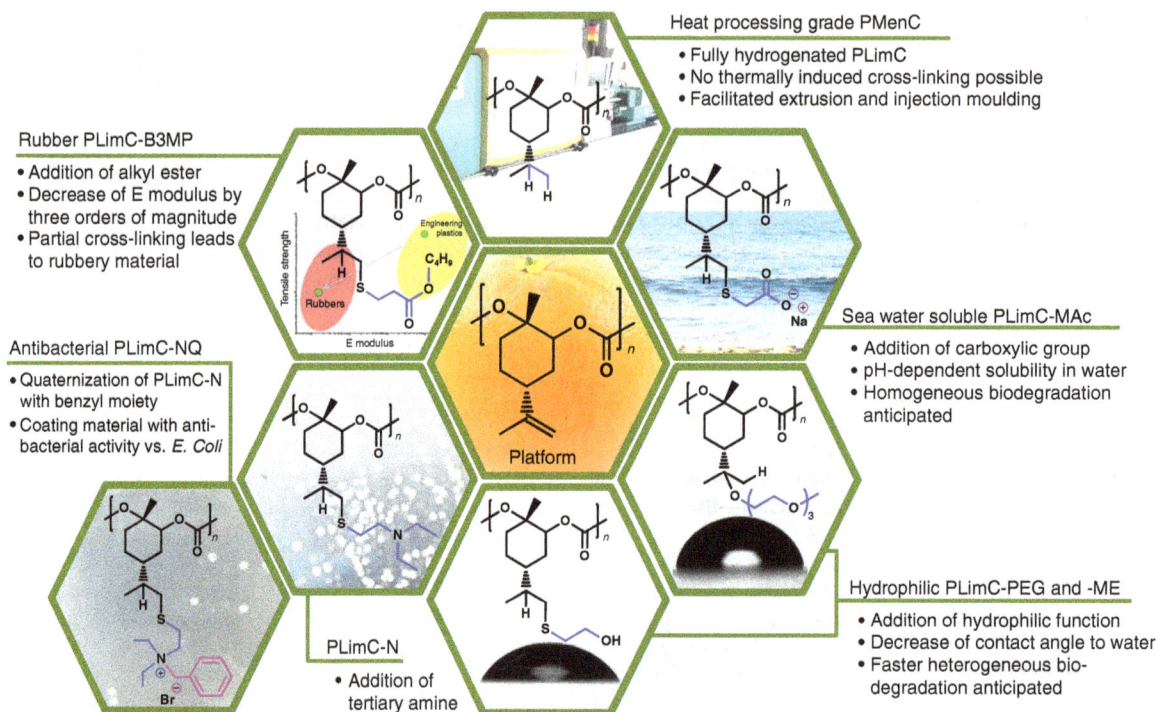

**Figure 2 | The valorization of PLimC.** The versatility of the platform polymer PLimC is illustrated in this hexagon cluster. The double bond is used for addition reactions, to induce dramatic changes of the properties of PLimC. The addition of an alkyl ester (PLimC-B3MP) leads to a drop of $T_g$ by 120 °C and of its Young's modulus by three orders of magnitude to give a PLimC rubber. The addition of a carboxylic group (PLimC-MAc) yields a pH-dependent solubility in water, whereas the functionalization with hydroxyl (PLimC-ME) or polyethylene glycol (PLimC-PEG) groups results in a higher hydrophilicity. The attachment of a tertiary amine (PLimC-N) and subsequent quaternization with an aromatic moiety (PLimC-NQ) leads to antibacterial activity against *E. coli*. Another possibility is the complete hydrogenation of the double bond to give PMenC, a superior material for heat processing.

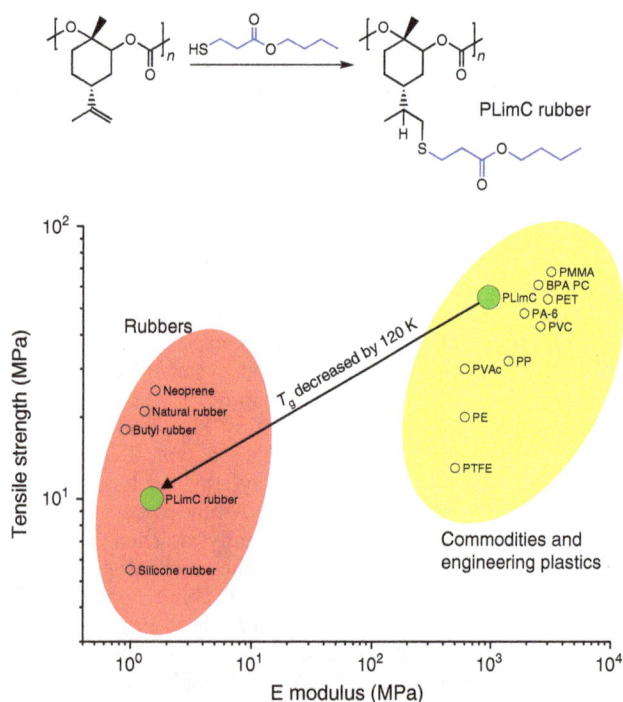

**Table 1 | Glass transition temperature and tensile properties of PLimC[22], PLimC rubber, BPA PC and silicone rubber.**

| Polymer | $T_g$ (°C) | Young's modulus (MPa) | $\sigma_s$ (MPa) | $\varepsilon$ (%) |
|---|---|---|---|---|
| PLimC | 130 | 950 | 55 | 15 |
| PLimC rubber | 5 | 1.0 | 6.8 | 228 |
| BPA PC | 145 | 2500 | 65 | 125 |
| Silicone rubber | −125 | 1.0 | 4.8–7.0 | 100–400 |

BPA PC, bisphenol A polycarbonate; PLimC, poly(limonene carbonate).
Data for PLimC rubber is taken from a cured sample of PLimC-B3MP with initially 2% unsaturation; for a more comprehensive table of polymers, see Supplementary Table 1.

**Figure 3 | The transformation into elastic PLimC.** The schematic illustrates the functionalization of PLimC with B3MP to give PLimC-B3MP, that is, a PLimC rubber. The Young's moduli of various engineering plastics, commodities (both yellow region) and rubbers (red region) are plotted against their tensile strength. The materials PLimC and PLimC rubber are highlighted as green circles, showing the dramatic change of mechanical properties on functionalization of pure PLimC with B3MP.

region to the rubber region, that is, the thermoplastic PLimC has become a PLimC rubber (PLimC-B3MP, red region in Fig. 3)[32].

This transition enables the application of the bio-based material in completely new areas, where elasticity and softness are required. Furthermore, we introduced a short alkyl chain ester into the repeating unit of PLimC. Addition of longer alkyl chains potentially leads to $T_g$ values well below 0 °C, which is another very important parameter to tune the performance of the resulting rubber. This is an example of the transformation of PLimC into a completely new material by simple polymer analogous click chemistry, while keeping the material based on natural resources. Further studies will focus on the reduction of $T_g$ and the control of mechanical properties to expand the coverage of the Ashby plot.

**The transformation into antibacterial PLimC.** In a recent publication we stated the thermal properties of PLimC, that is, a $T_g$ of 130 °C and a 5% decomposition temperature ($T_{5\%}$) of

240 °C, resulting in a rather narrow window for processing[22]. As an alternative to thermal processing, that is, extrusion or injection moulding, the employment of a PLimC solution for the application as a coating is self-evident. The high transparency and good scratch resistance are very promising properties that should yield high-value materials in combination with an extra functionality. We picked one out of many possible functionalizations, to show how PLimC can be transformed into an antibacterial material by rather simple means. The strategy involves the addition of a tertiary amine to PLimC via thiol-ene click chemistry (PLimC-N) and the subsequent quaternization of the amine with an aryl halide (PLimC-NQ) (Supplementary Figs 8–14 and Supplementary Tables 3 and 4). The functionalization was performed with different degrees of functionalization (DFs), although keeping it below 70% to keep the material insoluble in water. Resistance to water is of major importance, to make the material applicable as coatings in everyday life, where contact with water is inevitable. On the other hand, antibacterial activity rises exponentially, if the polymer—or part of it—is water soluble, as interaction of the charged amine with the bacteria's membrane is facilitated[33]. Therefore, a sample with 20% quaternized amine (PLimC-NQ20) was investigated, which does not disintegrate in contact with water and still shows antibacterial activity by inhibiting bacterial growth. For the evaluation of the antibacterial properties of the coating, films were placed in *E. coli* suspensions and the concentration of the Gram-negative bacteria was assessed after 0, 6, 12, 24 and 48 h (Supplementary Tables 5–7). Compared with pure PLimC, PlimC-NQ20 exhibited a strong inhibitory effect on bacteria growth after 24 h. The ratio of killed bacteria relative to PLimC samples is illustrated in Fig. 4. The inhibitory effect for the positive reference material polyhexamethylene guanidine hydrochloride is detected already after 6 h, when all bacteria were killed. The charged PLimC samples are less active. This is not surprising, as they are in condensed state and not dissolved such as polyhexamethylene guanidine hydrochloride. Still, the antibacterial activity could be observed after 12 h of contact with the bacteria suspension, indicating a successful valorization of PLimC into antibacterial PLimC.

Here, as proof of principle, we can demonstrate that PLimC is readily transformable into a material with antibacterial activity by rather simple and cost-effective means. We would assume though that lots of parameters are still to be optimized, that is, DF, length of spacer between thiol and amine, nature of the alkyl or aryl moiety on the amine and random distribution of quaternized amine along the backbone of the polymer versus block copolymer structure. Furthermore, the types of bacteria have to be selected in respect of the targeted application.

**The transformation into hydrophilic/water-soluble PLimC.** The idea to render PLimC more hydrophilic was born, when studies on the degradation behaviour of pure PLimC in highly active compost at 60 °C (positive reference poly(L-lactic acid) disintegrated within 1 week) had been stopped after 60 days, because no change, neither in the outer appearance nor in molecular weight, had been observed. The rather substantial biostability of this aliphatic PC is most probably explained by the three facts about PLimC. First of all, it has a very rigid backbone, resulting in a high $T_g$ of 130 °C, which is $\sim 100$ °C higher than that of readily biodegradable poly(propylene carbonate)[34,35]. The rigidity of the backbone prevents the polymer chain segments from moving and, therefore, no exposure of the carbonate groups to enzymes/bacteria is possible. Second, the polymer carries a very bulky 'side group', consisting of the cyclohexane ring connected to an isopropylene group and another methyl moiety vicinal to the carbonate. Thus, even if there are some exposed chains on the surface of such a film, the carbonate group is shielded against any attacking species. Eventually, PLimC is very hydrophobic, which is represented by its contact angle to water ($CA_W$) of 94°. This prevents not only enzymes from penetrating the polymer but also water. Hence, acid or basic hydrolysis— usually the major breakdown mechanism for long polymer chains—is inhibited; thus, nearly no fragmentation of PLimC for further degradation takes place. In consideration of the underlying circumstances, the hydrophilization of PLimC was assumed to be a likely enabler for biodegradation. Three different strategies were employed to achieve hydrophilic PLimC: two of

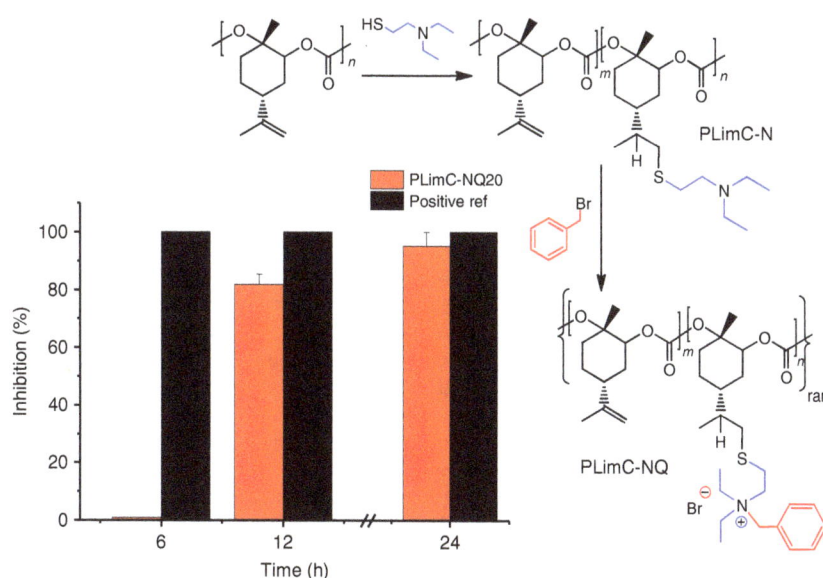

**Figure 4 | The transformation into antibacterial PLimC.** Schematic of the functionalization of PLimC with a tertiary amine (PLimC-N) and the subsequent quaternization with a benzyl moiety (PLimC-NQ). In the column diagram, the bacterial inhibition performances of PLimC-NQ20 and a positive reference material (polyhexamethylene guanidine hydrochloride, PHMG) relative to pure PLimC in a shaking flask experiment; tested on *E. coli* bacteria in buffer solution with polymer films (20 mg ml$^{-1}$) at 20 °C are illustrated. The inhibition was calculated by determination of the cfu of tenfold diluted dispersions spread on agar plates (Supplementary Table 7).

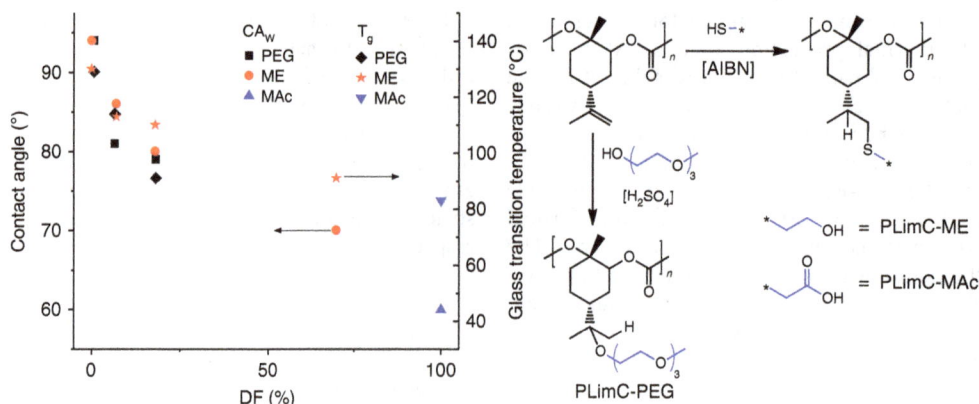

**Figure 5 | The transformation into hydrophilic/water-soluble PLimC.** Dependency of contact angle to water and $T_g$ of a PLimC film on the degree of functionalization with PEG-3-OH (black), ME (red) or MAc (blue), respectively. The schematic of the functionalization is shown for the thiol-ene addition of ME and MAc to give PLimC-ME and PLimC-MAc, respectively, and the electrophilic addition of PEG-3-OH to give PLimC-PEG.

them involve the well-known thiol-ene chemistry with ME (Supplementary Fig. 15 and Supplementary Table 8) or MAc as thiols, respectively (Supplementary Fig. 16)[36–40]. The other strategy is—an even simpler—acid-catalysed electrophilic addition of PEG-3-OH to the double bond (Supplementary Fig. 17 and Supplementary Table 9). The latter can also be acknowledged as a green reaction, hence keeping the bio-based character of PLimC, while grafting polar functionality. Indeed, the higher the DF, the smaller $CA_W$ becomes. Up to 18% conversion of the double bond was achieved by this electrophilic addition; however, as it is acid catalysed, hydrolysis of the carbonate is an immanent side reaction, which eventually breaks down the polymer chains (Supplementary Fig. 18). Thus, reaction times were kept short and a great excess of PEG-3-OH was maintained. The contact angle could be decreased below 80° with this technique. An even stronger decrease of the $CA_W$ could be achieved by radically adding ME to PLimC. Compared with the acid-catalysed addition, the advantage of thiol-ene chemistry is the absence of hydrolytic side reactions. Therefore, higher DFs are easily accessible. Here a DF of 70% resulted in a $CA_W$ of 70°. To hydrophilize the polymer even further, PLimC was functionalized with MAc, whereas 100% attachment of the acid yields a polymer with a $CA_W$ of 60°. A side effect of all above-mentioned modifications is the decrease of $T_g$ with increasing DF (Fig. 5 and Supplementary Figs 19–21). In terms of degradability, this should further promote the breakdown of the polymer. The PEG- and ME-modified PLimCs were assessed regarding their degradation behaviour in acidic (pH 4), basic (pH 9) and enzymatic environment (esterase) at elevated temperatures (37 °C). Within the timespan of 4 weeks, no degradation characteristics, that is, loss of mass, drop in molecular weight or surface alterations, could be observed (Supplementary Fig. 22 and Supplementary Table 10). These observations suggest that neither a bulk nor a surface erosion process is taking place for the samples in those environments on the time scale measured. Keeping in mind that hydrophilicity and the chain flexibility were augmented, but steric shielding of the carbonate group was unchanged (or even higher due to addition reactions), we noticed that the overall stability versus hydrolysis and/or enzymatic attack is still too high. However, on longer time scales, the backbone of PLimC is anticipated to be much more labile than that of a polyolefin and degradation should take place eventually. This makes it an interesting choice, wherever good stability against hydrolysis during application is required, but eventual disintegration on a reasonable timescale is desired.

The functionalization of PLimC with acid functionality renders the material not only hydrophilic ($CA_W = 60°$) but also pH

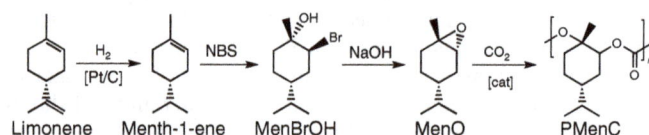

**Figure 6 | The saturation of PLimC.** The synthetic route to PMenC starts from the metal-catalysed hydrogenation of limonene, followed by the stereoselective epoxidation of Men via its bromohydrin (MenBrOH) to trans-MenO and the subsequent copolymerization with $CO_2$ to give the saturated PC PMenC.

responsive. A film of PLimC-MAc (DF = 100%) placed into a pH >8 buffer solution will dissolve within minutes. In the dissolved state, hydrolysis is of course highly accelerated compared with the condensed state. Therefore, the material would quickly disintegrate in seawater, which is usually slightly basic, and chain scission could readily occur. The exact degradation mechanism is yet to be studied, but this polymer could contribute in reducing the waste accumulation in the oceans[1].

**The saturation of PLimC.** Apart from the functionalization of PLimC, of course, it is also possible to hydrogenate the double bond, to render it unreactive, when it is heat processed or stored for a prolonged period of time. In contrast to the aforementioned modifications, this is not a polymer-analogous reaction, but the manipulation is performed on the pre-monomer (R)-limonene. Indeed, this hydrogenation is very regioselective, when a heterogeneous catalyst such as Pt on charcoal is used. Hence, a quantitative conversion to menth-1-ene (Men) can be achieved in reasonable time, whereas separation from the catalyst/carrier material is easy. We used the N-bromosuccinimide route for epoxidation (see Fig. 6) of Men to menthene oxide (MenO), as it is an established reaction for the stereo- and regio-selective epoxidation of limonene[22]. The route involves the formation of the bromohydrin MenBrOH and the subsequent ring closure in a basic medium to yield MenO. The conversion of the monomer MenO and pre-monomers Men and MenBrOH, respectively, were monitored by gas chromatography (GC) analysis (Supplementary Figs 23–25).

As the catalyst for the production of PMenC by copolymerization with $CO_2$ is also selective towards the trans-isomer of MenO (Supplementary Figs 26–28), we still recommend taking the detour via the bromohydrin, but plan to look into new reactions to perform a more economical oxidation of menthene. The properties of PMenC are very similar to PLimC, as both are high-$T_g$ amorphous PCs (Supplementary Fig. 29). The only, and of course anticipated, difference is the inability of the polymer to cross-link or to perform any undesired oxidation reactions at elevated temperatures. For PLimC, the addition of antioxidants, that is, butylated hydroxytoluene derivatives, helps to reduce those side reactions, but for PMenC no additives are needed. Indeed, a better processablilty, that is, extrusion and injection moulding, and a prolonged ultraviolet stability are anticipated. Both give extra value to this polymer. A combination of both MenO and LO for copolymerization with $CO_2$ is also possible, resulting in a defined number of available double bonds for postmodification reactions as shown above.

## Discussion

In summary, we could show the huge versatility of the green platform polymer PLimC. The valorization was achieved either by polymer analogous thiol-ene chemistry and acid-catalysed electrophilic addition, or by metal-catalysed hydrogenation of the pre-monomer. Thiol-ene chemistry proved to be the most versatile technique, adding not only hydrophilicity or pH-dependent solubility, but also antibacterial activity by functionalization with quaternary amines. Even a transformation of the high-$T_g$ thermoplastic PLimC into a rubbery material could be achieved by addition of a thiol. Not as adaptable is the acid-catalysed electrophilic addition of alcohols, although it proves to be more economical, as no costly functional thiols (employed in great excess) but only PEG and sulphuric acid are needed. Keeping the lability of the backbone of the aliphatic PC in mind, we note that this method is limited to short reaction times and hence only partial functionalization can be achieved. A quantitative conversion is possible for the regioselective hydrogenation of limonene, resulting in fully saturated PMenC after a few steps. This PC is oxidation resistant and thus exhibits improved processability for extrusion and injection moulding. The valorization of PLimC significantly broadens the range of applications, as PLimC-NQ is a viable antibacterial coating material and PLimC-ME, PLimC-PEG and PLimC-MAc could be employed as packaging materials with tuneable degradation/dissolution mechanism.

## Methods

**Instrumentation and characterization.** NMR spectra were recorded on a Bruker AMX-300 operating at 300 MHz. Chemical shifts $\delta$ are indicated in parts per million with respect to residual solvent signals. Thermogravimetric analysis was performed on a Netzsch TG 209 F1 Libra and differential scanning calorimetry on a Mettler Toledo DSC 821c, both at a heating rate of 10 K min$^{-1}$ under $N_2$ atmosphere. Infrared spectra of solids were recorded with an attenuated total reflection unit of a Digilab Excalibur FTS-3000. GC analysis was performed on a Shimadzu QP-5050 with $N_2$ as the carrier gas (temperature profile for GC studies: start at 40 °C hold for 5 min, heating to 80 °C with 5 °C min$^{-1}$, heating to 120 °C with 3 °C min$^{-1}$ and hold for 3 min, heating to 300 °C with 30 °C min$^{-1}$). Relative molecular weights and dispersities were determined by gel permeation chromatography on an Agilent 1200 system with chloroform as the eluent and polystyrene as the calibration standard. A Hazemeter BykGardner Haze-Gard Plus and a ultraviolet–visible spectrometer V-670 (JASCO) were employed for the testing of optical properties of solvent cast PLimC films having a thickness between 100 and 400 μm. A Zwick/Roell Z0.5 test equipment with testXpert II software was employed for the tensile testing. The tests were performed at 21 °C and a relative humidity of 20%. The strain rate was set to 5 mm min$^{-1}$, to test the tensile properties of cast polymer films that were die-cut into specimen (dumb-bell shaped) having a width of 2 mm, a length of 20 mm and a thickness of 100–200 μm. A BYK Pencil Hardness Tester and Derwent Graphic pencils were used to determine pencil hardness.

**Synthetic procedures.** All synthetic manipulations were carried out under exclusion of air in dry conditions, if not otherwise stated. The acid-catalysed electrophilic addition and the thiol-ene chemistry were carried out as polymer analogous reactions. The hydrogenation of the exo double bond of limonene was performed on the pre-monomer, which was subsequently epoxidized and copolymerized with $CO_2$, to give the PC PMenC. A detailed description of the synthetic procedures is given in the Supplementary Methods.

**Degradation tests in composting environment.** Composting tests were performed on cast films of PLimC ($M_n = 54.0$ kDa, $Đ = 1.11$) of 200 μm thickness fixed in slide mounts. The 3-month matured compost was supplied by an industrial composting plant (Mistelbach) and directly used for PLimC burial tests. Poly(L-lactic acid) (NatureWorks) was used as the positive reference material. During the test, the temperature was kept at 60 °C and the container was vented every 2 days (every 3 days after the first 2 weeks) and humidified if necessary (humidity was estimated by weighing of the container).

**Degradation tests in enzymatic environment.** The enzymatic tests were performed on cast films of PLimC, PLimC-ME (100% functionalized with ME) and BPA PC (reference material) having a thickness of 100 μm, which were cut into 40 mg pieces. As media water, pH 4 buffer, pH 9 buffer and an esterase in pH 9 buffer (Esterase EL-01, triacylglycerol lipase, ASA Spezialenzyme GmbH, one part of enzyme suspension mixed with four parts of buffer solution, replaced after 10 days) were selected to test the polymer samples' stability against. At a temperature of 37 °C, the samples were shaken in a mechanical shaker (50 r.p.m.) in 40 ml glass containers, which were filled to 75%. The mass loss (balance) and molecular weight (gel permeation chromatography) change were analysed after 3, 10 and 21 days in triplicate for each sample.

**Antibacterial activity tests.** *E. coli* (DSM no. 1077, K12 strain 343/113, DSMZ), as a Gram-negative test organism, was used to evaluate the antibacterial activity. We have chosen the Gram-negative bacteria *E. coli* for the antibacterial activity tests of PLimC-NQ, to evaluate the general activity of the polymer towards a very common bacterium. CASO-Boullion was used as nutrient for the *E. coli* (30 g l$^{-1}$ in distilled water for liquid nutrient; 15 g l$^{-1}$ agar-agar in addition for nutrient agar plates). The strain was preserved on nutrient agar plates and liquid cultures were grown by inoculation of liquid nutrient with a single bacteria colony using an inoculation loop. The inoculated broth was incubated with shaking at 37 °C until the optical density at 578 nm had risen to 0.125, indicating a cell density of $10^7$–$10^8$ cfu ml$^{-1}$. To obtain the final bacterial suspensions, the inoculated broth was diluted with buffer solution (phosphate-buffered solution, concentration of phosphate ions = 12 mM, pH 7.4) to an approximate cell density of $10^5$ cfu ml$^{-1}$. The antibacterial activity was determined by the shaking flask method: polymer films with a mass of 40 mg and a thickness of 100 μm were incubated with 2 mL of bacteria suspension at ambient temperature in microcentrifuge tubes with contact times of 6, 12 and 24 h. After the defined time intervals, 100 μl specimens were drawn and spread on nutrient agar plates. After incubation at 37 °C for 24 h, colonies were counted and the reduction was calculated relative to the unfunctionalized PLimC sample.

## References

1. Miller, S. A. Sustainable polymers: opportunities for the next decade. *ACS Macro Lett.* **2**, 550–554 (2013).
2. Tuck, C. O., Perez, E., Horvath, I. T., Sheldon, R. A. & Poliakoff, M. Valorization of biomass: deriving more value from waste. *Science* **337**, 695–699 (2012).
3. Bentley, R. W. Global oil and gas depletion: an overview. *Energ. Policy* **30**, 189–205 (2002).
4. Kale, G. *et al.* Compostability of bioplastic packaging materials: an overview. *Macromol. Biosci.* **7**, 255–277 (2007).
5. Mecking, S. Nature or petrochemistry?—Biologically degradable materials. *Angew. Chem. Int. Ed.* **43**, 1078–1085 (2004).
6. Mülhaupt, R. Green polymer chemistry and bio-based plastics: dreams and reality. *Macromol. Chem. Phys.* **214**, 159–174 (2013).
7. Babu, R. P., O'Connor, K. & Seeram, R. Current progress on bio-based polymers and their future trends. *Prog. Biomater.* **2**, 8 (2013).
8. Iwata, T. Biodegradable and bio-based polymers: future prospects of eco-friendly plastics. *Angew. Chem. Int. Ed.* **54**, 3210–3215 (2015).
9. Müller, H.-M. & Seebach, D. Poly(hydroxyalkanoates): Aa fifth class of physiologically important organic biopolymers? *Angew. Chem. Int. Ed. Engl.* **32**, 477–502 (1993).
10. Reichardt, R. & Rieger, B. Poly(3-hydroxybutyrate) from carbon monoxide. *Adv. Polym. Sci.* **245**, 49–90 (2011).
11. Dutta, S., Hung, W.-C., Huang, B.-H. & Lin, C.-C. Recent developments in metal-catalyzed ring-opening polymerization of lactides and glycolides: preparation of polylactides, polyglycolide, and poly(lactide-co-glycolide). *Adv. Polym. Sci.* **245**, 219–283 (2011).
12. Siegenthaler, K. O., Künkel, A., Skupin, G. & Yamamoto, M. Ecoflex® and Ecovio®: biodegradable, performance-enabling plastics. *Adv. Polym. Sci.* **245**, 91–136 (2011).
13. Firdaus, M., Montero De Espinosa, L. & Meier, M. A. R. Terpene-based renewable monomers and polymers via thiol-ene additions. *Macromolecules* **44**, 7253–7262 (2011).

14. Blanch, G. & Nicholson, G. Determination of the enantiomeric composition of limonene and limonene-1,2-epoxide in lemon peel by multidimensional gas chromatography with flame-ionization detection and selected ion monitoring mass spectrometry. *J. Chromat. Sci.* **36**, 19–22 (1998).

15. Zhao, J. & Schlaad, H. Synthesis of terpene-based polymers. *Adv. Polym. Sci.* **253**, 151–190 (2012).

16. Ciriminna, R., Lomeli-Rodriguez, M., Demma Carà, P., Lopez-Sanchez, J. A. & Pagliaro, M. Limonene: a versatile chemical of the bioeconomy. *Chem. Commun.* **50**, 15288–15296 (2014).

17. Wilbon, P. A., Chu, F. & Tang, C. Progress in renewable polymers from natural terpenes, terpenoids, and rosin. *Macromol. Rapid Commun.* **34**, 8–37 (2013).

18. Li, C., Sablong, R. J. & Koning, C. E. Synthesis and characterization of fully-biobased α,ω-dihydroxyl poly(limonene carbonate)s and their initial evaluation in coating applications. *Eur. Polym. J.* **67**, 449–458 (2015).

19. Harvey, B. G. *et al.* Sustainable hydrophobic thermosetting resins and polycarbonates from turpentine. *Green Chem.* **18**, 2416–2423 (2016).

20. Byrne, C. M., Allen, S. D., Lobkovsky, E. B. & Coates, G. W. Alternating copolymerization of limonene oxide and carbon dioxide. *J. Am. Chem. Soc.* **126**, 11404–11405 (2004).

21. Peña Carrodeguas, L., González-Fabra, J., Castro-Gómez, F., Bo, C. & Kleij, A. W. Al$^{III}$-catalysed formation of poly(limonene)carbonate: DFT analysis of the origin of stereoregularity. *Chem. A Eur. J.* **21**, 6115–6122 (2015).

22. Hauenstein, O., Reiter, M., Agarwal, S., Rieger, B. & Greiner, A. Bio-based polycarbonate from limonene oxide and $CO_2$ with high molecular weight, excellent thermal resistance, hardness and transparency. *Green Chem.* **18**, 760–770 (2016).

23. Koning, C. *et al.* Synthesis and physical characterization of poly(cyclohexane carbonate), synthesized from $CO_2$ and cyclohexene oxide. *Polymer* **42**, 3995–4004 (2001).

24. Hoyle, C. E. & Bowman, C. N. Thiol-ene click chemistry. *Angew. Chem. Int. Ed.* **49**, 1540–1573 (2010).

25. Artham, T. & Doble, M. Biodegradation of aliphatic and aromatic polycarbonates. *Macromol. Biosci.* **8**, 14–24 (2008).

26. Luinstra, G. A. & Borchardt, E. Material properties of poly(propylene carbonates). *Adv. Polym. Sci.* **245**, 29–48 (2012).

27. Wu, K.-J., Wu, C.-S. & Chang, J.-S. Biodegradability and mechanical properties of polycaprolactone composites encapsulating phosphate-solubilizing bacterium *Bacillus* sp. PG01. *Process Biochem.* **42**, 669–675 (2007).

28. Fuchs, A. D. & Tiller, J. C. Contact-active antimicrobial coatings derived from aqueous suspensions. *Angew. Chem. Int. Ed.* **45**, 6759–6762 (2006).

29. Sileika, T. S., Barrett, D. G., Zhang, R., Lau, K. H. A. & Messersmith, P. B. Colorless multifunctional coatings inspired by polyphenols found in tea, chocolate, and wine. *Angew. Chem. Int. Ed.* **52**, 10766–10770 (2013).

30. Burdock, G. A. *Encyclopedia of Food and Color Additives* Volume 1 (CRC Press, 1996).

31. Fried, J. R. *Polymer Science and Technology* Third Edition (Prentice Hall, 2014).

32. Brandrup, J., Immergut, E. H. & Grulke, E. A. *Polymer Handbook* Fourth Edition (Wiley, 1998).

33. Kenawy, E.-R., Worley, S. D. & Broughton, R. The chemistry and applications of antimicrobial polymers: a state-of-the-art review. *Biomacromolecules* **8**, 1359–1384 (2007).

34. Luinstra, G. A. Poly(propylene carbonate), old copolymers of propylene oxide and carbon dioxide with new interests: catalysis and material properties. *Polym. Rev.* **48**, 192–219 (2008).

35. Zhou, M., Takayanagi, M., Yoshida, Y., Ishii, S. & Noguchi, H. Enzyme-catalyzed degradation of aliphatic polycarbonates prepared from epoxides and carbon dioxide. *Polym. Bull.* **42**, 419–424 (1999).

36. Darensbourg, D. J. & Wang, Y. Terpolymerization of propylene oxide and vinyl oxides with $CO_2$: copolymer cross-linking and surface modification via thiol–ene click chemistry. *Polym. Chem.* **6**, 1768–1776 (2015).

37. Darensbourg, D. J., Chung, W., Arp, C. J., Tsai, F. & Kyran, S. J. Copolymerization and cycloaddition products derived from coupling reactions of 1,2-epoxy-4-cyclohexene and carbon dioxide. Postpolymerization functionalization via thiol–ene click reactions. *Macromolecules* **47**, 7347–7353 (2014).

38. Darensbourg, D. J. & Tsai, F.-T. Postpolymerization functionalization of copolymers produced from carbon dioxide and 2-vinyloxirane: amphiphilic/water-soluble $CO_2$-based polycarbonates. *Macromolecules* **47**, 3806–3813 (2014).

39. Gu, L., Qin, Y., Gao, Y., Wang, X. & Wang, F. Hydrophilic $CO_2$-based biodegradable polycarbonates: Synthesis and rapid thermo-responsive behavior. *J. Polym. Sci. A Polym. Chem.* **51**, 2834–2840 (2013).

40. Zhou, Q. *et al.* Biodegradable $CO_2$-based polycarbonates with rapid and reversible thermal response at body temperature. *J. Polym. Sci. A Polym. Chem.* **51**, 1893–1898 (2013).

## Acknowledgements

We thank H. Wang, H. Bakhshi and R. Schneider for discussing and carrying out antibacterial tests.

## Author contributions

A.G. directed the project. S.A. cosupervised the project. O.H. performed all the experiments, analysed the data and wrote the manuscript. All authors discussed the results and commented on the manuscript.

# 24

# Merging rhodium-catalysed C–H activation and hydroamination in a highly selective [4 + 2] imine/alkyne annulation

Rajith S. Manan[1] & Pinjing Zhao[1]

Catalytic C–H activation and hydroamination represent two important strategies for eco-friendly chemical synthesis with high atom efficiency and reduced waste production. Combining both C–H activation and hydroamination in a cascade process, preferably with a single catalyst, would allow rapid access to valuable nitrogen-containing molecules from readily available building blocks. Here we report a single metal catalyst-based approach for N-heterocycle construction by tandem C–H functionalization and alkene hydroamination. A simple catalyst system of cationic rhodium(I) precursor and phosphine ligand promotes redox-neutral [4 + 2] annulation between N–H aromatic ketimines and internal alkynes to form multi-substituted 3,4-dihydroisoquinolines (DHIQs) in high chemoselectivity over competing annulation processes, exclusive *cis*-diastereoselectivity, and distinct regioselectivity for alkyne addition. This study demonstrates the potential of tandem C–H activation and alkene hydrofunctionalization as a general strategy for modular and atom-efficient assembly of six-membered heterocycles with multiple chirality centres.

[1] Department of Chemistry and Biochemistry, North Dakota State University, Fargo, North Dakota 58102, USA. Correspondence and requests for materials should be addressed to P.Z. (email: pinjing.zhao@ndsu.edu).

n recent years, transition metal-mediated C–H bond activation has been increasingly explored in catalytic construction of heterocycles that involve tandem formation of carbon–carbon and carbon–heteroatom bonds[1–8]. A major strategy for such catalytic heterocycles synthesis via C–H activation is the intermolecular coupling between aromatic compounds and alkynes to form six-membered benzoheterocycles (Fig. 1a). These [4 + 2] annulations utilize a variety of heteroatom-based *ortho*-directing groups for aromatic C–H bond activation. The resulting metallacycle intermediates (**A**) undergo subsequent alkyne coupling and ring-closure steps to incorporate the directing groups into the heterocyclic product backbone[9–11]. A dominant majority of these domino processes occur in the form of oxidative annulation that retains a carbon–carbon double bond in the heterocycle structure (Fig. 1b)[8,12–34]. For heteroatom-based directing groups with a N–H or O–H moiety, stoichiometric amounts of Ag(I) or Cu(II) oxidants are commonly used for oxidative [4 + 2] annulations. This need for external oxidants can be eliminated either by developing aerobic oxidation[25–27,34] or dehydrogenative coupling conditions[30], or by using an 'oxidizing directing group' as the internal oxidant that releases a small molecule byproduct such as water[17,21], alcohol[18] or carboxylic acid[23,33]. In principle, redox-neutral [4 + 2] annulations with aromatic compounds and alkynes would provide direct access to

partially saturated benzoheterocycles in 100% atom efficiency and forms up to two new chirality centres (Fig. 1a). However, this strategy was only demonstrated in a single report in 2013 by Sun, Wang and coworkers[35], who developed a Re(I)–Mg(II) bimetallic catalyst for benzamide/alkyne coupling to synthesize 3,4-dihydroisoquinolinones with controlled *cis*- or *trans*-diastereoselectivity (Fig. 1c). This redox-neutral [4 + 2] annulation featured a ring-closure step of Mg-catalysed intramolecular alkene addition by the amide N–H bond, which represents an example of main group metal-catalysed alkene hydroamination[36].

Considering that late transition metal catalysts have been widely used in both C–H activation and alkene hydroamination processes[1,36,37], we envisioned that redox-neutral [4 + 2] annulations for N-heterocycle synthesis may be promoted by a single transition metal catalyst via a domino sequence of C–H bond activation, C–C bond formation by alkyne coupling and C–N bond formation by intramolecular alkene hydroamination. This 'one catalyst does it all' approach for redox-neutral annulations would complement existing methods with operationally simple procedures and provide opportunities for ligand-enabled control over chemo- and stereoselectivity for hydroamination. From the reaction mechanism perspective, our strategy can be compared with ruthenium- and gold-catalysed

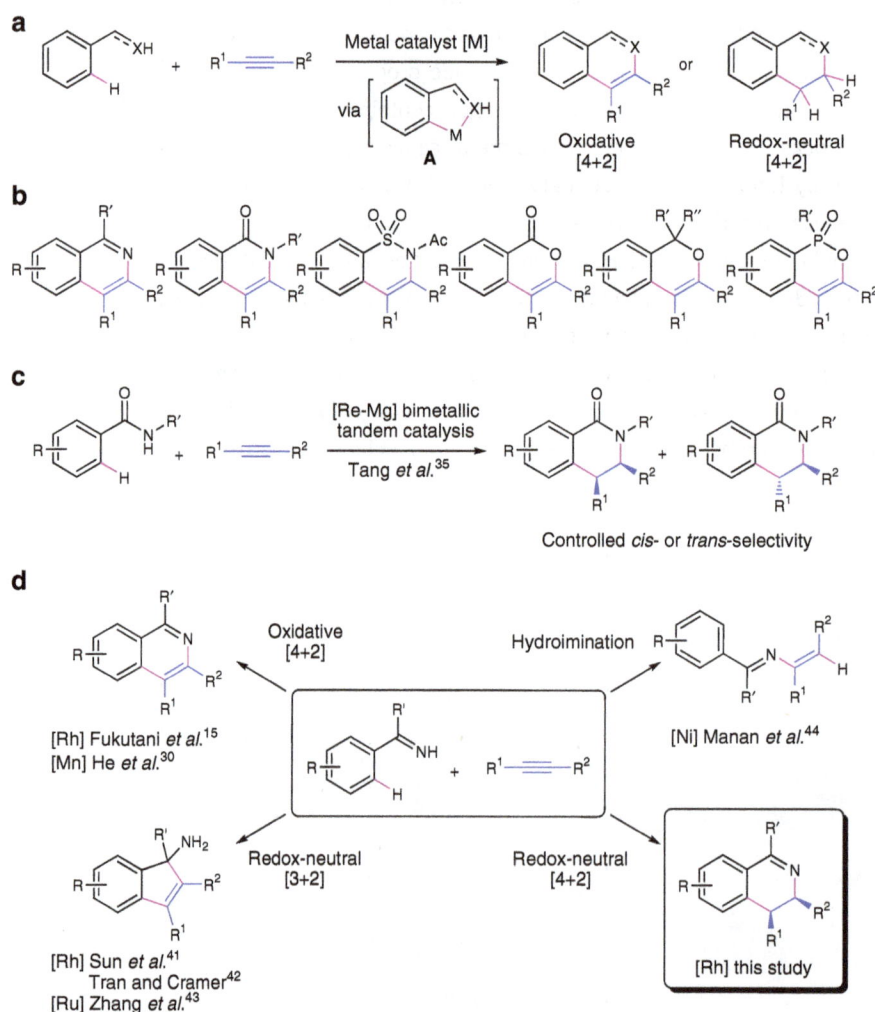

**Figure 1 | Transition metal-catalyzed [4 + 2] annulations and related strategies. (a)** Catalytic [4 + 2] annulations between aromatic compounds and alkynes via cyclometalated intermediates (**A**); XH = H-substituted σ-donating functional group to direct aromatic C–H activation at the *ortho*-position. **(b)** Reported benzoheterocycle products from oxidative [4 + 2] annulations with NH or OH directing groups. **(c)** A Re-Mg bimetallic catalyst system for redox-neutral [4 + 2] annulation between benzamides and alkynes. **(d)** Divergent catalytic couplings between N-H aromatic ketimines and alkynes.

synthesis of 1,2-dihydroquinoline derivatives by 3-component coupling between an aromatic amine and two alkynes[38–40]. This redox-neutral [3 + 2 + 1] annulation involves a tandem sequence of intermolecular alkyne hydroamination and aromatic C–H functionalization. In the current study, we have focused our attention on metal-catalysed intermolecular couplings between N–H aromatic ketimines and internal alkynes (Fig. 1d). We and others have previously reported three classes of couplings between these two reaction partners to selectively form isoquinolines by oxidative [4 + 2] N-heterocyclization[15,30], indene-based tertiary carbinamines by [3 + 2] carbocyclization[41–43], and 2-aza-1,3-butadienes by alkyne hydroamination with N–H imine nucleophile (hydroimination)[44]. These divergent catalytic processes attest to the challenge of targeting desired redox-neutral [4 + 2] annulation with high chemoselectivity.

We herein describe a mechanism-based development of a rhodium-catalysed redox-neutral [4 + 2] annulation with aromatic N–H ketimines and internal alkynes to form 3,4-dihydroisoquinolines (DHIQs), which are synthetic intermediates towards valuable 1,2,3,4-tetrahydroisoquinoline (THIQ) structures[45]. The strategic combination of a cationic Rh(I) catalyst precursor and a bis(phosphine) ligand enables this N-heterocyclization to proceed with high chemoselectivity over other possible coupling processes and exclusive diastereoselectivity for *cis*-3,4-disubstituted products. Regio- and stereochemistry results, as well as results from deuterium-labelling studies, are most consistent with a domino sequence of imine-directed aromatic C–H bond activation via oxidative addition, alkyne coupling, and a novel intramolecular alkene hydroimination. Synthetic utility of target DHIQ products are demonstrated with several stoichiometric and catalytic transformations including diastereoselective hydride reduction and Rh(III)-mediated regioselective C–H functionalizations.

## Results

**Initial observations and mechanistic implications.** Our study began with the detection of three annulation products from catalytic coupling between N–H aromatic ketimines and internal alkynes (Scheme 2a). In particular, Rh(I)-catalysed 1:1 coupling between benzophenone imine (**1a**) and diphenylacetylene (**2a**) led to formation of [3 + 2] carbocyclization product **3a** (refs 41–43), oxidative [4 + 2] N-heterocyclization product **4a** (refs 15,30), and the desired dihydroisoquinoline product **5aa** by redox-neutral [4 + 2] N-heterocyclization. The overall yield and chemoselectivity depended significantly on the choices of Rh(I) catalyst precursor, ancillary ligand and solvent for the reaction (*vide infra*). Thus, the major challenge for our catalyst development was to selectively promote formation of **5aa** over byproducts **3a** and **4a**. In particular, we expected that formation of isoquinoline byproduct **4a** would be highly competitive due to the strong thermodynamic driving force of aromatization. To this end, our effort was guided by an early observation that cationic Rh(I) precursors with non-coordinating counteranions, for example, [Rh(cod)$_2$]BF$_4$ (**6**), appeared to promote higher chemoselectivity for **5aa** than neutral Rh(I) precursors with anionic ligands such as [Rh(cod)$_2$(OH)]$_2$. In addition, **5aa** was detected exclusively as the *cis*-3,4-diphenyl diastereomer by $^1$H NMR spectroscopy with a relatively small H(3)–H(4) coupling in $^1$H NMR ($^3J \sim 6.0$ Hz), which supported gauche *cis*- H(3)–H(4) and gauche *cis*-3,4-disubstitution relationships[30,43,46]. Combined with proposed reaction pathways for mechanistically relevant annulation processes[8,15,30,35,41–43], these results led us to suggest an imine-directed C–H oxidative addition on cationic Rh(I)[47–49] to form a cyclometalated Rh(III) hydride **A1** (Fig. 2b, Path 1; see Supplementary Fig. 88 for a more detailed discussion)[9]. Alkyne insertion into the Rh–H linkage and

subsequent C–C reductive elimination gave an alkyne hydroarylation product **C**, which we envisioned as a key intermediate for [4 + 2] annulation products (*vide infra*)[47–49]. Alternatively, a neutral Rh(I) catalyst precursor may promote a deprotonation-type C–H activation pathway, leading to a cyclometalated Rh(I) complex **A2** (Path 2). Subsequent alkyne insertion into the Rh–C bond gave an imine-chelated Rh(I) alkenyl complex **D** (refs 10,11), which could lead to [3 + 2] carbocyclization product **3** by sequential intramolecular imine insertion into the Rh-alkenyl linkage and protonation of the resulting Rh(I) alkyl complex[41–43].

**Mechanism-based catalyst design.** The proposed acyclic intermediate **C** was not directly detected during our study. However, the involvement of similar alkyne hydroarylation intermediates was confirmed by Sun and Wang[35] in their recent report on Re(I)–Mg(II) bimetallic catalyst for redox-neutral [4 + 2] benzamide/alkyne annulation. In addition, Bergman, Ellman and coworkers have reported structurally analogous acyclic C–H alkenylation products in their reports on Rh(I)-catalysed N-heterocyclization with α,β-unsaturated N-benzyl imines and alkynes, which undergoes non-catalytic 6π-electrocyclization to form dihydropyridines as reactive intermediates for several one-pot procedures towards pyridine and tetrahydropyridine synthesis[49–51]. Aiming for a single catalyst-based domino procedure, we hypothesized that a cationic Rh(I) catalyst for the stage of C–H activation and alkyne coupling (Fig. 2b, Path 1) could also promote subsequent ring closure by intramolecular alkene hydroamination. As described in Fig. 2c, alkene π-complexation between **C** and a Lewis acidic Rh(I) centre would promote intramolecular nucleophilic attack by the N–H imine moiety in stereospecific 6-*endo-trig* fashion to give metal alkyl intermediate **F**. With the stereochemistry of C–N bond formation determined by such *anti*-aminometalation[36,37], **F** could undergo either *syn*-β-H elimination to give isoquinoline **4** or protonation of the Rh-alkyl linkage with stereospecific retention to give the *cis*-isomer of redox-neutral annulation product **5** (see Supplementary Fig. 88 for a more detailed description). Thus, we envisioned that an ideal catalyst system for selective synthesis of DHIQs should effectively promote a tandem sequence of C–H alkenylation (**1**→**C**), intramolecular alkene hydroamination (**C**→**F**), and selective protonation of a Rh-alkyl complex (**F**→**5**) over β-H elimination. This catalysis design is inspired by cationic Rh(I) catalysts that have been successfully explored by the Hartwig group for inter- and intramolecular alkene hydroamination[52–55]. For example, [Rh(cod)$_2$]BF$_4$ (**6**) was demonstrated as an effective catalyst precursor for *anti*-Markovnikov intramolecular hydroamination of vinylarenes with secondary aliphatic amines (Fig. 2d)[53]. Notably, selective formation of desired hydroamination products versus oxidative amination products in this report was significantly affected by the choice of chelating bis-phosphine ligands. Such ligand-controlled chemoselectivity can be attributed to the ligand bite angle effects on β-H elimination versus protonation of metal alkyl intermediates in several catalytic processes including alkene hydroamination versus oxidative amination, as well as Heck–Mizoroki olefinations versus the corresponding alkylation processes[56]. It is also noteworthy that we have recently reported a nickel-catalysed intermolecular alkyne hydroamination with N–H aromatic ketimines that proceeded by a proposed imine nulcleophilic attack on Ni(0)-coordinated alkyne for stereospecific *anti*-addition[44]. This result serves as an indirect evidence to support both the unconventional role of N–H imine moiety as a N-nucleophile for hydroamination and the proposed stereochemistry for C–N bond formation in current study.

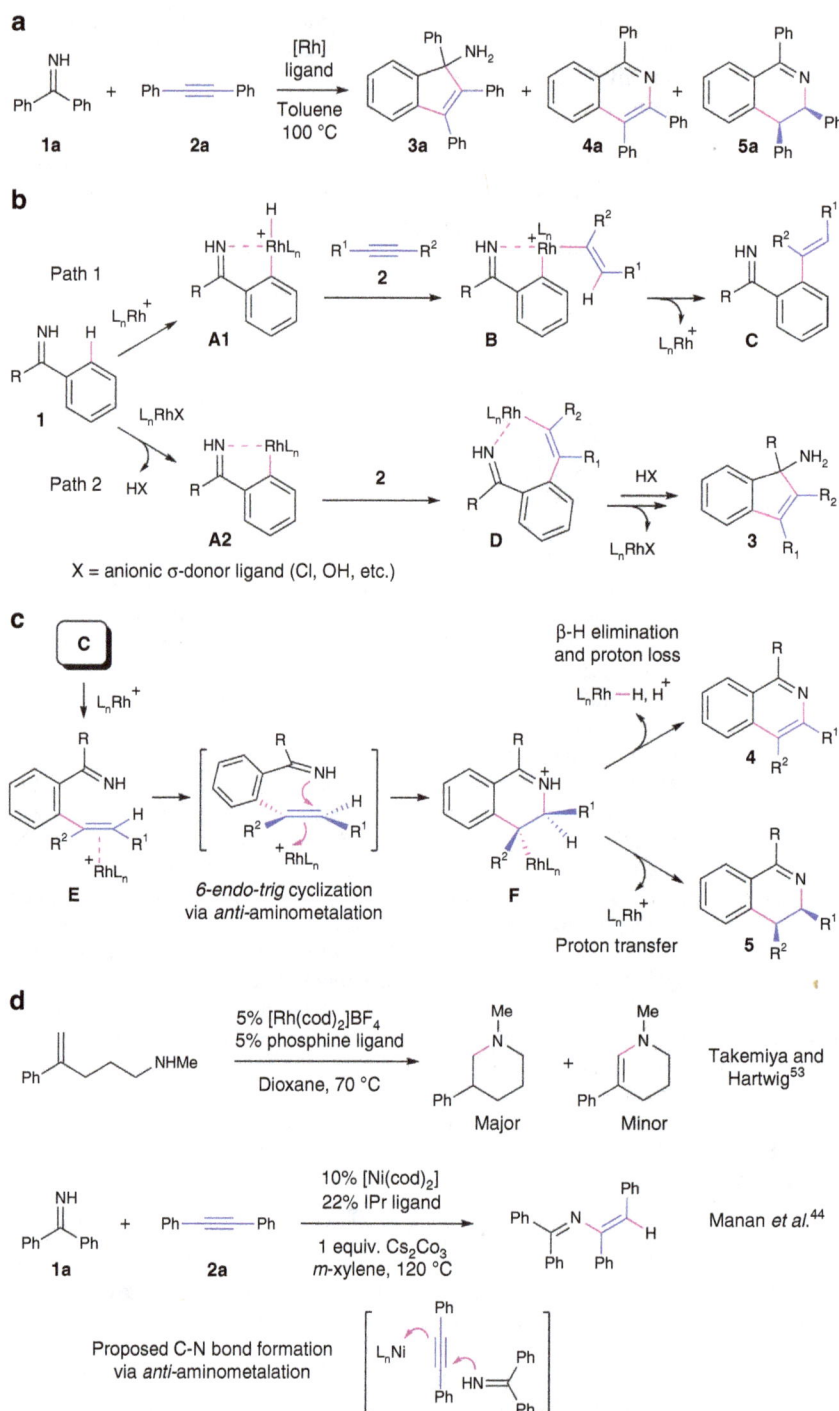

**Figure 2 | Catalyst design for redox-neutral [4 + 2] imine/alkyne coupling.** (a) Observation of three different annulation products with Rh(I) catalysts. (b) Proposed pathways for imine-directed aromatic C–H bond activation and subsequent alkyne coupling. (c) Proposed N-heterocyclization by intramolecular alkene hydroimination and subsequent competition between oxidative and redox-neutral [4 + 2] annulation product formations. (d) Mechanistically related reports on intramolecular alkene hydroamination and intermolecular alkyne hydroimination.

**Catalyst development.** With mechanistic insights described above, we have focused our attention on the combination of cationic Rh(I) precursors and chelating bis-phosphine ligands for catalyst development (Table 1). A particularly effective catalyst system was discovered using $[Rh(cod)_2]BF_4$ precursor (**6**) and DPEphos ligand (bis[(2-diphenylphosphino)phenyl] ether, **7**). It is noteworthy that DPEphos is a prominent example of bis-phosphine ligands with wide bite angles, whose significant ligand effects on chemo- and regioselectivity in metal-catalysed coupling reactions including hydroamination are well documented[52,56]. At 100 °C and in toluene solvent, a 1:1 coupling between **1a** and **2a** was promoted by 5 mol% $[Rh(cod)_2]BF_4$ (**6**) and 6 mol% DPEphos (**7**) to selectively form **5aa** in 93% yield over 24 h. Under these conditions, oxidative [4 + 2] annulation product **4a** was formed in 3% yield, and the [3 + 2] carbocyclization product **3a** was not detected (entry 1). Significantly reduced reactivity and chemoselectivity was observed when replacing DPEphos with various bis- and mono-phosphines (entries 2–8), or replacing **6** with several other Rh(I) catalyst precursors (entries 9–12). Switching from toluene to THF solvent led to similar overall

**Table 1 | Optimization of conditions for redox-neutral [4 + 2] imine/alkyne annulation.**

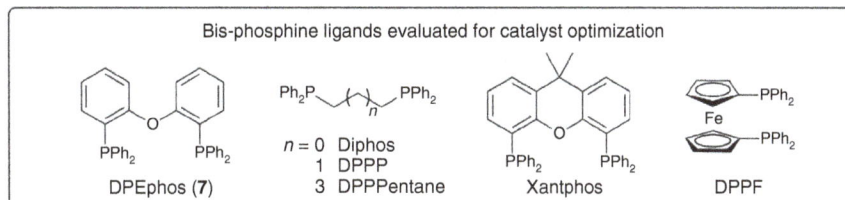

| Entry | Rh(I) catalyst | Ligand | Solvent | GC yield (%) | | |
|-------|----------------|--------|---------|------|------|------|
| | | | | **5aa** | **4a** | **3a** |
| 1 | [Rh(cod)$_2$]BF$_4$ (**6**) | DPEphos (**7**) | Toluene | 93 | 3 | 0 |
| 2 | **6** | Diphos | Toluene | 0 | 5 | 0 |
| 3 | **6** | DPPP | Toluene | 15 | 13 | 12 |
| 4 | **6** | DPPPentane | Toluene | 0 | 8 | 4 |
| 5 | **6** | DPPF | Toluene | 4 | 18 | 12 |
| 6 | **6** | Xantphos | Toluene | 0 | 11 | 0 |
| 7* | **6** | PPh$_3$ | Toluene | 0 | 11 | 2 |
| 8* | **6** | PCy$_3$ | Toluene | 6 | 9 | 12 |
| 9 | [Rh(cod)$_2$]OTf | **7** | Toluene | 5 | 35 | 2 |
| 10 | [Rh(nbd)$_2$]BF$_4$ | **7** | Toluene | 85 | 7 | 6 |
| 11 | [Rh(cod)(OH)]$_2$] | **7** | Toluene | 0 | 5 | 3 |
| 12 | [Rh(cod)(OH)]$_2$] | **7** | Toluene | 3 | 13 | 7 |
| 13 | **6** | **7** | THF | 89 | 8 | 2 |
| 14 | **6** | **7** | Hexane | 8 | 5 | 0 |
| 15 | **6** | **7** | Dioxane | 65 | 16 | 11 |
| 16 | **6** | **7** | DMF | 55 | 35 | 7 |
| 17 | **6** | **7** | CH$_2$Cl$_2$ | 11 | 22 | 0 |
| 18* | **6** | **7** | Toluene | 13 | 57 | 0 |
| 19† | **6** | **7** | Toluene | 32 | 48 | 0 |

Conditions: **1a** (0.17 mmol), **2a** (0.18 mmol), Rh catalyst (5.0 mol%), ligand (6.0 mol%), solvent (1.0 ml), 100 °C, 24 h.
*Using 11 mol% phosphine ligand.
†Using 2.0 mol% **6** and 2.4 mol% **7**.

reactivity but slightly lower selectivity for **5aa** (entry 13), while much lower reactivity and chemoselectivity was observed in hexane (entry 14) and several solvents of higher polarity (entries 15–17). Lastly, using an increased amount of DPEphos ligand (entry 18) or reduced catalyst loading (entry 19) both led to lower combined yield of **5aa** and **4a** but higher yield for **4a** formation. This observation suggested that the oxidative [4 + 2] annulation may also proceed in different pathways that involve non-catalytic ring-closure steps such as 6π-electrocyclization (see Supplementary Fig. 88 for a more detailed discussion on possible pathways for imine/alkyne coupling).

**Alkyne substrate scope.** With the standard reaction conditions established, various aromatic N–H ketimine (**1**) and internal alkyne (**2**) substrates were studied for Rh(I)-catalysed redox-neutral [4 + 2] annulation (Scheme 3). In general, 1,3,4-trisubstiuted DHIQs (**5**) were formed in exclusive cis-3,4-diastereoselectivity and high chemoselectivity, with only trace amounts (0–5%) of isoquinoline byproducts (**4**) and no detection of indenamine byproducts (**3**) as evidenced by gas chromatography (GC) and $^1$H NMR analysis of the unpurified reaction mixture. However, several DHIQ products appeared to undergo spontaneous dehydrogenation during the separation and purification procedures, which generated small amounts of isoquinoline byproducts **4** that could not be fully removed from the isolated DHIQ products **5** (vide

infra). Scope of the alkyne substrates was studied with benzophenone imine (**1a**) as the reaction partner, and high coupling yields were achieved with various symmetrical alkynes having aryl, 2-thienyl, and alkyl substituents (products **5aa–5am**). However, no coupling products were detected for bis(2-pyridyl)acetylene or terminal alkynes such as phenylacetylene. Reactions with non-symmetrical phenyl alkyl alkynes led to the exclusive formation of cis-3-alkyl-4-phenyl products (**5an–5ap**). By sharp contrast, most reported methods of oxidative [4 + 2] heterocyclization with non-symmetrical aryl alkyl akynes displayed the opposite regioselectivity[8,12–34]. For instance, regioselective formation of a 3-aryl-4-alkyl-substituted isoquinoline product was recently reported by the Wang group by manganese-catalysed dehydrogenative [4 + 2] annulation of N–H imines and alkynes[30]. The only example of similar regioselectivity with aryl alkyl alkynes was reported by the Cramer group on rhodium(III)-catalysed oxidative [4 + 2] annulation of N-acyl arylsulfonamides and alkynes, which promoted formation of a 3-alkyl-4-aryl-substituted benzosultam product in modest regioselectivity (2:1) (ref. 27). The reaction with 1-(2-thienyl)-2-phenylacetylene generated a 4:1 mixture of regioisomers, and the major isomer of 3-phenyl-4-(2-thienyl)-substituted DHIQ product **5aq** was isolated in 66% yield. This modest regioselectivity was likely affected by potential coordination between sulfur centre of the 2-thienyl moiety and the cationic Rh(I) centre during the akyne coupling process. Due to product decomposition by

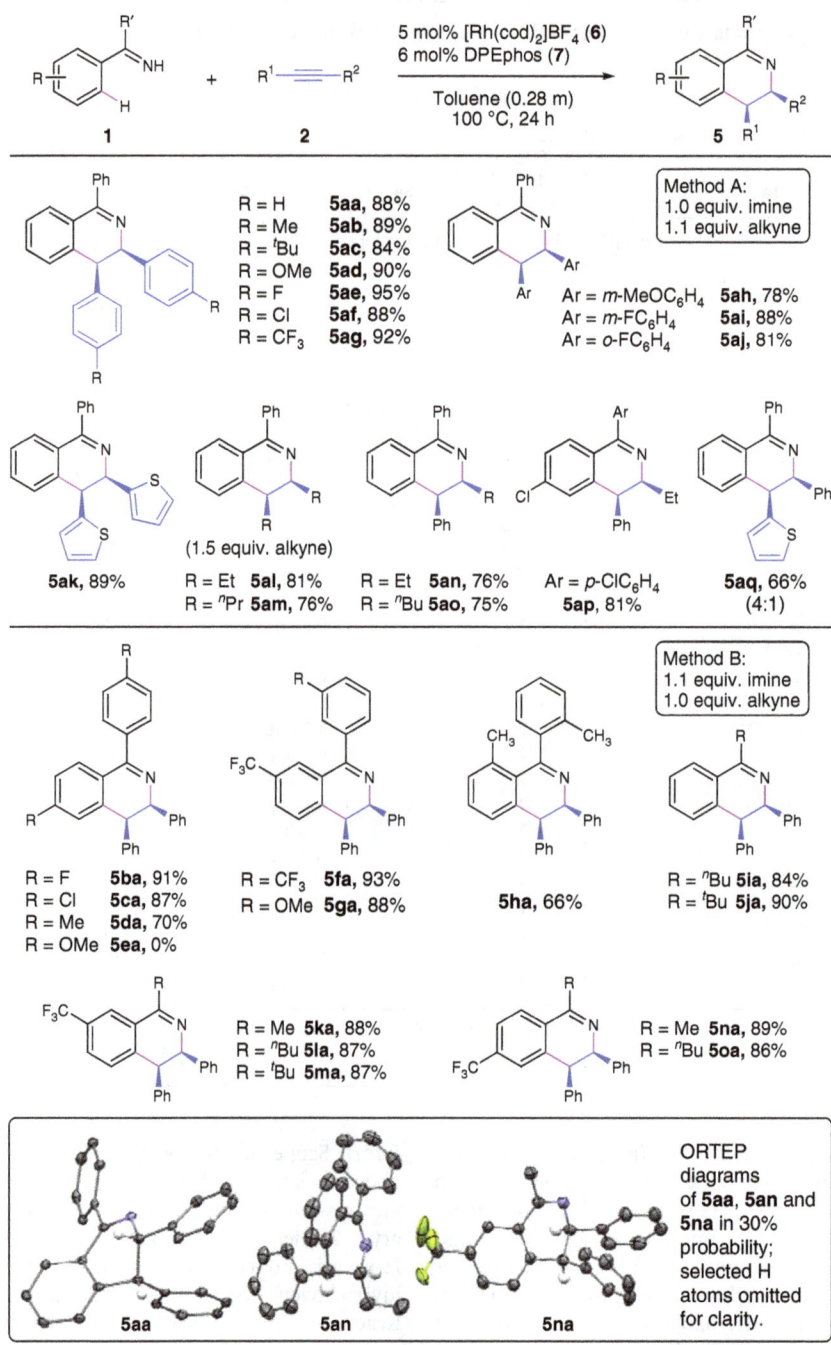

**Figure 3 | Scope of Rh-catalyzed redox-neutral [4 + 2] imine/alkyne annulation.** General reaction conditions: **1** (0.28 mmol for Method A, 0.31 mmol for Method B), **2** (0.31 mmol for Method A, 0.28 mmol for Method B); [Rh(cod)₂]BF₄ (**6**, 5.0 mol%), DPEphos (**7**, 6.0 mol%), toluene (1.0 ml), 100 °C, 24 h; averaged yield of isolated products from two runs.

dehydrogenation, isolated products **5am–5ao** were contaminated with ∼10–15% of the corresponding isoquinoline byproducts, and the reported yields were estimated by ¹H NMR analysis (see Supplementary Figs 29, 31 and 33 for more details).

**N–H ketimine substrate scope.** Scope of the ketimine substrates was studied by coupling with diphenylacetylene (**2a**), and high reactivity was observed for both diaryl and aryl alkyl ketimines with phenyl groups (**5ia** and **5ja**) or electron-poor aryl groups having F, Cl and CF₃ groups at *para* or *meta* positions (**5ba**, **5ca**, **5fa** and **5ka–5oa**). Electron-rich di(*p*-tolyl) N–H ketimine gave product **5da** in 70% yield, while di(*p*-anisyl) N–H ketimine failed to react with **2a** to give detectable coupling products. Such

electronic effect on ketimine reactivity was further demonstrated with product **5ga**, which was formed with exclusive regioselectivity for C–H functionalization of the electron-poor aryl group with *meta*-CF₃ substituent over the electron-rich one with *meta*-methoxy substituent. Notably, the sterically hindered di(*o*-tolyl) N–H ketimine did react with **2a** to give redox-neutral [4 + 2] adduct **5ha**. Product decomposition via dehydrogenation was also observed for **5ha**, and an isolated yield of 66% was calculated based on ¹H NMR analysis (see Supplementary Fig. 54 for details). The solid-state structures of products **5aa**, **5an** and **5na** were established by single-crystal X-ray diffraction analysis, which confirmed the *cis*-diastereoselectivity and 3-alkyl-4-aryl regioselectivity for corresponding DHIQ products.

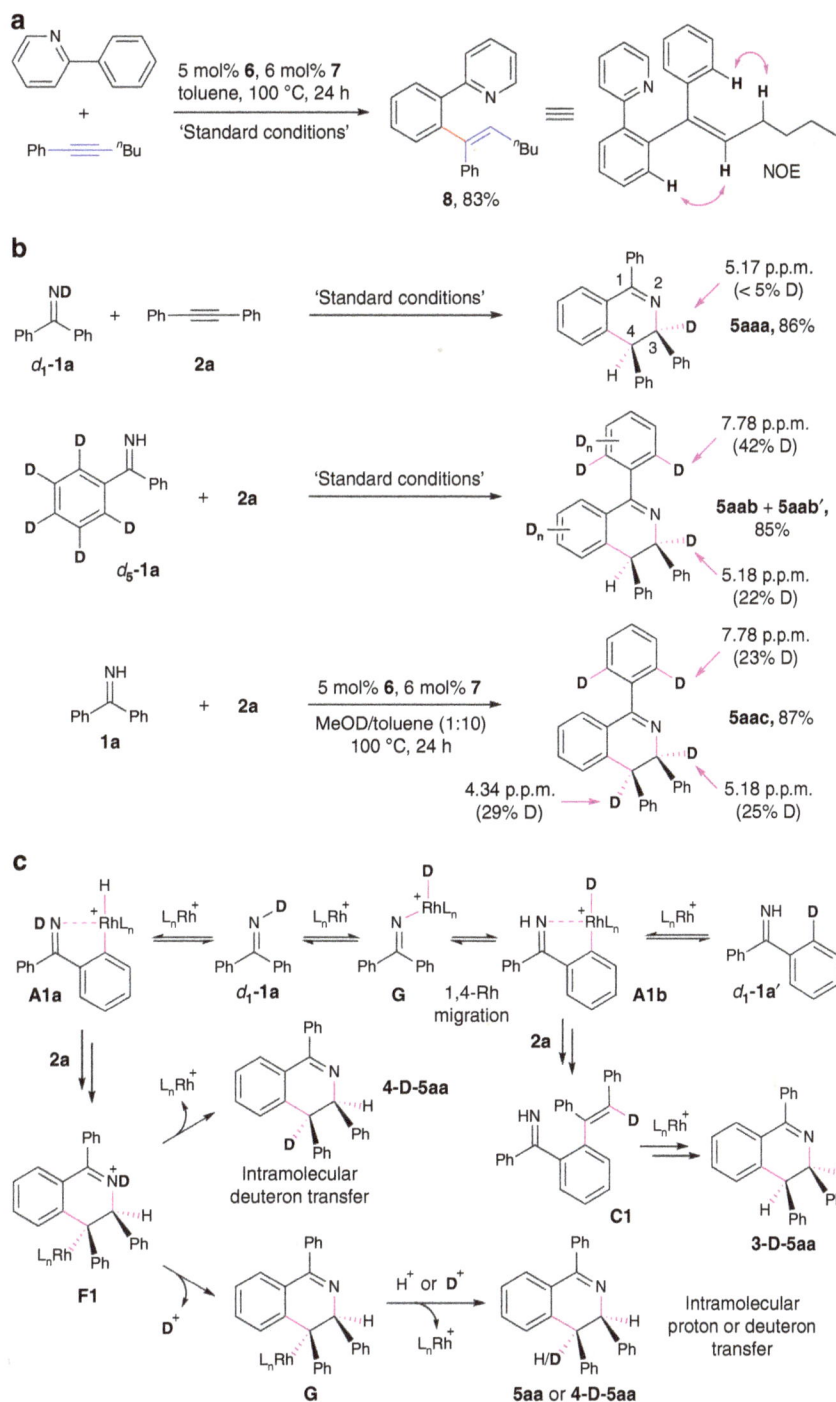

**Figure 4 | Reaction mechanism studies and analysis.** (**a**) Regioselective alkyne hydroarylation with 2-phenylpyridine under standard catalytic conditions for [4 + 2] imine/alkyne annulation. (**b**) Results from deuterium-labelling studies. (**c**) Proposed pathways for regioselective deuterium transfer and equilibrium processes for H/D scrambling.

## Discussion

Current results of the substituent effects on coupling reactivity and regioselectivity provide several mechanistic insights that were consistent with the proposed pathway for C–H alkenylation (Fig. 2b, Path 1). Firstly, the lack of significant reactivity dependence on symmetrically substituted alkyne substrates (products **5aa–5am**) suggested that the proposed C–C and C–N bond formation steps (**B→C, E→F**) were not rate determining. Secondly, the highly regioselective functionalization at the more electron-deficient aryl moiety in product **5ga** suggested a C–H activation pathway that does not involve electrophilic aromatic substitution[1]. Instead, these results were consistent with a proposed rate-determining step of C–H oxidative onto the Rh(I) centre of the catalyst (**1→A1**) (ref. 49). Thirdly, the distinct regioselectivity with non-symmetrical phenyl alkyl alkynes (products **5an–5ap**) was consistent with proposed alkyne insertion into a Rh–H linkage, which placed the Rh centre preferentially at the more stabilized, α-to-phenyl position in intermediate **B** ($R_1$ = alkyl, $R_2$ = phenyl). Subsequent C–C reductive elimination gave Murai-type hydroarylation product **C** (refs 47–49), whose regiochemistry was reflected in the 3-alkyl-4-phenyl substitution pattern of the corresponding DHIQ product **5**.

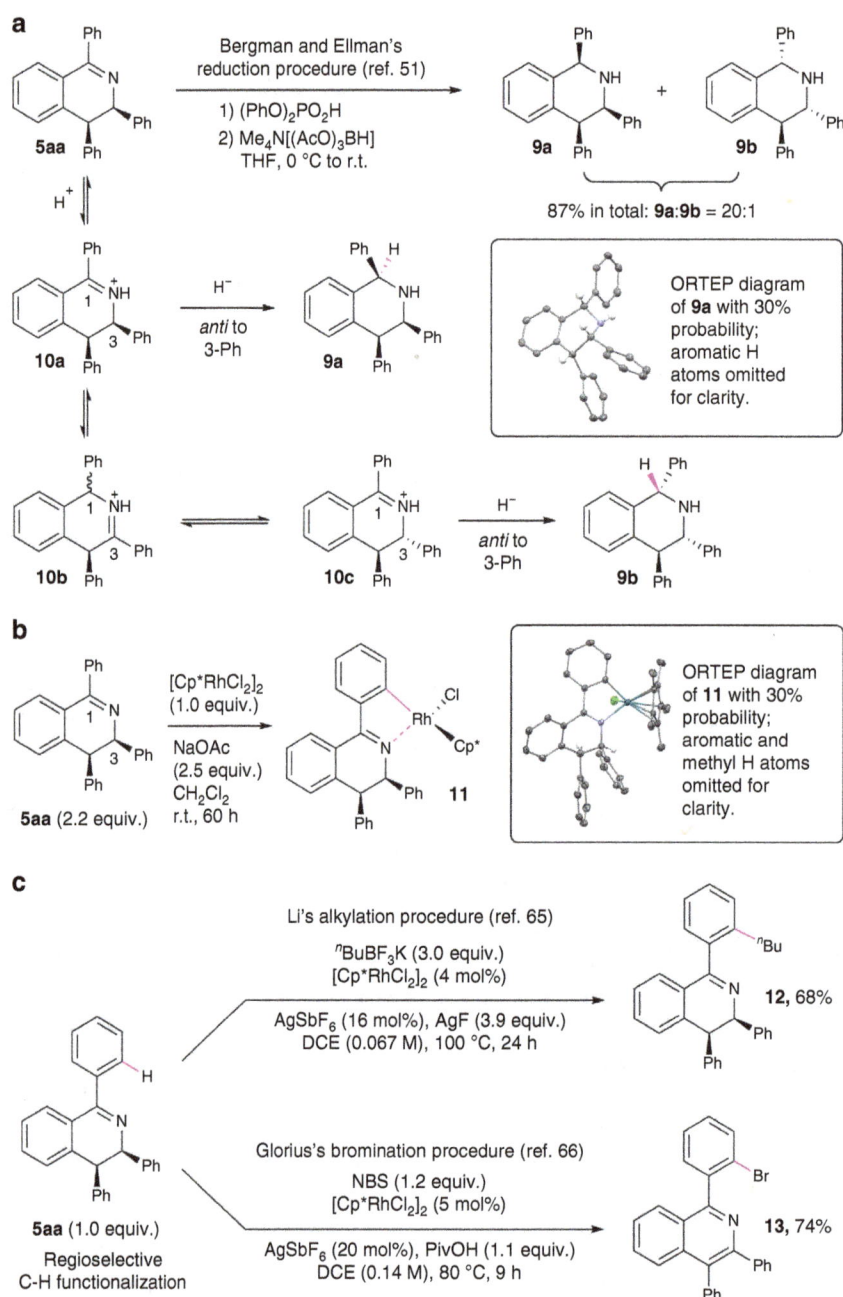

**Figure 5 | Demonstration of synthetic transformations of DHIQ products with the model compound 5aa. (a)** Diastereoselective hydride reduction. **(b)** Rh(III)-mediated regioselective cyclometalation. **(c)** Rh(III)-catalyzed aromatic functionalizations via directed C–H activation.

To further investigate regiochemistry of the proposed C–H alkenylation intermediate **C**, the current catalyst system was explored for a 1:1 coupling between 2-phenylpyridine and *n*-butylphenylacetylene that should not give an annulation product without oxidants (Fig. 4a)[10,57]. Under the standard conditions for catalytic redox-neutral [4 + 2] annulations, a (*E*)-1,1-diarylhexene product **8** was acquired in 83% yield and with exclusive regio- and stereoselectivity (see Supplementary Figs 75 and 77 for 1H NMR and NOESY spectra). This result is consistent with our proposed C–H alkenylation pathway and represents a relatively rare example of high selectivity towards 1,1-diarylalkene regioisomers for catalytic hydroarylation with aryl alkyl alkynes by the directed C–H activation strategy[21,58]. Lastly, the failure of produt formation for the highly electron-rich imine substrate di(p-anisyl) N–H ketimine (product **5ea**) suggested that C–H activation may be inhibited by strong

σ-complexation between Rh(I) and imine ligands. A similar lack of reactivity for electron-rich N–H aromatic imines was observed in our previous study on Ni-catalysed hydroimination of alkynes[44].

To better understand hydrogen atom transfer processes in the proposed mechanism for redox-neutral [4 + 2] annulation (Fig. 2b,c), we carried out deuterium-labelling studies on the formation of **5aa** by coupling between **1a** and **2a** under standard catalytic conditions and with various deuterium sources (Fig. 4b). Firstly, the reaction with $Ph_2C=ND$ ($d_1$-1a) led to <5% D incorporation at C3 position of the product **5aaa**. The significant deuterium loss suggested a rapid hydrogen/deuterium (H/D) srambling between the imine moiety and the reaction media, presumably due to traces of moisture or acid impurities. The regioselective D-transfer to C3, albeit in low conversion, suggested an intramolecular H/D exchange between the imine N centre and the *ortho* aromatic positions of $d_1$-1a. The resulting *ortho*-D atom would migrate to C3 position of **5aa** by the proposed pathways for C–H alkenylation (Fig. 2b) and intramolecular alkene hydroamination (Fig. 2c). Secondly, the reaction with (C₆D₅)PhC = NH (*d₅*-**1a**) led to an inseparable mixture of products **5aab** and **5aab′** via imine-directed C–H or C–D activation. This mixture displayed 22% D incorporation at C3 position, while 42% D could be measured at each *ortho*-position of the 1-phenyl group. This result further supported the proposed H/D exchange between the N centre and *ortho* aromatic positions of imine substrates. Thirdly, the reaction with non-deuterated **1a** and in a mixed solvent of 1:10 MeOD/toluene gave **5aac** with 25% D at C3, 29% D at C4 and 23% D at each *ortho*-position of the 1-phenyl group. The partial deuterium incorporation at C4 suggested that the cleavage of Rh-alkyl linkage (Fig. 2c, **F → 5**) likely occurred by both intra- and intermolecular proton/deuterium transfer processes.

Despite the complication by facile H/D scrambling, results from these deuterium-labelling experiments allowed us to gain further evidence and additional details for proposed reaction pathways as described in Fig. 4c. The proposed H/D exchange between the imine N atom and *ortho* aromatic positions likely occurs by reversible N–H(D) oxidative addition onto Rh(I) centre of the catalyst to form a Rh(III) hydrido iminyl intermediate (*d₁*-**1a → G**), which undergoes reversible 1,4-Rh migration[59] to activate an *ortho* aromatic C–H bond and form the cyclometalated intermediate **A1b** with a Rh–D linkage. Subsequent C–D reductive elimination generates *ortho*-deuterated imine (*d₁*-**1a′**) and completes the proposed H/D exchange. Notably, the cyclometalated Rh(III) intermediate with a Rh–D linkage that is analogous to **A1b** is also formed by imine-directed *ortho* C–D activation (that is, microscopic reverse of C–D reductive elimination) with *d₅*-**1a** (Fig. 2b). As described in Path 1 in Fig. 2b, a tandem sequence of alkyne insertion with **A1b** (or its analogue from *d₅*-**1a**) and C–C reductive elimination leads to regioselective D-transfer onto the mono-substituted alkenyl position in alkyne hydroarylation product **C1**, which followed the proposed hydroamination pathway (Fig. 2c, **C → 5**) to form the 3-deuterated compound **3-D-5aa** as observed in products **5aaa–5aac**. Besides the proposed intramolecular H/D exchange (*d₁*-**1a → d₁-1a′**), imine-directed *ortho* C–H oxidative addition forms the cyclometalated intermediate **A1a** that transforms into intermediate **F1** after the proposed alkyne coupling and ring-closure steps. With the D atom retained on the iminium N centre in **F1**, cleavage of the Rh-alkyl linkage by intramolecular deuteron transfer in stereospecific retention forms the 4-deuterated compound **4-D-5aa** as observed in product **5aac**. Alternatively, deuteron dissociation from **F1** and subsequent cleavage of the Rh-alkyl linkage by intermolecular proton or deuteron transfer forms the non-deuterated **5aa** or **4-D-5aa** respectively. The observed facile H/D scrambling can be attributed to the Brønsted acid behaviours by several proposed reactive intermediates such as cationic Rh(III) hydride/deuteride complexes (**A1a**, **A1b** and **G**) and the iminium species (**F1** and its protonated analogue), which could undergo fast and reversible H/D transfer with external proton or deuteron sources such as trace moisture, acid, or added MeOD in the reaction environment.

Product **5aa** was subjected to several stoichiometric and catalytic transformations to explore DHIQ products from the current study as valuable building blocks in chemical synthesis (Fig. 5). In particular, we expected that hydrogenation or nucleophilic addition of the imine moiety in these DHIQ products could lead to stereoselective formation of the corresponding THIQs, which are important structural motifs in biologically active compounds including natural alkaloids and drug molecules[45]. Thus, we studied the hydride reduction of **5aa** using a procedure reported by Bergman and Ellman (Fig. 5a)[51]. Upon acid-mediated activation, **5aa** underwent a borohydride

reduction to give a 20:1 mixture of two diastereomers of the corresponding THIQ product in 87% overall yield. ¹H NMR and X-ray crystallography indicated that the major product was the all-*cis*-diastereomer (**9a**), while the minor stereoisomer **9b** displayed *cis*-1,3 and *trans*-3,4 stereochemistry. The *cis*-1,3 relationship in both isomers should result from stereospecific, *anti* to 3-phenyl hydride transfer[51]. The formation of **9b** was likely due to acid-mediated epimerization at C3 via iminium intermediates[60,61]. This result highlights the pontential of redox-neutral [4 + 2] imine/alkyne annulation as a new approach towards stereoselective synthesis of poly-substituted THIQs that complements existing strategies such as intramolecular electrophilic aromatic substitutions[62] or metal-catalysed enantioselective hydrogenation of isoquinolines[63,64].

To explore 1-aryl-substituted DHIQs as aromatic imine analogues for imine-directed C–H functionalization, a 1:1 reaction between **5aa** and a Cp*-ligated Rh(III) complex [Cp*RhCl₂]₂ was carried out at room temperature with sodium acetate as an additive (Fig. 5b). Although this transformation occurred with incomplete conversion, we were able to isolate a Rh(III) product with a cyclometalated DHIQ ligand (**11**) and characterized its solid-state structure by single-crystal X-ray diffraction. This {Cp*RhCl[η²-(C,N)-DHIQ]} complex resulted from regioselective C–H activation of **5aa** at the *ortho*-position of 1-phenyl instead of 3-phenyl substituent. Notably, a preliminary test reaction between the oxidative [4 + 2] annulation product **4a** and [Cp*RhCl₂]₂ under similar conditions failed to generate cyclometalation products. Such reactivity difference between **5aa** and **4a** may result from the more rigid structure of isoquinoline than DHIQ, which led to more significant steric crowding by phenyl substituents and 1- and 3-positions that inhibits imine-directed aromatic C–H activation.

The Rh(III)-mediated regioselective C–H activation with **5aa** was further exploited in two catalytic transformations following reported procedures by Li[65] and Glorius[66], both using [Cp*RhCl₂]₂ as the catalyst precursor (Fig. 5c). A coupling between **5aa** and the organoboron reagent ⁿBuBF₃K gave alkylated DHIQ product **12** in 68% yield[65]. Consistent with the stoichiometric cyclometalation result (Fig. 5b), the *n*-butyl group was attached selectively at an *ortho*-position of the 1-phenyl group. Notably, the *cis*-3,4-diphenyl stereochemistry of the DHIQ backbone was retained in both compounds **11** and **12**. By contrast, **5aa** underwent C–H bromination with the NBS reagent in same regioselectivity but dehydrogenated under the reaction conditions to give brominated isoquinoline product **13** in 74% yield[66]. Although it is difficult to rationalize the observation of such dehydrogenation under non-oxidation conditions without a systematic investigation, this result resonates with the recent report by the Wang group on Mn-catalysed dehydrogenative [4 + 2] imine/alkyne annulation for isoquinoline synthesis[30]. In another preliminary reactivity evaluation, we found that the isoquinoline compound **4a** failed to undergo C–H bromination under similar reaction conditions to form **13**, presumably due to the steric hindrance against cyclometalation as previously discussed. Thus, DHIQ products from current study could also be explored as isoquinoline precursors by tandem functionalization-dehydrogenation strategy.

In summary, we have developed a single catalyst-based approach towards atom-efficient N-heterocycle construction by tandem C–H activation, alkyne coupling, and intramolecular alkene hydroamination. The mechanism-based catalyst development led to the combination of a cationic Rh(I) catalyst precursor and a bis(phosphine) ligand DPEphos, which promotes a redox-neutral [4 + 2] annulation between N–H aromatic ketimines and internal alkynes to form *cis*-3,4-disubstituted 3,4-dihydroisoquinolines (DHIQs) in high chemo-, regio- and stereoselectivity. With a proposed ligand-enabled intramolecular alkene hydroamination to introduce C3- and C4-chirality, this method can be potentially developed into an enantioselective version for asymmetric synthesis of poly-substituted chiral DHIQ building blocks towards highly valuable THIQ structures. The current strategy of combining metal-catalysed C–H functionalization and alkene hydrofunctionalization represents a unified synthetic approach towards various 6-membered benzoheterocycles by redox-neutral [4 + 2] annulation between aromatic compounds and alkynes.

## Methods

**General procedure for redox-neutral [4 + 2] imine/alkyne annulations.** Into a 4 ml scintillation vial equipped with a magnetic stir bar was placed $[Rh(cod)_2]BF_4$ (**6**, 5.6 mg, 0.014 mmol, 0.050 equiv.), DPEphos (**7**, 8.9 mg, 0.017 mmol, 0.060 equiv.), and 1.0 ml of toluene. Next, N–H ketimine **1** (0.28 mmol, 1.0 equiv.) and internal alkyne **2** (0.31 mmol, 1.1 equiv.) were added into the vial for the synthesis of products **5aa–5aq** (demonstration of alkyne substrate scope). For the synthesis of products **5ba–5oa** (demonstration of imine substrate scope), 0.28 mmol alkyne **2** (1.0 equiv.) and 0.31 mmol of imine **1** (1.1 equiv.) were added instead. The vial was sealed with a silicone-lined screw-cap, transferred out of the glovebox, and stirred at 100 °C for 24 h. After the reaction mixture was cooled to room temperature, all volatile materials were removed under reduced pressure. Further purification was achieved by flash-column chromatography using neutral alumina. Yields of the isolated products are based on the average of two runs under identical conditions. The exclusive cis-diastereoselectivity for products was determined by GC and $^1$H NMR analysis of the unpurified reaction mixture. See Supplementary Figs 1–87 and Supplementary Methods for full experimental details and analytical data for characterization of new compounds.

## References

1. Yu, J.-Q., & Shi, Z. (eds). *C–H Activation*. Topics in Current Chemistry, Vol. 292 (Springer, 2010).
2. Colby, D. A., Bergman, R. G. & Ellman, J. A. Rhodium-catalyzed C–C bond formation via heteroatom-directed C–H bond activation. *Chem. Rev.* **110**, 624–655 (2010).
3. Lyons, T. W. & Sanford, M. S. Palladium-catalyzed ligand-directed C–H functionalization reactions. *Chem. Rev.* **110**, 1147–1169 (2010).
4. Satoh, T. & Miura, M. Oxidative coupling of aromatic substrates with alkynes and alkenes under rhodium catalysis. *Chem. Eur. J* **16**, 11212–11222 (2010).
5. Mei, T.-S., Kou, L., Ma, S., Engle, K. M. & Yu, J.-Q. Heterocycle formation via palladium-catalyzed C–H functionalization. *Synthesis* **44**, 1778–1791 (2012).
6. Song, G., Wang, F. & Li, X. C–C, C–O and C–N bond formation via rhodium(iii)-catalyzed oxidative C–H activation. *Chem. Soc. Rev.* **41**, 3651–3678 (2012).
7. Ackermann, L. Carboxylate-assisted ruthenium-catalyzed alkyne annulations by C–H/Het-H bond functionalizations. *Acc. Chem. Res.* **47**, 281–295 (2014).
8. He, R., Huang, Z.-T., Zheng, Q.-Y. & Wang, C. Isoquinoline skeleton synthesis via chelation-assisted C–H activation. *Tetrahedron Lett.* **55**, 5705–5713 (2014).
9. Albrecht, M. Cyclometalation using d-block transition metals: fundamental aspects and recent trends. *Chem. Rev.* **110**, 576–623 (2010).
10. Li, L., Brennessel, W. W. & Jones, W. D. An efficient low-temperature route to polycyclic isoquinoline salt synthesis via C–H activation with $[Cp^*MCl_2]_2$ (M = Rh, Ir). *J. Am. Chem. Soc.* **130**, 12414–12419 (2008).
11. Li, L., Brennessel, W. W. & Jones, W. D. C–H activation of phenyl imines and 2-phenylpyridines with $[Cp^*MCl_2]_2$ (M = Ir, Rh): regioselectivity, kinetics, and mechanism. *Organometallics* **28**, 3492–3500 (2009).
12. Lim, S.-G., Lee, J. H., Moon, C. W., Hong, J.-B. & Jun, C.-H. Rh(I)-catalyzed direct ortho-alkenylation of aromatic ketimines with alkynes and its application to the synthesis of isoquinoline derivatives. *Org. Lett.* **5**, 2759–2761 (2003).
13. Ueura, K., Satoh, T. & Miura, M. An efficient waste-free oxidative coupling via regioselective C–H bond cleavage: Rh/Cu-catalyzed reaction of benzoic acids with alkynes and acrylates under air. *Org. Lett.* **9**, 1407–1409 (2007).
14. Umeda, N., Tsurugi, H., Satoh, T. & Miura, M. Fluorescent naphthyl- and anthrylazoles from the catalytic coupling of phenylazoles with internal alkynes through the cleavage of multiple C–H bonds. *Angew. Chem. Int. Ed.* **47**, 4019–4022 (2008).
15. Fukutani, T., Umeda, N., Hirano, K., Satoh, T. & Miura, M. Rhodium-catalyzed oxidative coupling of aromatic imines with internal alkynes via regioselective C–H bond cleavage. *Chem. Commun.* **2009**, 5141–5143 (2009).
16. Guimond, N. & Fagnou, K. Isoquinoline synthesis via rhodium-catalyzed oxidative cross-coupling/cyclization of aryl aldimines and alkynes. *J. Am. Chem. Soc.* **131**, 12050–12051 (2009).
17. Parthasarathy, K. & Cheng, C.-H. Easy access to isoquinolines and tetrahydroquinolines from ketoximes and alkynes via rhodium-catalyzed C–H bond activation. *J. Org. Chem.* **74**, 9359–9364 (2009).
18. Guimond, N., Gouliaras, C. & Fagnou, K. Rhodium(III)-catalyzed isoquinolone synthesis: the N–O bond as a handle for C–N bond formation and catalyst turnover. *J. Am. Chem. Soc.* **132**, 6908–6909 (2010).
19. Mochida, S., Umeda, N., Hirano, K., Satoh, T. & Miura, M. Rhodium-catalyzed oxidative coupling/cyclization of benzamides with alkynes via C–H bond cleavage. *Chem. Lett.* **39**, 744–746 (2010).
20. Song, G., Chen, D., Pan, C.-L., Crabtree, R. H. & Li, X. Rh-catalyzed oxidative coupling between primary and secondary benzamides and alkynes: synthesis of polycyclic amides. *J. Org. Chem.* **75**, 7487–7490 (2010).
21. Hyster, T. K. & Rovis, T. Pyridine synthesis from oximes and alkynes via rhodium(III) catalysis: Cp* and Cp$^t$ provide complementary selectivity. *Chem. Commun.* **47**, 11846–11848 (2011).
22. Morimoto, K., Hirano, K., Satoh, T. & Miura, M. Synthesis of isochromene and related derivatives by rhodium-catalyzed oxidative coupling of benzyl and allyl alcohols with alkynes. *J. Org. Chem.* **76**, 9548–9551 (2011).
23. Too, P. C., Chua, S. H., Wong, S. H. & Chiba, S. Synthesis of azaheterocycles from aryl ketone O-acetyl oximes and internal alkynes by Cu-Rh bimetallic relay catalysts. *J. Org. Chem.* **76**, 6159–6168 (2011).
24. Ackermann, L., Pospech, J., Graczyk, K. & Rauch, K. Versatile synthesis of isocoumarins and a-pyrones by ruthenium-catalyzed oxidative C–H/O–H bond cleavages. *Org. Lett.* **14**, 930–933 (2012).
25. Chinnagolla, R. K. & Jeganmohan, M. Regioselective synthesis of isocoumarins by ruthenium-catalyzed aerobic oxidative cyclization of aromatic acids with alkynes. *Chem. Commun.* **48**, 2030–2032 (2012).
26. Zhong, H., Yang, D., Wang, S. & Huang, J. Pd-catalysed synthesis of isoquinolinones and analogues via C–H and N–H bonds double activation. *Chem. Commun.* **48**, 3236–3238 (2012).
27. Pham, M. V., Ye, B. & Cramer, N. Access to sultams by rhodium(III)-catalyzed directed C–H activation. *Angew. Chem. Int. Ed.* **51**, 10610–10614 (2012).
28. Unoh, Y. *et al.* Rhodium(III)-catalyzed oxidative coupling through C–H bond cleavage directed by phosphinoxy groups. *Org. Lett.* **15**, 3258–3261 (2013).
29. Villuendas, P. & Urriolabeitia, E. P. Primary amines as directing groups in the Ru-catalyzed synthesis of isoquinolines, benzoisoquinolines, and thienopyridines. *J. Org. Chem.* **78**, 5254–5263 (2013).
30. He, R., Huang, Z.-T., Zheng, Q.-Y. & Wang, C. Manganese-catalyzed dehydrogenative [4 + 2] annulation of N–H imines and alkynes by C–H/N–H activation. *Angew. Chem. Int. Ed.* **53**, 4950–4953 (2014).
31. Li, J. & Ackermann, L. Ruthenium-catalyzed oxidative alkyne annulation by C–H activation on ketimines. *Tetrahedron* **70**, 3342–3348 (2014).
32. Nakanowatari, S. & Ackermann, L. Ruthenium(II)-catalyzed synthesis of isochromenes by C–H activation with weakly coordinating aliphatic hydroxyl groups. *Chemistry* **20**, 5409–5413 (2014).
33. Neufeldt, S. R., Jimenez-Óses, G., Huckins, J. R., Thiel, O. R. & Houk, K. N. Pyridine N-oxide vs pyridine substrates for Rh(III)-catalyzed oxidative C–H bond functionalization. *J. Am. Chem. Soc.* **137**, 9843–9854 (2015).
34. Warratz, S. *et al.* Ruthenium(II)-catalyzed C–H activation/alkyne annulation by weak coordination with $O_2$ as the sole oxidant. *Angew. Chem. Int. Ed.* **54**, 5513–5517 (2015).
35. Tang, Q., Xia, D., Zhang, Q., Sun, X.-Q. & Wang, C. Re/Mg bimetallic tandem catalysis for [4 + 2] annulation of benzamides and alkynes via C–H/N–H functionalization. *J. Am. Chem. Soc.* **135**, 4628–4631 (2013).
36. Mueller, T. E., Hultzsch, K. C., Yus, M., Foubelo, F. & Tada, M. Hydroamination: direct addition of amines to alkenes and alkynes. *Chem. Rev.* **108**, 3795–3892 (2008).
37. Huang, L., Arndt, M., Goossen, K., Heydt, H. & Goossen, L. J. Late transition metal-catalyzed hydroamination and hydroamidation. *Chem. Rev.* **115**, 2596–2697 (2015).
38. Yi, C. S. & Yun, S. Y. Scope and mechanistic study of the ruthenium-catalyzed ortho-C–H bond activation and cyclization reactions of arylamines with terminal alkynes. *J. Am. Chem. Soc.* **127**, 17000–17006 (2005).
39. Lium, X.-Y., Ding, P., Huang, J.-S. & Che, C.-M. Synthesis of substituted 1,2-dihydroquinolines and quinolines from aromatic amines and alkynes by gold(I)-catalyzed tandem hydroamination–hydroarylation under microwave-assisted conditions. *Org. Lett.* **9**, 2645–2648 (2007).
40. Zeng, X., Frey, G. D., Kinjo, R., Donnadieu, B. & Bertrand, G. Synthesis of a simplified version of stable bulky and rigid cyclic (Alkyl)(amino)carbenes, and catalytic activity of the ensuing gold(I) complex in the three-component preparation of 1,2-dihydroquinoline derivatives. *J. Am. Chem. Soc.* **131**, 8690–8696 (2009).
41. Sun, Z.-M., Chen, S.-P. & Zhao, P. Tertiary carbinamine synthesis by rhodium-catalyzed [3 + 2] annulation of N-unsubstituted aromatic ketimines and alkynes. *Chem. Eur. J* **16**, 2619–2627 (2010).
42. Tran, D. N. & Cramer, N. Enantioselective rhodium(I)-catalyzed [3 + 2] annulations of aromatic ketimines induced by directed C–H activations. *Angew. Chem. Int. Ed.* **50**, 11098–11102 (2011).
43. Zhang, J., Ugrinov, A. & Zhao, P. Ruthenium(II)/N-heterocyclic carbene catalyzed [3 + 2] carbocyclization with aromatic N–H ketimines and internal alkynes. *Angew. Chem. Int. Ed.* **52**, 6681–6684 (2013).
44. Manan, R. S., Kilaru, P. & Zhao, P. Nickel-catalyzed hydroimination of alkynes. *J. Am. Chem. Soc.* **137**, 6136–6139 (2015).

45. Scott, J. D. & Williams, R. M. Chemistry and biology of the tetrahydroisoquinoline antitumor antibiotics. *Chem. Rev.* **102**, 1669–1730 (2002).

46. Kwan, E. E. & Huang, S. G. Structural elucidation with NMR spectroscopy: practical strategies for organic chemists. *Eur. J. Biochem.* 2671–2688 (2008).

47. Murai, S. *et al.* Efficient catalytic addition of aromatic carbon-hydrogen bonds to olefins. *Nature* **366**, 529–531 (1993).

48. Kakiuchi, F., Yamamoto, Y., Chatani, N. & Murai, S. Catalytic addition of aromatic C-H bonds to acetylenes. *Chem. Lett.* **24**, 681–682 (1995).

49. Colby, D. A., Bergman, R. G. & Ellman, J. A. Synthesis of dihydropyridines and pyridines from imines and alkynes via C–H activation. *J. Am. Chem. Soc.* **130**, 3645–3651 (2008).

50. Duttwyler, S., Lu, C., Rheingold, A. L., Bergman, R. G. & Ellman, J. A. Highly diastereoselective synthesis of tetrahydropyridines by a C–H activation-cyclization-reduction cascade. *J. Am. Chem. Soc.* **134**, 4064–4067 (2012).

51. Duttwyler, S. *et al.* Proton donor acidity controls selectivity in nonaromatic nitrogen heterocycle synthesis. *Science* **339**, 678–682 (2013).

52. Utsunomiya, M., Kuwano, R., Kawatsura, M. & Hartwig, J. F. Rhodium-catalyzed anti-Markovnikov hydroamination of vinylarenes. *J. Am. Chem. Soc.* **125**, 5608–5609 (2003).

53. Takemiya, A. & Hartwig, J. F. Rhodium-catalyzed intramolecular, anti-Markovnikov hydroamination. Synthesis of 3-arylpiperidines. *J. Am. Chem. Soc.* **128**, 6042–6043 (2006).

54. Liu, Z. & Hartwig, J. F. Mild, rhodium-catalyzed intramolecular hydroamination of unactivated terminal and internal alkenes with primary and secondary amines. *J. Am. Chem. Soc.* **130**, 1570–1571 (2008).

55. Liu, Z., Yamamichi, H., Madrahimov, S. T. & Hartwig, J. F. Rhodium phosphine-π-arene intermediates in the hydroamination of alkenes. *J. Am. Chem. Soc.* **133**, 2772–2782 (2011).

56. van Leeuwen, P. W, Kamer, P. C. J., Reek, J. N. H. & Dierkes, P. Ligand bite angle effects in metal-catalyzed C–C bond formation. *Chem. Rev.* **100**, 2741–2769 (2000).

57. Zhang, G., Yang, L., Wang, Y., Xie, Y. & Huang, H. An efficient Rh/O₂ catalytic system for oxidative C–H activation/annulation: evidence for Rh(I) to Rh(III) oxidation by molecular oxygen. *J. Am. Chem. Soc.* **135**, 8850–8853 (2013).

58. Kitamura, T. Transition-metal-catalyzed hydroarylation reactions of alkynes through direct functionalization of C–H bonds. A convenient tool for organic synthesis. *Eur. J. Org. Chem.* **2009**, 1111–1125 (2009).

59. Ma, S. & Gu, Z. 1,4-migration of rhodium and palladium in catalytic organometallic reactions. *Angew. Chem. Int. Ed.* **44**, 7512–7517 (2005).

60. Pahadi, N. K., Paley, M., Jana, R., Waetzig, S. R. & Tunge, J. A. Formation of N-alkylpyrroles via intermolecular redox amination. *J. Am. Chem. Soc.* **131**, 16626–16627 (2009).

61. Deb, I., Das, D. & Seidel, D. Redox isomerization via azomethine ylide intermediates: N-alkyl indoles from indolines and aldehydes. *Org. Lett.* **13**, 812–815 (2010).

62. Stockigt, J., Antonchick, A. P., Wu, F.-R. & Waldmann, H. The pictet-spengler reaction in nature and in organic chemistry. *Angew. Chem. Int. Ed.* **50**, 8538–8564 (2011).

63. Zhao, D. & Glorius, F. Enantioselective hydrogenation of isoquinolines. *Angew. Chem. Int. Ed.* **52**, 9616–9618 (2013).

64. Wang, D.-S., Chen, Q.-A., Lu, S.-M. & Zhou, Y.-G. Asymmetric hydrogenation of heteroarenes and arenes. *Chem. Rev.* **112**, 2557–2590 (2012).

65. Wang, H., Yu, S., Qi, Z. & Li, X. Rh(III)-catalyzed C-H alkylation of arenes using alkylboron reagents. *Org. Lett.* **17**, 2812–2815 (2015).

66. Schroeder, N., Wencel-Delord, J. & Glorius, F. High-yielding, versatile, and practical [Rh(III)Cp*]-catalyzed ortho bromination and iodination of arenes. *J. Am. Chem. Soc.* **134**, 8298–8301 (2012).

## Acknowledgements

Financial support for this work was provided by NSF (CHE-1301409). We also thank NSF-CRIF (CHE-0946990) for funding the purchase of departmental X-ray diffractometer and Dr. Angel Ugrinov for solving the single-crystal XRD structures.

## Author contributions

R.S.M. performed the experiments and data analysis. P.Z. and R.S.M. designed the catalytic sequence and developed the reaction conditions. P.Z. prepared this manuscript with the assistance of R.S.M.

# Highly regio- and enantioselective multiple oxy- and amino-functionalizations of alkenes by modular cascade biocatalysis

Shuke Wu[1,2,3], Yi Zhou[3], Tianwen Wang[1], Heng-Phon Too[2], Daniel I.C Wang[2,4] & Zhi Li[1,2,3]

New types of asymmetric functionalizations of alkenes are highly desirable for chemical synthesis. Here, we develop three novel types of regio- and enantioselective multiple oxy- and amino-functionalizations of terminal alkenes via cascade biocatalysis to produce chiral α-hydroxy acids, 1,2-amino alcohols and α-amino acids, respectively. Basic enzyme modules 1-4 are developed to convert alkenes to (S)-1,2-diols, (S)-1,2-diols to (S)-α-hydroxyacids, (S)-1,2-diols to (S)-aminoalcohols and (S)-α-hydroxyacids to (S)-α-aminoacids, respectively. Engineering of enzyme modules 1 & 2, 1 & 3 and 1, 2 & 4 in *Escherichia coli* affords three biocatalysts over-expressing 4–8 enzymes for one-pot conversion of styrenes to the corresponding (S)-α-hydroxyacids, (S)-aminoalcohols and (S)-α-aminoacids in high e.e. and high yields, respectively. The new types of asymmetric alkene functionalizations provide green, safe and useful alternatives to the chemical syntheses of these compounds. The modular approach for engineering multi-step cascade biocatalysis is useful for developing other new types of one-pot biotransformations for chemical synthesis.

[1] Department of Chemical and Biomolecular Engineering, National University of Singapore, Singapore 117585, Singapore. [2] Singapore-MIT Alliance, National University of Singapore, Singapore 117583, Singapore. [3] Synthetic Biology for Clinical and Technological Innovation (SynCTI), Life Sciences Institute, National University of Singapore, Singapore 117456, Singapore. [4] Department of Chemical Engineering, Massachusetts Institute of Technology, Cambridge, Massachusetts 02139, USA. Correspondence and requests for materials should be addressed to Z.L. (email: chelz@nus.edu.sg).

Alkenes are readily available and excellent starting materials for chemical synthesis. Asymmetric functionalization of alkenes is of great importance in the synthesis of enantiopure chemicals for pharmaceutical manufacturing. Thus far, many metal catalyses have been developed for this, including the well-known Sharpless epoxidation[1], dihydroxylation[2] aminohydroxylation[3], Jacobsen epoxidation[4] and palladium-catalysed asymmetric alkene functionalization[5]. On the other hand, enzyme catalyses could provide green alternatives due to the non-toxicity, high selectivity and mild reaction conditions[6–8]. A number of enzyme-catalysed asymmetric alkene functionalizations have been reported, such as the epoxidation with monooxygenase or peroxidase[9], dihydroxylation with dioxygenase[10], hydration with hydratase[11], and more recently, cyclopropanation and aziridination with engineered P450 enzymes[12–14]. Despite of these achievements, the development of new types of reactions and catalysts for asymmetric alkene functionalizations is highly wanted and remains a significant challenge.

An attractive way of developing new asymmetric transformation is to develop novel cascade (tandem) catalysis for performing multi-step reactions sequentially or concurrently in one pot[15,16]. Recently, asymmetric hydroxy arylation and allylation of terminal alkenes was reported by combining platinum-catalysed diboration and palladium-catalysed cross-coupling[17] in a sequential manner due to divergent reaction conditions of diboration and cross-coupling. In comparison with cascade chemocatalysis or cascade hybrid catalysis[18], cascade biocatalysis is of advantages in combining multiple and complex reactions due to the natural compatibility and similar reaction condition of many enzymes[19–24], in addition to the green features of enzyme catalysis. Over the years, many types of non-natural biocatalytic cascades have been developed[25–36]. Nevertheless, the epoxidation–hydrolysis cascade for asymmetric *trans*-dihydroxylation of alkenes[37–39] recently developed by us is the only known biocatalytic cascade for asymmetric functionalization of alkenes. Thus far, most of the reported cascade biocatalysis enables only two to three relatively simple enzymatic reactions. It is very challenging to engineer the efficient cascade system containing more than four enzymatic reactions.

We have keen interest in developing new types of asymmetric alkene functionalizations and engineering enzyme cascade containing more than four enzymatic reactions. One-pot regio- and stereoselective multiple oxy- and amino-functionalizations of terminal alkenes to produce chiral α-hydroxy acids, 1,2-amino alcohols and α-amino acids, respectively, are designed as the target reactions (Fig. 1a). Enantiopure α-hydroxy acid[40], 1,2-amino alcohol[41] and α-amino acid[42] are the three very important groups of chiral chemicals with broad applications in chiral pharmaceutical and asymmetric syntheses. The designed new transformations could provide green, safe and complementary alternatives to the toxic cyanide-based asymmetric synthesis of chiral α-hydroxy acid (via cyanohydrin)[43] and α-amino acid (Strecker reaction)[44] and the osmium-based asymmetric synthesis of chiral amino alcohols[3,45]. Herein, we report the development of the three new types of asymmetric functionalizations of terminal alkenes, the modular approach for engineering efficient cascade biocatalysis containing more than four concurrent reactions, and the simple and green syntheses of α-hydroxy acids, 1,2-amino alcohols and α-amino acids in high enantiomeric excess (e.e.) and high yield.

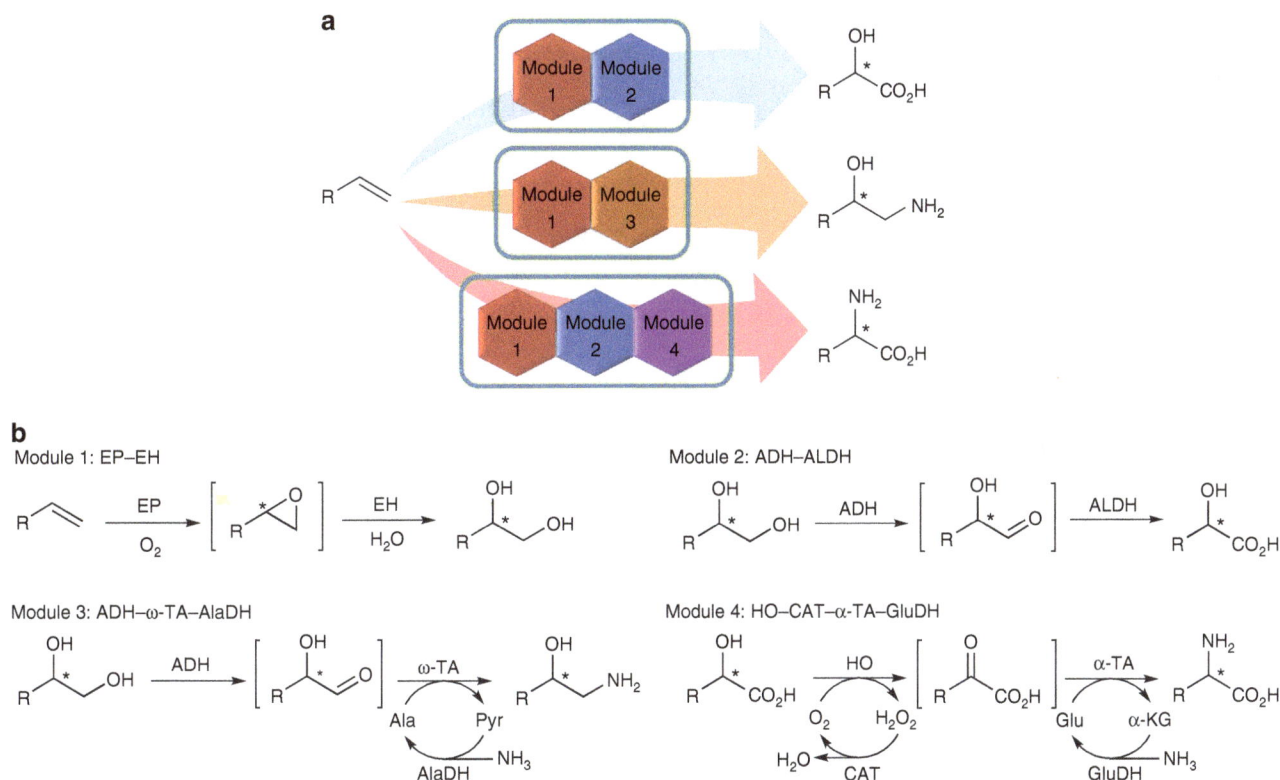

**Figure 1 | Regio- and enantioselective multiple oxy- and amino-functionalizations of terminal alkenes by modular cascade biocatalysis. (a)** One-pot conversion of terminal alkene to chiral α-hydroxy acid, 1,2-amino alcohol and α-amino acid with *E. coli* cells containing multiple basic enzyme modules, respectively. **(b)** Four general basic enzyme modules and their cascade biotransformations. Module 1: epoxidase (EP) and epoxide hydrolase (EH) for epoxidation–hydrolysis of terminal alkene to 1,2-diol; Module 2: alcohol dehydrogenase (ADH) and aldehyde dehydrogenase (ALDH) for terminal double oxidation of 1,2-diol to α-hydroxy acid; Module 3: ADH, ω-transaminase (ω-TA) and alanine dehydrogenase (AlaDH) for oxidation–transamination of 1,2-diol to 1,2-amino alcohol; Module 4: hydroxy acid oxidase (HO), α-transaminase (α-TA), catalase (CAT) and glutamate dehydrogenase (GluDH) for oxidation–transamination of α-hydroxy acid to α-amino acid.

## Results

### Design of modular biocatalysis for cascade reactions.

To realize the targeted asymmetric alkene functionalizations (Fig. 1a), we designed microbial cells containing two to three basic enzyme modules, each of them catalysing two to four enzymatic reactions (Fig. 1b), based on biocatalytic retrosynthesis analysis[46]. The basic modules were designed by using the following criteria: (a) each module utilizes a stable input, such as alkene, diol and hydroxy acid, and gives a stable output, such as diol, hydroxy acid, amino alcohol and amino acid; (b) each module enables fast conversion of unstable or toxic intermediates, such as epoxide, hydroxy aldehyde and keto acid, to minimize their accumulation and side reactions. Assemblies of module 1 and 2 in one cell, module 1 and 3 in one cell and module 1, 2 and 4 in one cell gave rise to whole-cell catalysts for one-pot transformations of terminal alkene to chiral α-hydroxy acid, 1,2-amino alcohol and α-amino acid, respectively (Fig. 1a). To demonstrate the concept, we chose the biotransformations of styrenes **1a–k** to (S)-α-hydroxy acids **5a–k**, (S)-1,2-amino alcohols **6a–k** and (S)-α-amino acids **8a–k**, respectively, as the representative examples of the three types of asymmetric reactions (Fig. 2). While the styrenes are easily available substrates, the (S)-α-hydroxy acids, (S)-1,2-amino alcohols and (S)-α-amino acids are highly valuable chiral chemicals with many applications (Supplementary Table 1).

### Engineering of basic enzyme modules.

Enzyme module 1 for the conversion of alkene to diol was engineered according to our previously reported method[39]. *Escherichia coli* (R-M1) containing gene module 1 on plasmid pRSFDuet-1 (Table 1) was constructed to coexpress styrene monooxygenase (SMO)[47] and epoxide hydrolase (SpEH)[48] (Fig. 2a). As shown in Fig. 3a, 5 g cdw l$^{-1}$ of *E. coli* (R-M1) cells efficiently transformed 50 mM styrene **1a** to 46 mM (S)-1-phenyl-1,2-ethanediol **3a** in 5 h, without significant

accumulation of (S)-styrene oxide **2a** (<1%). For further assembly of multiple basic modules and optimization of enzyme expression in one *E. coli* strain, gene module 1 was sub-cloned into other three different but compatible plasmids, pACYCDuet-1, pCDFDuet-1 and pETDuet-1, to generate three new recombinant plasmids, A-M1, C-M1 and E-M1, respectively (Table 1).

To engineer enzyme module 2 for the conversion of diol to α-hydroxy acid, many commercially available alcohol dehydrogenases (ADH), cloned ADHs and wild-type strains collected in our laboratory (Supplementary Table 2) were screened for the terminal oxidation of (S)-1-phenyl-1,2-ethanediol **3a** to identify a highly regioselective enzyme for the first reaction of the module (Fig. 2a). AlkJ from *Pseudomonas putida* GPo1 (ref. 49), a membrane-associated non-canonical ADH, was found to oxidize **3a** at the terminal position to give mandelaldehyde **4a** and mandelic acid **5a** with S-enantioselectivity. Phenylacetaldehyde dehydrogenase (EcALDH, encoded by padA) from *E. coli*[50] was then found to fully oxidize α-hydroxy aldehyde **4a** to give **5a**, the second reaction of the module. Thus, the genes of AlkJ and EcALDH were genetically engineered into a non-natural operon as gene module 2 on plasmid pRSFDuet-1 (R-M2, Table 1). *E. coli* (R-M2) cells expressed both AlkJ and EcALDH very well (Fig. 3b) and catalysed the highly regioselective terminal oxidation of 50 mM (S)-**3a** to give 49 mM (S)-**5a** in 8 h (Fig. 3b), without the accumulation of intermediate (S)-**4a**. Similarly, gene module 2 was also sub-cloned to the three plasmids to generate new recombinant plasmids, A-M2, C-M2 and E-M2, respectively (Table 1).

Enzyme module 3 is for the conversion of diol to α-amino alcohol. AlkJ-catalysed highly regioselective oxidation of (S)-**3a** to (S)-**4a** is the first reaction of module 3 (Fig. 2b). The ω-transaminase from *Chromobacterium violaceum* (CvωTA, encoded by cv_2025)[51] was chosen for the transamination of (S)-**4a**, the second reaction of the module. An *E. coli* strain was

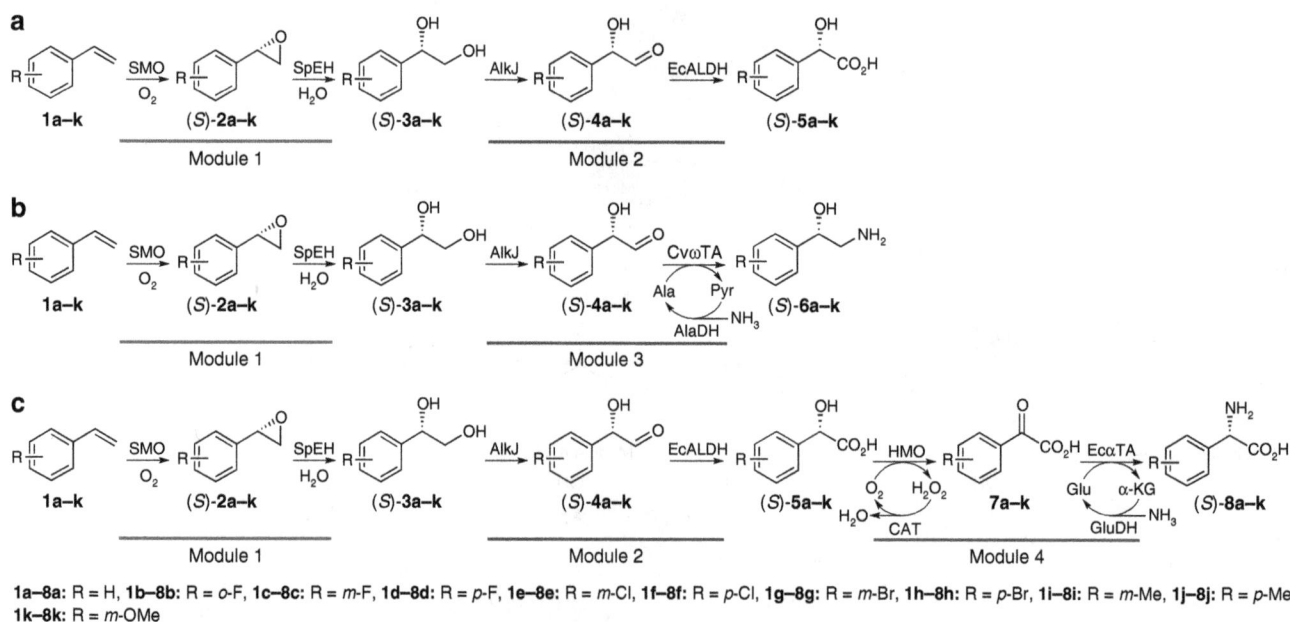

**1a–8a:** R = H, **1b–8b:** R = o-F, **1c–8c:** R = m-F, **1d–8d:** R = p-F, **1e–8e:** R = m-Cl, **1f–8f:** R = p-Cl, **1g–8g:** R = m-Br, **1h–8h:** R = p-Br, **1i–8i:** R = m-Me, **1j–8j:** R = p-Me, **1k–8k:** R = m-OMe

**Figure 2 | Regio- and enantioselective multiple oxy- and amino-functionalizations of styrenes by modular cascade biocatalysis.** (a) Conversion of styrenes to (S)-α-hydroxy acids with *E. coli* strains containing enzyme module 1 and 2. (b) Conversion of styrenes to (S)-1,2-amino alcohols with *E. coli* strains containing enzyme module 1 and 3. (c) Conversion of styrenes to (S)-α-amino acids with *E. coli* strains containing enzyme module 1, 2 and 4. SMO: styrene monooxygenase from *Pseudomonas* sp. VLB120; SpEH: epoxide hydrolase from *Sphingomonas* sp. HXN-200; AlkJ: alcohol dehydrogenase from *P. putida* GPo1; EcALDH: phenylacetaldehyde dehydrogenase from *E. coli*; CvωTA: ω-transaminase from *C. violaceum*; AlaDH: alanine dehydrogenase from *B. subtilis*; HMO: hydroxymandelate oxidase from *S. coelicolor* A3(2); EcαTA: branch chain amino acid transaminase from *E. coli*; GluDH: glutamate dehydrogenase from *E. coli*; and CAT: catalase from *E. coli*.

**Table 1 | Genetic construction of recombinant *E. coli* strains containing different enzyme modules.**

| Genetic construction of modules* | Plasmids containing modules† | Recombinant *E. coli* strains containing different modules |
|---|---|---|
| Module 1 (M1) `styA` `styB` `spEH` | M1 (A-M1), M1 (C-M1), M1 (E-M1), M1 (R-M1) | *E. coli* (A-M1), *E. coli* (C-M1), *E. coli* (E-M1), *E. coli* (R-M1) |
| Module 2 (M2) `alkJ` `padA` | M2 (A-M2), M2 (C-M2), M2 (E-M2), M2 (R-M2) | *E. coli* (A-M2), *E. coli* (C-M2), *E. coli* (E-M2), *E. coli* (R-M2) |
| Module 3 (M3) `alkJ` `cv_2025` `ald` | M3 (A-M3), M3 (C-M3), M3 (E-M3), M3 (R-M3) | *E. coli* (A-M3), *E. coli* (C-M3), *E. coli* (E-M3), *E. coli* (R-M3) |
| Module 4 (M4) `sco3228` `ilvE` `gdhA` `katE` | M4 (A-M4), M4 (C-M4), M4 (E-M4), M4 (R-M4) | *E. coli* (A-M4), *E. coli* (C-M4), *E. coli* (E-M4), *E. coli* (R-M4) |
| Module 1 + module 2 | A-M1_C-M2, A-M1_E-M2, A-M1_R-M2, C-M1_A-M2, C-M1_E-M2, C-M1_R-M2, E-M1_A-M2, E-M1_C-M2, E-M1_R-M2, R-M1_A-M2, R-M1_C-M2, R-M1_E-M2 | *E. coli* (A-M1_C-M2), *E. coli* (A-M1_E-M2), *E. coli* (A-M1_R-M2), *E. coli* (C-M1_A-M2), *E. coli* (C-M1_E-M2), *E. coli* (C-M1_R-M2), *E. coli* (E-M1_A-M2), *E. coli* (E-M1_C-M2), *E. coli* (E-M1_R-M2), *E. coli* (R-M1_A-M2), *E. coli* (R-M1_C-M2), *E. coli* (R-M1_E-M2) |
| Module 1 + module 3 | A-M1_C-M3, A-M1_E-M3, A-M1_R-M3, C-M1_A-M3, C-M1_E-M3, C-M1_R-M3, E-M1_A-M3, E-M1_C-M3, E-M1_R-M3, R-M1_A-M3, R-M1_C-M3, R-M1_E-M3 | *E. coli* (A-M1_C-M3), *E. coli* (A-M1_E-M3), *E. coli* (A-M1_R-M3), *E. coli* (C-M1_A-M3), *E. coli* (C-M1_E-M3), *E. coli* (C-M1_R-M3), *E. coli* (E-M1_A-M3), *E. coli* (E-M1_C-M3), *E. coli* (E-M1_R-M3), *E. coli* (R-M1_A-M3), *E. coli* (R-M1_C-M3), *E. coli* (R-M1_E-M3) |
| Module 1 + module 2 + module 4 | A-M1_E-M2_C-M4, A-M1_E-M2_R-M4, A-M1_R-M2_C-M4, A-M1_R-M2_E-M4, C-M1_E-M2_A-M4, C-M1_E-M2_R-M4, R-M1_E-M2_A-M4, R-M1_E-M2_C-M4 | *E. coli* (A-M1_E-M2_C-M4), *E. coli* (A-M1_E-M2_R-M4), *E. coli* (A-M1_R-M2_C-M4), *E. coli* (A-M1_R-M2_E-M4), *E. coli* (C-M1_E-M2_A-M4), *E. coli* (C-M1_E-M2_R-M4), *E. coli* (R-M1_E-M2_A-M4), *E. coli* (R-M1_E-M2_C-M4) |

*styA, styB and spEH are the genes of SMO and SpEH, respectively; alkJ and padA are the genes of AlkJ and EcALDH, respectively; alkJ, cv_2025 and ald are the genes of AlkJ, CvωTA and AlaDH, respectively; sco3228, ilvE, gdhA and katE are the genes of HMO, EcTA, GluDH and CAT, respectively.
†A-M1–4 using plasmid p**A**CYCDuet-1; C-M1–4 using plasmid p**C**DFDuet-1; E-M1–4 using plasmid p**E**TDuet-1; R-M1–4 using plasmid p**R**SFDuet-1.

engineered to coexpress AlkJ and CvωTA and used for the biotransformation of 45 mM (*S*)-**3a** with 200 mM L-alanine as amine donor. While the desired product (*S*)-phenylethanolamine **6a** was produced (50% yield; 22 mM), some (*S*)-**4a** (14 mM) and (*S*)-**5a** (5 mM) remained in the system (Supplementary Fig. 1). To increase the formation of (*S*)-**6a** in the reversible transamination reaction and utilize the cellular L-alanine and pyruvate, L-alanine dehydrogenase (AlaDH, encoded by *ald*, Genbank ID 936557) from *Bacillus subtilis* was used to regenerate L-alanine from cellular pyruvate with ammonia as amine donor[52]. A non-natural operon (gene module 3) containing the genes of AlkJ, CvωTA and AlaDH was cloned into the plasmid pRSFDuet-1 (R-M3, Table 1). The *E. coli* (R-M3) strain coexpressed the three enzymes well and catalysed the biotransformation of 50 mM (*S*)-**3a** to afford 35 mM (*S*)-**6a** in 8 h by using 200 mM ammonia with no addition of L-alanine (Fig. 3c). Substrate (*S*)-**3a**, intermediate (*S*)-**4a**, and by-product (*S*)-**5a** remained in relatively low amount (3, 5 and 2 mM, respectively). This *in vivo* amination with coexpressed ω-TA and AlaDH is complementary to the recently developed *in vitro* system[26,52]. Similarly, gene module 3 was sub-cloned into other three plasmids to generate A-M3, C-M3 and E-M3 (Table 1).

Enzyme module 4 is for the conversion of α-hydroxy acid to α-amino acid. To engineer this module, mandelate dehydrogenase[53] and hydroxymandelate oxidase (HMO, encoded by *sco3228*) from *Streptomyces coelicolor* A3(2)[54] were cloned into *E. coli* and examined for the oxidation of (*S*)-**5a**, respectively (Fig. 2c). HMO was found to be more efficient than the dehydrogenase (Supplementary Fig. 2), thus being chosen as the first enzyme of module 4. For the enantioselective amination of **7a** to (*S*)-**8a**, an amino acid dehydrogenase and four α-transaminases (α-TA) were screened with either ammonia or glutamate as amine donor

(Supplementary Fig. 3). EcαTA, the branch chain amino acid transaminase from *E. coli* (encoded by *ilvE*, Genbank ID 948278), was found to give the best results. Glutamate dehydrogenase (GluDH, encoded by *gdhA*, Genbank ID 946802) was then coexpressed with EcαTA in *E. coli* to enable the regeneration of glutamate by using ammonia during the transamination (Supplementary Fig. 4). Since HMO is a $H_2O_2$-generating oxidase, a catalase (CAT, encoded by *katE*, Genbank ID 946234) was used to decompose $H_2O_2$ to improve the biotransformation (Supplementary Fig. 5). The genes of HMO, EcαTA, GluDH and CAT were thus engineered on plasmid pRSFDuet-1 (R-M4, Table 1) to construct module 4. *E. coli* (R-M4) coexpressed the four enzymes well (Fig. 3d) and converted 50 mM (*S*)-**5a** to 45 mM (*S*)-**8a** within 26 h (Fig. 3d). Gene module 4 was sub-cloned into three plasmids to give A-M4, C-M4 and E-M4, respectively (Table 1).

**Engineering of catalyst to convert alkene to α-hydroxy acid.** Module 1 and 2 were assembled together in *E. coli* cells as the catalyst (Fig. 2a). To explore the optimal combination, four module 1 plasmids (A-M1, C-M1, E-M1 and R-M1) and four module 2 plasmids (A-M2, C-M2, E-M2 and R-M2) were combinatorially combined and transformed into *E. coli*. Since plasmids with the same backbone (for example, A-M1 and A-M2) are not compatible with each other, 12 *E. coli* strains were obtained (Table 1), each co-expressing SMO, SpEH, AlkJ and EcALDH. These *E. coli* strains grew well in M9-glucose medium and expressed the desired enzymes. The collected cells were examined for the biotransformation of 100 mM styrene **1a** in a two-liquid-phase system (buffer and *n*-hexadecane; 1:1) containing 0.5% glucose. As shown in Fig. 4a, all strains were able

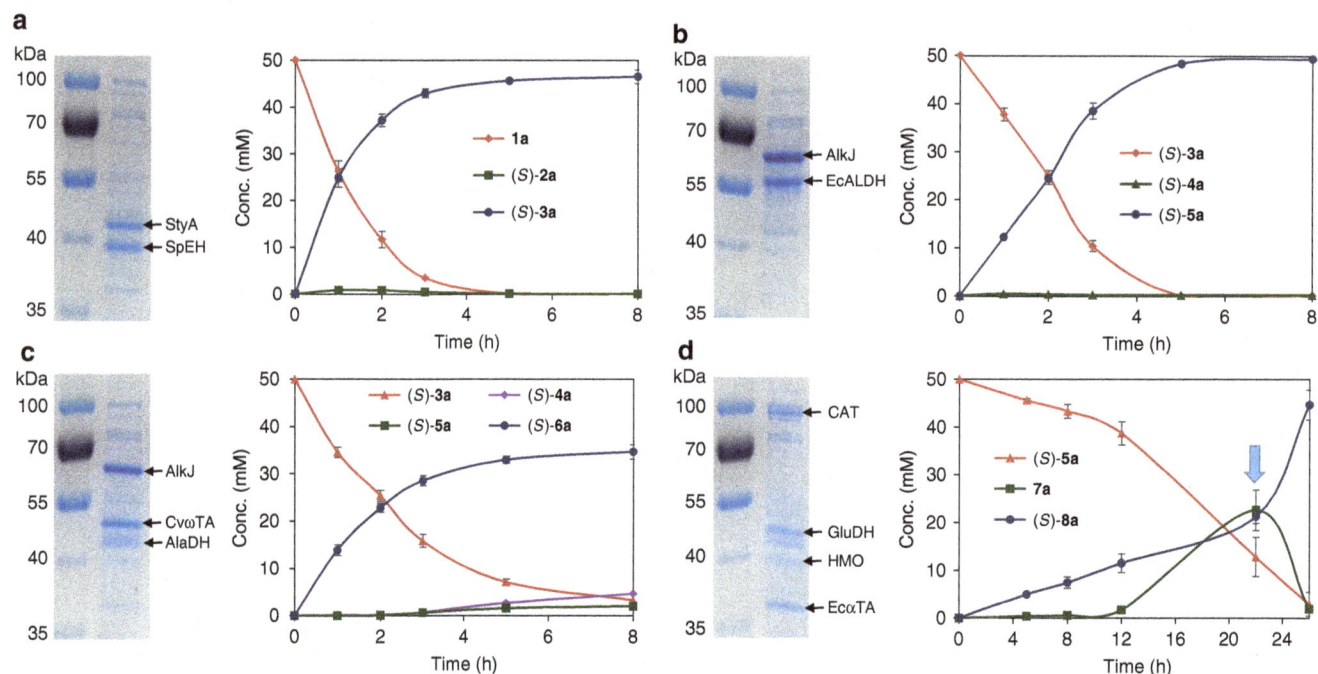

**Figure 3 | SDS–PAGE and biotransformation time course of *E. coli* strains containing individual enzyme modules.** (a) *E. coli* (R-M1) cells containing enzyme module 1 (SMO and SpEH); and biotransformation of styrene **1a** to (*S*)-1-phenyl-1,2-ethanediol **3a**. (b) *E. coli* (R-M2) cells containing enzyme module 2 (AlkJ and EcALDH); and biotransformation of (*S*)-1-phenyl-1,2-ethanediol **3a** to (*S*)-mandelic acid **5a**. (c) *E. coli* (R-M3) cells containing enzyme module 3 (AlkJ, CvωTA and AlaDH); and biotransformation of (*S*)-1-phenyl-1,2-ethanediol **3a** to (*S*)-phenylethanolamine **6a**. (d) *E. coli* (R-M4) cells containing enzyme module 4 (HMO, EcαTA, GluDH and CAT); and biotransformation of (*S*)-mandelic acid **5a** to (*S*)-phenylglycine **8a** (blue arrow: adding additional 0.5% glucose at 22 h). All biotransformations were performed in triplicate, and error bars show ± s.d.

to convert **1a** to (*S*)-**5a** (21–83 mM). Among them, three strains gave (*S*)-**5a** in 71–83 mM, and *E. coli* (A-M1_R-M2) is the best one to produce 83 mM (*S*)-**5a** (83% conversion) in 20 h together with 9 mM (*S*)-**3a**. SDS–PAGE analysis of the cell proteins of the 12 strains (Supplementary Fig. 6) revealed that the three good strains exhibited a relatively balanced expression of the four enzymes, whereas several strains with lower productivity expressed much less AlkJ and EcALDH (module 2) than SMO and SpEH (module 1). The whole-cell activities of *E. coli* (A-M1_R-M2) towards **1a**, (*S*)-**2a**, (*S*)-**3a** and (*S*)-**4a** were determined to be 43, 220, 29 and 42 U (g cdw)$^{-1}$, respectively. *E. coli* (A-M1_R-M2) was used to transform 100–150 mM **1a** to (*S*)-**5a**, and the highest product concentration was observed with 120 mM **1a** (Supplementary Fig. 7a). The time course of biotransformation of 120 mM **1a** with 15 g cdw l$^{-1}$ resting cells of *E. coli* (A-M1_R-M2) was shown in Fig. 4b. A total of 94 mM (14.2 g l$^{-1}$) (*S*)-**5a** was produced in 98% e.e. and 78% conversion in 22 h. The unreacted substrate **1a** and intermediate (*S*)-**3a** were found at a relatively low level (9 and 12 mM, respectively).

**Engineering of catalyst to convert alkene to amino alcohol.** Module 1 and module 3 were combined for the asymmetric aminohydroxylation of alkenes (Fig. 2b). Combinatorial assembly of module 1 plasmids (A-M1, C-M1, E-M1 and R-M1) and module 3 plasmids (A-M3, C-M3, E-M3 and R-M3) led to 12 different *E. coli* strains (Table 1), each co-expressing SMO, SpEH, AlkJ, CvωTA and AlaDH. Biotransformation of 50 mM **1a** was examined with resting cells of each *E. coli* strain in a two-liquid-phase system for 10 h (Fig. 4c). All strains produced (*S*)-**6a** (1–28 mM), and three of them produced (*S*)-**6a** in 26–28 mM. The best one, *E. coli* (A-M1_E-M3), gave 28 mM (*S*)-**6a** (56% conversion) with the accumulation of (*S*)-**3a** (2 mM), (*S*)-**4a**

(2 mM) and (*S*)-**5a** (5 mM). The reaction buffer and temperature were then optimized to improve the final product yield (Supplementary Fig. 8). *E. coli* (A-M1_E-M3) was chosen as the catalyst for this type of biotransformations. The specific activities of *E. coli* (A-M1_E-M3) towards **1a**, (*S*)-**2a**, (*S*)-**3a** and (*S*)-**4a** were 45, 280, 39 and 11 U (g cdw)$^{-1}$, respectively. The biotransformation was examined with styrene **1a** at 50–80 mM, and 60 mM substrate was found to give the highest concentration of (*S*)-**6a** (Supplementary Fig. 7b). Figure 4d depicted the time curve of the reaction of 60 mM **1a** with 15 g cdw l$^{-1}$ resting cells: 42 mM (5.8 g l$^{-1}$) (*S*)-**6a** was produced in 98% e.e. and 70% conversion in 12 h. Unreacted substrate **1a**, intermediates (*S*)-**3a** and (*S*)-**4a**, and by-product (*S*)-**5a** remained at low concentrations (0.2, 2, 0.1 and 4 mM, respectively). The cascade biocatalysis did not produce phenylglycinol, suggesting the excellent regioselectivity of the aminohydroxylation.

**Engineering of catalyst to convert alkene to α-amino acid.** Modules 1, 2 and 4 were assembled together as the catalyst (Fig. 2c). Instead of combinatorial assembly of the basic modules, module 4 on four different plasmids (A-M4, C-M4, E-M4 and R-M4) was transformed into the existing best four recombinant *E. coli* strains containing module 1 and 2, *E. coli* (A-M1_E-M2), *E. coli* (A-M1_R-M2), *E. coli* (C-M1_E-M2) and *E. coli* (R-M1_E-M2). This generated eight different *E. coli* strains, each containing module 1, 2 and 4 on different plasmids (Table 1). The eight strains were individually examined for biotransformation of 50 mM styrene **1a** to (*S*)-phenylglycine **8a** (Fig. 4e). All strains produced (*S*)-**8a** (15–40 mM), and five of them produced 37–40 mM (*S*)-**8a**. *E. coli* (A-M1_R-M2_C-M4) showed the highest productivity, generating 40 mM (*S*)-**8a** (80% conversion) in 24 h together with (*S*)-**3a**, (*S*)-**5a** and **7a** (1 mM each). This

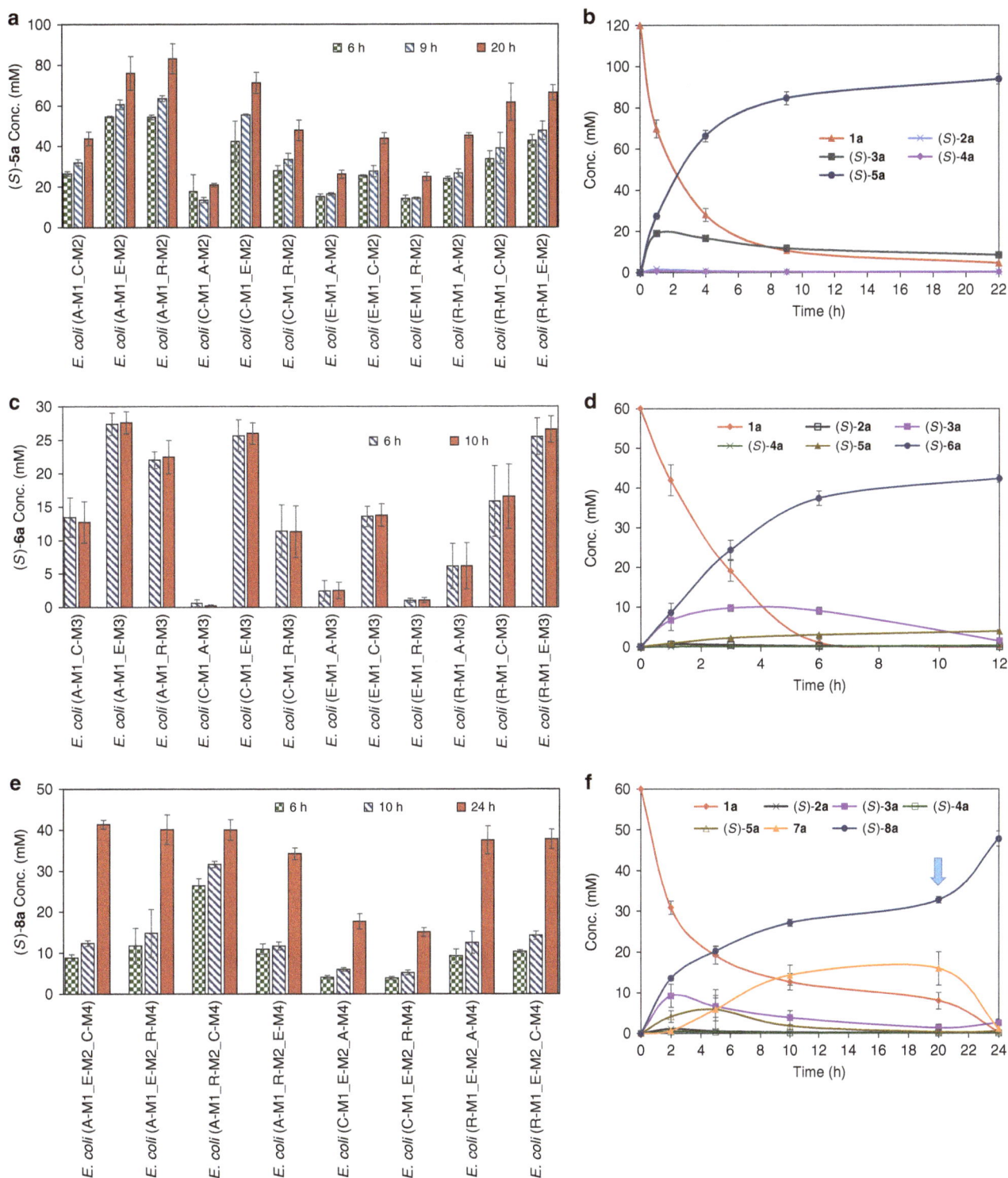

**Figure 4 | Regio- and enantioselective multiple oxy- and amino-functionalizations of styrene 1a with *E. coli* strains containing multiple enzyme modules.** (**a**) Product concentration of biotransformation of 100 mM **1a** to (*S*)-**5a** with twelve *E. coli* strains (10 g cdw l$^{-1}$), each containing both enzyme module 1 and 2, respectively. (**b**) Time course of biotransformation of 120 mM **1a** to (*S*)-**5a** with *E. coli* (A-M1_R-M2) cells (15 g cdw l$^{-1}$) in a two-liquid-phase system (KP buffer containing 0.25% glucose and *n*-hexadecane; 1:1) at 30 °C. (**c**) Product concentration of biotransformation of 50 mM **1a** to (*S*)-**6a** with twelve *E. coli* strains (10 g cdw l$^{-1}$), each containing both enzyme module 1 and 3, respectively. (**d**) Time course of biotransformation of 60 mM **1a** to (*S*)-**6a** with *E. coli* (A-M1_E-M3) cells (15 g cdw l$^{-1}$) in a two-liquid-phase system (NaP buffer containing 1% glucose and 200 mM NH$_3$/NH$_4$Cl and *n*-hexadecane; 1:1) at 25 °C. (**e**) Product concentration of biotransformation of 50 mM **1a** to (*S*)-**8a** with eight *E. coli* strains (10 g cdw l$^{-1}$), each containing enzyme module 1, 2 and 4, respectively. (**f**) Time course of biotransformation of 60 mM **1a** to (*S*)-**8a** with *E. coli* (A-M1_R-M2_C-M4) cells (15 g cdw l$^{-1}$) in a two-liquid-phase system (KP buffer containing 0.5% glucose and 100 mM NH$_3$/NH$_4$Cl and *n*-hexadecane; 1:1) at 30 °C (arrow: adding additional 0.5% glucose and 100 mM NH$_3$/NH$_4$Cl at 20 h). All biotransformations were performed in triplicate, and error bars show ± s.d.

strain was chosen for this type of biotransformations. The specific activities of *E. coli* (A-M1_R-M2_C-M4) towards **1a**, (*S*)-**2a**, (*S*)-**3a**, (*S*)-**4a**, (*S*)-**5a** and **7a**, were determined to be 20, 75, 11, 16, 10 and 16 U (g cdw)$^{-1}$, respectively. Biotransformations of 50–80 mM **1a** to (*S*)-**8a** were examined, and 60 mM **1a** was found to give the highest final product concentration (Supplementary Fig. 7c). As shown in Fig. 4f, biotransformation of 60 mM **1a** with 15 g cdw l$^{-1}$ resting cells gave 33 mM (*S*)-**8a** at 20 h, together with 8 mM **1a** and 16 mM **7a**. By the additional feeding of 0.5% glucose and 100 mM NH$_3$/NH$_4$Cl at 20 h, enantiopure (*S*)-**8a** was produced in 48 mM (7.3 g l$^{-1}$) and 80% conversion at 24 h, with intermediates (*S*)-**3a**, (*S*)-**5a** and **7a** at low concentrations (3, 0.5 and 1 mM, respectively).

**Bioconversion of alkenes *1a–k* to (*S*)-α-hydroxy acids *5a–k*.** To explore the substrate scope and synthetic potential of the asymmetric functionalization of alkenes to α-hydroxy acids, styrenes **1a–k** (20 mM) were biotransformed with resting cells of *E. coli* (A-M1_R-M2) (10 g cdw l$^{-1}$) for 12 h in a two-liquid-phase system (KP buffer and *n*-hexadecane; 1:1) (Table 2). Five (*S*)-α-hydroxy acids (**5a–d** and **5k**) were produced in 90–99% conversion, and six (*S*)-α-hydroxy acids (**5e–j**) were produced in 69–86% conversion. The (*S*)-configurations of **5a–k** were established by comparing the bioproducts with the commercially available enantiopure standards (**5a**, **5d–f** and **5j**) or derived from the previously established (*S*)-configurations of the diol intermediates (**3b**, **3c**, **3g–i** and **3k**)[39]. Ten chiral α-hydroxy acids

---

**Table 2 | Regio- and enantioselective functionalization of terminal alkenes 1a–k to α-hydroxy acids 5a–k with *E. coli* (A-M1_R-M2) via cascade biocatalysis.**

R—⟨benzene⟩—CH=CH₂  →  *E. coli* (A-M1_R-M2), O₂, Glucose, *n*-C₁₆H₃₄/buffer, 30 °C, 12 h  →  R—⟨benzene⟩—CH(OH)—CO₂H
**1a–k**                                                                  (*S*)-**5a–k**

| Substrate* | R group | Product | Conversion (%)† | e.e. (%)‡ |
|---|---|---|---|---|
| **1a** | H | (*S*)-**5a** | 95 | 98 |
| **1b** | *o*-F | (*S*)-**5b** | 90 | 98 |
| **1c** | *m*-F | (*S*)-**5c** | 99 | 99 |
| **1d** | *p*-F | (*S*)-**5d** | 94 | >99 |
| **1e** | *m*-Cl | (*S*)-**5e** | 83 | >99 |
| **1f** | *p*-Cl | (*S*)-**5f** | 83 | 99 |
| **1g** | *m*-Br | (*S*)-**5g** | 71 | 99 |
| **1h** | *p*-Br | (*S*)-**5h** | 69 | 99 |
| **1i** | *m*-Me | (*S*)-**5i** | 86 | 99 |
| **1j** | *p*-Me | (*S*)-**5j** | 78 | 96 |
| **1k** | *m*-OMe | (*S*)-**5k** | 97 | 98 |

*Reactions were conducted with **1a–k** (20 mM in organic phase) and resting cells of *E. coli* (A-M1_R-M2) (10 g cdw l$^{-1}$) in KP buffer (200 mM, pH 8.0, 0.5% glucose) and *n*-hexadecane (1:1) at 30 °C for 12 h.
†Conversion was determined by reversed phase HPLC analysis of the final product **5a–k**. The values are averages of two experiments
‡Enantiomeric excess (e.e.) was determined by chiral HPLC analysis. The values are averages of two experiments.

---

**Table 3 | Regio- and enantioselective functionalization of terminal alkenes 1a–k to 1,2-amino alcohols 6a–k with *E. coli* (A-M1_E-M3) via cascade biocatalysis.**

R—⟨benzene⟩—CH=CH₂  →  *E. coli* (A-M1_E-M3), O₂, NH₃, Glucose, *n*-C₁₆H₃₄/buffer, 25 °C, 24 h  →  R—⟨benzene⟩—CH(OH)—CH₂—NH₂
**1a–k**                                                                  (*S*)-**6a–k**

| Substrate* | R group | Product | Conversion (%)† | e.e. (%)‡ |
|---|---|---|---|---|
| **1a** | H | (*S*)-**6a** | 86 | 98 |
| **1b** | *o*-F | (*S*)-**6b** | 65 | >99 |
| **1c** | *m*-F | (*S*)-**6c** | 71 | 97 |
| **1d** | *p*-F | (*S*)-**6d** | 78 | 91 |
| **1e** | *m*-Cl | (*S*)-**6e** | 20 | 99 |
| **1f** | *p*-Cl | (*S*)-**6f** | 36 | >99 |
| **1g** | *m*-Br | (*S*)-**6g** | 16 | 99 |
| **1h** | *p*-Br | (*S*)-**6h** | 26 | 96 |
| **1i** | *m*-Me | (*S*)-**6i** | 81 | >99 |
| **1j** | *p*-Me | (*S*)-**6j** | 69 | 96 |
| **1k** | *m*-OMe | (*S*)-**6k** | 81 | 98 |

*Reactions were conducted with **1a–k** (20 mM in organic phase) and resting cells of *E. coli* (A-M1_E-M3) (10 g cdw l$^{-1}$) in NaP buffer (200 mM, pH 8.0, 1% glucose, 200 mM NH$_3$/NH$_4$Cl) and *n*-hexadecane (1:1) at 25 °C for 24 h (additional 0.5% glucose and 100 mM NH$_3$/NH$_4$Cl were added at 12 h).
†Conversion was determined by reversed phase HPLC analysis of the final product **6a–k**. The values are averages of two experiments.
‡Enantiomeric excess (e.e.) was determined by chiral HPLC analysis. The values are averages of two experiments.

(**5a–i** and **5k**) were produced in 98–99% e.e., and **5j** was obtained in 96% e.e.. The high product e.e. values were generated by the highly enantioselective epoxidation of **1a–k** with SMO, regioselective hydrolysis of (S)-**2a–k** with SpEH, and S-selective oxidation of **3a–k** with AlkJ.

**Bioconversion of alkenes _1a–k_ to (S)-amino alcohols _6a–k_.** The substrate scope and synthetic potential of the aminohydroxylation of alkenes were explored with _E. coli_ (A-M1_E-M3) $(10\,\mathrm{g\,cdw\,L^{-1}})$ to transform styrenes **1a–k** (20 mM) in a two-liquid-phase system (NaP buffer and _n_-hexadecane; 1:1) (Table 3). Good conversion (65–86%) was achieved for (S)-amino alcohols **6a–d** and **6i–k** in 24 h, while (S)-**6e–h** were produced in relatively low conversion (16–36%). (S)-configuration of **6a** was established by comparing the bioproduct with commercially available enantiopure standard, while (S)-configurations of **6b–k** were derived from the previously established (S)-configurations of diol intermediates **3b–k**[39]. Seven (S)-amino alcohols (**6a–b**, **6e–g**, **6i** and **6k**) were produced in 98–99% e.e., three (S)-amino alcohols (**6c**, **6h** and **6j**) were formed in 96–97% e.e. and (S)-**6d** was obtained in 91% e.e.. The high e.e. values were mainly generated by the high enantioselectivity of biotransformation with enzyme module 1.

**Bioconversion of alkenes _1a–k_ to (S)-α-amino acids _8a–k_.** The substrate scope and synthetic potential of the asymmetric functionalization of alkenes to (S)-α-amino acids were examined with resting cells of _E. coli_ (A-M1_R-M2_C-M4) $(10\,\mathrm{g\,cdw\,l^{-1}})$ to convert styrenes **1a–k** (20 or 5 mM) in a two-liquid-phase system (KP buffer and _n_-hexadecane; 1:1) for 24 h (Table 4). (S)-**8a** and (S)-**8c** were produced in excellent conversion of 88 and 91%, respectively. (S)-**8b** and (S)-**8h** were produced in moderate conversion of 55 and 28%, respectively. Seven other (S)-α-amino acids (**8d–g** and **8i–k**) were produced in good conversion of 60–76%. (S)-configuration of **8a** was established by comparing the bioproduct with the commercially available enantiopure standard, while (S)-configurations of **8b–k** were deduced from the L-selectivity of EcαTA on substrates **7b–k**. Remarkably, all produced (S)-α-amino acids **8a–k** were in enantiopure forms (e.e. ≥99%). These very high e.e. values were generated by the

highly enantioselective transamination of prochiral α-keto acids **7a–k** with EcαTA.

**Preparative biotransformations.** The synthetic application of the three new types of asymmetric functionalizations of alkenes was demonstrated by the preparative biotransformations to produce two (S)-α-hydroxy acids, two (S)-amino alcohols and two (S)-α-amino acids. The ratio of aqueous buffer and _n_-hexadecane of the two-phase system was examined for the biotransformations, and 1–5:1 ratios gave similar high conversion for all three types of reactions (Supplementary Fig. 9). Thus, the preparative biotransformations were performed at 5:1 ratio of aqueous buffer: _n_-hexadecane to reduce the use of organic solvent. Biotransformations of alkenes **1a** (100 mM) and **1d** (50 mM) with resting cells of _E. coli_ (A-M1_R-M2) $(20\,\mathrm{g\,cdw\,l^{-1}})$ gave (S)-α-hydroxy acids **5a** and **5d** in 83 and 80% conversion at 24 h, respectively. Simple extraction and crystallization gave (S)-**5a** (98% e.e.) and (S)-**5d** (98% e.e.) in 72 and 61% yield, respectively. Similarly, biotransformations of alkenes **1a** (50 mM) and **1i** (25 mM) with resting cells of _E. coli_ (A-M1_E-M3) $(20\,\mathrm{g\,cdw\,l^{-1}})$ afforded (S)-amino alcohols **6a** and **6i** in 71 and 63% conversion at 24 h, respectively. Extraction and flash chromatography afforded (S)-**6a** (98% e.e.) and (S)-**6i** (98% e.e.) in 62% and 55% yield, respectively. Finally, biotransformations of alkenes **1a** (50 mM) and **1d** (25 mM) were carried out with resting cells of _E. coli_ (A-M1_R-M2_C-M4) $(20\,\mathrm{g\,cdw\,l^{-1}})$ for 24 h to give (S)-α-amino acids **8a** and **8d** in 81 and 79% conversion, respectively. Simple extraction and evaporation afforded enantiopure (S)-**8a** and (S)-**8d** in 70 and 59% yield, respectively.

**Discussion**
Cascade biocatalysis is of great importance in green synthesis of chemicals, since it could avoid the waste-generating, time-consuming, and costly separation and purification of intermediates in traditional multi-step process. However, efficient non-natural cascades with more than four enzymatic reactions are rare, and their development is challenging. The modularization concept reported here provides a useful tool for the engineering of complex cascade biocatalysis, which is different

**Table 4 | Regio- and enantioselective functionalization of terminal alkenes 1a–k to α-amino acids 8a–k with _E. coli_ (A-M1_R-M2_C-M4) via cascade biocatalysis.**

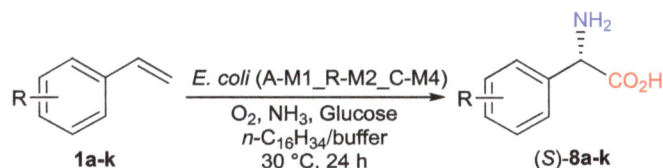

| Substrate* | R group | Product | Conversion (%)[†] | e.e. (%)[‡] |
|---|---|---|---|---|
| **1a** | H | (S)-**8a** | 88 | >99 |
| **1b** | _o_-F | (S)-**8b** | 55 | 99 |
| **1c** | _m_-F | (S)-**8c** | 91 | >99 |
| **1d** | _p_-F | (S)-**8d** | 76 | >99 |
| **1e** | _m_-Cl | (S)-**8e** | 73 | >99 |
| **1f** | _p_-Cl | (S)-**8f** | 63 | >99 |
| **1g** | _m_-Br | (S)-**8g** | 60 | >99 |
| **1h** | _p_-Br | (S)-**8h** | 28 | >99 |
| **1i** | _m_-Me | (S)-**8i** | 73 | >99 |
| **1j** | _p_-Me | (S)-**8j** | 63 | >99 |
| **1k** | _m_-OMe | (S)-**8k** | 68 | >99 |

*Reactions were conducted with **1a–k** (20 mM **1a–d** and **1i–k**, 5 mM **1e–h** in organic phase) and resting cells of _E. coli_ (A-M1_R-M2_C-M4) $(10\,\mathrm{g\,cdw\,l^{-1}})$ in KP buffer (200 mM, pH 8.0, 0.5% glucose, 100 mM NH₃/NH₄Cl) and _n_-hexadecane (1:1) at 30 °C for 24 h (additional 0.5% glucose and 100 mM NH₃/NH₄Cl were added at 20 h).
†Conversion was determined by reversed phase HPLC analysis of the final product **8a–k**. The values are averages of two experiments.
‡Enantiomeric excess (e.e.) was determined by chiral HPLC analysis. The values are averages of two experiments.

from the modularization in synthetic biology to build complex genetic circuits[55] and in metabolic engineering to optimize production of certain metabolites[56,57]. In the modular cascade biocatalysis, the appropriate basic modules are designed and engineered to ideally give full conversion with no accumulation of the intermediates. In these aspects, modules 1, 2 and 4 are excellent: they produced 90–98% of the final products in high e.e. from the starting materials, with no accumulation of the intermediates (Fig. 3a,b,d). These results were achieved by using enzymes having relatively high activities and using the second enzyme with much higher activity than the first enzyme (SpEH versus SMO; EcALDH versus AlkJ; EcαTA versus HMO) in the module (Supplementary Table 3). In module 3, the conversion of the starting material to the final product reached 70%. Overall, 10% of the intermediate remained, which is possibly due to the relatively low activity of CvωTA for the second reaction, the transamination (Fig. 3c). Further improvement of module 3 might be achieved by using other enzymes with higher amination activity.

Efficient cascade catalysis systems consisting of multiple basic modules were developed by the combinatorial assembly of the basic modules on different plasmids to adjust the expression level of the enzymes. This method led to the development of E. coli (A-M1_R-M2), E. coli (A-M1_E-M3) and E. coli (A-M1_R-M2_C-M4) as powerful catalysts for the asymmetric functionalizations of alkenes to (S)-α-hydroxy acids, (S)-1,2-amino alcohols and (S)-α-amino acids, respectively. These catalysts enabled the biotransformation of 120 mM 1a to (S)-5a in 78% conversion, 60 mM 1a to (S)-6a in 70% conversion and 60 mM 1a to (S)-8a in 80% conversion, respectively, with low accumulation of the intermediates (Fig. 4b,d,f). Since some oxidoreductases were used in the cascade biocatalysis, cofactor recycling has to be considered for efficient biotransformations. While SMO[47] and AlaDH[52] are nicotinamide adenine dinucleotide (NADH)-dependent, GluDH is nicotinamide adenine dinucleotide phosphate (NADPH)-dependent and EcALDH is NAD$^+$-dependent[50]. On the other hand, SpEH[48], HMO[54], EcαTA, CAT and CvωTA[51] are independent of nicotinamide cofactors, and AlkJ is a non-canonical ADH coupling to the bacterial respiratory chain instead of nicotinamide cofactors[49]. Therefore, there is no net consumption of the nicotinamide cofactor in the functionalization of styrenes to (S)-hydroxy acids (Fig. 2a). However, 2 moles NADH are needed for producing 1 mole (S)-amino alcohol from styrene (Fig. 2b), and 1 mole NADPH is required for producing 1 mole (S)-amino acid from styrene (Fig. 2c). For these two types of biotransformations with whole-cell biocatalysts, the regeneration of NAD(P)H was achieved via cell metabolism of glucose. This was clearly demonstrated in Fig. 4f: feeding of additional glucose at 20 h significantly improved the conversion of the final product (S)-8a. Future improvement of these reactions might be achieved by co-expressing a NAD(P)H-regenerating enzyme in the recombinant biocatalysts.

An important parameter in catalysis is the total turnover number (TTN). We calculated the TTN of individual enzymes in the biotransformations based on the amount of the enzymes inside three whole-cell biocatalysts estimated by separation and analysis of the proteins with SDS–PAGE and densitometer (Supplementary Fig. 10; Supplementary Table 4). SMO(StyA), SpEH, AlkJ and EcALDH were expressed in E. coli (A-M1_R-M2) at 10, 6.0, 23 and 14 mg protein (g cdw)$^{-1}$, respectively, and gave a TTN of 32,000, 51,000, 18,000 and 25,000, respectively, in the biotransformation of 1a to (S)-5a (Fig. 4b). SMO(StyA), SpEH, AlkJ and CvωTA were expressed in E. coli (A-M1_E-M3) at 13, 6.2, 15 and 7 mg protein (g cdw)$^{-1}$, respectively, and afforded a TTN of 13,000, 24,000, 13,000 and 21,000, respectively, in the biotransformation of 1a to (S)-6a (Fig. 4d). SMO(StyA), SpEH, AlkJ, EcALDH and EcαTA

were expressed in E. coli (A-M1_R-M2_C-M4) at 2.8, 1.6, 7.8, 4.6 and 4.6 mg protein (g cdw)$^{-1}$, respectively, and gave a TTN of 58,000, 95,000, 26,000, 38,000 and 24,000, respectively, in the biotransformation of 1a to (S)-8a (Fig. 4f). The good TTN values of these key enzymes indicate the high efficiency of their catalysis in the cascade biotransformations.

On the basis of the specific activities of the three whole-cell biocatalysts towards substrate 1a and the corresponding intermediates (Supplementary Table 5), the following bottlenecks could be deduced: alkJ-catalysed oxidation of (S)-3a in the transformation of 1a to (S)-5a; CvωTA-catalysed transamination of (S)-4a in the conversion of 1a to (S)-6a; and alkJ-catalysed oxidation of (S)-3a and HMO-catalysed oxidation of (S)-5a in the transformation of 1a to (S)-8a. These bottlenecks were also confirmed by some accumulation of (S)-3a (Fig. 4b), (S)-5a (Fig. 4d, possibly due to the oxidation of (S)-4a to (S)-5a by other enzymes inside the E.coli cells), and both (S)-3a and (S)-5a at early reaction stage (Fig. 4f) in the corresponding biotransformations. On the basis of the determined specific activities of AlkJ, EcALDH, CvωTA and EcαTA (0.19, 1.1, 0.22 and 14 U (mg protein)$^{-1}$, respectively) (Supplementary Table 3; Supplementary Fig. 11) and the reported specific activities of SMO[58], SpEH[48] and HMO[59] (2.1, 16 and 1.8 U (mg protein)$^{-1}$, respectively), AlkJ and CvωTA are not very active. Since their expression in the whole-cell catalysts is not low, these two enzymes might be replaced with more active enzymes selected from natural sources or enzyme engineering[60–62] to improve the efficiency of the cascade catalysis. On the other hand, HMO has a relatively high activity, but its expression level is low (too low to be estimated). Thus, future improvement might focus on the enhancement of the expression of HMO or replacement of HMO by other enzymes with higher activity and easier expression. In general, the expression of all involved enzymes might be fine-tuned to a high and balanced level by altering other genetic elements, such as promoters and ribosome-binding sites.

The engineered whole-cell biocatalysts accept a group of styrene derivatives as substrates for the three types of asymmetric alkene functionalizations. E. coli (A-M1_R-M2) catalysed the biotransformations of eleven alkenes (1a–k) to produce the corresponding (S)-α-hydroxy acids (5a–k) in good conversion (90–99% for five products and 69–86% for six products) and high e.e. (98–99% for ten products and 96% for one product). Biotransformation of the same eleven alkenes (1a–k) with E. coli (A-M1_E-M3) gave the corresponding (S)-1,2-amino alcohols (6a–k) in high e.e. (96–99% for ten products and 91% for one product) with good conversion (65–86%) for seven products (6a–d and 6i–k) and low conversion (16–36%) for four products (6e–h). E. coli (A-M1_R-M2_C-M4) catalysed the reaction of alkenes 1a–k to produce (S)-α-amino acids 8a–k in enantiopure form (all ≥99%e.e) with good conversion (88–91% for two products and 55–76% for eight products) except (S)-8h (28%). The regio- and enantioselectivity of three types of alkene functionalizations are outstanding. For the low-conversion biotransformations of 1e–h to (S)-6e–h and of 1h to (S)-8h, 42–63% of unreacted alkenes remained in the reaction mixture. Thus, the SMO-catalysed epoxidation is the main bottleneck in these reactions, which was possibly caused by (a) the relatively low epoxidation activity of SMO towards those styrenes containing a bulky or electron-withdrawing group, (b) the relatively low expression of SMO in E. coli (A-M1_E-M3) and E. coli (A-M1_R-M2_C-M4) and (c) the inefficient cofactor supply and regeneration in the biotransformations. The improvement of these bioconversions might be achieved by enhancing the expression of SMO, co-expressing a NADH-regenerating enzyme, and/or using an engineered SMO with higher activity towards the alkene substrates.

The synthetic application of the developed whole-cell biocatalysts and three new types of asymmetric functionalizations of alkenes were clearly demonstrated in the preparation of two (S)-hydroxy acids, two (S)-amino alcohols and two (S)-amino acids. Biotransformations, workup and purification are straightforward, affording 208–1,095 mg (55–72% yield) of (S)-**5a**, (S)-**5d**, (S)-**6a**, (S)-**6i**, (S)-**8a** and (S)-**8d** in 98–99% e.e.

The biocatalytic syntheses utilize non-toxic and biodegradable catalysts, consume inexpensive and green stoichiometric reagents ($O_2$, $NH_3$ or glucose) and operate under mild reaction conditions (25–30 °C, atmosphere pressure, etc), thus being greener than many chemical synthetic methods.

From organic synthesis perspective, the one-pot asymmetric functionalization of alkenes to give chiral α-hydroxy acids is unique and advantageous over the traditional multi-step synthesis involving many isolation and purification steps. The biocatalytic synthesis is greener and safer than the cyanide-based synthesis using cyanohydrin[43]. It gives high product yield and e.e. from the low-cost alkenes, being more attractive than the reported kinetic resolution (maximum yield: 50%) and asymmetric reduction of α-keto acids (substrates are not cheap)[63]. The one-pot asymmetric functionalization of alkenes to give chiral α-amino acids has also no chemical counterpart and is greener and safer than the cyanide-based Strecker synthesis of chiral α-amino acids[44]. It enables high product yield and e.e. from inexpensive alkenes, being advantageous over the kinetic resolution (maximum yield: 50%) and asymmetric hydrogenation (substrates are not cheap) appoaches[42]. The one-pot conversion of alkenes to chiral 1,2-amino alcohols is a new type of biotransformation and offers an alternative or even better method in some cases to the existing chemical asymmetric aminohydroxylation (oxyamination)[3,45]. The biocatalytic aminohydroxylation produces primary amines of the amino alcohols by utilizing ammonia as the nitrogen source, while chemical aminohydroxylation has difficulty in using ammonia. It could also provide much better regio- and enantioselectivity than the chemical methods. As an example for comparison, biotransformation of styrene **1a** afforded (S)-phenylethanolamine **6a** in 98% e.e. with 100% regioselectivity, while Sharpless asymmetric aminohydroxylation of styrene **1a** gave (R)-phenylethanolamine (as 4-toluenesulfonyl derivative) in 55% e.e. together with (S)-phenylglycinol (as 4-toluenesulfonyl derivative)[64].

From biochemical point of view, our synthetic routes from terminal alkenes to chiral α-hydroxy acids, 1,2-amino alcohols and α-amino acids are three novel non-natural pathways containing four to eight reactions, which are unambiguously distinguished from the natural aromatic or aliphatic alkene degradation pathways reported so far[47,65]. Nevertheless, the three synthetic pathways were successfully engineered in microbial cells by modular approach and catalysed well the desired non-natural reactions. In comparison with *in vitro* cascade biocatalysis, the whole-cell approach with a single recombinant strain enables the easy production of the multiple enzymes and the cost-effective biotransformation. This approach opens new possibility of engineering cells for one-pot multi-step biotransformations to manufacture different types of chemicals in a green, selective and cost-effective manner.

In summary, we successfully developed three new types of one-pot asymmetric oxy- and amino-functionalizations of terminal alkenes by cascade biocatalysis, simple and green syntheses of a group of useful and valuable (S)-α-hydroxy acids, (S)-1,2-amino alcohols and (S)-α-amino acids in high e.e. and high yields, and a modular approach for engineering efficient one-pot cascade biocatalysis containing more than four concurrent enzymatic reactions.

## Methods

**General procedure to engineer recombinant E. coli strains.** *E. coli* T7 expression strain (an *E. coli* B strain derivative) was purchased from New England Biolabs (#C2566I) and used as host strain for all molecular cloning and biocatalysis experiments. The gene module 1 comprising of *styA*, *styB* and *spEH* was constructed previously[39]. *AlkJ* gene was amplified from the OCT megaplasmid extracted from *P. putida* GPo1 as reported[49]. Genes of *padA*, *ilvE*, *gdhA* and *katE* were amplified from the genomic DNA extracted from *E. coli* K12 MG 1655 with genomic DNA Purification Kit (Thermo Scientific). *Ald* gene was amplified from the genomic DNA extracted from *B. subtilis* str.168 with genomic DNA Purification Kit. Codon-optimized *cv_2025* gene was synthesized from Genscript based on the sequence from *C. violaceum* DSM30191 (ref. 51). Codon-optimized *sco3228* gene was synthesized from Genscript based on the sequence from *S. coelicolor* A3(2)[54] (see Supplementary Methods for codon-optimized sequences).

All genetic constructions were carried out by using standard molecular biology techniques with Phusion DNA polymerase, FastDigest restriction enzymes and T4 DNA ligase (all from Thermo Scientific). PCR primers were synthesized from Integrated DNA Technologies (see Supplementary Table 6 for a full list of key primers). Purification of DNA after electrophoresis or enzyme digestion was performed with E.Z.N.A. Gel Extraction Kit (Omega Biotek), and extraction of plasmids was performed with Axyprep Plasmid Miniprep Kit (Axygen). Basic gene modules 1–4 were constructed on a set of compatible plasmids pACYCDuet-1, pCDFDuet-1, pETDuet-1 and pRSFDuet-1 (Novagen) as individual artificial operon under control of a T7 promoter with one ribosome-binding site before every gene (see Supplementary Methods for details and Supplementary Fig. 12 for enzyme expression). Gene modules were transformed into *E. coli* T7 competent cells to obtain the *E. coli* with individual basic modules. Further transformation of other basic genetic module(s) into a constructed *E. coli* strain containing one or two basic modules gave an *E. coli* strain containing two or three basic modules for the desired asymmetric alkene functionalizations (see Supplementary Methods for details and Supplementary Fig. 10 for enzyme expression).

**General procedure to grow E. coli strains.** Recombinant *E. coli* strain was first inoculated in 1 ml LB medium containing appropriate antibiotics ($50 \text{ mg l}^{-1}$ chloramphenicol, $50 \text{ mg l}^{-1}$ streptomycin, $100 \text{ mg l}^{-1}$ ampicillin, $50 \text{ mg l}^{-1}$ kanamycin or a combination of them) at 37 °C for 7–10 h. The culture was then transferred into 25 ml M9 medium containing glucose ($20 \text{ g l}^{-1}$), yeast extract ($6 \text{ g l}^{-1}$) and appropriate antibiotics in a 125 ml tri-baffled flask. The cells were grown at 37 °C and 300 r.p.m. for about 2 h to reach an $OD_{600}$ of 0.6, followed by the addition of IPTG to 0.5 mM to induce the enzyme expression. The cells were grown for 12–13 h at 22 °C to reach late exponential phase, and they were collected by centrifugation (3,500 g, 10 min). The cell pellets were resuspended in an appropriate buffer to the desired density as resting cells for biotransformation.

**General procedure to convert 1a–k to (S)-5a–k.** Overall, 2 ml suspension ($10 \text{ g cdw l}^{-1}$) of the freshly prepared *E. coli* (A-M1_R-M2) cells in KP buffer (200 mM, pH 8.0) containing glucose (0.5%, w/v) were mixed with 2 ml *n*-hexadecane containing one of alkene substrates **1a–k** (20 mM). The mixture was shaken at 300 r.p.m. and 30 °C for 12 h, and 150 μl aliquots of each phase were taken out at different time points for following the reaction. For organic phase, 100 μl *n*-hexadecane were separated after centrifugation (13,000 g, 2 min), diluted with 900 μl *n*-hexane (containing 2 mM benzyl alcohol as internal standard) and analysed by normal phase HPLC for quantifying alkenes **1a–k** and possible epoxides **2a–k**. For aqueous phase, 100 μl supernatant were separated after centrifugation (13,000 g, 2 min), diluted with 400 μl TFA solution (0.5%) and 500 μl acetonitrile (containing 2 mM benzyl alcohol as internal standard) and then analysed by reverse phase HPLC for quantifying hydrophilic products **3a–k**, **4a–k** and **5a–k**. To determine the e.e. of **5a–k**, the remaining aqueous phase at the end of reaction was separated after centrifugation (13,000 g, 2 min), acidified with TFA and saturated with NaCl, followed by extraction with ethyl acetate and dry over $Na_2SO_4$. After evaporation of ethyl acetate, the residue was dissolved in solvent (*n*-hexane: IPA = 9:1) for chiral HPLC analysis (see Supplementary Table 7; Supplementary Figs 19–29; Supplementary Methods for analytic methods).

**General procedure to convert 1a–k to (S)-6a–k.** Overall, 2 ml suspension ($10 \text{ g cdw l}^{-1}$) of freshly prepared *E. coli* (A-M1_E-M3) cells in NaP (sodium phosphate) buffer (200 mM, pH 8.0) containing glucose (1%, w/v) and $NH_3/NH_4Cl$ (200 mM, $NH_3$:$NH_4Cl$ = 1:10) were mixed with 2 ml *n*-hexadecane containing one of the alkene substrates **1a–k** (20 mM). The mixture was shaken at 300 r.p.m. and 25 °C for 24 h. At 12 h, additional glucose (0.5%, w/v) and $NH_3/NH_4Cl$ (100 mM) were added. Samples were taken at different time points and prepared for analysis according to the same procedure described above in the conversion of **1a–k** to **5a–k**. Alkenes **1a–k** and possible epoxides **2a–k** were analysed by normal phase HPLC, and hydrophilic products **3a–k**, **4a–k** and **6a–k** were determined by reverse phase HPLC. To determine the e.e. of **6a–k**, the remaining aqueous phase at the end of reaction was separated after centrifugation (13,000 g, 2 min), acidified with TFA, and 100 μl sample were separated and diluted with 900 μl TFA solution (0.1%) for chiral HPLC analysis (see Supplementary Table 7; Supplementary Figs 30–40; Supplementary Methods for analytic methods).

**General procedure to convert 1a–k to (S)-8a–k.** Overall, 2 ml suspension (10 g cdw l⁻¹) of freshly prepared *E. coli* (A-M1_R-M2_C-M4) cells in KP buffer (200 mM, pH 8.0) containing glucose (0.5%, w/v) and NH₃/NH₄Cl (100 mM, NH₃:NH₄Cl = 1:10) were mixed with 2 ml *n*-hexadecane containing one of alkene substrates **1a–k** (20 or 5 mM). The reaction mixture was shaken at 300 r.p.m. and 30 °C for 24 h. At 20 h, additional glucose (0.5%, w/v) and NH₃/NH₄Cl (100 mM) were added. 300 µl aliquots of the mixture (150 µl of each phase) were taken out at different time points for following the reaction. 150 µl HCl solution (0.8M) were mixed with the 300 µl sample, followed by centrifugation (13,000 g, 2 min) to separate the organic and aqueous phases. The procedures for the preparation of the analytic sample from organic phase and the quantification of alkenes **1a–k** and possible epoxides **2a–k** by normal phase HPLC are the same as the above mentioned ones for the conversion of **1a–k** to **5a–k**. For aqueous phase, 200 µl supernatant were diluted with 300 µl TFA solution (0.1%) and 500 µl acetonitrile (containing 2 mM benzyl alcohol as internal standard), and the samples were analysed by reverse phase HPLC for quantifying hydrophilic products **3a–k**, **4a–k**, **5a–k**, **7a–k** and **8a–k**. To determine the e.e. of **8a–k**, the sample preparation and analysis are the same as those described above for the analysis of the e.e. of **6a–k** (see Supplementary Table 7; Supplementary Figs 41–51; Supplementary Methods for analytic methods).

**Biotransformation of 1a or 1d to prepare (S)-5a or (S)-5d.** Overall, 100 ml suspension of *E. coli* (A-M1_R-M2) cells (20 g cdw l⁻¹) in KP buffer (200 mM, pH 8.0) containing glucose (0.5%, w/v) were mixed with 20 ml *n*-hexadecane containing **1a** (10 mmol, 1,042 mg) or **1d** (5 mmol, 611 mg). The reaction mixture was shaken at 300 r.p.m. and 30 °C, and 100 µl aliquots of the aqueous phase were taken out at different time points for reversed phase HPLC analysis to follow the reaction. After 24 h, the reaction mixture was subjected to centrifugation (4,000 g, 15 min) to remove the cells and organic phase. The aqueous phase was collected, saturated with NaCl, adjusted to pH > 12 with NaOH (10 M) and washed with ethyl acetate two times (2 × 25 ml) to remove trace *n*-hexadecane and other organic impurities. The aqueous phase was adjusted to pH < 2 with HCl (10 M) and extracted with ethyl acetate (3 × 100 ml). The organic phase was collected and dried over Na₂SO₄. After filtration, the organic phase was subjected to evaporation by using a rotary evaporator (Buchi Rotavapor R-215) to remove the solvent. The crude hydroxy acid was purified by crystallization in ethyl acetate through dissolving at 65 °C and slowly cooling down to − 20 °C. The crystals were taken by filtration, and the mother liquor was evaporated and subjected to crystallization again. The collected crystals were combined and dried overnight under vacuum. (S)-2-Hydroxy-2-phenylacetic acid **5a**: white crystal; 1,095 mg; yield: 72%; e.e.: 98%; [α]$_D^{20}$: + 146° (c 1.0, H₂O) {literature[66] [α]$_D^{20}$: + 148.8° (c 0.5, H₂O), 99% e.e.}. ¹H NMR (400 MHz, D₂O): δ = 7.40–7.36 (m, 5H), 5.22 (s, 1H) p.p.m.; ¹³C NMR (100 MHz, D₂O): δ = 176.2, 138.0, 129.1, 129.1, 127.1, 127.1, 72.9 p.p.m. (Supplementary Fig. 13). (S)-2-Hydroxy-2-(4-fluorophenyl)acetic acid **5d**: white crystal; 518 mg; yield: 61%; e.e.: 98%; [α]$_D^{20}$: + 141° (c 1.0, H₂O) {literature[67] [α]$_D^{28}$: + 137.3° (c 0.5, EtOH), 90% e.e.}. ¹H NMR (400 MHz, D₂O): δ = 7.41–7.37 (m, 2H), 7.12–7.08 (t, J = 8.8 Hz, 2H), 5.22 (s, 1H) p.p.m.; ¹³C NMR (100 MHz, D₂O): δ = 176.1, 163.9 and 161.5 (C–F), 134.0, 129.1, 129.0, 115.8, 115.7, 72.2 p.p.m. (Supplementary Fig. 14).

**Biotransformation of 1a or 1i to prepare (S)-6a or (S)-6i.** Overall, 100 ml suspension of *E. coli* (A-M1_E-M3) cells (20 g cdw l⁻¹) in NaP (sodium phosphate) buffer (200 mM, pH 8.0) containing glucose (1%, w/v) and NH₃/NH₄Cl (200 mM, NH₃:NH₄Cl = 1:10) were mixed with 20 ml *n*-hexadecane containing **1a** (5 mmol, 521 mg) or **1i** (2.5 mmol, 295 mg). The reaction mixture was shaken at 300 r.p.m. and 25 °C, and 100 µl aliquots of the aqueous phase were taken out at different time points for reversed phase HPLC analysis to follow the reaction. At 12 h, additional glucose (0.5%, w/v) and NH₃/NH₄Cl (100 mM) were added. After 24 h, the reaction mixture was subjected to centrifugation (4,000 g, 15 min) to remove the cells and organic phase. The aqueous phase was collected, saturated with NaCl, adjusted to pH < 2 with HCl (10 M), and washed with ethyl acetate (2 × 25 ml) to remove trace *n*-hexadecane and other organic impurities. The aqueous phase was adjusted to pH > 12 with NaOH (10 M), followed by extraction with ethyl acetate (3 × 100 ml). The organic phase was separated and dried over Na₂SO₄. After filtration, the organic phase was subjected to evaporation. The crude amino alcohol was purified by flash chromatography on a silica gel column with CH₂Cl₂:MeOH:NH₃(28% aqueous solution) of 100:10:1 as eluent (R$_f$≈0.2–0.3 for (S)-**6a** and (S)-**6i**).The collected fraction containing the product was dried over Na₂SO₄. After filtration, the organic solvent was removed by evaporation, and the product was dried overnight under vacuum. (S)-2-Amino-1-phenylethanol **6a**: white solid; 425 mg; yield: 62%; e.e.: 98%; [α]$_D^{20}$: + 46° (c 1.0, EtOH) {literature[68] [α]$_D^{20}$: + 48.6° (c 2.0, EtOH), 99% e.e.}. ¹H NMR (400 MHz, CDCl₃): δ = 7.37–7.23 (m, 5H), 4.61 (dd, J₁ = 8.0 Hz, J₂ = 4.0 Hz, 1H); 2.93–2.77 (m, 2H), 2.36 (br, 3H) p.p.m.; ¹³C NMR (100 MHz, CDCl₃): δ = 142.5, 128.4, 128.4, 127.5, 125.9, 125.9, 74.3, 49.2 p.p.m. (Supplementary Fig. 15). (S)-2-Amino-1-(*m*-tolyl)ethanol **6i**: light yellow syrup; 208 mg; yield: 55%; e.e.: 98%; [α]$_D^{20}$: + 41° (c 1.0, EtOH). ¹H NMR (400 MHz, CDCl₃): δ = 7.26–7.06 (m, 4H), 4.59 (m, 1H); 2.81 (m, 2H), 2.40 (br, 3H), 2.34 (s, 3H) p.p.m.; ¹³C NMR (100 MHz, CDCl₃): δ = 142.5, 138.1, 128.3, 128.3, 126.6, 122.9, 74.3, 49.1, 21.4 p.p.m. (Supplementary Fig. 16).

**Biotransformation of 1a or 1d to prepare (S)-8a or (S)-8d.** Overall, 100 ml suspension of *E. coli* (A-M1_R-M2_C-M4) cells (20 g cdw l⁻¹) in KP buffer (200 mM, pH 8.0) containing glucose (0.5%, w/v) and NH₃/NH₄Cl (100 mM, NH₃:NH₄Cl = 1:10) were mixed with 20 ml *n*-hexadecane containing **1a** (5 mmol, 521 mg) or **1d** (2.5 mmol, 305 mg). The reaction mixture was shaken at 300 r.p.m. and 30 °C, and 100 µl aliquots of the aqueous phase were taken out at different time points for reversed phase HPLC analysis to follow the reaction. At 12 h, additional glucose (1%, w/v) and NH₃/NH₄Cl (200 mM) were added. After 24 h, the reaction mixture was subjected to centrifugation (4,000 g, 15 min) to remove the cells and organic phase. The collected aqueous phase was filtered to further remove solid impurities, followed by washing with ethyl acetate (2 × 25 ml) to remove trace *n*-hexadecane and other organic impurities. After neutralization to pH = 7 with NaOH (10 M), the aqueous solution was concentrated to about 15 ml by evaporation to precipitate the amino acid. The solid was collected by filtration, washed with cold water and EtOH and dried overnight under vacuum. (S)-2-Amino-2-phenylacetic acid **8a**: white solid; 528 mg; yield: 70%; e.e.: 99%; [α]$_D^{20}$: + 148° (c 1.0, 1 M HCl) {literature[69] [α]$_D^{23}$: + 150° (c 1.0, 1 M HCl), 99% e.e.}. ¹H NMR (400 MHz, D₂O containing 2% H₂SO₄): δ = 7.36–7.30 (m, 5H), 5.04 (s, 1H) p.p.m.; ¹³C NMR (100 MHz, D₂O containing 2% H₂SO₄): δ = 170.7, 131.3, 130.4, 129.7, 129.7, 128.1, 128.1, 56.5 p.p.m. (Supplementary Fig. 17). (S)-2-Amino-2-(4-fluorophenyl)acetic acid **8d**: white solid; 250 mg; yield: 59%; e.e.: 99%; [α]$_D^{20}$: + 138° (c 1.0, 1M HCl) {literature[70] [α]$_D^{20}$: + 141° (c 1.0, 1M HCl), 99% e.e.}. ¹H NMR (400 MHz, D₂O containing 2% H₂SO₄): δ = 7.34–7.31 (m, 2H), 7.08–7.03 (m, 2H), 5.04 (s, 1H) p.p.m.; ¹³C NMR (100 MHz, D₂O containing 2% H₂SO₄): δ = 170.6, 164.6 & 162.2 (C–F), 130.4, 130.3, 127.4, 127.4, 116.7, 116.4, 55.8 p.p.m. (Supplementary Fig. 18).

**Chemicals.** Chemicals used in this study are listed in Supplementary Table 8.

## References

1. Katsuki, T. & Sharpless, K. B. The first practical method for asymmetric epoxidation. *J. Am. Chem. Soc.* **102**, 5974–5976 (1980).
2. Kolb, H. C., VanNieuwenhze, M. S. & Sharpless, K. B. Catalytic asymmetric dihydroxylation. *Chem. Rev.* **94**, 2483–2547 (1994).
3. Li, G., Chang, H. T. & Sharpless, K. B. Catalytic asymmetric aminohydroxylation (AA) of olefins. *Angew. Chem. Int. Ed.* **35**, 451–454 (1996).
4. Zhang, W., Loebach, J. L., Wilson, S. R. & Jacobsen, E. N. Enantioselective epoxidation of unfunctionalized olefins catalysed by salen manganese complexes. *J. Am. Chem. Soc.* **112**, 2801–2803 (1990).
5. McDonald, R. I., Liu, G. & Stahl, S. S. Palladium (II)-catalysed alkene functionalization via nucleopalladation: stereochemical pathways and enantioselective catalytic applications. *Chem. Rev.* **111**, 2981–3019 (2011).
6. Faber, K. *Biotransformations in Organic Chemistry: A Textbook* (Springer, 2011).
7. Bornscheuer, U. T. *et al.* Engineering the third wave of biocatalysis. *Nature* **485**, 185–194 (2012).
8. Reetz, M. T. Biocatalysis in organic chemistry and biotechnology: past, present, and future. *J. Am. Chem. Soc.* **135**, 12480–12496 (2013).
9. Li, A. & Li, Z. in *Biocatalysis in Organic Synthesis.* (eds Faber, K. *et al.*) **Vols. 2**, 479–506 (Thieme, 2014).
10. Boyd, D. R., Sharma, N. D. & Allen, C. C. Aromatic dioxygenases: molecular biocatalysis and applications. *Curr. Opin. Biotechnol.* **12**, 564–573 (2001).
11. Jin, J. & Hanefeld, U. The selective addition of water to C = C bonds; enzymes are the best chemists. *Chem. Commun.* **47**, 2502–2510 (2011).
12. Coelho, P. S., Brustad, E. M., Kannan, A. & Arnold, F. H. Olefin cyclopropanation via carbene transfer catalysed by engineered cytochrome P450 enzymes. *Science* **339**, 307–310 (2013).
13. Coelho, P. S. *et al.* A serine-substituted P450 catalyzes highly efficient carbene transfer to olefins *in vivo. Nat. Chem. Biol.* **9**, 485–487 (2013).
14. Farwell, C. C., Zhang, R. K., McIntosh, J. A., Hyster, T. K. & Arnold, F. H. Enantioselective enzyme-catalysed aziridination enabled by active site evolution of a cytochrome P450. *ACS Cent. Sci* **1**, 89–93 (2015).
15. Wasilke, J. C., Obrey, S. J., Baker, R. T. & Bazan, G. C. Concurrent tandem catalysis. *Chem. Rev.* **105**, 1001–1020 (2005).
16. Lohr, T. L. & Marks, T. J. Orthogonal tandem catalysis. *Nat. Chem.* **7**, 477–482 (2015).
17. Mlynarski, S. N., Schuster, C. H. & Morken, J. P. Asymmetric synthesis from terminal alkenes by cascades of diboration and cross-coupling. *Nature* **505**, 386–390 (2014).
18. Köhler, V. *et al.* Synthetic cascades are enabled by combining biocatalysts with artificial metalloenzymes. *Nat. Chem.* **5**, 93–99 (2013).

19. Riva, S. & Fessner, W.-D. (eds. *Cascade Biocatalysis Integrating Stereoselective and Environmentally Friendly Reactions* (Wiley, 2014).

20. Ladkau, N., Schmid, A. & Bühler, B. The microbial cell–functional unit for energy dependent multistep biocatalysis. *Curr. Opin. Biotechnol.* **30**, 178–189 (2014).

21. Simon, R. C., Richter, N., Busto, E. & Kroutil, W. Recent developments of cascade reactions involving ω-transaminases. *ACS Catal.* **4**, 129–143 (2014).

22. Köhler, V. & Turner, N. J. Artificial concurrent catalytic processes involving enzymes. *Chem. Commun.* **51**, 450–464 (2015).

23. Muschiol, J. *et al.* Cascade catalysis–strategies and challenges en route to preparative synthetic biology. *Chem. Commun.* **51**, 5798–5811 (2015).

24. Bayer, T., Milker, S., Wiesinger, T., Rudroff, F. & Mihovilovic, M. Designer microorganisms for optimized redox cascade reactions–challenges and future perspectives. *Adv. Synth. Catal.* **357**, 1587–1618 (2015).

25. Voss, C. V. *et al.* Orchestration of concurrent oxidation and reduction cycles for stereoinversion and deracemisation of sec-alcohols. *J. Am. Chem. Soc.* **130**, 13969–13972 (2008).

26. Sattler, J. H. *et al.* Redox self-sufficient biocatalyst network for the amination of primary alcohols. *Angew. Chem. Int. Ed.* **51**, 9156–9159 (2012).

27. Agudo, R. & Reetz, M. T. Designer cells for stereocomplementary de novo enzymatic cascade reactions based on laboratory evolution. *Chem. Commun.* **49**, 10914–10916 (2013).

28. Martin, C. H. *et al.* A platform pathway for production of 3-hydroxyacids provides a biosynthetic route to 3-hydroxy-γ-butyrolactone. *Nat. Commun.* **4**, 1414 (2013).

29. O'Reilly, E. *et al.* A regio-and stereoselective ω-transaminase/monoamine oxidase cascade for the synthesis of chiral 2, 5-disubstituted pyrrolidines. *Angew. Chem. Int. Ed.* **53**, 2447–2456 (2014).

30. Schrittwieser, J. H. *et al.* Deracemization by simultaneous bio-oxidative kinetic resolution and stereoinversion. *Angew. Chem. Int. Ed.* **53**, 3731–3734 (2014).

31. Thiel, D., Doknić, D. & Deska, J. Enzymatic aerobic ring rearrangement of optically active furylcarbinols. *Nat. Commun.* **5**, 5278 (2014).

32. Sattler, J. H. *et al.* Introducing an in situ capping strategy in systems biocatalysis to access 6-aminohexanoic acid. *Angew. Chem. Int. Ed.* **53**, 14153–14157 (2014).

33. Zhang, J., Wu, S., Wu, J. & Li, Z. Enantioselective cascade biocatalysis via epoxide hydrolysis and alcohol oxidation: one-pot synthesis of (R)-α-hydroxy ketones from meso-or racemic epoxides. *ACS Catal.* **5**, 51–58 (2015).

34. Schmidt, S. *et al.* An enzyme cascade synthesis of ε-caprolactone and its oligomers. *Angew. Chem. Int. Ed.* **54**, 2784–2787 (2015).

35. Parmeggiani, F., Lovelock, S. L., Weise, N. J., Ahmed, S. T. & Turner, N. J. Synthesis of d-and l-phenylalanine derivatives by phenylalanine ammonia lyases: a multienzymatic cascade process. *Angew. Chem. Int. Ed.* **54**, 4608–4611 (2015).

36. Mutti, F. G., Knaus, T., Scrutton, N. S., Breuer, M. & Turner, N. J. Conversion of alcohols to enantiopure amines through dual-enzyme hydrogen-borrowing cascades. *Science* **349**, 1525–1529 (2015).

37. Xu, Y., Jia, X., Panke, S. & Li, Z. Asymmetric dihydroxylation of aryl olefins by sequential enantioselective epoxidation and regioselective hydrolysis with tandem biocatalysts. *Chem. Commun.* 1481–1483 (2009).

38. Xu, Y., Li, A., Jia, X. & Li, Z. Asymmetric trans-dihydroxylation of cyclic olefins by enzymatic or chemo-enzymatic sequential epoxidation and hydrolysis in one-pot. *Green Chem.* **13**, 2452–2458 (2011).

39. Wu, S. *et al.* Enantioselective trans-dihydroxylation of aryl olefins by cascade biocatalysis with recombinant *Escherichia coli* coexpressing monooxygenase and epoxide hydrolase. *ACS Catal.* **4**, 409–420 (2014).

40. Coppola, G. M. & Schuster, H. F. *α-Hydroxy Acids in Enantioselective Syntheses* (Wiley, 2002).

41. Ager, D. J., Prakash, I. & Schaad, D. R. 1, 2-Amino alcohols and their heterocyclic derivatives as chiral auxiliaries in asymmetric synthesis. *Chem. Rev.* **96**, 835–876 (1996).

42. Soloshonok, V. A. & Izawa, K. (eds. *Asymmetric Synthesis and Application of α-Amino Acids* (American Chemical Society, 2009).

43. Brunel, J. M. & Holmes, I. P. Chemically catalysed asymmetric cyanohydrin syntheses. *Angew. Chem. Int. Ed.* **43**, 2752–2778 (2004).

44. Zuend, S. J., Coughlin, M. P., Lalonde, M. P. & Jacobsen, E. N. Scaleable catalytic asymmetric Strecker syntheses of unnatural α-amino acids. *Nature* **461**, 968–970 (2009).

45. Donohoe, T. J., Callens, C. K., Flores, A., Lacy, A. R. & Rathi, A. H. Recent developments in methodology for the direct oxyamination of olefins. *Chem. Eur. J.* **17**, 58–76 (2011).

46. Turner, N. J. & O'Reilly, E. Biocatalytic retrosynthesis. *Nat. Chem. Biol.* **9**, 285–288 (2013).

47. Panke, S., Witholt, B., Schmid, A. & Wubbolts, M. G. Towards a biocatalyst for (S)-styrene oxide production: characterization of the styrene degradation pathway of *Pseudomonas* sp. strain VLB120. *Appl. Environ. Microbiol.* **64**, 2032–2043 (1998).

48. Wu, S., Li, A., Chin, Y. S. & Li, Z. Enantioselective hydrolysis of racemic and meso-epoxides with recombinant *Escherichia coli* expressing epoxide hydrolase from *Sphingomonas* sp. HXN-200: preparation of epoxides and vicinal diols in high ee and high concentration. *ACS Catal.* **3**, 752–759 (2013).

49. Kirmair, L. & Skerra, A. Biochemical analysis of recombinant alkJ from pseudomonas putida reveals a membrane-associated, flavin adenine dinucleotide-dependent dehydrogenase suitable for the biosynthetic production of aliphatic aldehydes. *Appl. Environ. Microbiol.* **80**, 2468–2477 (2014).

50. Ferrández, A., Prieto, M. A., García, J. L. & Díaz, E. Molecular characterization of padA, a phenylacetaldehyde dehydrogenase from *Escherichia coli*. *FEBS Lett.* **406**, 23–27 (1997).

51. Kaulmann, U., Smithies, K., Smith, M. E., Hailes, H. C. & Ward, J. M. Substrate spectrum of ω-transaminase from *Chromobacterium violaceum* DSM30191 and its potential for biocatalysis. *Enzyme Microb. Technol.* **41**, 628–637 (2007).

52. Koszelewski, D. *et al.* Formal asymmetric biocatalytic reductive amination. *Angew. Chem. Int. Ed.* **47**, 9337–9340 (2008).

53. Resch, V., Fabian, W. M. & Kroutil, W. Deracemisation of mandelic acid to optically pure non-natural L-phenylglycine via a redox-neutral biocatalytic cascade. *Adv. Synth. Catal.* **352**, 993–997 (2010).

54. Li, T. L. *et al.* Characterisation of a hydroxymandelate oxidase involved in the biosynthesis of two unusual amino acids occurring in the vancomycin group of antibiotics. *Chem. Commun.* 1752–1753 (2001).

55. Purnick, P. E. & Weiss, R. The second wave of synthetic biology: from modules to systems. *Nat. Rev. Mol. Cell Biol.* **10**, 410–422 (2009).

56. Ajikumar, P. K. *et al.* Isoprenoid pathway optimization for taxol precursor overproduction in *Escherichia coli*. *Science* **330**, 70–74 (2010).

57. Sheppard, M. J., Kunjapur, A. M., Wenck, S. J. & Prather, K. L. Retro-biosynthetic screening of a modular pathway design achieves selective route for microbial synthesis of 4-methyl-pentanol. *Nat. Commun.* **5**, 5031 (2014).

58. Otto, K., Hofstetter, K., Röthlisberger, M., Witholt, B. & Schmid, A. Biochemical characterization of StyAB from *Pseudomonas* sp. strain VLB120 as a two-component flavin-diffusible monooxygenase. *J. Bacteriol.* **186**, 5292–5302 (2004).

59. Liu, S. P. *et al.* Heterologous pathway for the production of L-phenylglycine from glucose by E. coli. *J. Biotechnol.* **186**, 91–97 (2014).

60. Romero, P. A. & Arnold, F. H. Exploring protein fitness landscapes by directed evolution. *Nat. Rev. Mol. Cell Biol.* **10**, 866–876 (2009).

61. Turner, N. J. Directed evolution drives the next generation of biocatalysts. *Nat. Chem. Biol.* **5**, 567–573 (2009).

62. Reetz, M. T. Laboratory evolution of stereoselective enzymes: a prolific source of catalysts for asymmetric reactions. *Angew. Chem. Int. Ed.* **50**, 138–174 (2011).

63. Gröger, H. Enzymatic routes to enantiomerically pure aromatic α-hydroxy carboxylic acids: a further example for the diversity of biocatalysis. *Adv. Synth. Catal.* **343**, 547–558 (2001).

64. Andersson, M. A., Epple, R., Fokin, V. V. & Sharpless, K. B. A new approach to osmium-catalysed asymmetric dihydroxylation and aminohydroxylation of olefins. *Angew. Chem. Int. Ed.* **41**, 472–475 (2002).

65. Ensign, S. A. Microbial metabolism of aliphatic alkenes. *Biochemistry* **40**, 5845–5853 (2001).

66. Ma, B. D. *et al.* Increased catalyst productivity in α-hydroxy acids resolution by esterase mutation and substrate modification. *ACS Catal.* **4**, 1026–1031 (2014).

67. Yan, P. C. *et al.* Direct asymmetric hydrogenation of α-keto acids by using the highly efficient chiral spiro iridium catalysts. *Chem. Commun.* **50**, 15987–15990 (2014).

68. Cho, B. T., Kang, S. K. & Shin, S. H. Application of optically active 1, 2-diol monotosylates for synthesis of β-azido and β-amino alcohols with very high enantiomeric purity. Synthesis of enantiopure (R)-octopamine, (R)-tembamide and (R)-aegeline. *Tetrahedron: Asymmetry* **13**, 1209–1217 (2002).

69. Hassan, N. A., Bayer, E. & Jochims, J. C. Syntheses of optically active α-amino nitriles by asymmetric transformation of the second kind using a principle of O. Dimroth. *J. Chem. Soc., Perkin Trans.* 1, 3747–3758 (1998).

70. Beller, M. *et al.* Efficient chemoenzymatic synthesis of enantiomerically pure α-amino acids. *Chem. Eur. J.* **4**, 935–941 (1998).

## Acknowledgements

This work was financially supported by the Chemical and Pharmaceutical Engineering (CPE) Program from Singapore-MIT Alliance, a research grant (Project No. 1021010026) from Science & Engineering Research Council of A*STAR, Singapore, and the Synthetic Biology for Clinical and Technological Innovation (SynCTI) program from National University of Singapore. We thank Prof. Sven Panke from ETH Zurich, Switzerland for helpful discussion.

## Author contributions

S.W. designed and performed most of the experiments. S.W. and Y.Z. carried out preparative biotransformations, enzyme purification and enzyme activity determination. T.W. cloned AlkJ. H.-P.T. provided helpful discussion and lab facilities at the late stage of the project. D.I.C.W. co-advised some research work of S.W. Z.L. supervised the entire research project. S.W. and Z.L. wrote the manuscript.

# Metal-free intermolecular formal cycloadditions enable an orthogonal access to nitrogen heterocycles

Lan-Gui Xie[1], Supaporn Niyomchon[1], Antonio J. Mota[2], Leticia González[2] & Nuno Maulide[1]

Nitrogen-containing heteroaromatic cores are ubiquitous building blocks in organic chemistry. Herein, we present a family of metal-free intermolecular formal cycloaddition reactions that enable highly selective and orthogonal access to isoquinolines and pyrimidines at will. Applications of the products are complemented by a density functional theory mechanistic analysis that pinpoints the crucial factors responsible for the selectivity observed, including stoichiometry and the nature of the heteroalkyne.

[1] Institute of Organic Chemistry, University of Vienna, Währinger Strasse 38, 1090 Vienna, Austria. [2] Institute of Theoretical Chemistry, University of Vienna, Währinger Strasse 17, 1090 Vienna, Austria. Correspondence and requests for materials should be addressed to N.M. (email: nuno.maulide@univie.ac.at).

eteroarenes constitute one of the privileged core structural
motifs in organic chemistry[1]. Among them, isoquinolines
and pyrimidines represent two big families in
pharmaceutical agents, natural products and functional
materials[2–9]. Therefore, continued effort is devoted to the
exploration of new and efficient synthetic strategies for these
backbones.

The classical strategies to prepare isoquinolines (Fig. 1a)
generally focus on the crucial textbook disconnections C1-C8a
(Bischler-Napieralski and Pictet-Spengler syntheses) or C4-C4a
(Pomeranz-Fritsch synthesis). Recently developed routes centred
on the bond-forming events N2-C3 or N2-C3/C4-C4a, employing electrophile-triggered annulation and transition metal-
catalysed C–H or C–halogen bond activation, respectively[1,10–17]. A strategy relying on the simultaneous formation of N2-C3/
C1–C8a is much less documented[18].

Conversely, most of the known avenues towards pyrimidine
synthesis rely on the condensation of N-C-N subunits (mostly
amidines or guanidines) with 1,3-dicarbonyl derivatives or the
stoichiometric activation of carbonyl moieties with triflic
anhydride (Fig. 1a) (refs 19–23). Ynamides have recently shown
to be suitable candidates for regioselective cycloaddition
with nitriles in the presence of a gold catalyst, leading
to 4-aminopyrimidine cores[24]. Although the reactivity of
ynamides has received considerable recent attention[25–33],

analogous investigation of the potential enclosed in the triple
bond of thioalkynes is surprisingly rare[34–37], even though the
resulting sulfide is a useful[38] and versatile substituent[39–42].

Herein we report a family of reactions that enable a high
yielding, orthogonal access to either isoquinolines or pyrimidines
at will (Fig. 1b), by Brønsted acid-mediated regioselective formal
cycloaddition of ynamides and thioalkynes with nitriles (for a
review of transition-metal mediated [2 + 2 + 2] cycloadditions)
(ref. 43). Mechanistic studies reveal the subtle differences that are
responsible for selectivity.

## Results
### Synthesis.
Initial experiments involving the reaction of ynamide
1a with various Brønsted acids in the presence of varying
amounts of acetonitrile led to moderate yields of isoquinoline
3aa. After optimization of conditions (see Supplementary Table 1
for details), we found that essentially equimolar amounts of 1a, 2a
and TfOH in dichloroethane as solvent sufficed to enable
preparation of 3aa in 89% yield (for a discussion of stoichiometry
in these reactions, vide infra).

Holding suitable conditions in hand, we then examined several
nitriles 2a-j under the optimized conditions. As shown in Fig. 2a,
this direct formal cycloaddition is applicable to a broad range
of substrates, generally affording good to excellent yields
of isoquinoline products. Remarkably, alkyl nitriles bearing

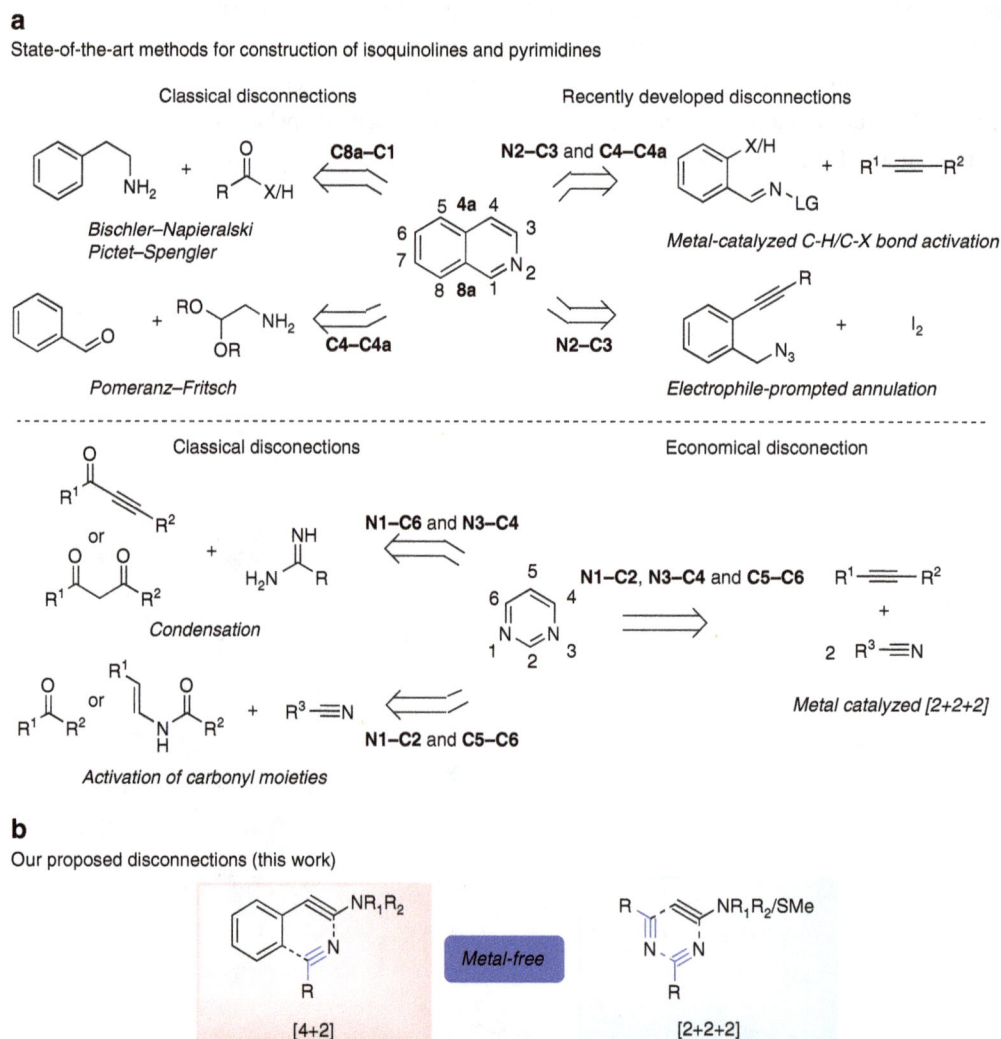

### a
State-of-the-art methods for construction of isoquinolines and pyrimidines

### b
Our proposed disconnections (this work)

**Figure 1 | Synthetic disconnections. (a)** Known synthetic disconnections for the isoquinoline and pyrimidine backbone. **(b)** Proposed direct disconnections through intermolecular metal-free alkyne/nitrile cycloadditions.

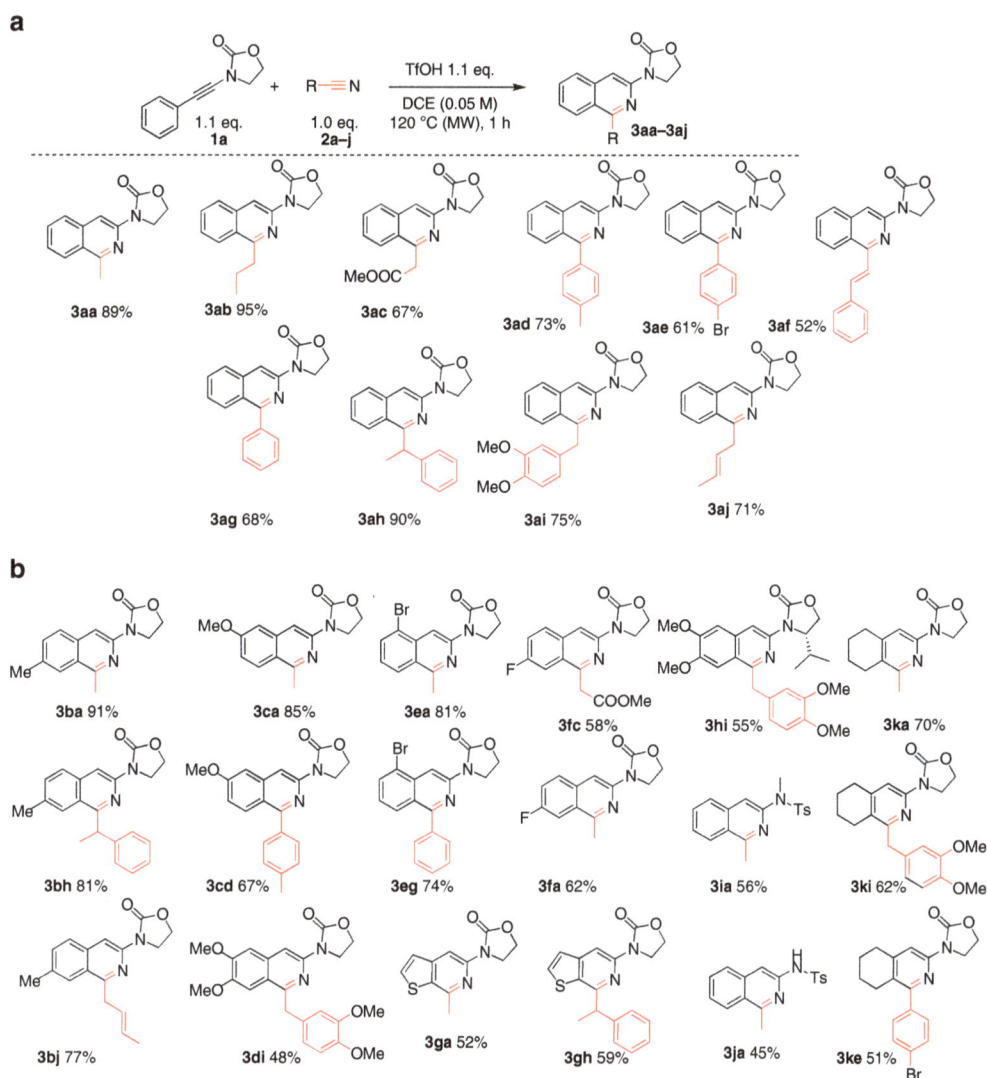

**Figure 2 | Scope of isoquinoline synthesis. (a)** Scope of nitriles and **(b)** Scope of ynamides for the synthesis of isoquinolines. Yields are for isolated products.

functional groups such as an ester (**2c**), aryl rings (**2h** and **2i**) or C–C double bonds (**2j**) are compatible with the reaction conditions. It is worth mentioning that the isolated double bond in product **3aj** does not migrate into conjugation with the isoquinoline ring under these conditions. Aryl nitriles (**2d–g**) and α,β-unsaturated nitrile **2f** are also viable partners delivering the corresponding substituted isoquinoline products in good to very good yields.

Subsequently, a broad range of ynamides were submitted to this protocol (Fig. 2b). In the event, both electron-donating (**1b–d**) and -withdrawing (**1f**) substituents were tolerated on the ynamide partner, leading to smooth isoquinoline assembly in good yields. Halogenated ynamides (**1e–f**) were also amenable to this reaction, delivering isoquinoline products ripe for subsequent divergent functionalization. Thienopyridine skeletons could be obtained in reasonable yields (**3ga** and **3gh**). Interestingly, N-tosyl-N-benzyl ynamide (**1j**) directly generated the corresponding debenzylated product: the tosyl-protected, pharmacologically relevant 3-aminoisoquinoline (**3ja**) (refs 44,45). Moreover, the use of an alkenyl-substituted ynamide (**1k**) led to the annulated pyridine product (**3ke**).

After this initial success, we hypothesized that other heteroatom-substituted alkynes might prove amenable to a similar modular assembly of isoquinolines. In particular, we were

drawn to the use of thioalkynes such as **4b**, with the expectation of obtaining an (alkylthio)-isoquinoline **6ba** where the sulfur residue could serve as a useful synthetic handle (Fig. 3a).

Much to our surprise, treatment of **4b** with acetonitrile **2a** under conditions identical to those employed previously led exclusively to the pyrimidine **5ba** in 52% yield (Fig. 3a). Remarkably, product **5ba** is the result of a formal, regioselective cycloaddition of one molecule of **4b** with two molecules of **2a**. This dramatic shift in product selectivity between ynamides and thioalkynes eventually presented us with a versatile cycloaddition route towards pyrimidines. Reaction optimization showed that this transformation proceeds most effectively at room temperature in the presence of an excess of acetonitrile (see Supplementary Table 2 for details).

Figure 3b depicts the full scope of nitriles **2b–v** compatible with this metal-free pyrimidine synthesis. Secondary aliphatic (**2k**) and alicyclic (**2m-2p**) carbonitriles smoothly coupled with thioalkyne **4a** under the reaction conditions. This formal cycloaddition was also tolerant of nitriles bearing triple (**2q**) and double bonds (**2j**), including conjugated olefins (**2f**). Both electron-rich (**2d** and **2s**) and electron-deficient (**2t** and **2u**) substituted benzonitriles could be employed, providing the desired pyrimidine products in good to excellent yields. It is worth noting that heteroarylnitriles such as 3-cyanothiophene (**2r**) were also tolerated. The possibility

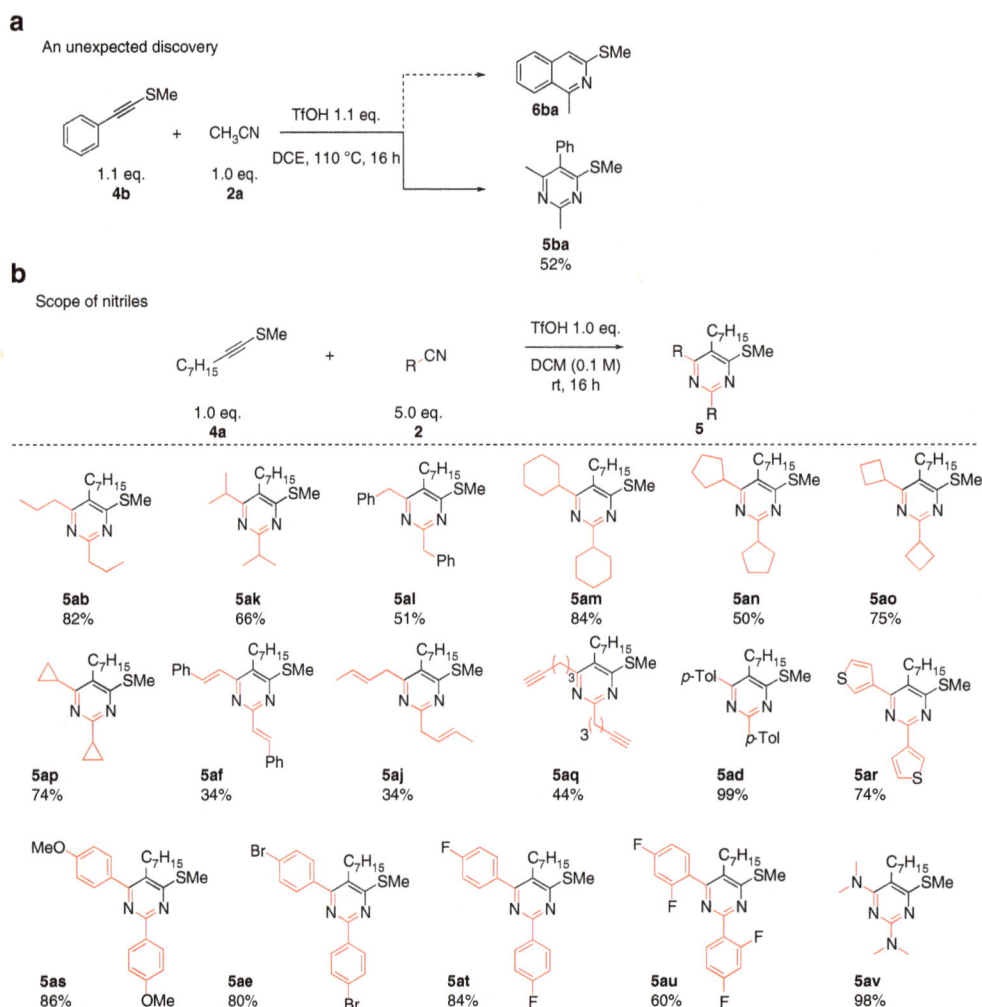

**Figure 3 | Synthesis of pyrimidine.** (**a**) Unexpected synthesis of pyrimidine **5ba**. (**b**) Scope of nitriles in the synthesis of pyrimidines **5**.

of using dimethylcyanamide (**2v**), delivering an aminated pyrimidine in excellent yield, further highlights the generality of this synthetic method. Pyrimidine **5ae** yielded crystals suitable for X-ray diffraction analysis, unambiguously confirming its structure (see Supplementary Fig. 64 and Supplementary Tables 4 and 5 for details).

Further studies focused on the scope of heteroalkynes for this pyrimidine synthesis (Fig. 4a). We were pleased to find that a cyclopropyl substituent (**4c**) was tolerated, as a cyclopropyl appended to a pyrimidine ring is a common feature in drug-like, biologically active cores[46–48]. Both electron-rich (**4e** and **4g**) and electron-poor (**4d** and **4f**) arylalkynes afforded the corresponding pyrimidine products in good yields. Furthermore, considerable flexibility can be exerted, concerning the location of substituents on the aryl ring (**4d–g**).

Strikingly, we found that 4-aminosubstituted pyrimidines can also be obtained by exposing ynamides (**1l**, **1a** and **1m**) to the standard conditions developed for pyrimidine synthesis. A distal nitrile group carried by the ynamide partner could be successfully introduced into the pyrimidine product (**7ma**). Remarkably, when phenyl-substituted ynamide **1a** was submitted to these conditions, a 4-amino-5-aryl pyrimidine product (**7aa**) was obtained in good yield (Fig. 4b). Together with the reactions described previously (*cf.* Figures 2–3), these results offer an entirely new orthogonal access to either isoquinoline or pyrimidine motifs at will, while unifying this novel, powerful family of formal cycloaddition reactions.

**Density functional theory study.** We approached the mechanistic study of this reaction performing density functional theory (DFT) calculations of two reaction manifolds: the first leading to isoquinoline products (by modelling the entire pathway introducing a single acetonitrile molecule, see Fig. 5a) and the second leading to pyrimidine adducts (by computing the mechanism with two acetonitrile molecules, see Fig. 5b). The first question that arises is what occurs when all these species are in the presence of the TfOH promoter, as there are many potential protonation sites. DFT calculations (see Computational details in the Supplementary Figs 65–67) show that protonation would take place preferably on the heteroalkyne partner. Indeed, calculated transition states for the oxazolidinone ($+7.6$ and $+6.0$) and methylthio ($+5.3$ and $+10.7$) derivatives (in the presence of either one or two acetonitrile molecules, $II_i$ and $II_p$, respectively) are much lower than the acetonitrile protonation ($+17.0 \, \text{Kcal mol}^{-1}$). Furthermore, we confirmed that this protonation takes place regioselectively β- to the heteroatom as anticipated, leading to either a keteniminium $III_i$ or ketenethionium $III_p$ species, as the TfO$^-$ anion is stabilized by acetonitrile (which in these reactions coincides with the nucleophilic species). In the second mechanistic step, a nucleophilic attack by acetonitrile takes place stereoselectively from the face opposite to the β-proton due to shielding by TfO$^-$ (see $IV_i$ and $IV_p$ in Fig. 5). In fact, in both cases the introduction of acetonitrile, giving respectively $V_i$ and $V_p$, is more stable than the corresponding TfO$^-$ bonded derivative by 6.1 and 7.7 Kcal mol$^{-1}$ for

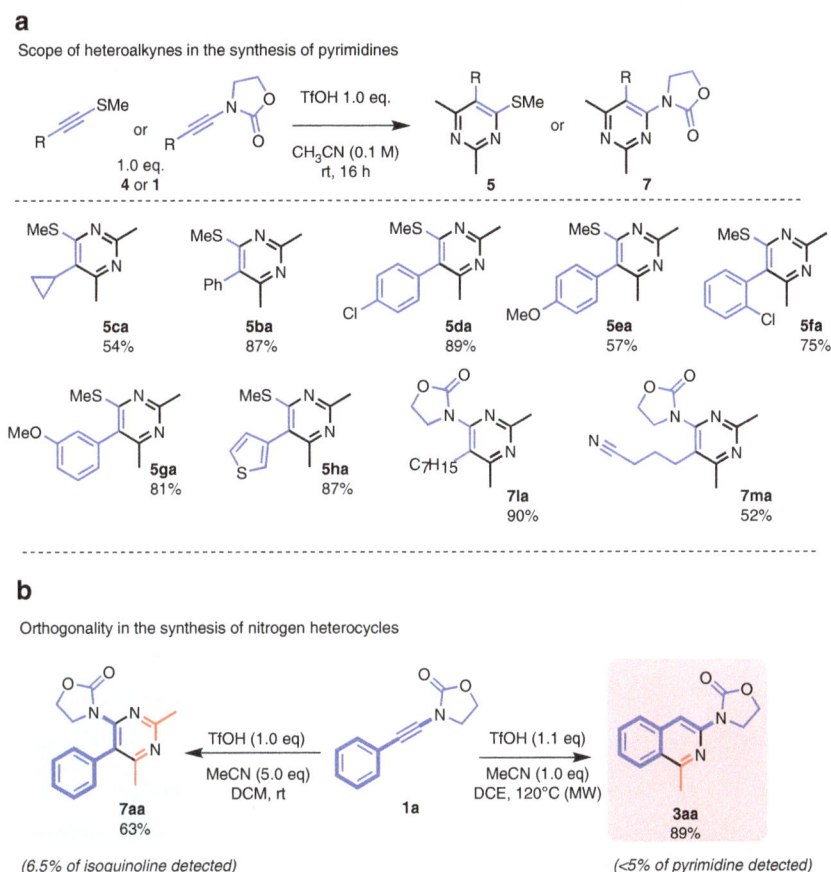

**Figure 4 | Scope and orthogonality of heterocycles.** (**a**) Scope of heteroalkynes for the synthesis of pyrimidines **5** or **7**. (**b**) Orthogonality in the synthesis of isoquinolines or pyrimidines from ynamides. See Supplementary Figs 62 and 63 for details.

the oxazolidinone and methylthio derivatives, respectively. Interestingly, and very important for the reaction outcome, in the absence of acetonitrile the TfO$^-$ species readily adds to the positively charged intermediate effectively blocking further reaction with acetonitrile. Moreover, we verified that nucleophilic attack by acetonitrile can only take place after the first protonation event as the highest occupied molecular orbital of acetonitrile ($-0.3264$ H) and the neutral ynamide's lowest unoccupied molecular orbital ($-0.0242$ H) are energetically too far apart. The protonation process, however, results in an alkyne-centred lowest unoccupied molecular orbital turned by 90° at $-0.2443$ H, whereas the highest occupied molecular orbital of the TfO$^-$ counteranion lies at $-0.0742$ H (see Supplementary Figs 65 in the Computational details section of the Supplementary Information). A similar trend is observed for the methylthio derivative. These two first processes (protonation + nucleophilic attack of acetonitrile) are common to both pathways. At this juncture, we can separately analyse the mechanisms leading to the isoquinoline and the pyrimidine scaffolds.

**Isoquinoline formation.** Following the addition of acetonitrile, the TfO$^-$ anion immediately adds to the resulting carbocation (as the former lost its prior stabilization by acetonitrile) delivering a neutral and highly stable imino-triflate $\mathbf{V_i}$. The last step consists of a Friedel–Crafts-like cyclization, $\mathbf{VII_i}$, with further elimination of TfOH giving rise to the isoquinoline skeleton ($\mathbf{VIII_i}$). This final addition process is characterized by high energy-transition states: $+30.6$ and $+34.5$ Kcal mol$^{-1}$, respectively, for the oxazolidinone and methylthio derivatives ($\mathbf{VI_i}$; Fig. 5a). This is why heating is necessary in this case. The higher enthalpic barrier for the methylthio derivative stands in agreement with the

experimental findings, as no isoquinoline product is observed for this derivative. In addition, nuclear magnetic resonance studies carried out with the starting heteroalkynes in the presence of TfOH suggest that the thermal stability of the methylthio-derivative is notably low. This could be a determining factor towards the experimental observations.

**Pyrimidine formation.** In this case, with two acetonitrile molecules (used for simplicity of the model, although the experimentally optimized molar ratio is higher), a different situation arises as a second addition becomes a more probable event. In fact, this process takes place through low energy-transition states ($+6.5$ and $+7.1$ Kcal mol$^{-1}$, respectively, for the oxazolidinone and methylthio derivatives), $\mathbf{VI_p}$, giving rise to rather stable intermediates ($\mathbf{VII_p}$; see Fig. 5b). Once the second molecule is added, the system could conceivably undergo a polymerization process with continued further addition of more acetonitrile molecules to the newly generated carbocationic species ($+0.546$ and $+0.529$ e$^-$, respectively, for the oxazolidinone and methylthio derivatives). Instead, the negatively charged ($-0.208$ and $-0.210$ e$^-$, for the oxazolidinone and methylthio derivatives, respectively) β-carbon atom can attack ($\mathbf{VIII_p}$) the newly generated carbocation to form an entropically favoured, six-membered pyrimidine ring (after a very exothermic re-aromatization promoted by TfO$^-$, $\mathbf{IX_p} \rightarrow \mathbf{X_p}$). This is the driving force for pyrimidine formation.

A comparative analysis of both calculated mechanisms reveals at a glance that those involving the oxazolidinone derivative proceed generally with lower energies than the thioalkyne one. The main reason for that is the greater stabilization of the positive charge in the former case. In addition, although pathways for

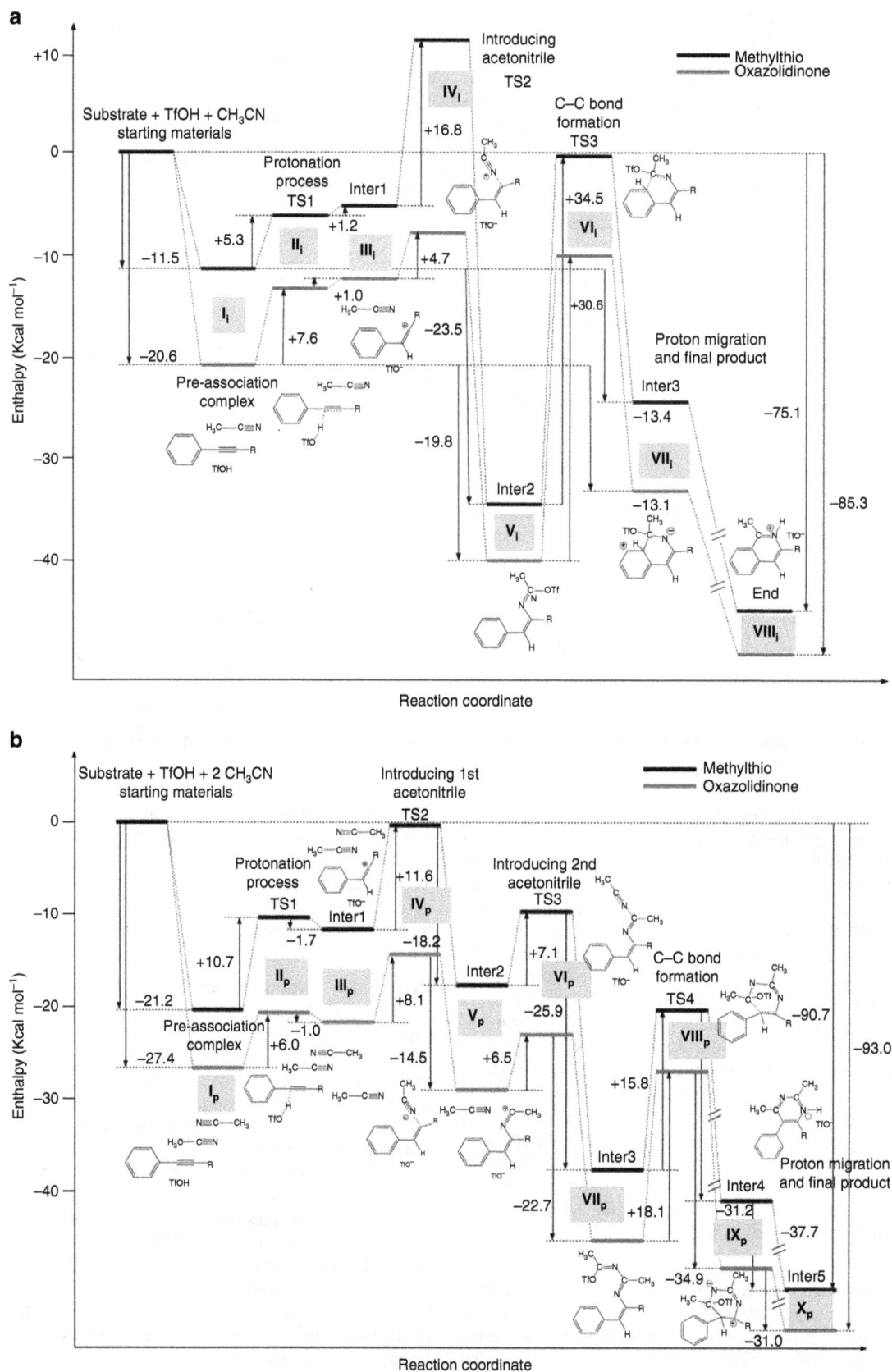

**Figure 5 | Energy profiles.** Isoquinoline (**a**, up, i subscript in roman numerals) and pyrimidine (**b**, down, p subscript in roman numerals) formation for the oxazolidinone (blue) and methylthio (red) derivatives. The corresponding three-dimensional structure sequence is exemplified for the methylthio derivative, for both **a** and **b** pathways, in Supplementary Figs 66 and 67, respectively

formation of either heterocycle could exist for both heteroalkynes, the pathway for isoquinoline formation through the thioalkyne derivative (Fig. 5a) has a prohibitive energy barrier when considering the addition of a single acetonitrile molecule. This barrier is much lower in the case of pyrimidine formation (Fig. 5b). This is due to the fact that in the latter case, there is a significant stabilization of the corresponding transition state introduced by the presence of the second acetonitrile molecule. Yet, the transition state for the first acetonitrile attack is 5.2 Kcal mol$^{-1}$ higher in the case of isoquinoline formation. In this value, 3.2 Kcal mol$^{-1}$ are purely due to the stabilization offered by the second acetonitrile molecule, as calculations made considering cationic structures (namely just the substrate and the

acetonitrile molecule(s) without the TfO$^{-}$ species) show this same energy difference. Therefore, the remaining 2 Kcal mol$^{-1}$ should derive from stabilization by the counteranion (which is present in our mechanistic studies) through cooperative cyclic $\delta^+ \cdot \delta^-$ interactions between the different molecular units (see Fig. 6). In fact, we expect this transition state to be very low-lying, taking into account more solvent molecules. The preceding mechanistic analysis also permits a rationalization of why isoquinoline synthesis (formal [4 + 2]; (refs 49–52)) requires high temperatures (highest energy barriers), whereas pyrimidine formation (formal [2 + 2 + 2]) typically occurs at room temperature.

In both cases, stoichiometry plays a crucial role and imposes the final result, as both pathways are irreversible. In the presence of several molecules of acetonitrile (Fig. 5b), the corresponding transition state for a real [2 + 2 + 2] approximation is either transition state $\mathbf{IV_p}$ or $\mathbf{VI_p}$, depending on the initial geometry conditions. These transition state also appear in a sequential pathway (as described in Fig. 5), thus indicating a natural direction for this molecular set. In the case of isoquinoline formation (Fig. 5a), the reaction of the unique acetonitrile molecule (imposed by stoichiometry) creates a positive charge that is readily neutralized by the negatively charged triflate present in the surrounding (as a remnant from the initial protonation event). The latter reaction should be faster than the time required for another acetonitrile molecule to approach, to

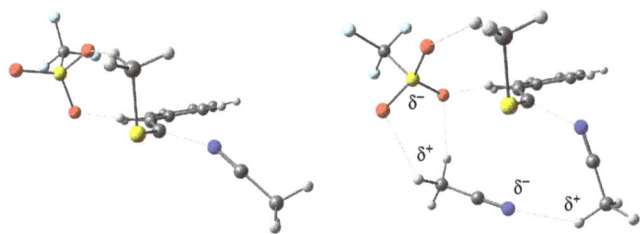

**Figure 6 | Transition states for the first acetonitrile addition.** Isoquinoline (left) and pyrimidine (right) pathways, the latter presenting additional cyclic, electrostatic stabilizing interactions.

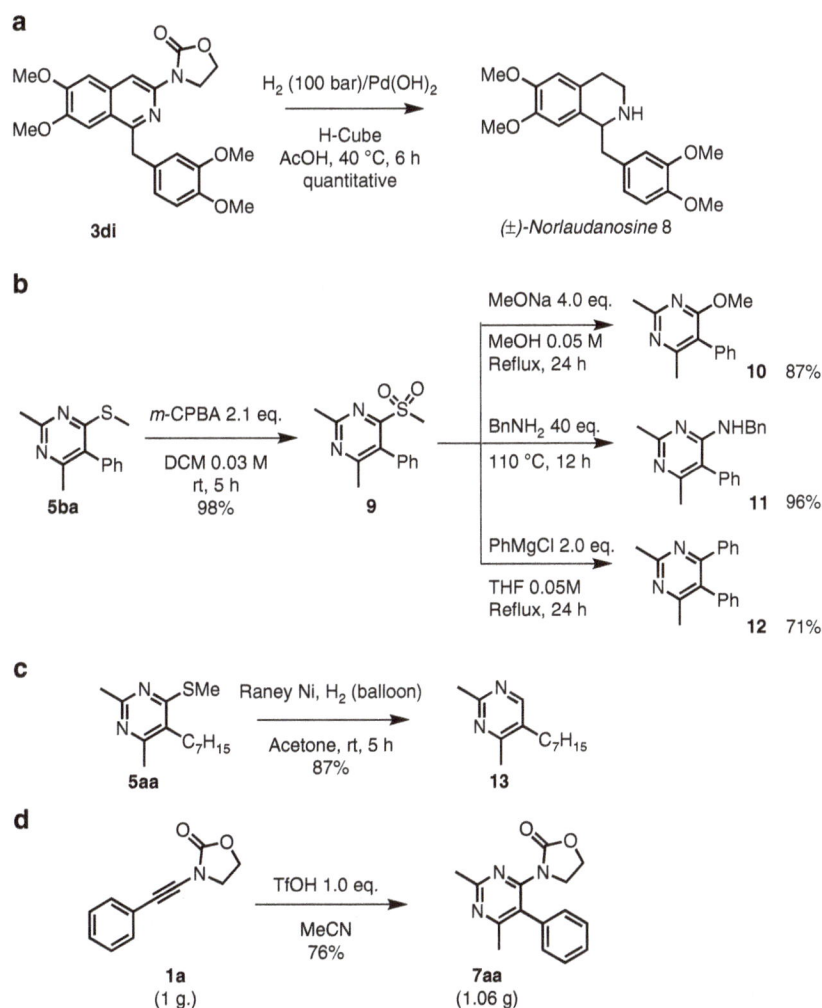

**Figure 7 | Synthetic application and modification.** (**a**) Preparation of norlaudanosine **8** by hydrogenation of **3di**. (**b**) Transformations of compound **5ba**. (**c**) Reductive desulfurization of compound **5aa**. (**d**) Pyrimidine **7aa** can be prepared in gram scale.

follow pathway B. Once triflate blocks this position, a quite stable intermediate ($V_i$) is formed and no additional acetonitrile molecules can be added.

It is noteworthy that although all the computed reaction pathways would only require a catalytic amount of TfOH to proceed (owing to its regeneration on aromatization, *vide supra*), the most stable final product in either pathway is the corresponding nitrogen-protonated heterocycle (readily converted into the experimentally isolated products following basic workup). This neatly accommodates the experimental need for stoichiometric amounts of acid, to obtain high yields.

**Further studies.** Given the prevalence of isoquinoline motifs in the core of bioactive molecules[53-55], we were eager to showcase the synthetic utility of our products (Fig. 7). The tetramethoxy adduct **3di**, carrying two electronically differentiated fused rings in its isoquinoline system, could be hydrogenated in acetic acid to deliver ( ± )-norlaudanosine **8** (Fig. 7a). Removal of the oxazolidinone takes place under these conditions analogously to previous work by Glorius *et al.*[56] on related pyridines and quinolones[57]. It is noteworthy that the use of a chiral 2-oxazolidone analogue, as in **3hi**, enabled the hydrogenation to proceed with some level of asymmetric induction (55% *e.e.*). See Supplementary Fig. 60 for more detail.

The (methylthio)pyrimidine **5ba** could be easily oxidized to the corresponding methylsulfonyl derivative **9**, in which the methylsulfonyl group is available for substitution. As shown in Fig. 7b, this can be achieved by the action of alcohols, amines or Grignard reagents, delivering substituted pyrimidines **10–12** in very good to excellent yields.

In addition, Raney-Ni-mediated hydrogenation of **5aa** smoothly excises the sulfide residue to afford the 2,5,6-trisubstituted pyrimidine **13** in 87% yield (Fig. 7c). These simple transformations outline the versatility and usefulness of the methods reported herein. Moreover, the reaction can be readily carried out in gram scale (Fig. 7d).

## Discussion

A family of reactions selectively leading to isoquinoline and pyrimidine motifs has been developed, by Brønsted acid-promoted regioselective merger of alkynes and nitriles. These methodologies benefit from the strategic use of readily available nitriles as the C–N sources. Most importantly, the orthogonality of the methods enables the preparation of either family of heterocycles from the same starting materials. The practicality of these metal-free formal cycloadditions is illustrated by the large scope of alkynes and nitriles that can be employed. DFT calculations reveal the crucial role of TfOH and the reaction stoichiometry in these processes. With one equivalent of acetonitrile, the preferred pathway leads to isoquinoline products through a Friedel–Crafts-like process; with larger amounts of nitrile, a second addition is allowed *en route* to the formation of a pyrimidine derivative. Furthermore, subtle differences between the classes of heteroalkynes employed control which products can be formed. We believe that the simple yet powerful heterocycle syntheses presented here will be eagerly adopted into the repertoire of synthetic chemistry.

## Methods

Full experimental details, characterization of compounds, Cartesian coordinates and energies of all the structures appearing in Supplementary Figs 66 and 67, and computational details can be found in the Supplementary Information (Supplementary Figs 1–67 and Supplementary Methods).

## References

1. Joule, J. A. & Mills, K. *Heterocyclic Chemistry* 5th edn 194–200 (John Wiley & Sons, Ltd, 2010).
2. Iranshahy, M., Quinn, R. J. & Iranshahi, M. Biologically active isoquinoline alkaloids with drug-like properties from the genus Corydalis. *RSC Adv.* **4**, 15900–15913 (2014).
3. Bentley, K. W. β-Phenylethylamines and the isoquinoline alkaloids. *Nat. Prod. Rep.* **23**, 444–463 (2006).
4. Su, Y. J. *et al.* Highly efficient red electrophosphorescent devices based on iridium isoquinoline complexes: remarkable external quantum efficiency over a wide range of current. *Adv. Mater.* **15**, 884–888 (2003).
5. Ho, C.-L. *et al.* Red-light-emitting iridium complexes with hole-transportingn 9-arylcarbazole moieties for electrophosphorescence efficiency/color purity trade-off optimization. *Adv. Funct. Mater.* **18**, 319–331 (2008).
6. Walker, S. R., Carter, E. J., Huff, B. C. & Morris, J. C. Variolins and related alkaloids. *Chem. Rev.* **109**, 3080–3098 (2009).
7. Lagoja, I. M. Pyrimidine as constituent of natural biologically active compounds. *Chem. Biodivers.* **2**, 1–50 (2005).
8. Köytepe, S., Paşahan, A., Ekinci, E. & Seçkin, T. Synthesis, characterization and $H_2O_2$-sensing properties of pyrimidine-based hyperbranched polyimides. *Eur. Polym. J.* **41**, 121–127 (2005).
9. Gompper, R., Mair, H.-J. & Polborn, K. Synthesis of oligo(diazaphenyls). Tailor-made fluorescent heteroaromatics and pathways to nanostructures. *Synthesis (Mass)* 696–718 (1997).
10. He, R., Huang, Z.-T., Zheng, Q.-Y. & Wang, C. Isoquinoline skeleton synthesis via chelation-assisted C–H activation. *Tetrahedron Lett.* **55**, 5705–5713 (2014).
11. Shi, Z., Koester, D. C., Boultadakis-Arapinis, M. & Glorius, F. Rh(III)-catalyzed synthesis of multisubstituted isoquinoline and pyridine N-oxides from oximes and diazo compounds. *J. Am. Chem. Soc.* **135**, 12204–12207 (2013).
12. Zhao, D., Lied, F. & Glorius, F. Rh(III)-catalyzed C–H functionalization/ aromatization cascade with 1,3-dienes: a redoxneutral and regioselective access to isoquinolines. *Chem. Sci.* **5**, 2869–2873 (2014).
13. Roesch, K. R., Zhang, H. & Larock, R. C. Synthesis of isoquinolines and pyridines by the palladium-catalyzed iminoannulation of internal alkynes. *J. Org. Chem.* **66**, 8042–8051 (2001).
14. Fischer, D. *et al.* Iodine-mediated electrophilic cyclization of 2-alkynyl-1-methylene azide aromatics leading to highly substituted isoquinolines and its application to the synthesis of Norchelerythrine. *J. Am. Chem. Soc.* **130**, 15720–15725 (2008) and references therein.
15. Gilmore, C. D., Allan, K. M. & Stoltz, B. M. Orthogonal synthesis of indolines and isoquinolines via aryne annulation. *J. Am. Chem. Soc.* **130**, 1558–1559 (2008).
16. Castillo, J.-C., Quiroga, J., Abonia, R., Rodriguez, J. & Coquerel, Y. The aryne aza-Diels–Alder reaction: flexible syntheses of isoquinolines. *Org. Lett.* **17**, 3374–3377 (2015).
17. Coppola, A., Sucunza, D., Burgos, C. & Vaquero, J. J. Isoquinoline synthesis by heterocyclization of tosylmethyl isocyanide derivatives: total synthesis of mansouramycin B. *Org. Lett.* **17**, 78–81 (2015).
18. Mrtínez, A. G., Fernández, A. H., Vilchez, D. M., Gutiérrez, M. L. L. & Subramanian, L. R. A new easy one-step synthesis of isoquinoline derivatives from substituted phenylacetic esters. *Synlett.* **1993**, 229–230 (1993).
19. Hill, M. D. & Movassaghi, M. New strategies for the synthesis of pyrimidine derivatives. *Chem. Eur. J.* **14**, 6836–6844 (2008) and references therein.
20. Movassaghi, M. & Hill, M. D. Single-step synthesis of pyrimidine derivatives. *J. Am. Chem. Soc.* **128**, 14254–14255 (2006).
21. Martínez, A. G. *et al.* On the mechanism of the reaction between ketones and trifluoromethanesulfonic anhydride. An improved and convenient method for the preparation of pyrimidines and condensed pyrimidines. *J. Org. Chem.* **57**, 1627–1630 (1992) and references therein.
22. Herrera, A., Martínez-Álvarez, R., Chioua, M., Chioua, R. & Sánchez, Á. On the regioselectivity in the reaction of aliphatic ketones and aromatic nitriles. Regiospecific synthesis of alkylarylpyrimidines. *Tetrahedron* **58**, 10053–10058 (2002).
23. Martínez, A. G. *et al.* Sterically hindered bases. Synthesis of 2,4,6-trisubstituted pyrimidine. *Synthesis (Mass)* 881–882 (1990) and references therein.
24. Karad, S. N. & Liu, R.-S. Regiocontrolled gold-catalyzed [2 + 2 + 2] cycloadditions of ynamides with two discrete nitriles to construct 4-aminopyrimidine cores. *Angew. Chem. Int. Ed.* **53**, 9072–9076 (2014).
25. Evano, G., Coste, A. & Jouvin, K. Ynamides: versatile tools in organic synthesis. *Angew. Chem. Int. Ed.* **49**, 2840–2859 (2010).
26. DeKorver, K. A. *et al.* Ynamides: a modern functional group for the new millennium. *Chem. Rev.* **110**, 5064–5106 (2010).
27. Wang, X.-N. *et al.* Ynamides in ring forming transformations. *Acc. Chem. Res.* **47**, 560–578 (2014).
28. Li, L. *et al.* Zinc-catalyzed alkyne oxidation/C–H functionalization: highly site selective synthesis of versatile isoquinolones and β-carbolines. *Angew. Chem. Int. Ed.* **54**, 8245–8249 (2015).

29. Shu, C. *et al.* Generation of α-imino gold carbenes through gold-catalyzed intermolecular reaction of azides with ynamides. *J. Am. Chem. Soc.* **137,** 9567–9570 (2015).

30. Theunissen, C. *et al.* Keteniminium ion-initiated cascade cationic polycyclization. *J. Am. Chem. Soc.* **136,** 12528–12531 (2014).

31. Peng, B., Huang, X., Xie, L.-G. & Maulide, N. A Brønsted acid-catalyzed redox arylation. *Angew. Chem. Int. Ed.* **53,** 8718–8721 (2014).

32. Xin, Z., Kramer, S., Overgaard, J. & Skrydstrup, T. Access to 1,2-dihydroisoquinolines through gold-catalyzed formal [4 + 2] cycloaddition. *Chem. Eur. J.* **20,** 7926–7930 (2014).

33. Minko, Y., Pasco, M., Lercher, L., Botoshansky, M. & Marek, I. Forming all-carbon quaternary stereogenic centres in acyclic systems from alkynes. *Nature* **490,** 522–526 (2012).

34. Ding, S. *et al.* Highly regio- and stereoselective hydrosilylation of internal thioalkynes under mild monditions. *Angew. Chem. Int. Ed.* **54,** 5632–5635 (2015).

35. Ding, S., Jia, G. & Sun, J. Iridium-catalyzed intermolecular azide–alkyne cycloaddition of internal thioalkynes under mild conditions. *Angew. Chem. Int. Ed.* **53,** 1877–1880 (2014).

36. Huang, K.-H. & Isobe, M. Highly regioselective hydrosilylation of unsymmetric alkynes using a phenylthio directing group. *Eur. J. Org. Chem.* 4733–4740 (2014) and references therein.

37. Frei, R. *et al.* Fast and highly chemoselective alkynylation of thiols with hypervalent iodine reagents enabled through a low energy barrier concerted mechanism. *J. Am. Chem. Soc.* **136,** 16563–16573 (2014).

38. Favre, A. & Fourrey, J.-L. Structural probing of small endonucleolytic ribozymes in solution using thio-substituted nucleobases as intrinsic photolabels. *Acc. Chem. Res.* **28,** 375–382 (1995).

39. Dubbaka, S. R. & Vogel, P. Organosulfur compounds: electrophilic reagents in transition-metal-catalyzed carbon–carbon bond-forming reactions. *Angew. Chem. Int. Ed.* **44,** 7674–7684 (2005).

40. Prokopcová, H. & Kappe, C. O. The Liebeskind–Srogl C–C cross-coupling reaction. *Angew. Chem. Int. Ed.* **48,** 2276–2286 (2009).

41. Pan, F. & Shi, Z.-J. Recent advances in transition-metal-catalyzed C − S activation: from thioester to (hetero)aryl thioether. *ACS Catal.* **4,** 280–288 (2014).

42. Melzig, L., Metzger, A. & Knochel, P. Pd- and Ni-catalyzed cross-coupling reactions of functionalized organozinc reagents with unsaturated thioethers. *Chem. Eur. J.* **17,** 2948–2956 (2011) and references therein.

43. Chopade, P. R. & Louie, J. [2 + 2 + 2] Cycloaddition reactions catalyzed by transition metal compleyes. *Adv. Synth. Catal.* **348,** 2307–2327 (2006).

44. Suzuki, H. & Abe, H. A simple cyclization route to some 4-substituted 3-aminoisoquinolines. *Synthesis (Mass)* 763–765 (1995) and references therein.

45. Neumeyer, J. L., Weinhardt, K. K., Carrano, R. A. & McCurdy, D. H. Isoquinolines 3. 3-aminoisoquinoline derivatives with central nervous system depressant activity. *J. Med. Chem.* **16,** 808–813 (1973).

46. Straub, A. *et al.* NO-independent stimulators of soluble guanylate cyclase. *Bioorg. Med. Chem. Lett.* **11,** 781–784 (2001).

47. Axon, J., Chakravarty, S., Dugar, S., Mcenroe, G. & Murphy, A. Inhibitors of TFGbeta. *PCT Int. Appl.* WO 2004/024159 A1 (2004).

48. Mciver, E. G. *et al.* Pyrimidine derivatives capable of inhibiting one or more kinases. *APCT Int. ppl.* WO 2009/122180 A1 (2009).

49. Barluenga, J., Fernández-Rodríguez, M. Á., García-García, P. & Aguilar, E. Gold-catalyzed intermolecular hetero-dehydro-diels − alder cycloaddition of captodative dienynes with nitriles: a new reaction and regioselective direct access to pyridines. *J. Am. Chem. Soc.* **130,** 2764–2765 (2008).

50. Wessig, P. & Müller, G. The dehydro-diels − alder reaction. *Chem. Rev.* **108,** 2051–2063 (2008).

51. Fernández-García, J. M., Fernández-Rodríguez, M. Á. & Aguilar, E. Catalytic intermolecular hetero-dehydro-Diels–Alder cycloadditions: regio- and diastereoselective synthesis of 5,6-dihydropyridin-2-ones. *Org. Lett.* **13,** 5172–5175 (2011).

52. Hoye, T. R., Baire, B., Niu, D., Willoughby, P. H. & Woods, B. P. The hexadehydro-Diels–Alder reaction. *Nature* **490,** 208–212 (2012).

53. Chrzanowska, M. & Rozwadowska, M. D. Asymmetric synthesis of isoquinoline alkaloids. *Chem. Rev.* **104,** 3341–3370 (2004).

54. Scott, J. D. & Williams, R. M. Chemistry and biology of the tetrahydroisoquinoline antitumor antibiotics. *Chem. Rev.* **102,** 1669–1730 (2002).

55. Kartsev, V. G. Natural compounds in drug discovery. Biological activity and new trends in the chemistry of isoquinoline alkaloids. *Med. Chem. Res.* **13,** 325–336 (2004).

56. Glorius, F., Spielkamp, N., Holle, S., Goddard, R. & Lehmann, C. W. Efficient asymmetric hydrogenation of pyridines. *Angew. Chem. Int. Ed.* **43,** 2850–2852 (2004).

57. Heitbaum, M., Frchlich, R. & Glorius, F. Diastereoselective hydrogenation of substituted quinolines to enantiomerically pure decahydroquinolines. *Adv. Synth. Catal.* **352,** 357–362 (2010).

## Acknowledgements

We are grateful to the European Research Council (ERC StG 278872 to N.M.) and the University of Vienna for support of this work. We acknowledge Dr Michael J. Fink and Professor Marko D. Mihovilovic (Vienna University of Technology) for hydrogenation reactions and helpful discussions. The computational results have been achieved in part using the Vienna Scientific Cluster (VSC).

## Author contributions

L.-G.X. and N.M. planned the project. L.-G.X. and S.N. carried out the experiments and analysed the data. A.J.M. and L.G. carried out the DFT analysis. L.-G.X., A.J.M., S.N., L.G. and N.M. wrote the manuscript.

# Permissions

All chapters in this book were first published in NC, by Nature Publishing Group; hereby published with permission under the Creative Commons Attribution License or equivalent. Every chapter published in this book has been scrutinized by our experts. Their significance has been extensively debated. The topics covered herein carry significant findings which will fuel the growth of the discipline. They may even be implemented as practical applications or may be referred to as a beginning point for another development.

The contributors of this book come from diverse backgrounds, making this book a truly international effort. This book will bring forth new frontiers with its revolutionizing research information and detailed analysis of the nascent developments around the world.

We would like to thank all the contributing authors for lending their expertise to make the book truly unique. They have played a crucial role in the development of this book. Without their invaluable contributions this book wouldn't have been possible. They have made vital efforts to compile up to date information on the varied aspects of this subject to make this book a valuable addition to the collection of many professionals and students.

This book was conceptualized with the vision of imparting up-to-date information and advanced data in this field. To ensure the same, a matchless editorial board was set up. Every individual on the board went through rigorous rounds of assessment to prove their worth. After which they invested a large part of their time researching and compiling the most relevant data for our readers.

The editorial board has been involved in producing this book since its inception. They have spent rigorous hours researching and exploring the diverse topics which have resulted in the successful publishing of this book. They have passed on their knowledge of decades through this book. To expedite this challenging task, the publisher supported the team at every step. A small team of assistant editors was also appointed to further simplify the editing procedure and attain best results for the readers.

Apart from the editorial board, the designing team has also invested a significant amount of their time in understanding the subject and creating the most relevant covers. They scrutinized every image to scout for the most suitable representation of the subject and create an appropriate cover for the book.

The publishing team has been an ardent support to the editorial, designing and production team. Their endless efforts to recruit the best for this project, has resulted in the accomplishment of this book. They are a veteran in the field of academics and their pool of knowledge is as vast as their experience in printing. Their expertise and guidance has proved useful at every step. Their uncompromising quality standards have made this book an exceptional effort. Their encouragement from time to time has been an inspiration for everyone.

The publisher and the editorial board hope that this book will prove to be a valuable piece of knowledge for researchers, students, practitioners and scholars across the globe.

# List of Contributors

**Pei-Qin Liao, Wei-Xiong Zhang, Jie-Peng Zhang and Xiao-Ming Chen**
MOE Key Laboratory of Bioinorganic and Synthetic Chemistry, School of Chemistry and Chemical Engineering, Sun Yat-Sen University, Guangzhou 510275, P. R. China

**Ai-Xin Zhu**
MOE Key Laboratory of Bioinorganic and Synthetic Chemistry, School of Chemistry and Chemical Engineering, Sun Yat-Sen University, Guangzhou 510275, P. R. China
Faculty of Chemistry and Chemical Engineering, Yunnan Normal University, Kunming 650092, P. R. China

**Hiroki Yokoi, Yuya Hiraoka, Satoru Hiroto and Hiroshi Shinokubo**
Department of Applied Chemistry, Graduate School of Engineering, Nagoya University, Nagoya 464-8603, Japan

**Daisuke Sakamaki and Shu Seki**
Department of Molecular Engineering, Graduate School of Engineering, Kyoto University, Kyoto 615-8510, Japan

**Xijian Li, Siyu Peng, Li Li and Yong Huang**
Key Laboratory of Chemical Genomics, Peking University, Shenzhen Graduate School, Shenzhen 518055, China

**Cheng Tao and Hongbin Zhai**
State Key Laboratory of Applied Organic Chemistry, Lanzhou University, Lanzhou 730000, China

**Xiaoming Chen and Shengguo Duan**
State Key Laboratory of Applied Organic Chemistry, Lanzhou University, Lanzhou 730000, China
Laboratory of Molecular Engineering, and Laboratory of Natural Product Synthesis, Guangzhou Institute of Biomedicine and Health, Chinese Academy of Sciences, 190 Kaiyuan Boulevard, The Science Park of Guangzhou, Guangzhou 510530, China

**Fayang G. Qiu**
Laboratory of Molecular Engineering, and Laboratory of Natural Product Synthesis, Guangzhou Institute of Biomedicine and Health, Chinese Academy of Sciences, 190 Kaiyuan Boulevard, The Science Park of Guangzhou, Guangzhou 510530, China

**Anne-Marie Caminade, Cédric-Olivier Turrin, Alexandrine Maraval, Jean-Pierre Majoral and Armelle Ouali**
Laboratoire de Chimie de Coordination du CNRS, UPR 8241, 205 route de Narbonne, BP 44099, 31077 Toulouse Cedex 4, France
Université de Toulouse, UPS, INP, LCC, F-31077 Toulouse, France

**Matteo Garzoni and Giovanni M. Pavan**
Department of Innovative Technologies, University of Applied Sciences and Arts of Southern Switzerland, Galleria 2, 6928 Manno, Switzerland

**Séverine Fruchon and Rémy Poupot**
Centre de Physiopathologie de Toulouse Purpan, F-31300 Toulouse, France
INSERM, U1043; CNRS, U5282; Université de Toulouse, UPS, Toulouse, France

**Mary Poupot**
Centre de Recherche en Cancérologie de Toulouse, F-31300 Toulouse, France
INSERM, U1037; CNRS, U5294; Université de Toulouse, UPS, Toulouse, France

**Marek Maly**
Faculty of Science, J.E. Purkinje University, Ceske mladeze 8, 400 96 Ústí nad Labem, Czech Republic

**Victor Furer**
Kazan State Architect and Civil Engineering University, Zelenaya 1, Kazan 420043, Russia

**Valeri Kovalenko**
A.E. Arbuzov Institute of Organic and Physical Chemistry of Kazan Scientific Center of Russian Academy of Science, Arbuzov Str., 8, Kazan 420088, Russia

**Anupam Bandyopadhyay, Kelly A. McCarthy, Michael A. Kelly and Jianmin Gao**
Department of Chemistry, Merkert Chemistry Center, Boston College, 2609 Beacon Street, Chestnut Hill, Massachuetts 02467, USA

**Chunji Li**
Department of Chemistry and Biotechnology, School of Engineering, The University of Tokyo, Hongo, Bunkyo, Tokyo 113-8656, Japan

**Joonil Cho, Kuniyo Yamada and Daisuke Hashizume**
RIKEN Center for Emergent Matter Science, 2-1 Hirosawa, Wako, Saitama 351-0198, Japan

**Fumito Araoka**
RIKEN Center for Emergent Matter Science, 2-1 Hirosawa, Wako, Saitama 351-0198, Japan
Department of Organic and Polymeric Materials, Tokyo Institute of Technology, 2-12-1-S8-42 O-okayama, Meguro, Tokyo 152-8552, Japan

**Hideo Takezoe**
Department of Organic and Polymeric Materials, Tokyo Institute of Technology, 2-12-1-S8-42 O-okayama, Meguro, Tokyo 152-8552, Japan

**Takuzo Aida**
Department of Chemistry and Biotechnology, School of Engineering, The University of Tokyo, Hongo, Bunkyo, Tokyo 113-8656, Japan

**Yasuhiro Ishida**
RIKEN Center for Emergent Matter Science, 2-1 Hirosawa, Wako, Saitama 351-0198, Japan
PRESTO, Japan Science and Technology Agency, 4-1-8 Honcho, Kawaguchi, Saitama 332-0012, Japan

**Masoud Kazemi and Johan Åqvist**
Department of Cell and Molecular Biology, Uppsala University, Biomedical Center, Box 596, SE-751 24 Uppsala, Sweden

**Zhenhua Jia and Qiang Liu**
Department of Chemistry, State Key Laboratory of Synthetic Chemistry and Centre of Novel Functional Molecules, Chinese University of Hong Kong, Shatin, New Territories, Hong Kong SAR, China

**Xiao-Shui Peng and Henry N.C. Wong**
Department of Chemistry, State Key Laboratory of Synthetic Chemistry and Centre of Novel Functional Molecules, Chinese University of Hong Kong, Shatin, New Territories, Hong Kong SAR, China
Shenzhen Center of Novel Functional Molecules and Shenzhen Municipal Key Laboratory of Chemical Synthesis of Medicinal Organic Molecules, Shenzhen Research Institute, Chinese University of Hong Kong, No.10, Second Yuexing Road, Shenzhen 518507, China

**Jisung Kim, Jinhee Lee, Sang Youl Kim, Hyungjun Kim, Sanghwa Lee and Yoon Sup Lee**
Department of Chemistry, Korea Advanced Institute of Science and Technology (KAIST), 291 Daehak-ro, Yuseong-gu, Daejeon 305-701, Korea

**Woo Young Kim**
Department of Mechanical Engineering, KAIST, Daejeon 305-701, Korea

**Hee Chul Lee**
Department of Electronic Engineering, KAIST, Daejeon 305-701, Korea

**Myungeun Seo**
Graduate School of Nanoscience and Technology, KAIST, Daejeon 305-701, Korea

**Miguel Garcia-Castro, Slava Ziegler and Christopher D. Reinkemeier**
Max-Planck-Institut für Molekulare Physiologie, Abteilung Chemische Biologie, Otto-Hahn-Strasse 11, 44227 Dortmund, Germany

**Lea Kremer and Kamal Kumar**
Max-Planck-Institut für Molekulare Physiologie, Abteilung Chemische Biologie, Otto-Hahn-Strasse 11, 44227 Dortmund, Germany
Technische Universität Dortmund, Fakultät für Chemie und Chemische Biologie, Otto-Hahn-Strasse 6, 44227 Dortmund, Germany

**Christian Unkelbach and Carsten Strohmann**
Technische Universität Dortmund, Fakultät für Chemie und Chemische Biologie, Otto-Hahn-Strasse 6, 44227 Dortmund, Germany

**Claude Ostermann**
Compound Management and Screening Center(COMAS), Max-Planck-Institut für Molekulare Physiologie, Otto-Hahn-Strasse 11, 44227 Dortmund, Germany

**Igor Pavlovic, Divyeshsinh T. Thakor, Philipp Anstaett, Laurent Bigler and Gilles Gasser**
Department of Chemistry, University of Zurich, Winterthurerstrasse 190, Zurich 8057, Switzerland

**Jessica R. Vargas, Colin J. McKinlay and Paul A. Wender**
Departments of Chemistry and Chemical and Systems Biology, Stanford University, Stanford, California 94305, USA

**Sebastian Hauke and Carsten Schultz**
European Molecular Biology Laboratory (EMBL), Cell Biology & Biophysics Unit, Meyerhofstrasse 1, 69117 Heidelberg, Germany

**Rafael C. Camuña**
Departamento de Química Orgánica, Facultad de Ciencias, Universidad de Málaga, Malaga 29071, Spain

**Henning J. Jessen**
Department of Chemistry and Pharmacy, Albert-Ludwigs University Freiburg, Albertstrasse 21, 79104 Freiburg, Germany

**René-Chris Brachvogel, Frank Hampel and Max von Delius**
Department of Chemistry and Pharmacy, Friedrich-Alexander University Erlangen-Nürnberg (FAU), Henkestrasse 42, 91054 Erlangen, Germany

**Antonio Leyva-Pérez and Avelino Corma**
Instituto de Tecnología Química, Universidad Politécnica de Valencia-Consejo Superior de Investigaciones Científicas, Avda. de los Naranjos s/n, 46022 Valencia, Spain

**Antonio Doménech-Carbó**
Departament de Química Analítica, Universitat de Valencia, Dr Moliner, 50, 46100 Burjassot, Valencia, Spain

**Cristian Pezzato and Leonard J. Prins**
Department of Chemical Sciences, University of Padova, Via Marzolo 1, 35131 Padova, Italy

**Xisen Hou, Chenfeng Ke, Carson J. Bruns, Paul R. McGonigal and J. Fraser Stoddart**
Department of Chemistry, Northwestern University, Evanston, Illinois 60208-3113, USA

**Roger B. Pettman**
Cycladex, c/o Innovation and New Ventures Office, Northwestern
University, 1800 Sherman Avenue, Suite 504, Evanston, Illinois 60201-3789, USA

**Ji-Wei Zhang, Jin-Hui Xu, Dao-Juan Cheng, Chuan Shi, Xin-Yuan Liu and Bin Tan**
Department of Chemistry, South University of Science and Technology of China, Shenzhen 518055, China.

**Weijiang Guan, Si Wang and Chao Lu**
State Key Laboratory of Chemical Resource Engineering, Beijing University of Chemical Technology, 15 Beisanhuan East Road, PO Box 98, Beijing 100029, China

**Ben Zhong Tang**
Department of Chemistry, Hong Kong Branch of Chinese National Engineering Research Center for Tissue Restoration and Reconstruction,
Hong Kong University of Science and Technology, Clear Water Bay, Hong Kong 999077, China

**Nicola Casati**
Paul Scherrer Institute, WLGA/229, CH-5232 Villigen, Switzerland

**Annette Kleppe**
Diamond light source Ltd., Harwell Science and innovation Campus, Didcot OX110DE, UK

**Andrew P. Jephcoat**
Institute for Study of the Earth's interior, Okayama University, Yamada 827, Misasa, Tottori 682-0193, Japan

**Piero Macchi**
Department of Chemistry and Biochemistry, University of Bern, Freiestrasse 3, Bern CH-3012, Switzerland

**Xi Lu**
Hefei National Laboratory for Physical Sciences at the Microscale, iChEM, CAS Key Laboratory of Urban Pollutant Conversion, Anhui Province Key Laboratory of Biomass Clean Energy, University of Science and Technology of China, Hefei 230026, China
Department of Chemistry, Tsinghua University, Beijing 100084, China.

**Bin Xiao, Zhenqi Zhang,Tianjun Gong, Wei Su, Jun Yi and Yao Fu**
Hefei National Laboratory for Physical Sciences at the Microscale, iChEM, CAS Key Laboratory of Urban Pollutant Conversion, Anhui Province Key Laboratory of Biomass Clean Energy, University of Science and Technology of China, Hefei 230026, China

**Lei Liu**
Department of Chemistry, Tsinghua University, Beijing 100084, China

**Thilo Krause, Sabrina Baader, Benjamin Erb and Lukas J. Goo_en**
FB Chemie-Organische Chemie, Technische Universität Kaiserslautern, Erwin Schrädinger Strasse Geb. 54, 67663 Kaiserslautern, Germany

**Pilar García-García1, José María Moreno1, Urbano Díaz1,**
Instituto de Tecnología Química, UPV-CSIC, Universidad Politécnica de Valencia, Avenida de los Naranjos s/n, E-46022 Valencia, Spain

**Marta Bruix**
Instituto de Química Física Rocasolano, CSIC, Serrano 119, 28006 Madrid, Spain

**Avelino Corma**
Instituto de Tecnología Química, UPV-CSIC, Universidad Politécnica de Valencia, Avenida de los Naranjos s/n, E-46022 Valencia, Spain
King Fahd University of Petroleum and Minerals, PO Box 989, 31261 Dhahran, Saudi Arabia

**O. Hauenstein, S. Agarwal and A. Greiner**
Macromolecular Chemistry II and Center for Colloids and Interfaces, University of Bayreuth, Universita¨tsstrasse 30, 95440 Bayreuth, Germany

**Rajith S. Manan and Pinjing Zhao**
Department of Chemistry and Biochemistry, North Dakota State University, Fargo, North Dakota 58102, USA

**Yi Zhou**
Synthetic Biology for Clinical and Technological Innovation (SynCTI), Life Sciences Institute, National University of Singapore, Singapore 117456, Singapore

**Tianwen Wang**
Department of Chemical and Biomolecular Engineering, National University of Singapore, Singapore 117585, Singapore

**Heng-Phon Too**
Singapore-MIT Alliance, National University of Singapore, Singapore 117583, Singapore

**Daniel I.C Wang**
Singapore-MIT Alliance, National University of Singapore, Singapore 117583, Singapore
Department of Chemical Engineering, Massachusetts Institute of Technology, Cambridge, Massachusetts 02139, USA

**Shuke Wu and Zhi Li**
Department of Chemical and Biomolecular Engineering, National University of Singapore, Singapore 117585, Singapore
Singapore-MIT Alliance, National University of Singapore, Singapore 117583, Singapore
Synthetic Biology for Clinical and Technological Innovation (SynCTI), Life Sciences Institute, National University of Singapore, Singapore 117456, Singapore

**Lan-Gui Xie, Supaporn Niyomchon and Nuno Maulide**
Institute of Organic Chemistry, University of Vienna, Währinger Strasse 38, 1090 Vienna, Austria

**Antonio J. Mota and Leticia González**
Institute of Theoretical Chemistry, University of Vienna, Währinger Strasse 17, 1090 Vienna, Austria

# Index

**A**

Acetylenes, 170, 175, 203
Aerobic Oxidization, 1-2
Aldol Condensation, 27-29, 32, 145
Alkyne Annulation, 193, 195, 197-199, 201-202
Alkyne Homocoupling, 112
Amide Coupling Protocol, 170
Amide Synthesis, 170-171, 174-175
Amine-presenting Lipids, 44
Amino-functionalizations, 204-206, 209, 213
Aromatic Compound, 154
Aromaticity, 154-161
Arrhenius Plots, 62-63, 65-66
Axially Chiral Urazoles, 137-138, 143-144
Azacorannulene, 9-10, 16

**B**

Bacteria, 44-50, 52, 187, 189, 191
Bio-based Polycarbonate, 186, 192
Biocatalysis, 181, 204-206, 208, 210-216
Brute Force, 62
Buckybowl, 9-10, 13, 16

**C**

C-h Activation, 16, 118, 193-195, 201-203, 224
Carbon-carbon Bond, 32, 69-70, 113, 162, 164, 166, 168, 225
Carboxylic Acids, 34-35, 170-172, 174-176, 215-216
Cellular Delivery, 97, 99
Chemical Reaction Mechanisms, 62
Circularly Polarized Light (cpl), 58
Complex Compounds, 18
Concentration-dependent Activation, 120
Coordination Polymers, 7-8, 177
Cross-coupling, 10, 52, 69-75, 118-119, 145, 162-163, 165, 168-169, 202, 205, 214, 225
Cryptographic Algorithm, 128

**D**

De Novo Branching, 84-86, 90, 92
Dendrimer Nanodrugs, 33
Dendritic Cells, 33
Diels-alder Reactions, 18-19, 24, 26
Digold Catalyst, 112
Distal Size Selectivity, 112-113, 115-118
Dynamic Orthoester Cryptates, 105

**E**

Electron Density, 16, 19, 68, 78, 154-155, 157-158, 160
Electron Microscopy, 56, 61, 77-78, 81, 147-148, 151
Enantiomeric Excess, 27, 82-83, 144, 184, 205
Entropic Components, 62-63
Ethylene Coupling, 69, 72-73

**F**

Fluorescence Microscopy, 49, 51-52, 102, 147-148, 150, 152
Fluorescent Materials, 128
Functional Diversity, 84, 86, 93

**G**

Gelsemine Skeleton, 27, 32
Green Platform Polymer, 186
Guest Molecules, 7, 53-54, 57-58

**H**

Helical Nanostructures, 53, 76
Helical Pores, 53-54, 58, 60
Heterocyclic Compounds, 26, 138
Human Monocytes Activation, 33
Hydroamination, 145, 168, 171, 193-196, 201-203

**I**

Iminoboronate Chemistry, 44-45, 49-50
Inositol Pyrophosphates, 97-98, 104
Intermolecular Formal Cycloadditions, 217
Irreversible Deconstruction, 105
Isopropyllithium, 69, 72-73, 75

**M**

Macroscopic Ordering, 53
Metal-organic Frameworks, 7-8, 58, 152, 177, 184
Microscale Dispersion, 147
Molecular Crystal, 154, 159-160
Molecular Sensing, 105

**N**

Nanocarbon Materials, 9
Nanosystem, 120
Natural Signalling Pathways, 120, 123-124
Nickel Catalyst System, 162
Nitrogen Heterocycles, 26, 217, 221
Nitrogen-containing Molecules, 193
Non-cycloaddition Reactions, 18, 25

## O

Olefin Hydrocarbonation, 162-165, 167
One-pot Biotransformations, 204
Optical Chromophore, 53
Organic Compounds, 70, 105
Organic Halides, 69-70
Organic Linker, 1, 5, 7
Organic Reactions, 118, 168, 177, 181
Organic-inorganic Composites, 147-149
Organocatalytic Tyrosine Click Reaction, 137-138
Organolithium Compounds, 69-71, 74-75, 169
Orthogonal Access, 217-218, 220
Oxidative Homocoupling, 112-113

## P

Ph-domain Translocation, 97-98, 100-102
Photochemical Release, 97
Photopolymerization, 76-82
Polychromic Images, 128
Porous Crystal, 1-2, 5-7

## R

Rationalization, 62, 223
Retrosynthetic Analysis, 28-29, 162

## S

Signal Intensity, 120-121
Small Molecules, 51-52, 82, 84, 91-95
Solid Catalyst, 177
Sorption Behaviours, 1
Stereochemical Information, 137, 139
Supramolecular Chirality, 59, 76-81
Supramolecular Encryption, 128, 133-135
Synthetic Molecules, 44-45, 49
Synthetic Toolbox, 186

## T

Tetrabenzocarbazole, 9-10
Thermoplastic Polymer, 186
Total Synthesis, 18, 25-28, 32, 75, 95, 145, 224
Transient Signal Generation, 120-123, 125

www.ingramcontent.com/pod-product-compliance
Lightning Source LLC
Chambersburg PA
CBHW080251230326
41458CB00097B/4263